# Physicomimetics

William M. Spears · Diana F. Spears

Editors

# Physicomimetics

## Physics-Based Swarm Intelligence

Springer

Editors
Dr. William M. Spears
Swarmotics LLC
Laramie, WY
USA
wspears@swarmotics.com

Dr. Diana F. Spears
Swarmotics LLC
Laramie, WY
USA
dspears@swarmotics.com

ISBN 978-3-642-44863-8          ISBN 978-3-642-22804-9 (eBook)
DOI 10.1007/978-3-642-22804-9
Springer Heidelberg Dordrecht London New York


ACM Classification (1998): I.2, I.6, J.2

Springer is part of Springer Science+Business Media (www.springer.com)

*To Daisy*

# Preface

We are used to thinking of the world in a centralized and hierarchical manner. Governments and businesses rely on organizations with someone "at the top" who collects information and issues orders that trickle down the hierarchy until they reach the rest of us. Even portions of our economic system work in this fashion. The reason organizations exist is because they work well in many situations. But there is another view of the world that is entirely different. This view starts at the "bottom," and realizes that much of the organization that we see does not stem from centralized control, but emerges from the local interactions of a multitude of entities (as with insects, people, vehicles, and the movement of money). These multitudes are swarms. Standard approaches to understanding swarms rely on inspiration from biology. These are called "biomimetic" approaches. In this book, we focus on a different inspiration, namely, physics. We refer to physics-based swarm approaches as "physicomimetics." Both approaches are complementary, but physics-based approaches offer two unique advantages. The first is that these approaches capture the notion that "nature is lazy." This means that physical systems always perform the minimal amount of work necessary. This is very important for swarm robotics, because robots are always limited by the amount of power they have at their disposal. The second advantage is that physics is the most predictive science, and it can reduce complex systems to amazingly simple concepts and equations. These concepts and equations codify emergent behavior and can be used to help us design and understand swarms.

This book represents the culmination of over 12 years of work by numerous people in the field of swarm intelligence and swarm robotics. We include supplemental material, such as simulation code, simulation videos, and videos of real robots. The goal of this book is to provide an extensive overview of our work with physics-based swarms. But we will not do this in the standard fashion. Most books are geared toward a certain level of education (e.g., laymen, undergraduates, or researchers). This book is designed to "grow with the reader." We start with introductory chapters that use simple but powerful graphical simulations to teach elementary concepts in physics and swarms. These are suitable for junior and senior high school students. Knowledge of algebra and high school physics is all you need to understand this material. However, if you are weak in physics, we provide a chapter, complete with simulations, to bring you back up to speed. In fact, even if you have had physics already, we recommend that you read this chapter—because the simulations provide insights into physics that are difficult to achieve using only equations and standard high school physics textbooks.

All you need is a computer to run the simulations. They can be run directly on your machine or through your web browser. You do not need to have had any programming courses. But if you have had a programming course, you will be ready to modify the simulations that come with this book. We provide an introductory chapter to explain the simple simulation language that we use throughout. Suggestions for modifications are included in the documentation with the simulations.

The middle of the book is most appropriate for undergraduates who are interested in majoring in computer science, electrical computer engineering, or physics. Because we still use simulations, these chapters generally require only algebra and an understanding of vectors. A couple of chapters also require basic calculus (especially the concepts of derivatives and integrals) and elementary probability. If you don't know anything about electrical computer engineering, that is fine. You can merely skim over the hardware details of how we built our robots and watch the videos that also come with this book. But if you have had a course or two, much of this material will be quite accessible.

The final sections contain more advanced topics suitable for graduate students looking for advanced degree topics, and for researchers. These sections focus on how to design swarms and predict performance, how swarms can adapt to changing environments, and how physicomimetics can be used as a function optimizer. However, even here most of the chapters require little mathematics (e.g., only two require knowledge of linear algebra and calculus).

It is important to point out that this is a new and rapidly developing field. Despite the fact that the authors share a surprisingly consistent vision, we do not always view swarms in precisely the same way. This should not be a cause for concern. Unlike Newtonian physics or mathematics, which have been developed for hundreds of years, physicomimetics is relatively new and does not yet have entirely consistent terminology and notation. But all this means is that not everything is cast in stone—we are at the beginning of an adventure. You are not learning old knowledge. Instead, you are seeing how science progresses in the here and now. We hope you enjoy the journey.

Laramie, Wyoming                                                    *William M. Spears*
September 2011                                                        *Diana F. Spears*

# Acknowledgements

This book would not have been possible without the hard work, creativity, and perseverance of our numerous students. Quite a few of them are contributors to this book, and we thank them all. Suranga Hettiarachchi, Paul M. Maxim, and Dimitri Zarzhitsky earned Ph.D.'s working on the projects described in this book. In addition, Dimitri Zarzhitsky provided valuable proofreading of portions of this book.

Thanks to Adam Sciambi for our first graphical user interface for artificial physics (physicomimetics). Adam was a high school student who participated in the Science and Engineering Apprentice Program at the Naval Research Laboratory. During July 2000, he converted our original C code into Java, including two-dimensional and three-dimensional simulations. Brian Zuelke and Vaibhav Mutha subsequently extended Adam's simulations.

We appreciate the collaboration with Oleg Sokolsky and Insup Lee of the University of Pennsylvania on an early project that consisted of combining artificial physics with monitoring and checking.

We are grateful to David Walrath and Ben Palmer of the University of Wyoming for the creation of a spheroidal radio-controlled vehicle that we tested for outdoor use. Ben did this for his senior design project. Thanks also to Ella Wellman, who helped us with our early simulation experiments on chain formations.

We thank Alan Schultz of the Naval Research Laboratory for the use of his Lego robotics kits from the KISS Institute of Practical Robotics. Thanks to Paul Hansen for using these kits to provide us with the first robots that we used for our initial experiments. These early robots were invaluable for testing artificial physics, and they served as platforms for tests of our trilateration hardware and for preliminary chemical plume tracing experiments. John Schabron of the Western Research Institute provided valuable guidance on the calibration and testing of our chemical sensors.

Furthermore, we would like to thank the students who took our course on artificial intelligence and robotics: Nathan Horter, Christoph Jechlitschek, Nathan Jensen, Nadya Kuzmina, Glenn McLellan, Vaibhav Mutha, Markus

Petters, Brandon Reed, and Peter Weissbrod. They worked hard, pushing the Lego robots to their limits, and provided a lot of data for us on the proper use of Sharp infrared sensors and shaft encoders. One student thought it was "by far one of the most entertaining projects I've ever worked on." Another said, "I'll probably remember more about this class than most other classes I've taken, in part because I got to work in a group and discuss how to proceed in solving problems in both software and hardware designs which I really don't get a chance to do in other classes, and, it was just a really fun project to do." Students found that working with actual hardware provided them with a very real appreciation for how difficult it is to sense an environment accurately and to respond appropriately. Given that modern operating systems require almost one gigabyte of memory, it was exciting to see how much could be accomplished with only 32 kilobytes! Student enthusiasm and energy always provided us with a deep sense of satisfaction.

We appreciate the guidance from Jerry Hamann of the University of Wyoming on designing the trilateration hardware, which was key to all of our later experiments. Numerous students were instrumental in building and improving the hardware, especially Rodney Heil, Thomas Kunkel, and Caleb Speiser. These three students earned master's degrees while working on these projects. Rodney Heil also solved the initial trilateration equations, thus providing an extremely elegant solution.

Special thanks to William Agassounon and Robert Welsh from Textron Defense Systems for funding and support for our chemical plume tracing research and our next-generation trilateration system.

We wish to thank James Partan, who was a visiting Ph.D. student from the University of Massachusetts, Amherst. Despite the fact that robotics was not Jim's area of expertise, he shared an interest with respect to disruption-tolerant wireless networks. He always provided excellent insights every time we used him as a sounding board.

Most of the NetLogo simulations provided with this book were originally written in StarLogo, which is a similar language. We thank Eric Klopfer of the Massachusetts Institute of Technology for his advice and feedback on StarLogo.

Finally, we extend our appreciation to our supportive European colleagues, especially Erol Şahin, Alan Winfield, and Alcherio Martinoli. Those of you who are interested in learning still more about swarms will find that the proceedings of the First and Second International Workshops on Swarm Robotics will provide you with a thorough introduction to the field [202, 203].

# Contents

xi

## Part III  Physicomimetics on Hardware Robots

## 10  What Is a Maxelbot? .......................................... 301
Paul M. Maxim

# List of Contributors

Richard Anderson-Sprecher
Department of Statistics, University of Wyoming, Laramie, Wyoming, USA,
e-mail: sprecher@uwyo.edu

Thomas B. Apker
NRC/NRL Postdoctoral Fellow, Washington DC, USA, e-mail:
apker@aic.nrl.navy.mil

Andrea Bravi
Dynamical Analysis Laboratory, University of Ottawa, Ottawa, Ontario,
Canada, e-mail: a.bravi@uottawa.ca

Paolo Corradi
Human Space Flight and Operations, European Space Agency, Noordwijk,
The Netherlands, e-mail: paolo.corradi@sssup.it

Chris Ellis
Department of Electrical Engineering and Computer Science, University of
Central Florida, Orlando, Florida, USA, e-mail: chris@cs.ucf.edu

Charles Lee Frey
Harbor Branch Oceanographic Institute, Ft. Pierce, Florida, USA, e-mail:
cfrey@hboi.fau.edu

Derek T. Green
Department of Computer Science, University of Arizona, Tucson, Arizona,
USA, e-mail: dtgreen@arizona.edu

Suranga Hettiarachchi
Department of Computing Science, Indiana University Southeast, New
Albany, Indiana, USA, e-mail: suhettia@ius.edu

Christer Karlsson
Department of Mathematical and Computer Sciences, Colorado School of
Mines, Golden, Colorado, USA, e-mail: ckarlsso@mines.edu

Sanza Kazadi
Jisan Research Institute, Alhambra, California, USA, e-mail:
skazadi@jisan.org

Wesley Kerr
Department of Computer Science, University of Arizona, Tucson, Arizona,
USA, e-mail: wkerr@email.arizona.edu

Aleksey Kletsov
Department of Physics, East Carolina University, Greenville, North Carolina,
USA, e-mail: kletsov@gmail.com

Paul Maxim
Wyoming Department of Transportation, Cheyenne, Wyoming, USA, e-mail:
paul.maxim@dot.state.wy.us

Arianna Menciassi
CRIM Laboratory, Scuola Superiore Sant'Anna, Pisa, Italy, e-mail:
arianna.menciassi@sssup.it

Mitchell A. Potter
US Naval Research Laboratory, Washington DC, USA, e-mail:
mpotter@aic.nrl.navy.mil

Antons Rebguns
Department of Computer Science, University of Arizona, Arizona, USA,
e-mail: anton@email.arizona.edu

Florian Schlachter
Applied Computer Science – Image Understanding, University of
Stuttgart, Stuttgart, Germany, e-mail: Florian.Schlachter@
ipvs.uni-stuttgart.de

Diana F. Spears
Swarmotics LLC, Laramie, Wyoming, USA, e-mail: dspears@
swarmotics.com

William M. Spears
Swarmotics LLC, Laramie, Wyoming, USA, e-mail: wspears@
swarmotics.com

Ying Tan
Taiyuan University of Science and Technology, Taiyuan, Shanxi, China,
e-mail: tanying1965@gmail.com

David Thayer
Physics and Astronomy Department, University of Wyoming, Laramie,
Wyoming, USA, e-mail: drthayer@uwyo.edu

R. Paul Wiegand
Institute for Simulation and Training, University of Central Florida,
Orlando, Florida, USA, e-mail: wiegand@ist.ucf.edu

Edith A. Widder
Ocean Research and Conservation Association, Ft. Pierce, Florida, USA,
e-mail: ewidder@oceanrecon.org

Liping Xie
Taiyuan University of Science and Technology, Taiyuan, Shanxi, China,
e-mail: xieliping1978@gmail.com

Dimitri V. Zarzhitsky
Jet Propulsion Laboratory, California Institute of Technology, Pasadena,
California, USA, e-mail: Dimitri.Zarzhitsky@jpl.nasa.gov

Jianchao Zeng
Taiyuan University of Science and Technology, Taiyuan, Shanxi, China,
e-mail: zengjianchao@263.net

# Acronyms

| | |
|---|---|
| AP | Artificial physics |
| APO | Artificial physics optimization |
| ASV | Autonomous surface vehicle |
| AUV | Autonomous underwater vehicle |
| BP | Bathyphotometer |
| BPT | Biological plume tracing |
| CFD | Computational fluid dynamics |
| CFL | Chain formation list |
| CPT | Chemical plume tracing |
| DAEDALUS | Distributed agent evolution with dynamic adaptation to local unexpected scenarios |
| DG | Density gradient |
| DMF | Divergence of chemical mass flux |
| EA | Evolutionary algorithm |
| EL | Evolutionary learning |
| IP | Internet protocol |
| EEPROM | Electrically erasable programmable read-only memory |
| FOFE | First-order forward Euler integration |
| GDMF | Gradient of the divergence of the mass flux |
| GMF | Gradient of the mass flux |
| GPS | Global Positioning System |
| GUI | Graphical user interface |
| HABS | Harmful algal blooms |
| HBOI | Harbor Branch Oceanographic Institute |
| HLS | Hyperbolic localization system |
| I2C | Inter-Integrated Circuit |
| IR | Infrared |
| KL | Kullback–Leibler |
| KT | Kinetic theory |
| LCD | Liquid crystal display |
| LOPF | Local oriented potential fields |

| LOS | Line-of-sight |
| MAV | Micro-air vehicle |
| MAXELBOT | Robot built by Paul M. Maxim and Tom Kunkel |
| MDS | Multi-drone simulator |
| MST | Minimum spanning tree |
| MSE | Mean squared error |
| NAVO | Naval Oceanographic Office |
| NRL | Naval Research Laboratory |
| NSF | National Science Foundation |
| ONR | Office of Naval Research |
| PCB | Printed circuit board |
| PIC | Programmable intelligent computer |
| PSO | Particle swarm optimization |
| PSP | Paralytic shellfish poisoning |
| PVC | Polyvinyl chloride |
| RF | Radio frequency |
| RL | Reinforcement learning |
| SRF | Sonic range finder |
| TCP/IP | Transmission control protocol / Internet protocol |
| UAV | Unmanned aerial vehicle |
| VOC | Volatile organic compound |
| VSP | Voith–Schneider propeller |
| XSRF | Experimental sonic range finder |

# NetLogo Syntax Glossary

This glossary provides a useful (albeit incomplete) glossary of NetLogo syntax.

| | |
|---|---|
| **ask patch x y [a]** | Ask patch at $(x, y)$ to execute a. |
| **ask patches [a]** | Ask all patches to execute a. |
| **ask turtle n [a]** | Ask turtle $n$ to execute a. |
| **ask turtles [a]** | Ask all turtles to execute a. |
| **ca** | Kills all turtles and resets all variables to zero. |
| **clear-all** | Kills all turtles and resets all variables to zero. |
| **crt n** | Create $n$ turtles, numbered 0 to $n-1$. |
| **end** | The keyword that ends the definition of NetLogo procedures. |
| **home** | Send all turtles to $(0, 0)$. |
| **if (a) [b]** | If condition a is true execute b. |
| **ifelse (a) [b] [c]** | If condition a is true execute b, else execute c. |
| **pd** | Set the pen to be down. |
| **pen-down** | Set the pen to be down. |
| **pen-up** | Set the pen to be up. |
| **pu** | Set the pen to be up. |
| **random n** | Returns a random integer uniformly from 0 to $n-1$. |
| **set a b** | Set the value of variable a to b. |
| **set color white** | Set the color of a turtle to white. |
| **set heading a** | Set the heading of a turtle to a. |
| **to** | The keyword that starts the definition of NetLogo procedures. |
| **[xcor] of turtle 1** | The $x$-coordinate of turtle number one. |

# NetLogo Parameters Glossary

This glossary provides a useful (albeit incomplete) glossary of parameters used in our NetLogo simulations.

**angular_mom**     The angular momentum of the system.

**breed**     The breed of an agent.

**center_of_mass_x**     The $x$-component of the center of mass of the system.

**center_of_mass_y**     The $y$-component of the center of mass of the system.

**color**     The color of an agent.

**dv**     The change in velocity of an agent.

**dvx**     The $x$-component of the change in velocity of an agent.

**dvy**     The $y$-component of the change in velocity of an agent.

**DeltaT**     The time granularity of the simulation.

**deltax**     The $x$-component of distance between two agents.

**deltay**     The $y$-component of distance between two agents.

**F**     The force on an agent.

**FMAX**     The maximum pair-wise force magnitude.

**FR**     The amount of friction in the system.

**Fx**     The $x$-component of the force on an agent.

**Fy**     The $y$-component of the force on an agent.

**G**     The gravitational constant in the system.

| | |
|---|---|
| **hood** | The neighbors (neighborhood) of an agent. |
| **heading** | The heading of an agent. |
| **k** | The spring constant. |
| **ke** | The kinetic energy of an agent. |
| **lm** | The linear momentum of an agent. |
| **lmx** | The $x$-component of the linear momentum of an agent. |
| **lmy** | The $y$-component of the linear momentum of an agent. |
| **mass** | The mass of an agent. |
| **omega** | The initial amount of rotation in the system. |
| **pcolor** | The color of a patch. |
| **pe** | The potential energy of an agent. |
| **r** | The $r$ distance between two agents. |
| **total_energy** | The total energy of the system. |
| **total_ke** | The total kinetic energy of the system. |
| **total_lm** | The total linear momentum of the system. |
| **total_lmx** | The $x$-component of the total linear momentum of the system. |
| **total_lmy** | The $y$-component of the total linear momentum of the system. |
| **total_pe** | The total potential energy of the system. |
| **v** | The velocity of an agent. |
| **VMAX** | The maximum velocity of an agent. |
| **vx** | The $x$-component of the velocity of an agent. |
| **vy** | The $y$-component of the velocity of an agent. |
| **who** | Returns the id of the turtle. |
| **xcor** | The $x$-coordinate of an agent or patch. |
| **ycor** | The $y$-coordinate of an agent or patch. |

# Part I
# Introduction

# Chapter 1
# Nature Is Lazy

William M. Spears

> *"...sudden I wav'd My glitter falchion, from the sanguine pool Driving th' unbody'd host that round me swarm'd"* (1810) [29]

> *"...fince then it is called a fwarm of bees, not fo much from the murmuring noife they make while flying, as the manner in which they conneɛt, and join themfelves together at that remarkable time of fwarming"* (1783) [135]

## 1.1 What Are Swarms?

No one knows the exact origin of the word "swarm." Most sources refer to the Old English word "swearm." Samuel Johnson's 1804 dictionary defined a swarm as "a great number of bees; a crowd" [108]. Similarly, in 1845 a swarm was defined as a large number of persons or animals in motion [178]. This agrees with the use of the word in the two quotations above. The first is from a translation of Homer's Odyssey [29]. The second is from a book on English etymology [135].

The most frequent connotation of "swarm" is with respect to bees, animals and people. And, not surprisingly, much of the swarm research is inspired by biology. This research is referred to as "biomimetics" or "biomimicry." These words are from the Greek *bios* ($\beta\iota o\varsigma$), meaning "life," and *mimesis* ($\mu\iota\mu\eta\sigma\iota\varsigma$), which means "imitation." Hence much of swarm research focuses on the imitation of live organisms such as birds, fish, insects, and bacteria.

This book examines a different approach to swarms, namely a "physicomimetics" approach. This word is derived from *physis* ($\varphi\upsilon\sigma\iota\varsigma$), which is Greek for "nature" or "the science of physics." Physics involves "the study of matter and its motion through spacetime, as well as all related concepts, including energy and force" [265]. Hence we will focus on swarms of matter particles that are subject to forces. Depending on the application the particles can act as robots in an environment, agents solving a task, or even as points in a high dimensional space. We will show how various forms of energy play a role in how the particles move.

William M. Spears
Swarmotics LLC, Laramie, Wyoming, USA, e-mail: wspears@swarmotics.com

W.M. Spears, D.F. Spears (eds.), *Physicomimetics*,
DOI 10.1007/978-3-642-22804-9_1,

## 1.2 Why Do We Care About Swarms?

Why are we interested in swarms? Let's answer this question from the point of view of robotics. It has been traditional to build expensive robotic platforms that are highly capable and sophisticated. Yet, these individual platforms also represent a single point of failure. When one Predator unmanned aerial vehicle (UAV) is lost, the mission is over and US $4,500,000 is gone. The same can be said of planetary rovers and other highly automated large vehicles.

Such expensive platforms are sometimes the only way to accomplish a task because they have the power and load-bearing capacity to carry out various missions. However, other tasks require cooperation and coordination over a wide geographic area. Expensive platforms are not appropriate for such tasks because they are simply not fast enough to watch everything at the same time. Example tasks include surveillance, sentry duty, reconnaissance and scenarios that require simultaneous measurements at different locations.

There are a number of characteristics natural to swarms that are advantageous:

- **Self-organization**: Swarms have the ability to organize themselves. The individual units (platforms) can be placed at a central location, and then disperse to perform their task.
- **Robustness**: Unlike individual expensive platforms, swarms may lose individual units but still maintain a level of functionality. Usually there must be a massive loss before the swarm can no longer function.
- **Self-repair**: Furthermore, swarms are able to adapt to the loss of units by automatically reorganizing the placement of the remaining units in order to perform a task more efficiently.
- **Scalability**: It is easy to add more units to the swarm as they become available.
- **Noise tolerance**: Swarms generally are quite resilient to "noise." Such noise can be a result of faulty sensors, faulty actuators, or even communication errors. In fact, as we will see, noise often improves the behavior of swarms.
- **Local communication**: One of the reasons that swarms exhibit these characteristics is because they are designed around the premise that units interact mostly with their immediate neighbors. This turns a pragmatic constraint (i.e., long range communication is expensive) into a behavioral advantage. Local communication is a major reason why swarms scale well and are robust.

A comparison of swarms to more traditional robotics is provided in Table 1.1. Traditional robots are often very complex entities that must perform various functions. A good example is a rover that must move, manage its own power, collect and analyze samples, monitor its health, and perform a myriad of functions. Their architecture is complex, and their software is generally based on a combination of control algorithms and heuristics.[1] They can utilize extensive communication and sensing, if available. A traditional robot can include algorithms to help it adapt. Behavior is engineered. Analysis of these systems is difficult, due to their complexity.

Table 1.1: A comparison of swarms with traditional robotics

|  | Swarms | Traditional Robotics |
| --- | --- | --- |
| Agents | Particles/Critters/Simple Robots | Complex Robots |
| Motivation | Biology/Physics | Control Algorithms |
| Software | Simple | Complex Heuristics |
| Architecture | Simple | Complex |
| Communication | Minimal | Extensive |
| Sensing | Local | Local and Global |
| Distributed | Fully | Uses Central Controller |
| Robustness | Automatic | Hard to Achieve |
| Learning | No/Yes | Yes |
| Behavior | Engineered/Emergent | Engineered |
| Organization | Self-organized | Engineered |
| Analysis | Possible | Difficult |
| Assurance | Possible | Difficult |

In swarms the agents are usually modeled as simple biological entities or even particles. The motivation for their use is based on analogies to biology, ethology, or physics. The algorithms, software, and architecture are relatively simple. Communication and sensing are usually local and minimal. The swarm is fully distributed, without centralized control. It is often robust by virtue of being a swarm. It is usually adaptive enough that learning is not necessary, although new distributed learning algorithms are being created for swarms, as we will show in Chap. 14. Although the behavior of each entity can be engineered or designed from first principles, the behavior of the swarm is far more "emergent" and acts as a whole. A swarm often self-organizes. Because the algorithms are simple, they are often subject to analysis and behavioral assurance (i.e., guarantees that the swarm will perform as expected).

As one can see, traditional robots and swarms are the opposite ends of a spectrum. Both approaches are valuable and serve different purposes. A "middle of the road" approach is to use a team of traditional robots. These are often seen in robotic soccer competitions. We can also contrast these more traditional teams of robots with swarms. Figure 1.1 provides one such contrast. The horizontal axis refers to the amount of communication, with increasing communication being to the right. The vertical axis refers to the behavioral heterogeneity, with heterogeneity increasing upwards. Robot teams utilize the same complex software that normal traditional robots use. Each

---

[1] Heuristics are strategies using readily accessible, though loosely applicable, information to control problem solving in human beings and machines [177].

member of the team can perform a different role (goalie, defender, forward, etc.). They can communicate extensively. This represents the area of the figure that is above the diagonal line. Swarm robots usually have the same software, or only very minor variations, and generally do not have different roles (although recent approaches break swarms into castes that play different roles, as shown in Chap. 16). They usually do not communicate as much as the robots in robot teams. This represents the area below the diagonal line.

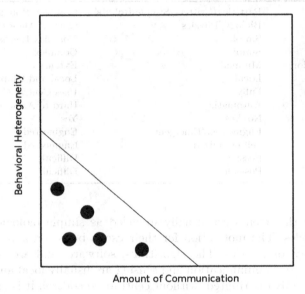

Fig. 1.1: The types of swarms we will be examining in this book

Even below the diagonal line we can make distinctions between two different types of swarms. In the first, the individual robots interact minimally, usually to simply avoid each other. But this doesn't mean that this form of swarm is not worthwhile. In fact, for some tasks, this is precisely what we want. For example, Chap. 11 shows how this type of swarm is used to uniformly search an environment. The technique works for any number of simple robots. This form of swarm has an advantage over one expensive robot due to the multiplicity of units. If the robots are acting as sentries, the swarm is more likely to catch an intruder, due to the large numbers of sentries. Another example is surveillance, as shown in Chap. 7.

We can describe the first type of swarm mathematically as follows. Suppose there are $N$ robots and each robot $i$ has probability $p_i$ of completing a task. Then the probability that none will complete the task is $\Pi_{i=1}^{N}(1 - p_i)$, where the $\Pi$ symbol is shorthand for a product. Then the probability that at least one will complete the task is $P = 1 - \Pi_{i=1}^{N}(1 - p_i)$. This formalism

assumes that the robots are largely independent of each other, which is true
of this type of swarm because the robots interact minimally. Note that the
$p_i$ values tell us how homogeneous the swarm is. If they are all roughly the
same value, then the robots are behaviorally similar and the swarm is homogeneous. If the values differ considerably, the swarm is heterogeneous. Note
also that $P$ provides a baseline level of performance for a swarm with minimal
interactions.

The second type of swarm relies on interactions to perform tasks that
are difficult (or impossible) for a single unit. In this case, the interactions
create emergent behavior, and the swarm acts more like a high-level entity. Example tasks involve cooperative pulling or pushing of an object (see
video "pull.avi"), resource protection (Chap. 16), chemical and biological
plume tracking (Chaps. 6, 8 and 9), tracking a moving optimum while
minimizing noise (Chap. 19), moving as a formation through an obstacle
course (Chap. 14), and creating multi-robot organisms by having swarms
self-assemble (Chap. 5). Experimentally, we can estimate the probability of
task completion as some value $Q$. If $Q \neq P$ we have a swarm of robots that are
interacting. If $Q < P$ the interactions are poorly designed and hurt performance. If $Q > P$ the interactions are well designed and improve performance.
Clearly, we are interested in designing swarms where $Q \geq P$, and that is the
focus of this book.

## 1.3 What Is the Modern History of Swarms?

The modern computer science swarm literature can be subdivided into *swarm
intelligence*, *behavior-based*, *rule-based*, *control-theoretic*, and *physics-based*
techniques. Swarm intelligence [12, 13] techniques are often ethologically motivated and have had excellent success with task allocation, foraging, and
division of labor problems [18, 86]. Hence, biomimetics is closely associated
with swarm intelligence.

Behavior-based approaches [7, 22, 69, 152, 175, 211] and rule-based [210,
276] systems have been quite successful in demonstrating a variety of swarm
behaviors. Behavior-based and rule-based techniques use heuristics for changing the velocity vectors of agents. Control-theoretic approaches have also been
applied effectively (e.g., [3, 60, 137]). Some approaches assume that there are
leaders and followers [48, 49, 64]. We do not make this assumption.

One of the earliest physics-based techniques was the *artificial potential
fields* approach (e.g., [83, 124]). The early literature dealt with a small number of robots (typically just one) that navigate through a field of obstacles
to get to a target location, using global virtual environmental forces. The
environment, rather than the agents, exerts forces. Obstacles exert repulsive
forces while goals exert attractive forces.

This earlier work was extended to larger numbers of robots, with the addition of virtual inter-agent forces. A proliferation of terminology also simultaneously ensued, resulting in "potential fields" and "virtual force fields" [20], "harmonic potential functions" [125], "potential field methods" [129], "artificial potential functions" [194], "social potential fields" [190], and "potential functions" [255]. One can see that an overarching framework is required, since these methods are very similar to each other. They are all physicomimetics.

Carlson et al. [27] investigated techniques for controlling swarms of microelectromechanical agents with a global controller. This global controller imposes an external potential field that is sensed by the agents. Researchers then moved away from global controllers, due to the brittleness of such an approach. Howard et al. [98] and Vail and Veloso [253] extended potential fields to include inter-agent repulsive forces. Although this work was developed independently of physicomimetics, it affirms the feasibility of a physics force-based approach.

Another physics-based method is the "Engineered Collective" work by Duncan at the University of New Mexico and Robinett at Sandia National Laboratory. Their technique has been applied to search-and-rescue and other related tasks [209]. Shehory et al. [217] convert a generic, goal-oriented problem into a potential energy problem, and then automatically derive the forces needed to solve the problem. The *social potential fields* framework by Reif and Wang [190] is highly related to our work in physicomimetics. They use a force-law simulation that is similar to our own, allowing different forces between different agents. Their emphasis is on synthesizing desired formations by designing graphs that have a unique potential energy embedding.

Although biomimetics is generally used for flocking [193], there have also been physics-based approaches. Central to this work is "velocity matching," wherein each agent attempts to match the average velocity of its neighbors. Vicsek provides a point particle approach, but uses velocity matching (with random fluctuations) and emphasizes biological behavior [43, 89, 256]. His work on "escape panic" with Helbing utilizes an $F = ma$ model, but includes velocity matching [89]. Toner and Tu provide a point particle model, with sophisticated theory, but again emphasize velocity matching and flocking behavior [251].

Regardless of the task, the physics-based techniques described above can all be considered to be physicomimetics. One of the goals of this book is to pull the related work together under one umbrella. Physics-based swarm concepts have been around for a while, but they are dispersed in the literature, leading to a subsequent lack of direction and vision for the field. This is unfortunate because physics obeys emergent properties that are desirable for swarms. There is often a tendency, however, to forget the properties and simply use the equations. Also forgotten are the assumptions made in physics that lead to the emergent properties. *But the assumptions and properties are as important as the equations.* By understanding these assumptions and properties more fully we can start to develop a vision for the future.

## 1.4 Why Do We Need Physicomimetics?

Given the success of other approaches, why do we need physicomimetics? The simplest answer is that different approaches are best for different applications, although they are complementary and can overlap. For example, biomimetic swarm approaches have been used for routing (using "ants") and foraging (using "bees"), as well as for exploring the dynamics of river formations. Biomimetic swarms excel at portraying flocks of birds, schools of fish, herds of animals, and crowds of people in computer graphics game and movie applications. A number of biomimetic techniques have also been applied to optimization problems, including ant colony optimization, particle swarm optimization, and bacteria foraging optimization.

Physicomimetic approaches are also used for computer graphics simulations. These are referred to as "particle simulations," which use physics forces to imitate the movement of water, explosions, fire and smoke. Physics "engines" have been built in hardware that utilize "physics processing units" to speed up particle systems and rigid body simulations. From the robotics point of view, physicomimetics has been successful at applications that require coverage of an environment or use moving formations that act as distributed sensor meshes. As stated in our original paper [235]:

> The example considered here is that of a swarm of MAVs whose mission is to form a hexagonal lattice, which creates an effective sensing grid. Essentially, such a lattice will create a virtual antenna or synthetic aperture radar to improve the resolution of radar images.

At that time we were considering using micro-air vehicles (MAVs) as our inexpensive (roughly US $1000) robotic platforms. In a later paper on chemical plume tracing we state [242]:

> In our approach, each robot acts as a computational fluid dynamics grid point, thereby enabling the robotic lattice to behave as a distributed computer for fluid computations. Although robots only gather information about the fluid locally, the desired direction of movement for the overall lattice emerges in the aggregate, without any global control.

Although we do show examples of both kinds of swarms in this book (those that interact minimally and those that interact collaboratively and cooperatively), it is the latter that is of greatest interest to us. Hence, we generally view swarms as a mobile mesh of distributed computers that sense environmental information, process it, and respond appropriately (e.g., in Chaps. 5, 6, 8, 9, 12, 16, 14, 18, and 19). The swarm *is* an entity unto itself. In fact, if the distributed computer is a neural network, the swarm is a moving, learning organism.

*The goal of physicomimetics is to synthesize efficient distributed multi-agent systems that contain tens to thousands of mobile agents, each carrying one or more sensors. We want to design these systems from first principles using scientific methods amenable to analysis.* There are numerous reasons why we think that a physics-based approach can help us attain our goal:

- Physics is the derivation of macroscopic behavior from microscopic interactions.
- Physics is a reductionist approach, allowing us to express the macroscopic behavior elegantly and simply.
- Physics is the most predictive science.
- Physical systems are naturally robust and fault-tolerant. Low-level physical systems can also self-organize and self-repair.
- The physical world is lazy.

An example of self-organization (and self-assembly) is shown in Fig. 1.2. Thirty nine magnets were placed on a board, in the same magnetic orientation. Hence they repelled each other. Left that way, nothing would have happened. The system was in equilibrium. But when I dropped the fortieth magnet from above, the equilibrium was broken. Individual magnets moved, jumped, and realigned themselves until all forty achieved a new equilibrium. They self-organized into a linear structure. There was no global controller, and the self-organization occurred via local interactions between the magnets. A video of this self-organization is shown in "selforg.avi".

Physics is a rich field, providing an enormous number of behaviors that are worth emulating to achieve various swarm-related tasks. Furthermore, physics has an excellent track record at showing how these behaviors emerge from simple interactions between the constituent parts of nature, *and* of predicting the behavior of the emergent system. The most important reason, however, for physicomimetics is the statement that "Nature is lazy." This was the statement that I remembered hearing over and over when I was learning physics. However, I never saw it again until I found an excellent new physics book by Cristoph Schiller, who states:

> Everyday motion is described by Galilean physics. It consists of only one statement: all motion minimizes change. In nature, change is measured by physical action $W$. More precisely, change is measured by the average difference between kinetic and potential energy. In other words, motion obeys the so-called *least action principle...* In other terms, nature is as lazy as possible. [207]

> In nature, motion happens in a way that minimizes change. Indeed, in classical physics, the principle of least action states that in nature, the motion of a particle happens along that particular path—out of all possible paths with the same end points—for which the action is minimal. [208]

The principle of least action is one that we intuitively rely upon all the time. Figure 1.3 shows some examples. The top left picture is a piece of steel. When pushed (top right) it bends in such a manner that each portion changes shape as little as possible. We know this is because the steel is rigid, but it is not something we think about most of the time. The same thing happens if we pull the steel (lower left). Finally, if we take another piece of steel and bend it into an arc, each portion of the arc makes a minimal change.

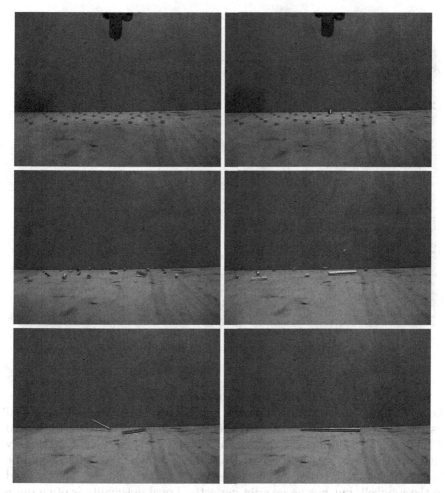

Fig. 1.2: The self-organization of 40 magnets

You may have heard that water moves along the *path of least resistance*. This is exactly the same concept. Figure 1.4 shows that not only does a fluid move downhill, but it does it in the easiest way possible. The concept is impossible to understate—it pervades all of classical physics, and holds even in quantum physics [208].

This deeper view of physics has two important consequences. First, the realization that motion is minimal is an excellent match for swarm robotic systems. Robots always have power constraints. We can either attempt to explicitly compute the action that minimizes change, *or* we can achieve this automatically by using physicomimetics. Well designed physicomimetic swarms obey the principle of least action. They are efficient.

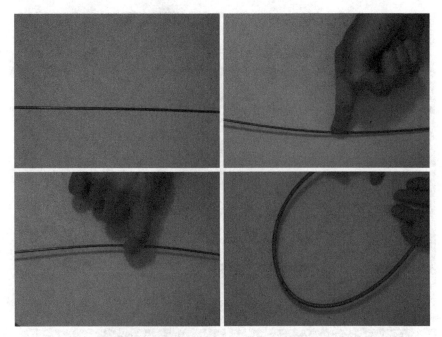

Fig. 1.3: Principle of least action in real life

Second, this minimality of change occurs only at the local level. Hence, the system is locally optimal. However, this does *not* indicate that the global system is optimal. Many researchers focus on this issue, and then work hard at forcing global optimality. But, in my opinion, this goes against the very nature of the basic swarm concept. Instead of optimality, swarms should be "satisficing" (a word that combines "satisfy" with "suffice"). They should do the job well, but not necessarily perfectly. Local optimality is often quite sufficient. Perfection requires far too many unrealistic expectations, such as a lack of robot failure, or global knowledge of the environment. Perfect systems are brittle because changes occur that aren't anticipated. Organic and physical systems, however, are not perfect. They never are. Hence they are more robust. There are systems where we need precise guarantees of behavior— such as the dynamic control of an airplane wing, to ensure that the plane does not stall and crash. But for other tasks, perfection and optimality are not even relevant.[2] We can make greater strides by simply accepting the fact that errors can and will occur.

---

[2] The Internet is a wonderful example of a robust system that at any moment has failing nodes, yet continues to operate.

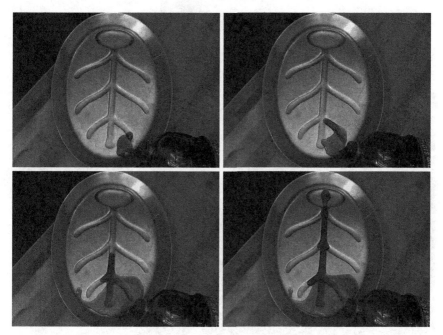

Fig. 1.4: Principle of least resistance with fluids in real life

## 1.5 Where Did Physicomimetics Come From?

This sentence, paragraph, chapter, book, and a whole decade of research literally arose from one idea. A colleague at the Naval Research Laboratory (NRL) wondered if micro-air vehicles could be programmed to fly in a triangular lattice formation. After some thought, the compass construction of triangular (hexagonal) lattices came to my mind.

Figure 1.5 shows how this can be done. First, draw a circle with a compass. The center of the circle is where one vehicle resides. Anywhere on the circle is where a neighbor should be located. Place the compass somewhere on the first circle and draw a second circle. This is shown in the top left picture of Fig. 1.5, which has two vehicles. Now a third vehicle should be placed at one of the two points of intersection of the two circles. This is done in the top right figure. If we keep adding vehicles to the locations where circles intersect, a triangular lattice can be built. The lower right picture shows a possible result with seven vehicles.[3] Nothing prohibits us from adding more and more vehicles. This compass construction is scalable to as many vehicles as we want. Note that removing a vehicle will not disturb the formation either, so it is robust.

---

[3] Of course, other triangular formations can be built, depending on the order and placement of the new vehicles.

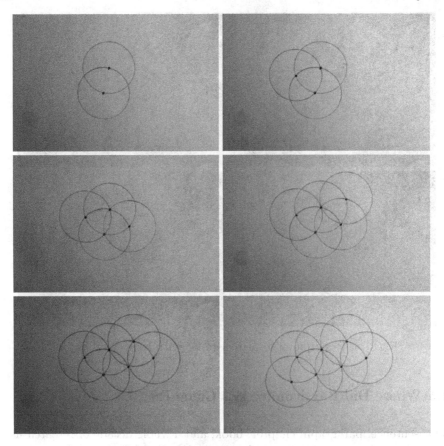

Fig. 1.5: Compass construction of a triangular lattice

Figure 1.5 provides a geometric interpretation of triangular lattices. But vehicles aren't really going to get into formation by the use of compass and paper. So, let's provide an interpretation that is closer to physics. This interpretation is called the "rubber sheet model." Usually it is used to explain gravity wells, but it will help us explain how lattices can organize from a physics perspective. Figure 1.6 illustrates the rubber sheet model. We start with one vehicle (particle), in the top left picture. The vehicle is at the tip. Other vehicles near the tip will want to move away and move down towards the trough. Vehicles that enter from the perimeter will also want to move down towards the trough. The bottom of the trough represents the circle drawn around a particle in Fig. 1.5. It is a "potential well" where new vehicles would like to reside.

Once a new vehicle is added (upper right picture), it wants to move to the trough around the original vehicle. But this new vehicle also deforms the rubber sheet, so the original vehicle wants to move to the trough around the

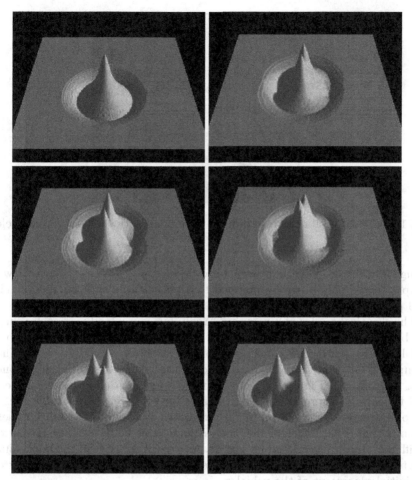

Fig. 1.6: Gravity well depiction of triangular lattice incremental construction

new vehicle! Both particles are deforming the rubber sheet, which in turn causes both to move. When the two particles come to a rest, they are each at a position corresponding to the original troughs that both particles created. But now there are new troughs. And, due to a concept called "constructive interference," the two troughs intersect in two areas to create even deeper deformations of the rubber sheet. This corresponds to the two intersection points of the two circles in the top left picture in Fig. 1.5. When a new particle comes in (middle right), it moves to one of the two intersection points, as shown in the lower right picture.

This process continues until all seven particles have been introduced, as shown in Fig. 1.7. Note the resemblance between this end result and that shown in Fig. 1.5. A video of this process is shown in "selforg_inc.avi". Al-

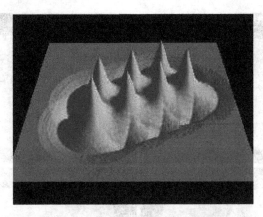

Fig. 1.7: Gravity well depiction of final triangular lattice with seven particles

though the details have not been fully fleshed out, we have shown how a compass construction can be mapped to a physics rubber sheet model. The final procedure, using physics forces, is shown in Chap. 3.

The skeptical reader will now say, "Yeah, but you introduced each vehicle one by one! That isn't reasonable. It is better to introduce them all at once, and let them self-organize." The skeptical reader is indeed correct. What is impressive is that particles *can* self-organize when introduced all at once (i.e., in a batch). Figure 1.8 shows the self-organization of seven vehicles into formation. There is no need to introduce the vehicles one by one. A video of this process is shown in "selforg_batch.avi". It is this ability to self-organize that drives much of the motivation for this book! Yes, the formation is different, but it is a triangular lattice. As we will see later in the book, since Newtonian physics is deterministic, the final formation depends entirely on the initial placement of the vehicles.

After remembering the triangular construction via the compass, I decided to write down my thoughts as generally as possible. Figure 1.9 shows the first two pages (of six) as I wrote them down on September 28, 1998. It outlines my original $F = ma$ Newtonian physics version of physicomimetics. Most importantly, however, was my vision of the goal of the work, especially with respect to the last sentence (Fig. 1.9, right):

> The goal of this work is to be able to clarify how local interactions between agents lead to global phenomena. Since physics often identifies macroscopic behavior (i.e., empirical laws) from microscope entities, we use physics to provide enlightenment. In this framework agents are particles and local interactions are described by the forces acting on those particles.
> We will not restrict ourselves to any particular form of physics.

As you will see (Fig. 1.9, left), the first name I gave to this work was "artificial physics." However, the physicists at NRL disliked the phrase. Whereas I thought of "artificial" as meaning "made by human skill," they interpreted

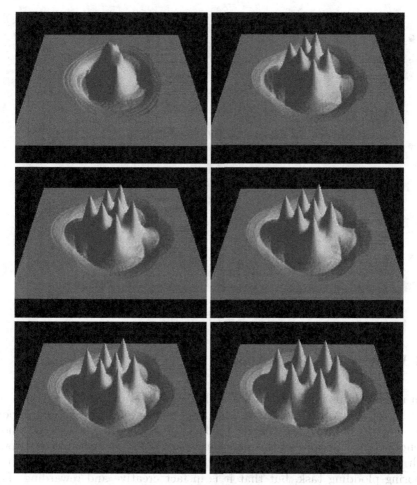

Fig. 1.8: Gravity well depiction of triangular lattice batch construction

it as "imitation, sham" and recommended that I find another way to describe the approach. After some thought it seemed quite reasonable that "physicomimetics" described the work precisely, just as "biomimetics" was describing the biologically-motivated swarm research. You will see both phrases in this book, depending on the context.

It is the last sentence, however, that I think is the most important. It stresses the notion that physicomimetics is not simply $F = ma$. It is a way of thinking about problems. As stated earlier, physics provides new metaphors for approaching these problems.

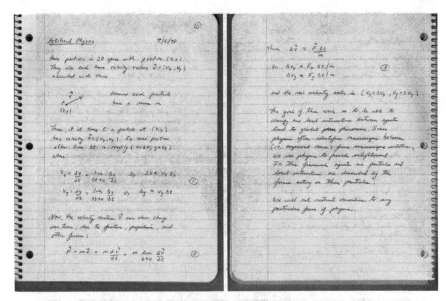

Fig. 1.9: My notes from September 28, 1998

## 1.6 What Is This Book About?

In a nutshell, this book provides a well-documented discussion of how a research program starts and grows. This is an unusual perspective, but important, because it provides younger readers the opportunity to see just how
science is done. Because of this perspective the book is structured to "grow"
with the student. One of our goals is to show that scientific research is not
a boring plodding task, but that it is in fact creative and rewarding. The
beginning of this book is suitable for high school students, especially when
coupled with the simulations. Even if you have no programming experience,
you can run the simulations. All you need is your web browser. If you find
this material interesting you may want to consider majoring in physics, computer science or electrical computer engineering in college. The middle of the
book is tailored for an undergraduate in computer science or electrical computer engineering. Numerous senior design projects can be obtained from this
material, especially by expanding the simulations or by building the hardware we describe. The final sections contain more advanced topics suitable
for graduate students looking for advanced degree topics and for researchers
in the swarm community.

## 1.6.1 A Holistic View

The structure of this book mirrors the real-world process. Chapters build upon each other, with increasing levels of sophistication. For example, a simple idea led to an algorithm (a piece of software). The algorithm determined the hardware requirements for our robots. We then built the necessary hardware. From there various tasks were defined, which led to new algorithms and additional hardware. Strengths and weaknesses became more apparent. Strengths were further exploited, while weaknesses became avenues for continued research. Most research papers hide weaknesses, because of the nature of the research community (which only rewards success). But the identification of weaknesses is often the moment when an important concept is learned and appreciated. Eventually, as the pieces fell together, more insights were gained, providing a level of understanding that we never imagined originally. The hope is that this book allows you to glimpse some of the excitement we felt during the journey.

This book is unusual also in that it is both deeper and broader than others in this area. It is deeper because we not only use physics to build swarm algorithms, but we also explore the assumptions underlying physics, and then carefully modify those assumptions when necessary, if it helps us achieve the emergent behavior that we desire. A careful investigation of symmetry, locality, and various conservation laws aids us in this process. What we will learn is that *we can change assumptions, yet still retain the properties that we want. We don't have to precisely copy physics from the real world, but can modify it to suit our needs.* The book is broader because of the variety of concepts that are explored:

- Not only do we build swarm formations that act as solids, but we also have swarms that act as gases and liquids. In addition, we explore different forms of physics, including kinetic theory and Brownian motion. This allows us to provide algorithms for a wide variety of applications, including pre-assembly alignment, uniform coverage, chemical plume tracing, monitoring chemical and biological concentrations on the way to a goal, snaking through narrow passageways, surveillance in corridors, and optimization. When appropriate, these algorithms are compared to the leading biomimetic approaches.
- Two of these applications are tested on our own robots, which were specifically designed around the requirements of the algorithms. This book provides a thorough discussion of our enabling hardware technology.
- Predictive theory is provided, which serves three purposes. First, it verifies the correctness of our algorithms (i.e., the theory and simulation must match). Second, it helps the user to set algorithm parameters. Third, it allows a subset of a swarm to predict the success of a whole swarm on some tasks.

- Although physics-based swarms are quite adaptive, they cannot anticipate dramatic changes in the environment. In the past, this was handled by having a central robot recompute the swarm algorithms, utilizing global information of the environment. This process is too slow and requires too much information. We provide an algorithm that allows swarms to adapt immediately to changes in an environment.
- We show how physics-based swarms can be used for function optimization and the tracking of optima in noisy domains, carrying our work to the domain of swarm intelligence.
- We address the vision of "swarm engineering." Here the goal is to describe the desired behavior and then automate the process of determining the interactions between the robots (agents) that will create that behavior.

### 1.6.2 An Atomistic View

Now let's provide an overview, chapter by chapter. The book is broken into five parts. Part I provides the introductory material. Part II starts to push the algorithms towards applications. Part III provides details on our robots and shows how we used them for two applications. Part IV discusses more advanced topics, such as learning, prediction and swarm engineering. Part V shows two different artificial physics algorithms for optimization problems.

Part I is composed of this chapter, up through Chap. 4. Chapter 2 provides an introduction to Newtonian physics. Most importantly, it also explains some emergent properties of Newtonian systems, such as the conservation of linear momentum, angular momentum, and energy. These emergent properties occur because of the assumption of "equal and opposite" forces. We use simulations of a spring and a simple solar system to explore these concepts in detail. The simulations are written in NetLogo, which is simple to understand and modify if you wish to pursue further explorations.

Chapter 3 shows how we can utilize the concepts from physics to create swarm systems that self-organize into formation, and move through obstacle courses while moving towards a goal. This chapter also introduces some simple physics theory that helps the user to set simulation parameters appropriately. We show that these formations scale to a large number of agents, but are robust in the face of agent loss. We show that physicomimetics is not only noise tolerant, but in fact can demonstrate improved performance when noise is present. Finally, the chapter summarizes the hardware requirements of the algorithms.

These early chapters assume that sensors and communication are both local and very limited in capability and bandwidth. Chapter 4 pushes physicomimetics forward in four ways. First, we allow a small amount of global information. Second, we permit slightly more communication. Third, we start to allow the algorithms to violate some standard physics assumptions. Fourth, we generalize to new forms of physics. This allows us to expand the repertoire of physicomimetics behaviors. We examine four tasks: (1) creating perfect formations, (2) creating chain formations, (3) uniformly exploring a region, and (4) using a swarm of robots to "sweep" an area. The latter task introduces a new version of physicomimetics based on kinetic theory. In addition, theory is used to optimize the performance of uniform exploration and to verify the correctness of the kinetic theory simulation.

Part II examines applications of physicomimetics. Chapter 5 utilizes a different form of physicomimetics called "local oriented potential fields." The emphasis is on swarm robots that can physically dock with each other, to share energy and computational resources within a single artificial life form. Using four different deployment strategies, numerous swarm behaviors are investigated, including safe wandering, following, dispersion, aggregation, homing and angularity.

Chapter 6 examines the use of physicomimetics for the control of mobile aquatic sensor networks. Biological organisms produce light from chemical reactions. There are numerous applications that require monitoring and tracking these organisms. By using a swarm of aquatic vehicles as a distributed sensing network, the chapter shows how the network can be utilized for both ocean monitoring and submarine stealth. The chapter starts with a NetLogo simulation, but then proceeds to discuss a more complex multidrone simulator and the "water helicopter" vehicles that are being built to perform these tasks.

Chapter 7 examines a new form of physicomimetics based on the behavior of gases, called kinetic theory, expanding on the work shown in Chap. 4. Applications include surveillance and sweeping of an obstacle-filled environment to provide continuous monitoring. Theoretical laws are derived, which make predictions about the stochastic behavior of the swarms. A comparison indicates that the kinetic theory approach achieves comparable performance to a biomimetic approach, but in roughly one-third the time.

Chapters 8 and 9 provide very thorough discussions of our work in chemical plume tracing. As opposed to predicting where a toxic substance will go, this work uses swarms as distributed sensor meshes to find the source of the emission. This is obviously important for military and manufacturing activities. Based on principles of fluid dynamics, a physicomimetics algorithm called "fluxotaxis" was created. This algorithm uses sensor mesh computations to drive the swarm lattice to a chemical source emitter. Simulations indicate that fluxotaxis is superior to biomimetic approaches. Results are also shown on our first MAXELBOT prototype robots, indicating that a small swarm of three robots (with one chemical sensor per robot) outperforms one robot with four sensors. This is an excellent demonstration of an application that is perfect for swarms.

Part III provides a summary of our research with our more advanced MAXELBOTS. Chapter 10 provides a thorough summary of the robot itself. Our algorithms require that robots know the range and bearing to neighboring robots. This is accomplished via specialized trilateration hardware, that was built in-house. Chapter 10 explains the mathematics underlying trilateration and shows how we built the hardware. Then it expands the capability of the robot by adding more components, such as an obstacle detection module, temperature sensor, liquid crystal display, and magnetic compass. All modules communicate with each other over an $I^2C$ bus.

Chapter 11 expands on the uniform coverage section that was briefly covered in Chap. 4. A robot-faithful simulation tool is used to build environments and test the quality of the uniform coverage algorithm. The algorithm is discussed in more detail, including a proof of correctness. The chapter also shows the derivation of the mean free path that provides optimal performance, and concludes with three different experiments using the MAXELBOTS.

Chapter 12 expands on the chain formation section that was also briefly covered in Chap. 4. Again a robot-faithful simulation tool is used to build environments and test the performance of the chain formation algorithm. A more thorough description of the chain formation algorithm is provided, along with performance metrics and simulation results. The effects of sensor and motor noise are explored, showing that the algorithm is quite robust. Finally, the chapter concludes with details of our MAXELBOT implementation. The robots were able to organize into a chain, maintain a communication network over this chain, and report the shape of the environment back to the user.

In the previous chapters, we have assumed that our robots can "turn on a dime." This makes robot implementations much easier. Chapter 13 shows how to apply physicomimetics to vehicles whose motions are physically constrained. It is best to think of physicomimetics as establishing waypoints for a vehicle. A lower-level controller must compute the best way to get to these waypoints, while not imposing too much stress on the vehicle. This is especially important for fixed-wing aircraft, which are subject to minimum turn rate and stall speed limitations. Prior to this chapter we have considered an agent to be a point particle. Chapter 13 provides a more sophisticated view of an agent. By allowing an agent to be composed of multiple particles, physicomimetics can be applied to physically constrained vehicles. This is accomplished by introducing a new form of physicomimetics that uses virtual *torsion* springs between agents.

Part IV introduces more advanced topics. Chapter 14 is a recognition that although swarms are adaptive and robust, environmental changes can be sufficiently severe that swarm performance can suffer. This chapter provides a distributed learning algorithm for swarms. It is not a physics-based algorithm and can be used for almost any kind of swarm. In fact, the algorithm is inspired by distributed evolutionary algorithms, where neighbors exchange information to help them improve performance in a changed environment. Chapter 14 provides a detailed study of the application of this algorithm to physicomimetics, in the context of the movement of formations towards a goal, when the obstacle density increases. Through adaptation the swarm is able to cope with the increased difficulty of the problem.

Chapter 15 presents a novel framework for using a small subset of a swarm (i.e., a group of scouts) to predict the probability that the remainder of the swarm will succeed at a given task. Two predictive formulas are provided within the framework, one based on Bayesian reasoning. The framework is applied to the physicomimetic control of agents that need to travel through a field of obstacles to get to a goal location. Interestingly, we note a difference in the accuracy of the predictions depending on whether the formations act as liquids or solids. This intriguing synergy between physicomimetics and predictability is explained in the chapter.

Chapters 16 and 17 both focus on the important issue of how to design swarm systems, a topic that is called "swarm engineering." Chapter 16 describes a graph-based method for designing interaction models in physicomimetic systems. This method allows engineers to construct graphs that clearly define what interactions are possible. By using this method for condensing subgraphs, engineers can think more modularly about the design process and produce reusable behavioral modules, giving the engineer the ability to directly control the scalability of the system. Chapter 16 uses the graph-based model to design heterogeneous, modular solutions to a resource protection problem.

Chapter 17 provides a swarm design methodology that follows two steps. First, a mathematical equation is created that defines the goal of the swarm in terms of properties of the system that the agents are capable of interacting with. This equation is used to derive a swarm condition that indicates what the agents must be able to do to achieve the goal. Then the swarm engineer creates the specific model that accomplishes what the swarm condition indicates. The chapter illustrates how this methodology can be used with physicomimetics to engineer desired swarm behavior.

Part V concludes the book by introducing two methods for using artificial physics for function optimization. Chapter 18 introduces an artificial physics optimization (APO) algorithm that focuses on changing the mass of particles based on their fitness. The better the fitness the larger the mass. Due to inertia this provides a bias (preference) for particles to stay near areas of the search space with better fitness. Chapter 18 provides a suite of APO algorithms and convergence analyses. The best of the APO algorithms outperform particle swarm optimization on numerous problems.

Chapter 19 also provides an APO algorithm, but for a different purpose. As with our earlier work on chemical plume-tracing, the goal now is to surround an optimum even when the fitness function is noisy, and then track the optimum if it moves. The rigid distributed formation serves two purposes. First, the algorithm never prematurely converges to a small area of the search space, because the swarm is always in formation. Also, the rigidity of the formation allows the swarm to move in a direction that is obtained via an implicit emergent consensus of the particles. This helps to minimize the effects of noise. Particle swarm optimization has great difficulty with this task. The chapter concludes with a discussion of the importance of large-scale and small-scale structures in fitness landscapes, and how artificial physics optimization can adapt to these structures.

## 1.7 Where Do We Go From Here?

In 1979, when I was a teenager, I wrote a letter to A. K. Dewdney, asking for a copy of his manuscript "Two Dimensional Science and Technology." Although I was just a high school student, he kindly replied, and sent me his manuscript. I read it and I was hooked. It is one of the most influential pieces I have ever read—not merely because he rederived physics for two dimensions specifically, but because he was willing to reinvent physics for a new situation. In fact, Dewdney has sections devoted to physics, chemistry, astronomy, planetary sciences, biology, and technology. Is it surprising that I later came up with the term "artificial physics?"[4]

A recent book called "The Microverse" discusses the world that is invisible to the naked eye [182]. Using a series of essays coupled with science fiction stories, the book delves deeper and deeper into the structure of the universe. Starting with cells, the book proceeds to the genetic code, molecules, the nuclear world, electrons, particle physics, quarks and the four forces. Then, coming full circle, we are brought back to the link between fundamental physics and cosmology.

What one should notice is that *each* level provides metaphors for swarms. Biomimetics is at the very highest level. The goal of physicomimetics is to capture the deepest levels. Can metaphors from statistical mechanics or classical thermodynamics help us? Perhaps quantum physics, superconductivity, or relativity might help. Can we make use of the Pauli exclusion principle or the Heisenberg uncertainty principle? These are topics worth exploring in far more detail.

Finally, there are intermediary levels that warrant much more exploration. Chemistry has been largely overlooked in the swarm community. But what if one really wanted swarms that could move like water, become solid, or become a gas? Then mimic $H_2O$ molecules as closely as possible. Self-assembly via covalent bonds is an interesting possibility. And why limit ourselves to standard chemistry? Dewdney's manuscript discusses two-dimensional chemistry, creating a table of elements that is quite different from what we have been taught. What is the behavior of molecules built from these elements?

The main point is that there are countless new metaphors for us to explore with respect to swarms. But how do we adapt these metaphors to problems that we want to solve? Dewdney proposes two principles for the planiverse, the "principle of similarity" and the "principle of modification" [53]:

---

[4] Dewdney's manuscript is now available as a book entitled "The Planiverse: Computer Contact with a Two-Dimensional World" [53].

Subject to the principle of modification, a phenomenon in the planiverse shall be as similar as possible to its steriversal counterpart.

In the case of contradictory conclusions it will be the more superficial theory which is modified, i.e., the one less deeply rooted in an invocation of the principle of similarity.

Physicomimetics is about the imitation of physics. Thus we also follow a principle of similarity and a principle of modification. But there is one important difference. Dewdney is restricted to providing a whole two-dimensional universe where there are no (or few) inconsistencies. We are not trying to do that. We are interested in solving problems. Hence we can create a new physicomimetics instantiation for any problem. All that we ask is that each instantiation be internally consistent and obey its own laws. Our two guiding principles can be summarized as:

- Subject to the principle of modification, a phenomenon in the swarmiverse shall be as similar as possible to its natural counterpart.
- In the case of contradictory or competing theories it will be the poorer performing theory that is modified or abandoned.

The first principle is almost identical to Dewdney's, and captures the idea that we want to be as close to real-world physics as possible, because it possesses so many desirable swarm qualities (e.g., the principle of least action). However, we are *not* trying to build a consistent artificial world. We are task driven. As a consequence, there must be performance metrics associated with our physicomimetics algorithms. If changes to the physics break some assumptions but yield superior performance, this is the path we will take. It is a cautious path, but as you will see in this book, there are large payoffs for pursuing this route.

# Chapter 2
# NetLogo and Physics

William M. Spears

## 2.1 A Spring in One Dimension

This chapter will introduce you to NetLogo and Newtonian physics. Let's
start with NetLogo first. NetLogo is a tool that allows us to simulate distributed, non-centralized systems of agents. As we discussed in Chap. 1,
agents are autonomous entities (either in software or hardware) that observe, react and interact with other agents and an environment. People and
other living creatures are agents. Robots are agents. Some agents are computer programs that exist only in computer networks. NetLogo allows us to
simulate many types of agents. There are two ways you can interact with the
NetLogo simulations in this book. The simplest way is to run them as Java
applets in your browser. In this case, you will need to make sure that Java 5
(or a later version) is installed on your computer. You will be able to run the
simulations in your browser, but you will not be able to modify the code. If
you have trouble with one browser, try another. Browsers that work well are
Firefox, Internet Explorer, Flock, Chrome, Opera and Safari.

Alternatively, you can run the NetLogo simulations directly on your computer. This is what we will assume throughout this book. In this case, you
will have to install NetLogo. We will be using NetLogo 4.1.2. Currently, the
NetLogo web page is at http://ccl.northwestern.edu/netlogo/. If
this link does not work, use a search engine to locate NetLogo 4.1.2. Then
follow the download and installation instructions. NetLogo will run on a Windows machine, a Macintosh or under Linux.[1] When you are done, download the NetLogo code that comes with this book. We are going to start
with a spring simulation called "spring1D.nlogo" (nlogo is the suffix used for

William M. Spears
Swarmotics LLC, Laramie, Wyoming, USA, e-mail: wspears@swarmotics.com

[1] The NetLogo code in this book has been tested on Debian and Ubuntu Linux, Mac OS
X, as well as Windows Vista and Windows 7.

W.M. Spears, D.F. Spears (eds.), *Physicomimetics*,
DOI 10.1007/978-3-642-22804-9_2,
© Springer-Verlag Berlin Heidelberg 2011

NetLogo programs). On a Windows or Macintosh, simply download this file
and double-click on it. If you are using Linux, run "netlogo.sh". If everything works properly, one window with three tabs should open. One tab is
for the NetLogo code (which we will talk about later). Another tab is for the
documentation. The third is the graphical window. Select File, then Open...,
and select "spring1D.nlogo" (you may have to navigate your way to the file,
depending on where you placed it on your computer). Now your graphical
window should look like Fig. 2.1.

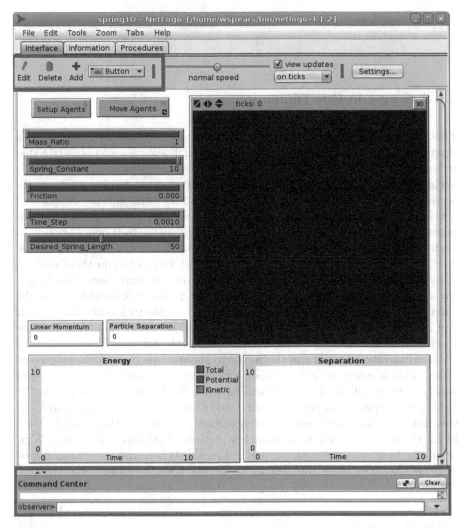

Fig. 2.1: A picture of the NetLogo graphical user interface for the one-
dimensional spring simulation

If you are using a web browser, you will click on "spring1D.html" instead. In this case, the graphical window shown in Fig. 2.1 should appear in your browser. As you scroll down you will also see the documentation about this simulation, and the actual code will be shown at the bottom.

A full explanation of NetLogo is beyond the scope of this book. For example, we will not use the upper left portion and bottom portion (highlighted in red) of the graphical window shown in Fig. 2.1. However, we will explain enough to get you started. It is very intuitive, and once you start to learn NetLogo you will gain confidence very quickly. The graphical interface to most of the simulations shown in this book contains several components. Violet buttons allow you to start and stop the simulation. For example, Setup Agents initializes the agents. The other button, Move Agents, tells the agents to start running. If you click it once, the agents start. If you click it again the agents stop. If you look carefully you will notice single capital letters in the buttons. Initially, they are light gray in color. If you mouse click in the white area of the graphical window, the letters will become black. When they become black the letters represent keys you can type on your keyboard to control the simulation. In this case, you can type "S" or "s" to set up the agents (NetLogo is not case-sensitive in this situation). Then you can type "M" or "m" to start and pause the simulation. These action keys are not available, however, if you are running the simulation in your browser.

Green sliders allow you to change the values of parameters (variables) in the simulation. Some sliders can be changed when the simulation is running, while others can be changed only before initialization. The "initialization sliders" and "running sliders" depend on the particular simulation. The light brown bordered boxes are "monitors" that allow you (the observer) to inspect values of parameters as the simulation runs. Similarly, graphs (shown at the bottom) also allow you to monitor what is happening in the simulation. Graphs show you how values change over time, whereas monitors only give you values at the current time step. One other useful feature is the built-in speed slider at the top of the graphical user interface. The default value is normal speed, but you can speed up or slow down the simulation by moving the slider. This slider is available when running the simulation in your browser.

So what precisely is "spring1D"? It is a one-dimensional simulation of a spring that is horizontal. There is a mass connected to each end of the spring. There is a slider called Desired_Spring_Length that represents the length of the spring when no force is applied. It is set to 50 (with NetLogo simulations such as this, units are in terms of screen pixels) in the simulation, but you are free to change that. Consider the following Gedanken (thought) experiment. Imagine that the two masses are pulled apart beyond 50 units (this is not something you are doing with NetLogo yet; this is an experiment you perform in your mind). What happens? The spring pulls the masses closer. On the other hand, if the masses are closer than 50 units, the spring pushes them back out.

Another important parameter that is controlled by a slider is the Spring_Constant. This is simply the stiffness of the spring. If the spring constant is higher, the spring is stiffer. For now we will focus on the Particle Separation monitor and Separation graph. The monitor indicates the distance between the two weights at any moment in time. The graph shows how this separation changes over time.

Let's run the simulation. Start the code, but do not change any parameters. First, click Setup Agents. You will see two small white dots. These represent the weights at the ends of the spring. You will also see a small red dot, which is the center of mass of the spring system. We will explain this more later. For now you will note that the red dot is halfway between the two white dots. This is because we are currently assuming that the two weights at the ends of the spring are of equal mass. Click Setup Agents a couple of times. You will notice that the initial positions change each time. This is because NetLogo uses a random number generator. If you have any randomness in the system, the simulation will run differently each time. In the case of this particular simulation, the only elements of randomness are the initial positions of the two weights. The initial position of the red dot representing the center of mass also changes with each experiment, since it depends on the initial positions of the two weights.

Now click Move Agents and the simulation will start. At any time you can click on it again to pause the system, and then click again to continue. Watch the two white dots. Because of the randomized initialization, the two white dots are highly unlikely to be placed initially such that they are precisely 50 units apart. If they are closer than 50 units, they immediately move away from each other. If they are farther than 50 units, they move towards each other. Then they oscillate back and forth, as we would expect from a spring.

Examine the Particle Separation monitor. You will note that the separation increases and decreases around 50. In fact, the amount of deviation from 50 is the same in both directions, i.e., if the maximum separation is $50 + x$, then the minimum separation is $50 - x$. You can also see this oscillation clearly on the Separation graph. An example is shown in Fig. 2.2. The $x$ (horizontal) axis represents time. The $y$ (vertical) axis represents the amount of separation between the weights on the ends of the spring.

Feel free to play with the system. Stop everything, change the desired spring length, and rerun the system. Similarly, decrease the spring constant (to one, for example). What happens? Precisely what you would expect. The spring is weaker and less stiff, so the oscillations are longer. In fact, change the desired spring length and/or spring constant while the system is running! This is analogous to modern automotive suspension systems that can be adjusted by the driver—pushing a switch inside a car changes the characteristics of the spring. Very soon you will understand the dynamics of a spring.

To continue with the automotive analogy, the movement of each wheel is damped by a spring connected to the car. But the wheel weighs much less than the car. This is essentially the Mass_Ratio slider in the simulation. The

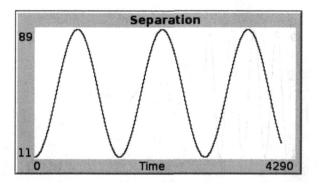

Fig. 2.2: A graph of the length of the spring over time in the one-dimensional spring simulation

mass on one end of the spring has a fixed weight of one. However, during initialization (and only then) we can adjust the weight of the other mass to be one or greater. For example, set it to 10 and rerun the system. What you will notice is that one dot hardly moves, while the other moves a lot. The dot that is almost stationary is the heavier mass. Simply put, the car doesn't move much while the wheel does.

You will also notice that the red dot representing the center of mass of the system is no longer halfway between the two white masses, but is very close to the "heavy" dot. Although we will discuss this more precisely later in the chapter, it is the location where the system is balanced. If you replaced the spring with a bar, and attached the two masses to the ends, the center of mass is where you would pick up the bar so that it would be balanced properly.

At this point you will have noticed that the system continues to oscillate forever. This is an "ideal" spring that doesn't exist in the real world. Real springs are damped by friction and eventually come to rest (assuming they are not perturbed again). And in fact, this is precisely what we would want a car spring to do. We do not want our car to bounce forever because of one pothole. Shock absorbers provide friction that damp the spring. We can simulate this easily by changing the Friction slider from zero to 0.001. First, start the simulation with zero friction, and let it oscillate for a while. Then increase the friction to 0.001. Very quickly the oscillations will damp and the separation between the two masses will settle at the desired separation distance again. An example is shown in Fig. 2.3.

Up to this point we have not discussed the Time_Step slider, the Linear Momentum monitor or the Energy graph. These are somewhat more technical and require a deeper discussion of Newtonian physics. As we discuss these topics next we will also take the opportunity to indicate how these concepts are programmed in NetLogo.

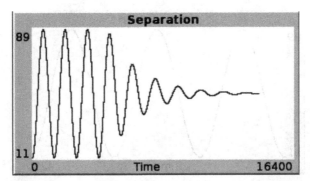

Fig. 2.3: A graph of the length of the spring over time in the one-dimensional spring simulation, when friction is applied after 5,000 time units

## 2.1.1 NetLogo Details and Newtonian Physics

In addition to the graphical Interface tab, NetLogo has a Procedures tab for the code. NetLogo assumes that there are two *conceptual* perspectives for the code, the agent perspective and the observer perspective. The agent perspective is from the agent's point of view. For historic reasons agents were called "turtles" a long time ago, and this convention still holds. We will use the terms "turtles," "agents," "particles," and "robots" synonymously throughout this book, because in all cases we are discussing a set of distributed cooperating entities. The observer perspective is a more global perspective and generally deals with higher-level concepts. Stated another way, the observer perspective is used to monitor the emergent behavior of the system, while the turtle perspective deals with the individual agents themselves. The distinction between the two perspectives is not always clear, and NetLogo itself does not care much about the difference. But it allows you, the programmer, to more easily maintain different perspectives on the system.

Let's look at the code in Listing 2.1. At the top of the code is a list of global variables. These are variables that we (the observer) are interested in examining or controlling. Some of the variables refer to the total linear momentum, total kinetic energy, and total potential energy of the whole system. If you are unfamiliar with these concepts, do not worry. We will define them shortly. Other variables refer to the desired separation D between the two masses at the ends of the spring, the spring constant k and the location of the center of mass.

In NetLogo, procedures start with the keyword to and conclude with the keyword end. The glossaries at the beginning of this book summarize the syntax of NetLogo and give a list of important variables used in our NetLogo simulations. The procedure setup is called when you click on the Setup Agents button. To see this more clearly, start the NetLogo simulation.

```
globals [total_lm total_ke total_pe total_energy
         center_of_mass_x k FR DeltaT D S]

to setup
   clear-all crt 2
   ask turtles [setup-turtles]
   set S abs(([xcor] of turtle 1) - ([xcor] of turtle 0))

   set center_of_mass_x
      (([mass] of turtle 0) * ([xcor] of turtle 0) +
      ([mass] of turtle 1) * ([xcor] of turtle 1)) /
      (([mass] of turtle 0) + ([mass] of turtle 1))
   ask patch (round center_of_mass_x) 0
      [ask patches in-radius 4 [set pcolor red]]
end
```

Then "select" the Setup Agents button. This can be done by dragging a rectangle around it with your mouse. If it has been selected properly a gray rectangle is displayed around the button. You can resize the button by dragging the small black squares. You can also "right-click" with your mouse and select Edit... with your mouse. A new window is opened, describing what the button does. In this case, it calls the setup procedure of the code.[2] Sliders, monitors and graphs can be selected and edited in the same manner, showing you what is being controlled, monitored, and displayed.

Now let's examine what the setup procedure does. First, the clear-all command clears everything. Then crt 2 creates two turtles. They are given unique identifiers, namely 0 and 1. Then the turtles are asked to run their own setup-turtles routine. The separation S between the two turtles is computed with set S abs(([xcor] of turtle 1) - ([xcor] of turtle 0)). At this point the procedure computes the center of mass of the system. Mathematically, in a one-dimensional system, the center of mass is:

$$\frac{\sum_i m_i x_i}{\sum_i m_i}, \qquad (2.1)$$

where the sum is taken over all particles in the system. Since there are only two in this particular simulation, set center_of_mass_x (([mass] of turtle 0) * ([xcor] of turtle 0) + ([mass] of turtle 1) *

---

[2] NetLogo states that you should use control-click rather than right-click if using a Macintosh. See http://ccl.northwestern.edu/netlogo/requirements.html for more information.

Listing 2.2: A portion of the turtle code for the one-dimensional spring

```
turtles-own [hood deltax r F v dv mass ke lm]

to setup-turtles
  set color white home
  set shape "circle" set size 5

  ifelse (who = 0)
    [set mass 1 set heading 90 fd (1 + random (D / 2))]
    [set mass Mass_Ratio set heading 270 fd (1 + random (D / 2))]
end
```

([xcor] of turtle 1)) / (([mass] of turtle 0) + ([mass] of turtle 1)) computes the center of mass.

Finally, the center of mass is drawn on the graphics screen. In addition to turtles, NetLogo includes the concept of "patches." Patches represent pieces of the environment that the turtles inhabit. For example, each patch could represent grass for rabbit agents to munch on. NetLogo was modeled on the premise that much of the interaction between agents occurs indirectly through the environment, a process referred to as "stigmergy." Stigmergy is an excellent way to model various biological and ethological systems, and hence biomimetics relies heavily on this form of communication. Physicomimetics generally does not make use of stigmergy, because of the numerous difficulties in reliably changing the environment when performing research in swarm robotics. The spring simulation does not rely on the center of mass computation—it is merely a useful mechanism for us (the observer) to understand the dynamics of the simulation. Hence, the final step in the setup procedure is to make the patch red at the coordinates of the center of mass by saying ask patch (round center_of_mass_x) 0 [ask patches in-radius 4 [set pcolor red]].

The code for the turtles is shown in Listing 2.2. At the top of the code is a list of "turtle" variables. These are variables that each turtle maintains. They can also be shared with other turtles. Each turtle knows its neighbors, called the hood. In this case, since there are only two turtles, each turtle has only one neighbor. Each turtle also knows the $x$-displacement to the neighbor (which can be positive or negative), the range to the neighbor, the force felt on itself, its velocity, change in velocity, mass, kinetic energy and linear momentum. We will discuss these concepts further below.

Both turtles execute the setup-turtles procedure. First, each turtle is given a white color. Then they are placed at the home position, which is at the center of the graphics pane, and are drawn as large circles. Each turtle

Listing 2.3: Remainder of observer code for the one-dimensional spring

```
to run-and-monitor
    ask turtles [ap]
    ask turtles [move]

    set center_of_mass_x
        (([mass] of turtle 0) * ([xcor] of turtle 0) +
        ([mass] of turtle 1) * ([xcor] of turtle 1)) /
        (([mass] of turtle 0) + ([mass] of turtle 1))
    ask patch (round center_of_mass_x) 0
        [ask patches in-radius 4 [set pcolor red]]

    set total_lm sum [lm] of turtles
    set total_ke sum [ke] of turtles
    set total_pe (k * (S - D) * (S - D) / 2)
    set total_energy (total_ke + total_pe)
    tick
end
```

is identified by a unique identifier given by who. One turtle is given a mass of one by default, is turned 90° clockwise to face to the right (in NetLogo a heading of 0° points up), and moves randomly forward an amount determined by the desired spring length D.[3] The other turtle is given a mass determined by the Mass_Ratio slider, is turned 270° clockwise to face to the left, and moves randomly forward. The effect is to create a horizontal spring with a random spacing.

As stated before, after initialization, you start the simulation by clicking on the Move Agents button. If you "open" the button using the instructions given earlier (select the button and then click on Edit...) you will see that the Observer procedure run-and-monitor is executed. Note that the Forever box is checked, indicating that the simulation will run forever, unless you pause it by clicking the Move Agents button again.

The run-and-monitor procedure is shown in Listing 2.3. This procedure, which is looped infinitely, performs numerous actions. First, it tells each turtle to execute a procedure called ap. This calculates where each turtle should move. Then each turtle is asked to move. The center of mass is recomputed and redrawn, to see if *it* has moved. At this point the global variables store the total linear momentum, total kinetic energy, total potential energy, and total energy of the system. Finally, tick is called, incrementing the system counter by one. The energy variables are displayed on the Energy graph in the simulation. However, fully understanding these concepts means

---

[3] The discrepancy between Cartesian and NetLogo coordinates will arise more than once in this book—code that is wrong from a Cartesian standpoint is correct for NetLogo.

```
to ap                                    ; Artificial Physics!
   set v (1 - FR) * v

   set hood [who] of other turtles
   foreach hood [
      set deltax (([xcor] of turtle ?) - xcor)
      set r abs(deltax)
      set S r
      ifelse (deltax > 0)
         [set F (k * (r - D))]
         [set F (k * (D - r))]
   ]
   set dv DeltaT * (F / mass)
   set v  (v + dv)
   set deltax DeltaT * v
end

to move
   set xcor (xcor + deltax)
   set ke (v * v * mass / 2)
   set lm (mass * v)
end
```

we need to carefully examine the core procedures ap and move, as shown in
Listing 2.4.

The first thing that procedure ap does is apply friction. The parameter
FR is controlled by the Friction slider. If it is set to zero then the velocity v
does not change. Otherwise the velocity is decreased. It should be pointed
out that friction is quite difficult to model properly (i.e., in accordance with
real-world observations). In fact, there are numerous forms of friction, and
they are not well understood. Because we are less interested in perfect fidelity
we are free to use a form that provides behavior that is quite reasonable, as
shown in Fig. 2.3.

Next the turtle looks for its neighbors. Since there are only two turtles
in this system, each turtle has one neighbor. The turtle sees whether the
neighbor is to the left or right of itself. If the neighbor is to the right, then
the force on the turtle is computed with set F (k * (r - D)). If they
are separated by a distance greater than D, the turtle feels a positive force to
the right. Otherwise it feels a force to the left. If the neighbor is to the left,
then the force on the turtle is computed with set F (k * (D - r)). If
they are separated by a distance greater than D, the turtle feels a force to
the left. Otherwise it feels a force to the right. Note that this obeys Newton's
third law, which states that the reaction between two bodies is equal and

opposite. This "force law" is referred to as Hooke's law, after Robert Hooke,
a 17th century British physicist.

Each turtle obeys Newton's second law, namely, $\boldsymbol{F} = m\boldsymbol{a}$. This is stated
in vector notation. However, since this simulation is in one dimension we can
write it simply as $F = ma$ or $F/m = a$. What does this mean? It means that
a force of magnitude $F$ imparts an acceleration $a$ to an object. Note that
as the mass $m$ of the object is increased, the same amount of applied force
results in decreased acceleration.

What precisely is the acceleration $a$? Mathematically it is equivalent to
$dv/dt$. This is the instantaneous change in velocity. However, we must always
keep in mind that we wish to implement our equations on computer hardware.
Nothing is instantaneous in this domain—everything takes some amount of
time. Hence we can approximate $a$ with the discrete time equation $\Delta v/\Delta t$
(the change in velocity over some short interval of time). As $\Delta t$ approaches
zero we get closer to the continuous form $dv/dt$. Similarly, velocity $v$ itself
can be expressed as the instantaneous change in position $dx/dt$, with the
approximation $\Delta x/\Delta t$. Hence $\Delta x = v\Delta t$.

Let us write Newton's second law as $F = m\Delta v/\Delta t$. We can reorder the
terms so that $\Delta v = \Delta t F/m$. The following NetLogo code computes the
change in velocity: `set dv DeltaT * (F / mass)`. Then the new velocity is established with `set v (v + dv)`. This leads to the change in
position, computed with `set deltax DeltaT * v`, and finally the position of the turtle is updated in the move procedure with `setxcor (xcor
+ deltax)`. In the simulation $\Delta t$ is represented with `DeltaT`. The variable
`DeltaT` is controlled by the Time_Step slider. Try increasing the Time_Step.
You will notice that the simulation runs much faster. However, the behavior
is similar (Fig. 2.4). This indicates that this approach can work even if our
computer hardware has delays in processing.

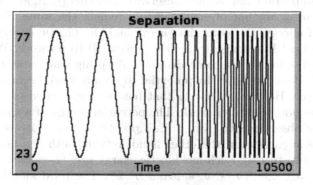

Fig. 2.4: A graph of the length of the spring, where $\Delta t$ is increased

## 2.1.2 Conservation Laws

A nice aspect of using physics for agent control is that the emergent system obeys various laws. This serves as a mechanism for verifying that the agent algorithms are working correctly, because if the laws are broken, something is wrong with our algorithms. Also, as we will see later, we can use our understanding of physics to appropriately set some parameters a priori! This is in contrast to other techniques where it is hard to verify that the system is indeed correct, because there are no known laws governing the emergent behavior.

The first example is conservation of linear momentum, which means that the velocity of the center of mass of a system will not change in a "closed system." A closed system is one that experiences no external forces. Our one-dimensional spring is a "closed system" (when we do not modify any sliders). Due to this, the total linear momentum of the system should not change. The linear momentum $p$ of each turtle is $p = mv$. The total linear momentum is simply the sum of the linear momentum of both particles. In Listing 2.4 we see that each turtle computes its own linear momentum as set lm (mass * v). Then Listing 2.3 shows that the total linear momentum is calculated with set total_lm sum [lm] of turtles. Because our two turtles are initialized with zero velocity, the total linear momentum should always be zero. Reset the Time_Step to 0.001 and run the simulation again. The Linear Momentum monitor should remain at zero. The fact that it does so indicates that, with respect to this aspect at least, the algorithm is written correctly. The result is a direct consequence of the fact that we are obeying Newton's third law. Another way to state this is that the system should not move as a whole in any direction across the screen. This is indeed what we observe.

Another important law is the conservation of energy. Again, in a closed system, the total amount of energy should remain the same. We focus on two forms of energy, the potential energy and the kinetic energy. What is potential energy? It reflects the ability (potential) to do work. Work is the application of a force over some distance. A spring that is at the desired spring length does not have the potential to do any work because it cannot exert any force. However, a spring that has been expanded or compressed does have the potential to do work. The potential energy of a spring is computed by the observer with set total_pe (k * (S - D) * (S - D) / 2).[4] Kinetic energy, on the other hand, reflects both the mass and velocity of the two turtles. Each has kinetic energy $mv^2/2$, calculated by each turtle with set ke (v * v * mass / 2). The total kinetic energy is computed by the observer with set total_ke sum [ke] of turtles. In the absence of friction, the sum of the potential energy and kinetic en-

---

[4] Let $s = S - D$. Then the potential energy is $-\int_0^s -k\,x\,\mathrm{d}x = k\,s^2/2$.

ergy should remain constant. We compute this in the observer with `set total_energy (total_ke + total_pe)`.

This does not say that both the potential energy and kinetic energy remain constant, merely that the sum remains constant. In fact, the amount of both forms of energy change with time. Rerun the simulation again and note the Energy graph (Fig. 2.5 provides an example). Again the $x$ (horizontal) axis represents time. The $y$ (vertical) axis represents the amount of kinetic energy (green curve), potential energy (blue curve) and total energy (brown curve). What behavior do we see? There is a constant trade-off between kinetic energy and potential energy. However, the total remains constant (be sure to set the Time_Step to 0.001 and Friction to zero). The fact that the total energy remains quite constant provides more evidence that our simulation is correct.

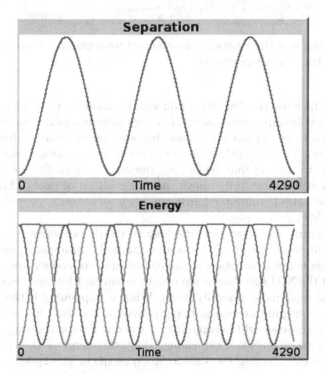

Fig. 2.5: A graph of the kinetic, potential and total energy of the spring, over time, when there is no friction

You will also notice something that you probably noticed before—the turtles appear to move slowly when the spring is maximally compressed or maximally expanded, while moving most quickly when they are D apart. The explanation for this can be seen in the Energy graph. When maximally compressed or expanded the potential energy is maximized. Hence the kinetic energy is minimized and the turtles move slowly. However, when they are D

apart, potential energy is minimized and kinetic energy is maximized. The
correlation between the energy graph and the separation graph is shown in
Fig. 2.5.

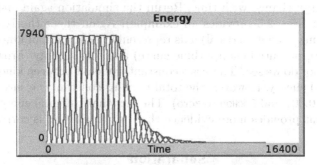

Fig. 2.6: A graph of the kinetic, potential and total energy of the spring, over
time, when friction is introduced

However, increase the Time_Step. You will see small fluctuations in the total
energy. This is because the conservation laws assume continuous time rather
than discrete time. As our simulation becomes more "coarse" from a time
point of view, it can slightly violate the conservation laws. However, what
is important to note is that (thus far), the errors are really not that large.
The system still almost fully obeys the conservation of energy law. Hence,
when we use actual computer hardware (such as robots), we will find that
the physics laws still apply.

It is instructive to reset the Time_Step to 0.001, but raise Friction. What
happens? The total energy of the system gradually disappears, as shown in
Fig. 2.6. This is not a violation of the conservation of energy law. Instead it
reflects that the NetLogo simulation is not measuring frictional energy (which
in real-world systems is generally heat). What is happening is that the total
energy of the original system is being converted to heat, in the same way
that a shock absorber will become hotter when it damps a spring.

There are other ways you can change the total energy of the system. Two
examples are by changing the Desired_Spring_Length or the Spring_Constant. In
this case, the total energy can increase or decrease. Is this a violation of the
conservation of energy? No, because changing the spring length or spring
constant requires the observer to "open" the system and modify the spring
itself. Hence, when a car adjusts the suspension system, it must do so by
adding or removing energy to that suspension system. However, do note that
once the change has been made, the total energy remains constant again,
albeit at a different level.

Finally, due to the nature of NetLogo itself, there is one additional way
to obtain unusual results. The graphics pane in NetLogo is toroidal. This
means that when an agent moves off the pane to the north (south), it re-

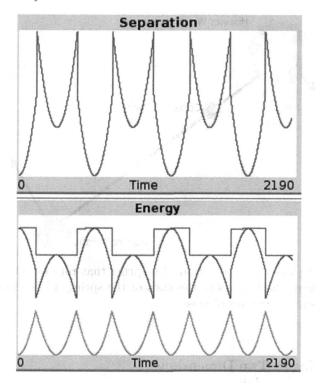

Fig. 2.7: Separation and energy graphs for the one-dimensional spring simulation when the particles cross the world boundary

enters from the south (north). Similarly when an agent moves off the pane to the east (west), it re-enters from the west (east). Newtonian physics assumes Euclidean geometry. The toroidal nature of the NetLogo world breaks the Euclidean geometry assumption. To see an example of the results, open the Desired_Spring_Length slider. Set the maximum value to 500 and the current value to 500. Then run the simulation. Figure 2.7 illustrates some example results for both the Separation and Energy graphs. The Separation graph is odd because when the particle leaves one side of the world, it re-enters from the other, changing the separation dramatically. The Energy graph reflects two different levels of total energy, depending on which side the particle is on. The results are understandable, but do not reflect standard physics. In the real non-toroidal world, with robots, this will never be a problem. But we do need to be careful when using NetLogo for physics simulations to ensure that particles do not cross the environment boundaries.

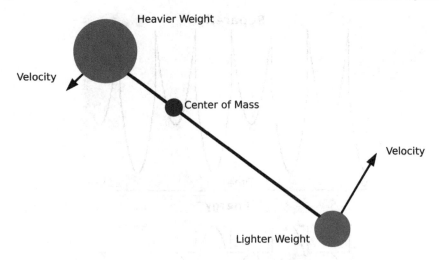

Fig. 2.8: A pictorial representation of a spring that rotates. The light gray circles represent the weights at the ends of the spring, while the dark gray circle represents the center of mass

## 2.2 A Spring in Two Dimensions

Although the one-dimensional simulation shown above is excellent for a tutorial, we do not expect our robots to be limited to one dimension. Hence, in this section we generalize the simulation to two dimensions (the file is "spring2D.nlogo"). We do this by imparting a rotation to the spring. This is shown in Fig. 2.8. This figure represents the spring with a black line (which can expand or contract). The light gray circles represent the masses on the ends of the spring. The upper mass has greater weight than the lower mass, so the center of mass (a dark gray circle) is closer to the upper mass. At the beginning of the simulation the spring is given an initial rotation around the center of mass. In the case of Fig. 2.8 that rotation is counterclockwise. Note that, as expected, the lower mass moves more quickly (the velocity arrow is longer) because it is farther from the center of mass.

Generalizing the one-dimensional simulation to two dimensions is not very difficult. A picture of the graphical interface is shown in Fig. 2.9. There is one additional monitor for the linear momentum, as both the $x$- and $y$-components need to be monitored now. There is one additional slider, called Angular_Motion, representing the amount of initial rotation. Finally, there is a monitor for the angular momentum.

For the two-dimensional spring, the core code is again contained in two procedures, ap and move. The procedure ap calculates where each particle should move, as shown in Listing 2.5. It is essentially the same as in List-

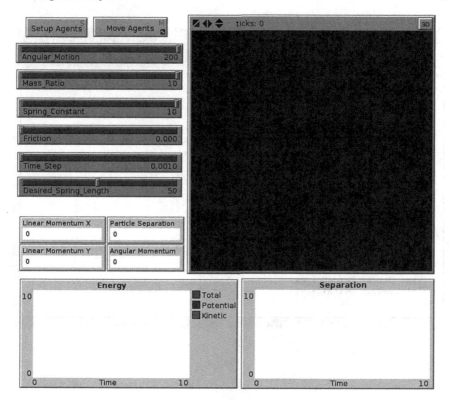

Fig. 2.9: A picture of the graphical user interface for the two-dimensional spring simulation

ing 2.4, except that all operations need to be performed on both the $x$- and $y$-components of the system. This is a nice aspect of physics. Motion in two dimensions can be computed by decomposing two-dimensional vectors into their $x$- and $y$-components and performing the computations at that level. So, as an example, instead of one line of code to describe how friction changes velocity, we now have two lines of code. There is one line for the $x$-component of velocity vx and one line for the $y$-component of velocity vy.

The procedure move (shown in Listing 2.6) moves each particle and computes information pertaining to the momenta and energy. The only large change to the code is at the bottom, where the angular momentum is computed. Angular momentum is the amount of resistance to changing the rate of rotation of a system. The equation is given as $L = r \times p$, where $L$ is the angular momentum vector, $r$ is the position vector of the particle with respect to the origin (in our case this is just the center of mass), $p$ is the linear momentum vector, and $\times$ represents the cross product. In two dimensions this equation can be written with scalars instead of vectors to obtain

Listing 2.5: The main turtle computation code for the two-dimensional spring

```
to ap
    set vx (1 - FR) * vx
    set vy (1 - FR) * vy

    set hood [who] of other turtles
    foreach hood [
        set deltax (([xcor] of turtle ?) - xcor)
        set deltay (([ycor] of turtle ?) - ycor)
        set r sqrt (deltax * deltax + deltay * deltay)
        set S r
        set F (k * (r - D))
        set Fx (F * (deltax / r))
        set Fy (F * (deltay / r))
    ]
    set dvx DeltaT * (Fx / mass)
    set dvy DeltaT * (Fy / mass)
    set vx  (vx + dvx)
    set vy  (vy + dvy)
    set deltax DeltaT * vx
    set deltay DeltaT * vy
end
```

Listing 2.6: The main turtle movement and monitor code for the two-dimensional spring

```
to move
    set xcor (xcor + deltax)
    set ycor (ycor + deltay)

    set lmx (mass * vx)
    set lmy (mass * vy)

    set v sqrt (vx * vx + vy * vy)
    set ke (v * v * mass / 2)
    set lever_arm_x (xcor - center_of_mass_x)
    set lever_arm_y (ycor - center_of_mass_y)
    set lever_arm_r sqrt (lever_arm_x * lever_arm_x +
                          lever_arm_y * lever_arm_y)
    set theta (atan (mass * vy) (mass * vx)) -
              (atan lever_arm_y lever_arm_x)
    set angular_mom (lever_arm_r * mass * v * (sin theta))
end
```

```
ifelse (who = 0)
    [set mass 1 set vx Angular_Motion]
    [set mass Mass_Ratio
     set vx (- Angular_Motion) / Mass_Ratio]
```

$L = r\,m\,v\,\sin\theta$, where $L$ is the magnitude of $\boldsymbol{L}$ and $r$ is the magnitude of $\boldsymbol{r}$. In our simulation each mass has a velocity. Each velocity can be decomposed into two quantities, the velocity that is along the line of the spring, and the velocity that is at a right angle to the ends of the spring. The former quantity does not impart spin to the system, so we focus only on the latter. This quantity is computed as $v\,\sin\theta$, where $\theta$ is the angle of the particle velocity in comparison to the line of the spring. The magnitude of the vector $\boldsymbol{p}$ that is at a right angle to the spring is given by $m\,v\,\sin\theta$. For example, if the particle was moving directly outwards along the line of the spring, $\theta = 0$. Then $\sin\theta = 0$ and $m\,v\,\sin\theta = 0$. Hence $L = 0$ for this particle, because it is not rotating the system. However, if $\theta = 90°$, then $\sin\theta = 1$, $m\,v\,\sin\theta = m\,v$, and $L = r\,m\,v$. This represents a particle that is rotating the system as much as possible.

It is useful to examine $L = r\,m\,v$ more closely. Angular momentum depends on the weight of the masses at the ends of the spring, the distance $r$ from the center of mass, and the velocity. Also, just as with linear momentum, a closed system obeys the conservation of angular momentum! This explains something we have all observed. When an ice skater is spinning, and then brings in her arms, she spins much faster. This is because $r$ has decreased. Since her mass $m$ is constant, $v$ must increase. For example, if $r$ is decreased to one-third of the initial value, velocity must increase three-fold.

The Angular_Motion slider controls the initial amount of rotation, as shown in Listing 2.7. For the lighter particle vx is set to Angular_Motion. To ensure that both masses of the spring are initialized properly, the $x$-component of the velocity of the heavier particle is -Angular_Motion / Mass_Ratio, where Mass_Ratio is set by the Mass_Ratio slider. This creates a system with zero linear momentum.

Run the simulation with the default parameter settings. You will notice that the spring rotates around the center of mass, while expanding and contracting. The Linear Momentum monitors remain at zero, indicating that the center of mass is not moving. The Angular Momentum monitor also stays constant (or very close to constant—a small amount of error can occur due to numerical precision issues in the simulation). Again, this serves to verify that the simulation is indeed written correctly. By convention a negative value for the angular momentum indicates a clockwise rotation. A positive value indicates a counterclockwise rotation.

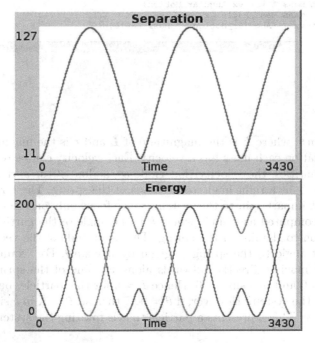

Fig. 2.10: Separation and energy graphs for the two-dimensional spring simulation

Figure 2.10 shows both the Separation graph and the Energy graph. Although still quite cyclic, the Separation graph shows two new behaviors. First, the separation can exceed twice the desired spring length (which is 50). Also, one can clearly see that the average separation is greater than the desired spring length. What is happening? We are seeing the effects of "centrifugal force" due to the rotation. Centrifugal force is an outward force from the center of rotation. The Energy graph is also more interesting than in the one-dimensional simulation, again due to the effects of the rotation of the system. Although the graph shows cycles, they are more complex than the simple sinusoidal cycles shown earlier.

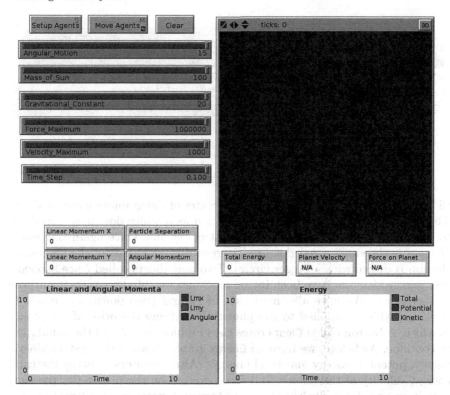

Fig. 2.11: A picture of the graphical user interface for the solar system simulation

## 2.3 A Simple Solar System

The two-dimensional spring looks quite a bit like a solar system, with the lighter particle being a planet, and the heavier particle being the sun. However, solar system dynamics are governed by Newton's gravitational force law $F = G\,m_1\,m_2/r^2$ (where $G$ is the gravitational constant), not by Hooke's law. The nice aspect of our simulation is that it is simple to modify the simulation to be more like a one-planet solar system, as shown in Listing 2.8. However, note that this particular force law has a feature that Hooke's law does not—the magnitude of the force goes to infinity as $r$ goes to zero. This section will explore how this feature can be problematic for both simulations and real robots, and how the problems can be overcome.

Figure 2.11 shows the graphical user interface for the solar system model, which looks very similar to those we have seen before ("solar_system.nlogo"). The gravitational constant $G$ is controlled by the Gravitational_Constant slider. The mass of the sun, shown as a yellow dot, is controlled with the Mass_of_Sun

Listing 2.8: The portion of code that implements the Newtonian gravitational force law in the solar system model

```
set F (G * mass * ([mass] of turtle ?) / (r * r))
```

slider. The sun is initialized to be at the center of the graphics pane, at $(0,0)$. The planet has a mass of one and is shown as a white dot. It is initialized to be at location $(15,0)$, just to the right of the sun. Once again the center of mass is shown as a red dot (and is generally obscured by the sun, since the sun is drawn with a bigger circle). Two new sliders called Force_Maximum and Velocity_Maximum have been introduced, and we will discuss their purpose further down. We have also made use of the pd (pen down) command in NetLogo, which is applied to the planet. This draws the orbit of the planet in white. A button called Clear erases the graphics, but allows the simulation to continue. As before, we have an Energy graph, showing the system kinetic energy, potential energy and total energy.[5] Also, monitors showing the total energy, velocity of the planet, and force exerted on the planet (by the sun) have been added. Finally, the Linear and Angular Momenta graph allows us to see if the momenta are changing with time. Blue is used for angular momentum, red for the $x$-component of linear momentum, and green for the $y$-component of linear momentum. Recall from before that in a properly designed system, the laws of conservation of both momenta and energy should hold.

The slider Angular_Motion once again controls the amount of rotation of the system. Specifically, the planet is initialized with an upward velocity of Angular_Motion, while the sun is initialized with a downward velocity of -Angular_Motion / Mass_of_Sun. The result of this is to create a rotating system that has no linear momentum. Unlike the previous simulations, there is no random component in the code. This maximizes the likelihood that the simulation will run similarly for you, making it easier to discuss the behavior and results. For example, let's show the orbits when Angular_Motion is 15 and 10. Let's also change the mass of the sun, using the values of 100 and 95. The resulting orbits are shown in Fig. 2.12. The higher initial velocity given when Angular_Motion is 15 results in a large elliptical orbit. The orbit is smaller when Angular_Motion is 10. The effect of mass on the orbits is qualitatively the same for both values of Angular_Motion. As mass decreases the orbit increases in size. This is because the gravitational force is weaker. The effect is bigger when the orbit is bigger. For all experiments the

---

[5] The computation for potential energy is shown in [240].

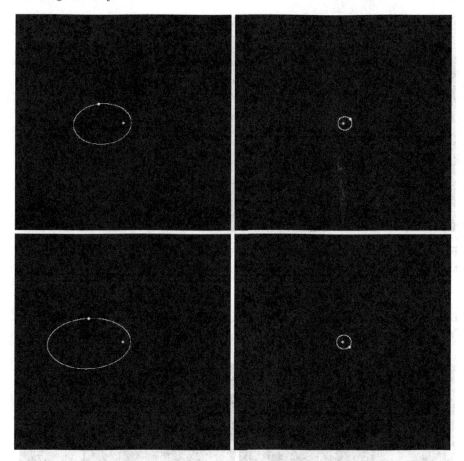

Fig. 2.12: The orbits when Angular_Motion is 15 (left) and 10 (right) and
the mass of the sun is 100 (top) and 95 (bottom)

linear momenta stay at zero, while the angular momentum and total energy
remain very steady. Feel free to try different values of Angular_Motion,
Mass_of_Sun and gravity. Just remember that if the planet moves beyond
the perimeter of the environment, the orbit will be very odd!

Reset the mass of the sun to 100, and let's examine the orbits when
Angular_Motion is five and zero (see Fig. 2.13). These results look unusual. When Angular_Motion is five the planet breaks away from the sun.
When Angular_Motion is zero the results make no sense and the total energy changes radically. What is happening? When Angular_Motion is zero
there is no angular momentum in the system. Both planet and sun are initialized as stationary objects in space. Hence, due to gravitation, they are
attracted to one another. Initially, the planet is 15 units to the right of the
sun. Assuming the planet can plunge through the sun without any interac-

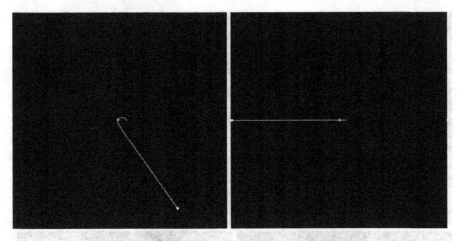

Fig. 2.13: The orbits when Angular_Motion is five (left) and zero (right), the mass of the sun is 100, and $\Delta t$ is 0.1

tion, it should end up being no farther than 15 units to the left of the sun. Yet this is not the case.

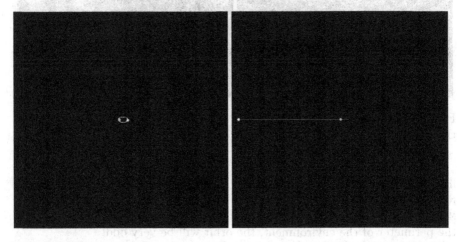

Fig. 2.14: The orbits when Angular_Motion is five (left) and zero (right), the mass of the sun is 100, and $\Delta t$ is 0.001

Could the cause be that $\Delta t$ is too large? After all, the physics of real solar systems assume continuous time, not discrete time. Reset the mass of the sun back to 100. Set Time_Step to 0.001 and try again. Figure 2.14 shows the results. When Angular_Motion is five, the orbit now looks reasonable, and both momenta and energy are conserved. However, when Angular_Motion

is zero the results are still wrong—the planet continues to move to the left
indefinitely. The reason for this behavior lies in the very nature of the gravitational force, which is proportional to $1/r^2$. This means that when the planet
is extremely close to the sun, the force becomes huge. This yields a large
acceleration and large velocity. In the real world of continuous time, this is
not a problem. But with discrete time the particle moves somewhat farther
than it should (i.e., the velocity is too large) and the physics is broken. Even
when $\Delta t = 0.001$ the simulation is not sufficiently fine-grained.

This also creates a difficulty for implementation on real robots. Again,
everything takes time when implemented in hardware, and the faster the
hardware the more expensive it becomes. So, this raises an important issue. Can we find a solution that "fixes" the physics in a reasonable manner,
without the use of expensive hardware?

One possibility would be to limit the velocity of the robots. This is a
natural solution, given that robots have a maximum velocity. In fact, the
reason for the planet velocity monitor is to show the velocity that a robot
would need to reach. This might not be physically possible. Hence it is natural
to wonder if a simple solution of limiting the velocity to that presented by
the hardware would solve some of these issues.

Let's reexamine the cases where Angular_Motion is 15, the mass of the
sun is 100, and the Time_Step is 0.1. The maximum velocity of the planet
is roughly 15. Suppose our robot can only move 12 units per time step.
Move the Velocity_Maximum slider to 12 and run the simulation.[6] This seems to
run quite well, with only a slight change in the linear momentum. However,
move the Velocity_Maximum slider to nine and try again. Almost immediately
the planet plunges into the sun and remains nearby. The linear momenta
oscillate wildly, and the total energy varies a lot. Now try the case where
Angular_Motion is zero and Velocity_Maximum is set to 12. This is again
a disaster, with large changes in linear momenta and total energy. This is
also true with Velocity_Maximum set to nine (or one, for an extreme example).
Apparently, capping velocity explicitly does not work well in many cases.

There is an alternative solution. Physics works with force laws, and an
implicit way to decrease velocity is to cap the force magnitude. Lower forces
lead to lower velocities. This is especially important when using a force law
that has the form $1/r^2$, since the force magnitude has no limit. Interestingly,
no one really knows how the Newtonian force law behaves at small distances,
and modifications have been proposed [226]. The slider Force_Maximum allows
us to place a maximum on the magnitude of the force. Lower Force_Maximum
to one. Because the maximum attractive force is now reduced, we need to
use different values of Angular_Motion to achieve effects similar to those
we obtained before.

The results are shown in Fig. 2.15, with Angular_Motion set to nine,
four, one and zero. The biggest change in behavior is the precession of

---

[6] If you are ever unable to achieve a particular value using the slider, open the slider and
set the value manually.

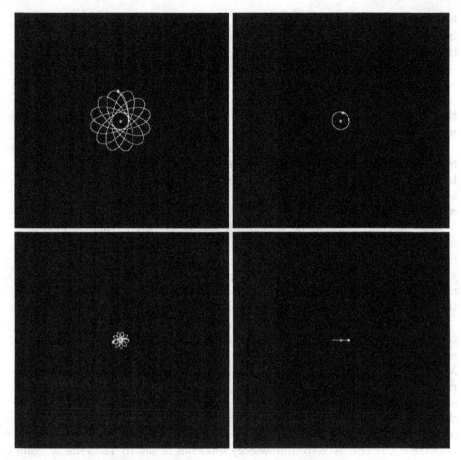

Fig. 2.15: The orbits when Angular_Motion is nine, four, one and zero, while the maximum force is one

the orbits, which did not occur before. Interestingly, this flower-petal shape occurs in real-world orbits, mainly due to perturbations from other planets. Of course, it is occurring in our system because of the imposition of Force_Maximum, which perturbs the elliptical orbit. Force_Maximum has caused a change in behavior for the system, although it is not unreasonable. The planet is still orbiting the sun, and is not breaking away from it.

A nice aspect is that for all these experiments the conservation of energy and momenta hold quite well. The deviations that remain can be reduced further by lowering the Time_Step. The behavior when Angular_Motion is zero is especially nice, with the separation ranging from zero to 15, as we had hoped for originally. The Linear and Angular Momenta graph has values at zero. Also, we have succeeded in our original goal, which was to limit the velocity. For the four experiments velocity is limited to 5.6, 5, 4 and 9 (when Angular_Motion is zero, one, four and nine).

Ultimately, if our robots are fast enough and have very fast sensors, we may not need to impose any force magnitude restrictions in the force laws. However, to further extend the range of behaviors that the robots can perform, using Force_Maximum is a reasonable option, as long as we understand the consequences. In practice, for the problems that we have examined, orbital precession is not an issue, and the use of Force_Maximum has worked quite well. In fact, as we will show in the next chapter, it can be used to our advantage, to help set parameters theoretically.

# Chapter 3
# NetLogo and Physicomimetics

William M. Spears

## 3.1 Biomimetic Formations

In Chap. 2 we examined NetLogo simulations of the physics of two agents in motion. Naturally, a swarm consists of more than two agents. In this chapter we generalize to large numbers of agents. One topic of interest to us is the application of swarm formations, and we will show how the code from the previous chapter can be generalized to many agents, via physicomimetics.

However, for the sake of contrast, it is instructional to examine a biomimetics solution to formations of flying flocks of birds or swimming schools of fish, originated by Craig Reynolds [192]. Reynolds generically refers to such agents as "boids." NetLogo [269] provides a flocking simulation inspired by the original boids model [268]. Generally, the flocking models assume that all boids are moving with the same constant velocity but in different directions (headings). Boids turn by changing their heading, depending on the positions and headings of the neighboring boids. There are three behaviors that specify how a boid turns—separation, alignment, and cohesion. A description of these behaviors is included in the NetLogo flocking code, with a priority for the separation rule (using the term "bird" instead of "boid") [268]:

> "Alignment" means that a bird tends to turn so that it is moving in the same direction that nearby birds are moving.
> "Separation" means that a bird will turn to avoid another bird which gets too close.
> "Cohesion" means that a bird will move towards other nearby birds (unless another bird is too close).
> When two birds are too close, the "separation" rule overrides the other two, which are deactivated until the minimum separation is achieved.

I have rewritten much of the original NetLogo simulation, in order to help explain the semantics of the flocking model ("flock.nlogo"). Figure 3.1 shows

William M. Spears
Swarmotics LLC, Laramie, Wyoming, USA, e-mail: wspears@swarmotics.com

W.M. Spears, D.F. Spears (eds.), *Physicomimetics*,
DOI 10.1007/978-3-642-22804-9_3,
© Springer-Verlag Berlin Heidelberg 2011

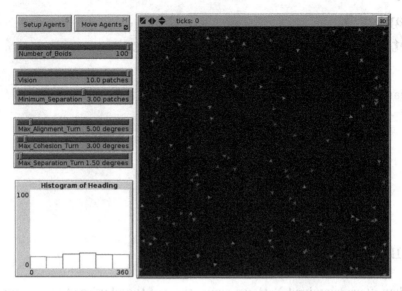

Fig. 3.1: The picture of the graphical user interface for the flocking simulation in NetLogo after initialization

the interface, after **Setup Agents** has been clicked. Let's provide an explanation of the features that are shown. The first thing that you will notice is that the agents are not just small circles. Each boid is represented with an arrow, where the arrow points in the direction that the boid is moving (i.e., its heading). The boid is given one of six colors, depending on its heading. When the simulation is initialized the boids are randomly placed in the pane, with random headings.

The buttons **Setup Agents** and **Move Agents** have the same function as in our prior NetLogo simulations. **Setup Agents** creates and initializes the agents. **Move Agents** is used to run the system when clicked, and to pause the system when clicked again. Since we are now working with more than two agents, the slider **Number_of_Boids** allows you create up to a hundred boids at initialization. The **Vision** slider is the distance that a boid can see in all directions. The **Minimum_Separation** slider allows you to set how closely the boids can approach one another before the separation behavior is activated (as mentioned above).

There are three other sliders in the simulation. Regardless of the turns actually computed by the separation, alignment, and cohesion procedures (to be discussed in more detail below), the simulation will not exceed various limits in the turns. The **Max_Alignment_Turn**, **Max_Cohesion_Turn**, and **Max_Separation_Turn** are initially set to 5°, 3° and 1.5°, respectively. One way to interpret these settings is to think in terms of importance. Higher settings mean that behavior is more important. With the initial settings as they are, the most important behavior is alignment, followed by cohesion, and ending with separation. Naturally, different settings yield different behavior by the flock. We will explore some extreme examples soon.

The last feature is a new type of graph called a "histogram." This graph has six vertical bars. The height of the first bar represents the number of boids that have their heading between 0° and 59°. The second bar represents the number of boids with heading between 60° and 119°. The last bar represents the number of boids with heading between 300° and 359°. The colors of the bars correspond to the colors of the boids. When the system starts, the bars have similar heights, because the headings were initialized randomly (see Fig. 3.1).

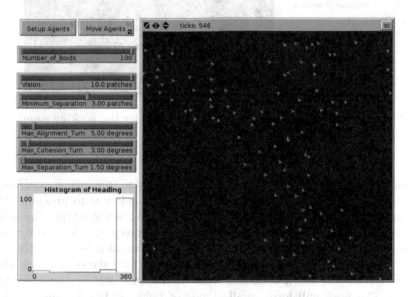

Fig. 3.2: The picture of the graphical user interface for the flocking simulation in NetLogo after 546 steps

Click Setup Agents and Move Agents to run the system As the boids start to flock, certain headings start to dominate. Figure 3.2 shows an example after 546 steps, where the boids are exhibiting flocking behavior and are heading relatively consistently up and towards the left. Of course, you are very likely to see something different, due to the random initialization of the boids. However, you should always see the boids self-organizing into a common direction.

An interesting experiment to perform is to set Minimum_Separation to zero. The boids quickly collapse into a long linear formation, as shown at 1,309 steps in Fig. 3.3. These particular boids do not mind colliding with each other!

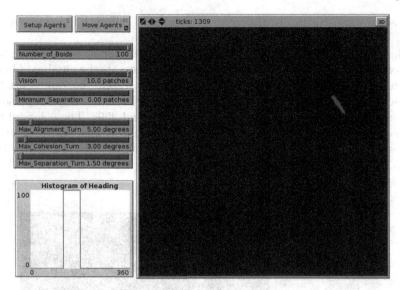

Fig. 3.3: The picture of the graphical user interface for the flocking simulation in NetLogo when the minimum separation is zero

Now let's try to better understand the separation, alignment, and cohesion behaviors. The easy way to examine the behaviors is to maximize the maximum turn slider associated with the behavior, while setting the other maximum turn sliders to zero. For example, set Max_Alignment_Turn to 40, while setting the other two to zero. Keep Minimum_Separation at zero. Rerun the system. The flock will align very quickly. Figure 3.4 (left) shows the flock after only 78 steps. Now set Max_Cohesion_Turn to 40 while setting the other two to zero. The flock will form small clouds of boids within 42 steps, as can be seen in Fig. 3.4 (right). The boids appear to collide with each other and they never align. The histogram shows that all headings remain in a state of flux. Finally, set Max_Separation_Turn to 40, while setting the other two to zero. Reset Minimum_Separation to three. The behavior is hard to show with a picture, but is easy to spot in the simulation. Look for two boids that are approaching each other (generally they will have different colors). When they get close they suddenly change headings (and colors) as they avoid each other. However, the boids never show cohesion or alignment—the histogram shows that all headings remain in a state of flux.

Let's examine the main portions of the NetLogo code. Listing 3.1 provides the core boid procedure, flock. The procedure find-neighbors serves two purposes. First, it finds the boids within distance Vision and stores them in the list hood. Second, it records the nearest-neighbor. Now each boid examines how many neighbors it has. If it has no neighbors, it merely moves forward one step (fd 1). If it does have neighbors, the boid

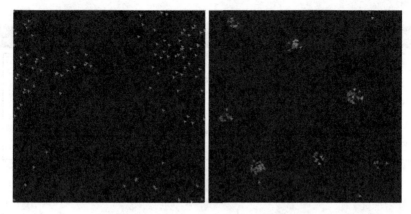

Fig. 3.4: Depictions of the alignment behavior (left) and cohesion behavior (right) in the flock simulation

Listing 3.1: The main flock procedure for the flocking model

```
to flock
    find-neighbors
    if ((length hood) > 0) [
        ifelse (min_r < Minimum_Separation)
            [separation]
            [alignment cohesion]
    ]
    monitor
    fd 1
end
```

determines whether the distance min_r to the nearest neighbor is less than the Minimum_Separation. If it is, the boid calls the separation procedure. Otherwise it calls the alignment and cohesion procedures. The monitor procedure updates heading information for the histogram.

The separation procedure is shown in Listing 3.2. This code is very simple. The procedure turn-away calls the turn-at-most procedure. This in turn calls subtract-headings, which computes the minimum angular difference between the boid heading and the heading of the nearest neighbor. The minimum angular difference is computed so that the agent makes the minimum turn. For example, 5° minus 355° is best described as a 10° turn, not a −350° turn. Then the procedure turn-at-most performs the turn or the Max_Separation_Turn (determined by the slider), whichever is less.

**Listing 3.2: The separation procedure for the flocking model**

```
to separation
    turn-away ([heading] of turtle nearest-neighbor)
            Max_Separation_Turn
end

to turn-away [new-heading max-turn]
    turn-at-most (subtract-headings heading new-heading) max-turn
end
```

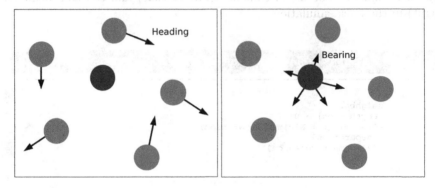

Fig. 3.5: Depictions of the alignment computation (left) and cohesion computation (right) in the flock simulation

The `alignment` and `cohesion` procedures are more complex. Figure 3.5 shows an illustration of both behaviors. Alignment (shown on the left) shows that the dark gray boid will compute a new heading for itself based on the headings of the neighbors. Cohesion (shown on the right) indicates that the dark gray boid will compute a new heading for itself based on the *bearings* to the neighbors.[1] Let's consider the `alignment` procedure first. The idea is to somehow compute the average heading of the neighbors. But how is this done? Let the heading of boid $j$ be denoted as $\theta_j$. Assuming $N$ neighbors, the "obvious" equation is as follows:

$$\bar{\theta} = \frac{\sum_{j=1}^{j=N} \theta_j}{N}. \tag{3.1}$$

---

[1] Bearing and heading are often confused. Heading is the direction you are pointing. Bearing is the direction to some other location (in this case, another boid).

Listing 3.3: The alignment procedure for the flocking model

```
to alignment
    let x 0 let y 0
    foreach hood [
        set y (y + (sin ([heading] of turtle ?)))
        set x (x + (cos ([heading] of turtle ?)))
    ]
    ifelse ((x = 0) and (y = 0))
        [set newh heading]
        [set newh (atan y x)]  ; The mathematical function atan2(y,x)
    turn-towards newh Max_Alignment_Turn
end
```

However, what is the average of 10° and 350°? According to Eq. 3.1, the average would be 180°. But we realize that the correct answer should be 0°! Why are we getting an incorrect answer? Recall that in Chap. 2 we discussed how a toroidal world breaks standard Newtonian physics, because Newtonian physics assumes Euclidean geometry. Similarly, we are again dealing with a non-Euclidean space, due to the circular nature of an angle (e.g., $360° = 0°$, $-180° = 180°$). So, standard statistical metrics do not work. The NetLogo community clearly understands this because they state [268]:

> "Alignment" means that a bird tends to turn so that it is moving in the same direction that nearby birds are moving. We can't just average the heading variables here. For example, the average of 1 and 359 should be 0, not 180. So we have to use trigonometry.

But what does "use trigonometry" mean and what precisely is being computed? The code for the alignment procedure is shown in Listing 3.3. This procedure examines the headings of each neighbor to produce a new heading called newh. The code is computing the following:

$$\text{newh} = \text{atan2}\left(\sum_{j=1}^{j=N} \sin\theta_j, \ \sum_{j=1}^{j=N} \cos\theta_j\right). \tag{3.2}$$

What is Eq. 3.2 computing? Interestingly, the headings are being treated as unit vectors (vectors of length one), with the heading of boid $j$ being $\theta_j$. Then the $x$- and $y$-components of the vectors are being summed. Finally, the angle of the summed vector is stored as newh.[2]

---

[2] Mathematicians will recognize this also as $\arg(\sum_{j=1}^{j=N} e^{i\theta_j}/N)$ from complex analysis. Also, atan2$(y, x)$ returns the angle from the positive $x$-axis to the point $(x, y)$.

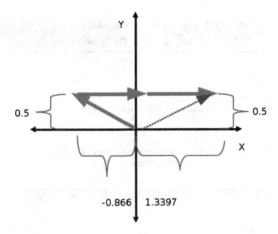

Fig. 3.6: An example of computing the angle of the sum of three unit vectors representing angles 150°, 0° and 0°

A nice aspect of Eq. 3.2 is that it does not suffer from the circular nature of angles, and hence does not produce nonsensical answers. However, how well does it agree with Eq. 3.1 under more normal circumstances? There is no simple answer—it depends on the circumstances. Given the three angles of 0°, 45° and 90°, both equations yield the same value of 45°, as we would hope for. However, let's try another example with three angles of 150°, 0° and 0°, as shown in Fig. 3.6. Our intuitive response would be that the average angle should be 50°. However, if we consider this as the sum of three unit vectors, the answer is quite different. First, the tip of the vector for 150° is at $(-0.8660254, 0.5)$. When we add the second unit vector depicting 0° the tip has moved to $(0.1339746, 0.5)$. Finally, when we add the third unit vector depicting 0°, the tip moves to $(1.1339746, 0.5)$. The angle of this vector (shown as a dashed green arrow) is atan2$(0.5, 1.1339746)$ which is approximately 23.794°. This is a discrepancy of roughly 26.206°, which is considerable.

The cohesion procedure also uses Eq. 3.2, where $\theta_j$ now refers to the bearing to neighboring boid $j$. This is shown in Listing 3.4. Hence, both alignment and cohesion transform the calculation of averaging angles into the calculation of finding the angle of the sum of unit vectors. This may not always produce results that agree with our intuition, but the transformation does yield behaviors that look reasonable. As the NetLogo community states [268]:

> The exact details of the algorithm tend not to matter very much—as long as you have alignment, separation and cohesion, you will usually get flocking behavior resembling that produced by Reynolds' original model.

Listing 3.4: The cohesion procedure for the flocking model

```
to cohesion
    let x 0 let y 0 let b 0
    foreach hood [
        set deltax (([xcor] of turtle ?) - xcor)
        set deltay (([ycor] of turtle ?) - ycor)
        set b (atan deltax deltay)
        set y y + (sin b)
        set x x + (cos b)
    ]
    ifelse ((x = 0) and (y = 0))
        [set bearing heading]
        [set bearing (atan y x)]
    turn-towards bearing Max_Cohesion_Turn
end
```

This raises an important scientific issue, however. Given that there are different possible implementations, how does one choose which one to use? There is an important concept in the scientific method referred to as the "ground truth." The ground truth represents the objective reality, as opposed to a subjective observation. What is the objective reality underlying flocking? Ultimately it would have to be measurements pertaining to the flocking of actual entities, such as birds. Different birds may yield different measurements. Although the flocking model certainly *looks* reasonable, there is no way to actually gauge the quality of the model until the ground truth is defined. This is one reason for the development of physicomimetics. We prefer to define the ground truth a priori (if possible), design the model, and then assess how well the model is performing in relationship to the ground truth.

In addition to our desire for a ground truth to evaluate our models, another reason we prefer physicomimetics for formations is that biomimetic formations, such as boids, are not stable or rigid enough for our purposes. We are interested in having our swarms act as distributed sensor meshes or distributed computers. This requires more rigid lattice-like formations than the biomimetics solutions appear to offer. This book will discuss three applications of these lattices in detail. Two applications are bioluminescence and chemical plume tracing (Chaps. 6 and 8). The other is the optimization of noisy functions (Chap. 19). For all three applications it is necessary to have formations that are more stable. Hence we now introduce our physics-based approach to formations, called "artificial physics" or "physicomimetics."

There are several advantages to our approach. First, we can specify the formations we want a priori and then measure how well the algorithm is performing. Second, we can use theory to help set parameters to achieve the proper behavior. Finally, because the model is based on physics, it obeys a concept in physics referred to as the principle of least action (also known as the principle of least effort or the path of least resistance). Robots have limited power sources. This principle indicates that robots will perform the minimum action required to respond to the environment. The principle applies at a local level, not globally. However, as stated in Chap. 1, our central premise is that we believe that building swarm systems that are globally optimal is simply not feasible for most applications. The best we can realistically hope for is local optimality.

There is also a very pragmatic issue that arises when we think about how to implement the boids flocking model on real robots. It is not hard for robot sensors to measure the bearing and distance to another robot, and we will show our own hardware implementation in Chap. 10. However, separation and alignment require that a robot know the headings of the neighbors. How is this to be determined? This is not a trivial issue, generally requiring multiple sensor readings or communication.

In fact, early work dispensed with the idea of measuring headings and instead explicitly computed the centroid of the neighbors to provide a location to avoid or approach [153]. The centroid is identical to the definition of the center of mass given in Chap. 2 when the masses are identical. One nice feature of physicomimetics is that it provides similar behavior to the "centroid" techniques, but without the explicit computation of the centroid. All that is required is that each agent make pairwise force computations, just as with our physics simulations in Chap. 2. Concepts such as the center of mass (and the centroid) are emergent properties of the system, as opposed to being part of the algorithm.

Due to hardware constraints, the physicomimetics algorithms discussed throughout most of this book do not require the computation of the heading of neighbors. However, we do not rule out the use of heading information. Advances in hardware are starting to allow the use of heading information to be feasible [168]. As an example, Turgut et al. [252] describes a "Virtual Heading Sensor" that requires that each robot has a digital compass to provide a common coordinate system, coupled with a wireless communication system. In recognition of these developments, Chap. 7 introduces a very different form of physicomimetics where computations are performed using the velocity vectors of neighbors. Unlike boids, however, this variation of physicomimetics still has a ground truth. This ground truth is "kinetic theory" and hence obeys physics laws. The details of how this is implemented, in addition to an explanation of kinetic theory, are provided in Chap. 7.

## 3.2 Physicomimetic Formations

The previous section showed a biomimetic approach to formations. However, as indicated above, we want our formations to be more geometric and stable than those shown with flocking. In addition, our initialization scenario is different. As opposed to placing agents uniformly randomly throughout the environment, the situations we are interested in employ a central point of dispersion for the initial location of the agents. For example, a container with a swarm of micro-air vehicles may be dropped from an airplane. Or, a swarm of land robots is launched from a mother vehicle. From this central point the autonomous vehicles should self-organize and move as a formation.

This section serves to accomplish two purposes. First, the "moving formation" task will be used to introduce physicomimetics to you via a concrete, real-world problem. The task itself will be broken into two subtasks. First, we would like the group of initially unorganized agents to self-organize into formation. Second, the formation of agents should move towards a goal and avoid obstacles.

The second purpose is to demonstrate several desirable aspects of physicomimetics. One useful analogy is to consider the $H_2O$ molecule. A "swarm" of $H_2O$ molecules creates water. In fact, water represents the self-organization of $H_2O$ molecules. The molecules are not reasoning about the nature of water. Similarly, when you examine the physicomimetics code, you will not see an approach that reasons about the number of neighbors, the angular separation (i.e., $60°$) between neighbors, and so forth. The lattice will self-organize and emerge from the local interactions of the particles in a manner that is similar to water.

We can continue the water analogy. Water is still water when some of it is removed or more is added. Similarly, water generally does not change when it is shaken. It moves and changes shape, but it is still water. However, under certain circumstances it can change form, when subject to certain temperatures. The transformations are referred to as "phase transitions." Similarly, lattice formations built using physicomimetics are robust when agents are removed, scale well when more agents are added, and are noise tolerant. In fact, they can also exhibit behavior similar to a phase transition. As we will show, such phase transitions can be explained and predicted because the system is based on physics. These concepts will be highlighted as the chapter continues.

In the meantime, let's focus on the task itself. Chapter 1 gave a high-level pictorial view showing how circles intersect to create triangular formations. Now we need to map this to a physics-based approach. Let the center of each circle represent an agent. The radius of the circle is the desired separation $D$ between two agents. The key is to define a virtual force law between two agents that has certain obvious characteristics. First, the force should be repulsive when the agents are too close. Second, the force should be attractive when the agents are too far. Third, the force law should be "circularly

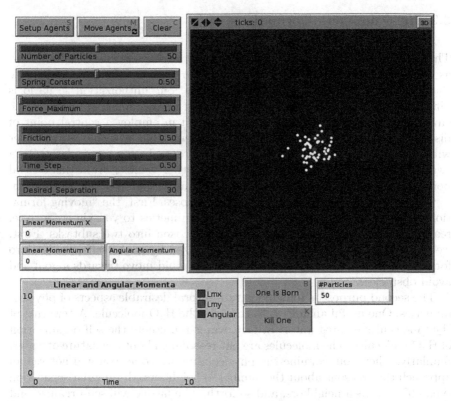

Fig. 3.7: A picture of the graphical user interface for the self-organization of triangular lattices using Hooke's law, after initialization

symmetric"—which means the force is the same regardless of the angle of rotation of the system. Finally, given the inherent limitation of agent sensors, we do not expect (or necessarily even desire) an agent to detect all other agents. As with the flocking simulation, there is a maximum sensing distance. Beyond this distance there can be no force between agents.

Chapter 2 introduced you to two force laws. The first was Hooke's law for a spring. The second was Newton's gravitational force law. Both laws are circularly symmetric since they depend only on the distance between two agents. At first blush, Hooke's law would seem more appropriate, because it has both attractive and repulsive components. However, will it allow agents to move enough to create large formations, since the attractive force grows as the distance increases (when the distance is greater than the desired separation)? Newton's law decreases with distance, possibly allowing agents to move farther apart to create larger formations. However, it is purely attractive. Can it be modified to have a repulsive component? We will examine the

Listing 3.5: The portion of observer code that initializes the Hooke's law formation simulation

```
to setup
  clear-all
  create-particles Number_of_Particles [setup-particles]
  update-info

  set center_of_mass_x (sum [xcor * mass] of particles) /
                       (sum [mass] of particles)
  set center_of_mass_y (sum [ycor * mass] of particles) /
                       (sum [mass] of particles)
  ask patch (round center_of_mass_x) (round center_of_mass_y)
     [ask patches in-radius 4 [set pcolor red]]
end
```

use of both force laws in this section. In fact, both can work quite well, but exhibit different behavioral characteristics.

## 3.2.1 Hooke's Law for Triangular Formations

We first examine the self-organization of triangular formations using Hooke's law ("formation_hooke.nlogo"). Figure 3.7 shows the graphical user interface, which looks much like our previous simulations. One item that has been omitted is the Energy graph. The reason for this becomes clear when one realizes that a stable formation requires some friction. Without friction the particles would either oscillate or move beyond sensor range. Once friction is included, potential and kinetic energy decrease to zero as the initial energy is converted to heat (see Fig. 2.6 in the previous chapter). However, the Linear and Angular Momenta graph is still part of the simulation. If the particles are motionless at initialization, linear and angular momenta will be zero. When the simulation is run, linear and angular momenta should remain zero if the simulation is implemented properly.

Go ahead and run the simulation. As usual, click Setup Agents and then click Move Agents. The initialization procedure called by Setup Agents is shown in Listing 3.5. First, the observer clears everything with clear-all. Then Number_of_Particles are created. The update-info procedure updates the parameters controlled by the sliders. Finally the center of mass is computed using Eq. 2.1 and displayed as a red patch. The setup-particles procedure, as shown in Listing 3.6, is run by each particle at initialization.

**Listing 3.6: The portion of turtle code that initializes the Hooke's law formation simulation**

```
to setup-particles
    setxy (random-normal 0 20.0) (random-normal 0 20.0)
    set heading random 360 set vx 0 set vy 0 set mass 1
    set color white set shape "circle" set size 5
end
```

**Listing 3.7: The core portion of turtle code that computes Hooke's law**

```
to ap-particles
    ...
    if (r < 1.5 * D) [                    ; Sensor range
      set F (k * (r - D))
      if (F > FMAX) [set F FMAX]          ; Maximum force bound
      set Fx (Fx + F * (deltax / r))
      set Fy (Fy + F * (deltay / r))
    ]
    ...
end
```

Each turtle (particle) is given an initial position at the origin, with some additional Gaussian noise to form a cloud around the origin. The particles all have a mass of one, no velocity (and hence, no momenta), and are displayed as white circles.

The core piece of code is within the ap-particles procedure, as shown in Listing 3.7. The sensor range is $1.5D$, where $D$ is the desired separation between particles. If a neighboring particle is within view then that neighbor exerts a force on the particle. If the neighbor is not in view it exerts no force on the particle. As with the solar system simulation in Chap. 2, the force is capped via FMAX. The stiffness of the spring, represented by the variable k, is controlled by the Spring_Constant slider.

To understand where the sensor range "1.5 * D" comes from, examine Fig. 3.8. In a triangular lattice, the neighbors nearest to the central particle are $D$ away. The next closest neighbors are $\sqrt{3}D$ away.[3] We want our algo-

---

[3] This requires just a bit of trigonometry to prove.

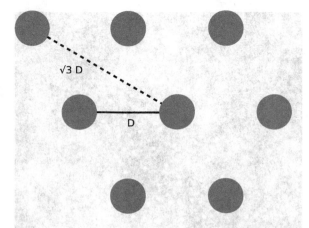

Fig. 3.8: A picture of an ideal triangular lattice

rithms to use minimal sensor information, so a sensor range of $1.5D$ allows a particle to respond to the nearest neighbors, but not the rest. In fact, a limited sensor range is key to the self-organization of the lattice. If you extend the sensor range (e.g., change 1.5 to 2.5 in the code), you will see that the particles cannot properly expand into a lattice. This is a perfect example of a minimalist principle referred to as "less is more," where greater functionality is obtained by removing extraneous information. Although we tend to think that more information leads to better results, this is not always true in science, and is very often not true in distributed systems.

The simulation generally yields nice triangular formations, as shown in Fig. 3.9. Note that the formation is not globally perfect. Is that a problem? No, because as stated earlier in this book, global perfection is generally not our goal. Since we are utilizing local information and forces with limited range, defects can and will occur in the formation. In fact, natural crystals (which are lattices composed of atoms or molecules) always have imperfections, and our formations often have similar imperfections. For example Fig. 3.11 shows a picture of bismuth crystal. One can easily see the lattice structure and the imperfections.

The overall quality of the formation shown in Fig. 3.9 is quite good and is more than adequate for the applications we will discuss later in this book. In terms of the correctness of our simulation, note that the center of mass does not move (at least not noticeably). Also, Fig. 3.10 shows that the linear and angular momenta do not deviate from zero, providing further confirmation. It is important to note that NetLogo automatically provides the vertical scale between $-1$ and 10. The momenta monitors confirm that the momenta remain close to zero.

There are two additional buttons in this simulation, namely, One is Born and Kill One. Listing 3.8 shows what these buttons do. The Kill One button calls

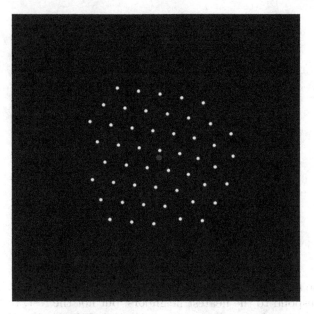

Fig. 3.9: A picture of the triangular lattice formed by 200 time steps using Hooke's law

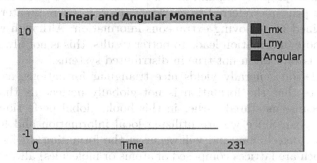

Fig. 3.10: The momenta graph of the simulation stays at zero

the one-dies procedure. This code asks the particles to choose one of themselves to die, if there are multiple particles remaining. Once a particle dies it is removed entirely from the simulation. The One is Born button calls the one-is-born procedure. This code is quite interesting. The hatch command *clones* one of the particles, copying all the variables associated with the original particle. Once the new particle is created it executes several commands. The first two commands (set deltax 0 set deltay 0) ensure that the new particle makes no initial move, because movement is done with the fd sqrt (deltax * deltax + deltay * deltay) command in the move procedure (as shown in the full NetLogo code). Then the position

Fig. 3.11: A picture of a piece of bismuth crystal, which has a spiral stair structure

Listing 3.8: The portion of observer code that creates new particles and kills current particles

```
to one-dies
    if (count (particles with [mass = 1]) > 1) [
        ask one-of particles with [mass = 1] [die]
        clear-drawing
    ]
end

to one-is-born
    if (count (particles with [mass = 1]) > 0) [
        ask one-of particles with [mass = 1]
            [hatch 1 [set deltax 0 set deltay 0 pd
                        setxy xcor + (random-normal 0 (D / 2))
                              ycor + (random-normal 0 (D / 2))]]
    ]
end
```

of the newly cloned particle is randomly offset from the original. Finally, the pd command puts the pen down for the particle, allowing us to see the trajectory that the new particle takes as it settles into position in the lattice. The Kill One button allows us to see how robust the system is with respect to the complete failure of individual agents. If you kill a lot of particles you will note that holes appear in the formation, but the formation stays remarkably stable. Similarly, the One is Born button shows how the system responds to the insertion of new particles. You will notice that the formation smoothly adapts to incorporate the new particle.

You may notice that linear and angular momenta are not preserved when you remove and add particles. Does this indicate that there is an error in the simulation? No, because (as explained in Chap. 2) the laws of conservation of momenta are valid only for closed systems. Adding and removing particles require actions by an external observer. This opens the system and allows laws to be violated. However, what is important to note is that after the change has been made, the system becomes closed again. Hence, linear and angular momenta should return to zero. The Clear button has been added to the simulation to allow you to clear both the graphics pane and the graph. After you make changes, click Clear, and you will see that the deviations from zero become extremely small once again.

Our earlier concern was that Hooke's law may have difficulty creating large formations. This was not evident when FMAX was set to one. However, changing Force_Maximum to three creates a formation that does not look like a triangular lattice. As shown in Fig. 3.12 (left), the initial cloud of particles has trouble expanding. Worse, setting Force_Maximum to five creates a very unstable formation that is always in a state of flux (Fig. 3.12, right). Similarly, reset Force_Maximum to one and change the Spring_Constant slider to 0.25. This does not work well either. Hence, although Hooke's law *can* work well, it is sensitive to the parameter settings.

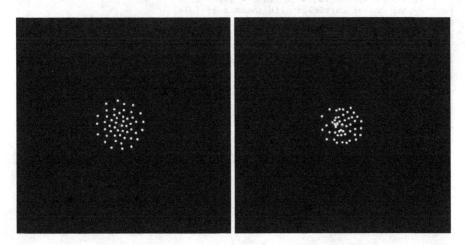

Fig. 3.12: A picture of the formations created using Hooke's law when FMAX is three (left) and five (right)

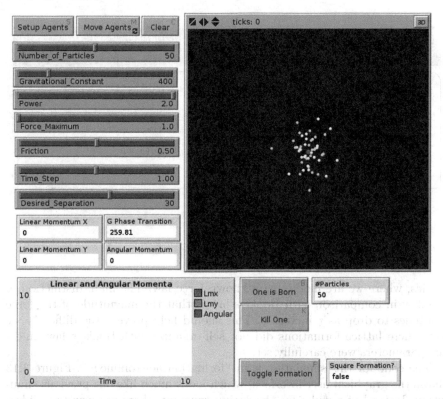

Fig. 3.13: A picture of the graphical user interface for the self-organization of triangular lattices using the "split Newtonian" force law, after initialization

## 3.2.2 Newton's Gravitational Force Law for Triangular Formations

Now let's try Newton's gravitational force law instead. This law is attractive, and must be modified to include a repulsive component. We also take the opportunity to generalize the force between two masses $m_1$ and $m_2$ as follows:

$$F_{ij} = \begin{cases} -Gm_1m_2/r^p & \text{repulsive when } 0 < r \leq D \ , \\ Gm_1m_2/r^p & \text{attractive when } D < r \leq 1.5D \ , \\ 0 & \text{out of sensor range when } r > 1.5D \ . \end{cases} \quad (3.3)$$

We refer to this force law as the "split Newtonian" force law because it changes sign at the desired separation $D$. Note also that we have generalized the denominator from $r^2$ to $r^p$. This is our first real example of physicomi-

```
if (r < view * D) [
    set F (G * mass * ([mass] of turtle ?) / (r ^ p))
    if (F > FMAX) [set F FMAX]
    ifelse (r > D)
        [set Fx (Fx + F * (deltax / r))     ; Attractive force
         set Fy (Fy + F * (deltay / r))]
        [set Fx (Fx - F * (deltax / r))     ; Repulsive force
         set Fy (Fy - F * (deltay / r))]
]
```

metics, where we are modifying a known force law. The main distinction of this law in comparison with Hooke's law is that the magnitude of the force continues to drop as $r$ increases. This should help prevent the difficulty we saw where lattice formations did not self-organize with Hooke's law unless the parameters were carefully set.

The code for this simulation is in "formation_newton.nlogo". Figure 3.13 shows the graphical user interface, which looks much like our previous simulation. Instead of a slider for the spring constant, there are now two sliders for the Gravitation_Constant and Power. The parameter $p$ in the force law is determined by the Power slider. The main change to the NetLogo code is in the ap-particles procedure, as shown in Listing 3.9. Note that FMAX still puts a bound on the force magnitude, as with the Hooke's law simulation.

Go ahead and run the simulation with the default settings. Then rerun the system with Force_Maximum set to five. Typical results are shown in Fig. 3.14. If you look carefully you will note that some particles are white, while others are yellow. The reason for this will become apparent later, but you can ignore this distinction for now. The center of mass (drawn as a red dot) does not move, which indicates that linear momentum is conserved. The momenta graph also indicates that the conservation of linear and angular momenta still hold, despite the discontinuity (at $r = D$) in the force law. Both values of Force_Maximum create triangular formations. However, the formation created with Force_Maximum set to one contains clusters of particles at the "nodes" in the lattice. Clustering was not seen with Hooke's law. Why is it happening now?

To explain why clustering is occurring, see Fig. 3.15. Two particles are clustered inside a ring of light gray particles. Because the force law is repulsive at small distances, the two central particles repel each other. Hence the dark gray particle tries to move away from the center of the formation. However,

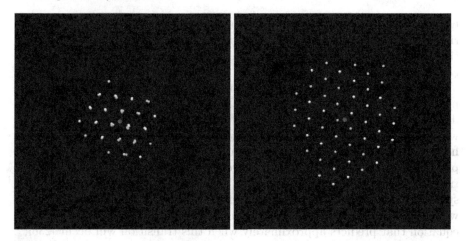

Fig. 3.14: A picture of the formations created using the split Newtonian law when FMAX is one (left) and five (right)

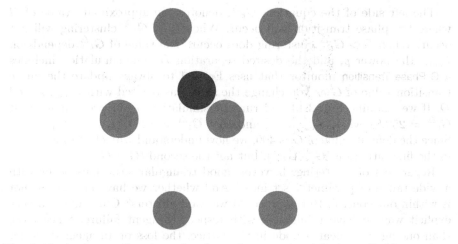

Fig. 3.15: A picture showing why clustering occurs with the split Newtonian force law

the outer light gray particles try to keep the dark particle near the center. The two closest light gray particles try to *push* the dark particle towards the center. The other four light gray particles try to *pull* the dark particle back towards the center.

Assume the dark particle is very close to the center. If Hooke's law was used, the outer ring of particles would exert weak forces on the dark particle, because it is almost at the separation distance $D$. The central light gray particle would succeed at pushing the dark particle away. But the split Newtonian force is not negligible at distance $D$. Hence it is possible for the outer ring of particles to exert more force on the dark particle (towards the center) than the repulsive force exerted by the central light gray particle. This causes clustering.

As we can see from Fig. 3.14, clustering does not always occur. It depends on the values of the parameters. In fact, rerun the simulation with the default parameters. Then slowly lower the Gravitational_Constant. The clusters will start to lose cohesion, but will not totally separate. At a certain point, however, a phase transition occurs, and the clusters fall apart. In physics a phase transition represents a change from one state to another, as with the transition of water to ice. However, it also refers generally to a qualitative system change caused by a small modification to a parameter. This is what we are seeing. One advantage of physicomimetics is that we can derive an equation that predicts approximately when this transition will occur [236]:

$$G_t{}^\triangle \approx \frac{F_{\max} D^p}{2\sqrt{3}} . \tag{3.4}$$

The left side of the equation, $G_t{}^\triangle$, denotes the approximate value of $G$ where the phase transition will occur. When $G < G_t{}^\triangle$ clustering will not occur. When $G > G_t{}^\triangle$ clustering does occur. The value of $G_t{}^\triangle$ depends on $F_{\max}$, the power $p$, and the desired separation $D$. The simulation includes a G Phase Transition monitor that uses Eq. 3.4 to always update the phase transition value of $G$ as you change the sliders associated with $F_{\max}$, $p$, and $D$. If we examine Fig. 3.14 and run the simulation again, we will see that $G_t{}^\triangle = 259.81$ when $F_{\max} = 1$, and that $G_t{}^\triangle = 1299.04$ when $F_{\max} = 5$. Since the default value of $G$ is 400, we now understand why clustering occurs in the first situation $(G > G_t{}^\triangle)$, but not the second $(G < G_t{}^\triangle)$.

Regardless of clustering, however, good triangular situations occur with a wide range of parameter settings.[4] And whether we have clusters or not is within our control. But why would we want clusters? Clusters are a very explicit way to have robustness with respect to agent failure. Since more than one agent is near a node in the lattice, the loss of an agent does not necessarily create a hole in the formation. Furthermore, if a hole does appear the surplus agents can move to fill a hole. Figure 3.16 illustrates this nicely. The left picture shows a clustered formation with 50 agents. These agents occupy 21 nodes in the lattice. Then 25 agents were removed. The formation is quite intact and still occupies 19 nodes in the lattice (right picture).

The nice aspect of this is that the system user can decide a priori whether clusters are desirable, and set the parameters appropriately. With the current implementation, clusters can be formed only during the initial self-organization of the formation. If $G$ is too low, clusters will not appear, and raising $G$ later will not create clusters once the formation is stable. However, raising $G$ and introducing new particles can create clusters. Alternatively,

---

[4] An analysis of formation quality is provided in Spears and Gordon [235].

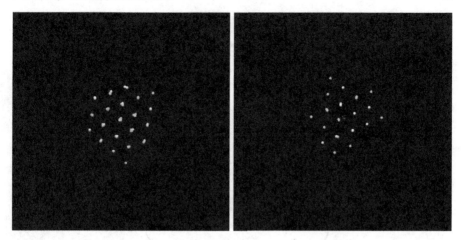

Fig. 3.16: A picture of the formations created using the split Newtonian law with 50 agents (left) and after 25 agents have been removed

repulsion could be turned off, allowing the formation to collapse, and then repulsion could be turned back on, with a higher value of $G$.

### 3.2.3 Newton's Gravitational Force Law for Square Formations

Given the success with triangular lattices, can other formations self-organize? The obvious alternative is a square lattice, since this also tiles a two-dimensional environment. But how is this to be accomplished? While some neighbors are $D$ away, others are $\sqrt{2}D$ away, as shown in Fig. 3.17. How do we distinguish between the two different types of neighbors?

One elegant solution is to create two types of particles. In this case we have hashed particle and dark particle *types*. The closest neighbors of unlike type should be $D$ away, while the closest neighbors of like type should be $\sqrt{2}D$ away. The NetLogo code sets the types of particles at the initialization stage, depending on their who identification number, as shown in Listing 3.10. Particles with even identification numbers are given a white color, while those with odd identification numbers are yellow. Note that the NetLogo code uses the colors white and yellow, instead of hashed and dark. All that matters is that we have two different labels for the particles, regardless of their depiction [235].

The only question remaining is how to modify the force law to accommodate the two types of particles. This turns out to be a very simple modification, as shown in Listing 3.11. There are two important components to this

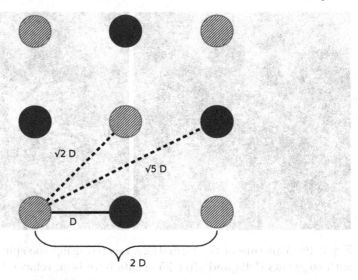

Fig. 3.17: A picture showing how a square lattice can be built, using two types of particles, hashed and dark gray

Listing 3.10: The initialization of the two types of particles needed to create square lattices

```
to setup-particles
    ...
    ifelse ((who mod 2) = 0) [set color white] [set color yellow]
end
```

code. The first involves a renormalization of the distance $r$. The code examines whether two particles both have even (or odd) identification numbers. If so, they should be $\sqrt{2}D$ away from each other. But we accomplish this in a subtle manner—as opposed to changing the desired distance $D$, we set r (r / (sqrt 2)). This renormalizes (i.e., divides) the distance by $\sqrt{2}$. This allows us to continue to require that the force between any two particles (regardless of their particle type) be repulsive if $r < D$ and attractive if $r > D$ (and if within the sensor view).

All that remains is to recalculate the sensor view. We will follow the same reasoning that was used to derive the sensor view of $1.5D$ for triangular lattices in Fig. 3.8. Consider particles of unlike type first. As shown in Fig. 3.17, the closest neighbors should be $D$ away. But the second closest unlike neigh-

**Listing 3.11:** The additional code needed in the `ap-particles` procedure to create square lattices

```
set view 1.5
if (square?) [                         ; For square lattice
ifelse ((who mod 2) = (? mod 2))
   [set view 1.3 set r (r / (sqrt 2))] ; Like color/type
   [set view 1.7]                      ; Unlike color/type
]
```

bors are $\sqrt{5}D \approx 2.236D$ away. Since we would like our sensors to be as local as possible, we choose $1.7D$ for our sensor view in that situation (see Listing 3.11). Now consider particles of like type. The closest should be $\sqrt{2}D$ away (before renormalization). The next closest are $2D$ away (again before renormalization). The ratio is $2D/\sqrt{2}D = \sqrt{2} \approx 1.414$. Since we are already utilizing sensor views of $1.5D$ for triangular lattices and $1.7D$ for unlike particles in square lattices, a sensor view of $1.3D$ is reasonable for like particles in a square lattice. The behavior of the system is not very sensitive to the precise sensor views, and you should feel free to experiment with the code if you wish.

The simulation includes a Toggle Formation button and assumes that a triangular formation is the default. However, you can create a square formation by clicking Setup Agents, Move Agents, and Toggle Formation.[5] Go ahead and run the simulation with the default settings. Then rerun the system with Force_Maximum set to five. Typical results are shown in Fig. 3.18. If you look carefully you will see alternating patterns of yellow and white particles, as we expected. As with triangular lattices, clustering occurs when Force_Maximum is set to one, but not when Force_Maximum is set to five. As before, a phase transition will occur if there is clustering and the Gravitational_Constant is lowered. Hence, we can again derive an equation that predicts approximately when this transition will occur [236]:

$$G_t{}^{\square} \approx \frac{F_{\max}D^p}{2 + 2\sqrt{2}}. \tag{3.5}$$

Self-organization of square lattices is more difficult than triangular lattices because there are two types of particles in the square lattice, and the initial distribution of both types has a large effect. Hence the lattices can contain larger defects than before. But there are ways to improve these lattices. Defects represent suboptimal lattices. These suboptima can be removed through

---

[5] Alternatively, you can change the NetLogo code to make a square formation the default.

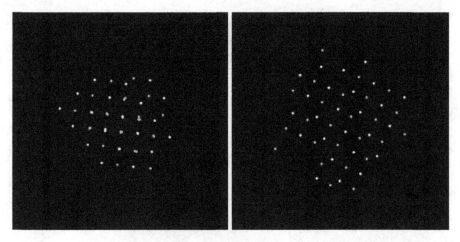

Fig. 3.18: A picture of the square formations created using the split Newtonian law when FMAX is one (left) and five (right)

the addition of noise, which "unsticks" the particles from their suboptima positions. This technique is used in metallurgy, where high temperatures create thermal noise that allow atoms to reposition. Lowering the temperature later then creates a metal with fewer defects. There are numerous ways to introduce noise into our simulation. One way is to raise the Time_Step even higher than 1.00. Interestingly, simulation inaccuracy helps to remove defects! Another way is to reduce Friction to a low value (e.g., 0.1) and then raise it. Figure 3.19 (left) shows the results when applied to Fig. 3.18 (left). Another way is to add more particles to the system. This perturbs the system and can help create better formations. As an example, 10 more particle were added to the formation in Fig. 3.18 (right). The results are shown in Fig. 3.19 (right). Note that the although the local shape of the formation is good, the global shape need not be. This is to be expected, since we are using only local information. However, globally perfect formations are shown in Chap. 4.

Another technique stems from the observation that clusters contain like particles (all yellow or all white). If a particle changes color occasionally, it will leave that cluster and move to another location. This movement helps to remove defects.[6] One consequence, however, is that clusters eventually disappear (resulting in individual particles at lattice node locations). An alternative mechanism for introducing noise is to hit Toggle Formation several times. Note how easily the formation switches from triangular to square, and vice versa. Repeated transformations tend to improve lattice quality. The main lesson to be learned is that, as with biomimetics, physicomimetics systems are not only noise tolerant, but can actually improve performance when noise is added. Since real robots and hardware exhibit sensor and actuator

---

[6] Formation quality results are shown in [236].

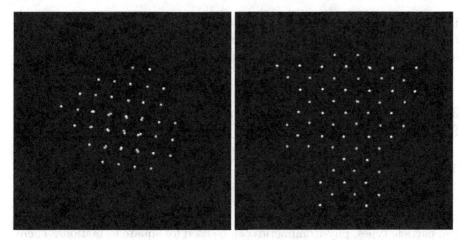

Fig. 3.19: A picture of the improved square formations created using additional noise in terms of lower friction (left) and the additional insertion of particles (right)

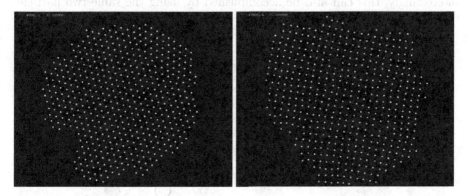

Fig. 3.20: Improved formations for 500 particles created using additional noise by lowering friction

noise, this is an extremely important observation. Generally, we have been taught to regard noise as something bad that should be removed or reduced through the application of various hardware and software techniques. It is rather difficult to change to the viewpoint that a certain amount of noise can be beneficial! In our experiments with real robots using physicomimetics, we often fell into the trap of worrying about noise. However, except in the cases of truly unreliable hardware, we found that it was best to ignore noise and let the system respond naturally. The end result was almost always better.

Although we have discussed robustness earlier in this section, we have not yet examined scalability. It is necessary for swarm techniques to scale to large numbers of robots. Does our technique scale well? Yes, as shown in Fig. 3.20.

By using the technique mentioned above where friction is lowered and then raised, 500 particles can self-organize into good quality triangular and square lattices! This is true of all of the simulations we show in the book.

### 3.2.4 Newton's Gravitational Force Law for Honeycomb Formations

Thus far this chapter has shown how to use physicomimetics so that particles can self-organize into triangular lattices. It is important to note that the term "hexagonal lattice" is synonymous with "triangular lattice" and they are used interchangeably throughout this book. We also showed that by introducing two particle types, physicomimetics can be used for square formations. There is one last formation of interest, namely, the honeycomb lattice. The honeycomb lattice is useful because, for a given separation distance $D$, it provides the maximum area coverage of any regular lattice pattern, given $N$ particles. Interestingly, this can also be accomplished by using the same two particle types, but in a slightly different manner [157].

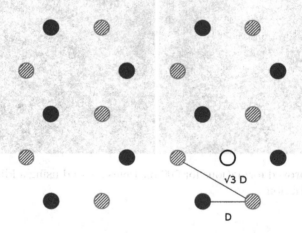

Fig. 3.21: A picture showing how a honeycomb lattice can be built, using two types of particles [157], hashed and dark gray

Figure 3.21 shows how this is accomplished. In a square lattice, unlike particle types are distance $D$ apart, while like particle types are $\sqrt{2}D$ apart. For a honeycomb lattice, the only difference is that like particle types are $\sqrt{3}D$ apart. Hence, the only change that is needed is to renormalize the distance by set r (r / (sqrt 3)) when two particles have the same type. Each honeycomb cell is composed of three particles of one type and three of another type. There is one difficulty, however, that does not occur with the square lattices. Suppose a particle is at the location denoted by the open circle in Fig. 3.21. Regardless of the type of particle, the forces acting upon it by its six neighbors constrain this extra particle to remain in this location. Simply put, it is caught in a potential well. Clearly, if particles end up in the centers of every honeycomb cell, the lattice will become the standard triangular lattice.

If the honeycomb is perfect, then there is nothing to do. However, if a cell has a center particle, this can be detected and repaired. In a perfect honeycomb, the neighbors of every particle (at distance $D$) have an angular separation of 120°. When a particle is at the center of a cell, this is no longer true. Some particles will have neighbors that are separated by 60° (see Fig. 3.21). Such particles are then perturbed, in an effort to break them out of their potential well [157].

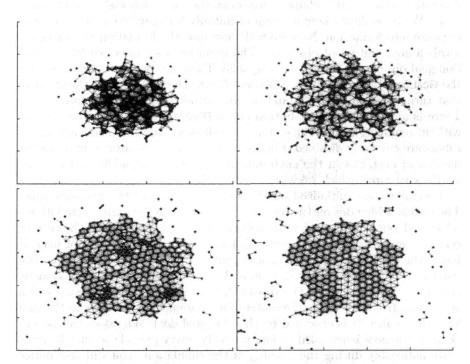

Fig. 3.22: The self-organization of a honeycomb lattice at time steps 5, 10, 100, and 1,800 (courtesy of James McLurkin)

Figure 3.22 shows four snapshots from the video "honeycomb.avi", showing the self-organization of a honeycomb lattice [157]. The initial conditions are similar to what we have seen earlier in this chapter—the particles start in a small cluster. Due to repulsion, this cluster expands (top right). At 100 time steps, a very clear honeycomb lattice has emerged. After 1,800 steps, utilizing the repair mechanism discussed in [157], the honeycomb has been greatly improved.

As opposed to detecting and repairing honeycomb lattice defects, an alternative technique is to use lattice transformations to remove the potential wells. The key is to note that square lattices do not have these potential wells. By periodically switching between square and honeycomb lattices, particles at the centers of honeycombs can escape. The NetLogo code "honeycomb.nlogo" provides the details on this technique.

## 3.3 Physicomimetics for Moving Formations

As stated earlier in this chapter, we broke the formation task into two subtasks. We have shown how a group of initially unorganized agents can self-organize into formation. Now we will show how the formation can move towards a goal and avoid obstacles. The graphical user interface for "formation_goal_obs.nlogo" is shown in Fig. 3.23. There are two new sliders, one for the Goal_Force and one for the Obstacle_Force. There is a Toggle Goal button that turns the goal force on and off. At initialization the goal force is off. There is another additional button called Disable One that disables a robot without removing it from the system, as well as a monitor to show how many robots are currently disabled. Finally, there is a new monitor to indicate the number of obstacles in the environment, as well as one additional monitor for the goal force, which we will discuss further below.

The robots are initialized on the right hand side of the graphics pane. The goal is a blue dot on the left hand side. As with the earlier simulations, white and yellow dots depict the robots, and a red dot denotes the current center of mass. Click Setup Agents. You will note that the center of mass is inside the original cluster of particles (robots). Now click Move Agents. The particles will move into a formation, and the center of mass will not change. But now click Toggle Goal. The robots start moving towards the goal, which is an attractive force. Also, the center of mass immediately moves to the goal location (in fact, it is obscured by the blue goal dot). Why does this occur? Because the goal is modeled as an extremely heavy particle so that it won't move noticeably during the running of the simulation. You will also notice that the momenta graph deviates from zero when you turn the goal force on. This occurs because you "opened" the system when you turned on the goal force. If you click Clear and examine the momenta graph you will see that the momenta quickly move back to zero after the goal force has been turned on.

Figure 3.24 (left) shows the formation centered around the goal, when the goal force is 0.25 (the Goal_Force slider is at 0.25). The formation takes quite a few time steps to reach the goal. Figure 3.24 (right) shows what happens when the goal force is 2.00. The formation moves much more quickly to the goal, but then is compressed somewhat by the goal force.

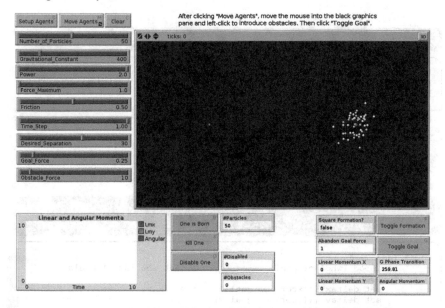

Fig. 3.23: A picture of the graphical user interface for the self-organization and movement of lattices toward a goal and around obstacles, after initialization

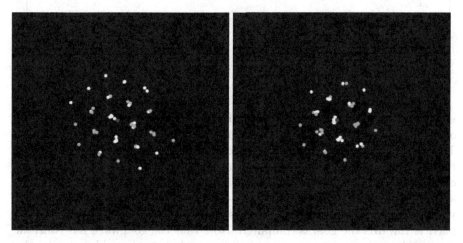

Fig. 3.24: A picture of the triangular lattice surrounding the goal when the goal force is 0.25 (left) and 2.00 (right)

The changes in the NetLogo code needed to include goals and obstacles are not dramatic. First, as shown in Listing 3.12, we take advantage of a feature in NetLogo that allows turtles to be of different breeds. Each breed has all of the parameters defined by turtles-own. For this simulation it

Listing 3.12: The code assumes there are three different breeds of turtles

```
breed [particles particle]
breed [goals goal]
breed [obstacles obstacle]
```

Listing 3.13: The code needed in `ap-particles` to respond to the attractive goal force

```
if (goal?) [
    set hood [who] of goals                ; Get the ids of goals
    foreach hood [
        set deltax (([xcor] of goal ?) - xcor)
        set deltay (([ycor] of goal ?) - ycor)
        set r sqrt (deltax * deltax + deltay * deltay)
        set F goalF
        set Fx (Fx + (F * (deltax / r)))   ; Attractive force
        set Fy (Fy + (F * (deltay / r)))
    ]
]
```

is conceptually elegant to have three breeds: particles (robots), goals, and obstacles. One agent of a breed is referred to as a particle, goal, or obstacle. The additional code needed in the procedure ap-particles to compute the attractive force exerted by the goal is given in Listing 3.13. First, the code checks to see whether the goal has been toggled on. Then it examines all the goals (in this simulation we use only one goal, but we are allowing you to experiment with more if you wish). Finally, the attractive force of the particle towards the goal is computed. Note that the goal force has constant magnitude and can be seen at any distance. As a real-world example, the goal could be a predetermined GPS location.

Another feature added to this simulation is the ability to disable a robot. In our earlier simulations a robot could die and be removed from the system. But a more realistic scenario is that the robot malfunctions and can barely move. However, it is still in contact with its neighbors. To try this feature, click **Setup Agents** and **Move Agents** again. Allow the formation to self-organize. Now click **Disable One**. You will see a huge jump in the momenta graph. This occurs because we are emulating a broken robot by making it extremely heavy (so that it can only move extremely slowly). You can click Clear to remove the spike in the graph, showing you that the momenta drop to zero again very quickly. Again, this spike occurred because we briefly opened the system to disable the robot. A disabled robot is displayed as a purple dot.

Now click Toggle Goal and make sure that your Goal_Force slider is at 0.25. You will notice that the center of mass moves towards the center of the graphics pane. This is because both the goal particle and the disabled particle have identically heavy masses. In terms of the behavior of the formation, usually most of the particles will move towards the goal. If not, increase the Goal_Force slider slowly until they do. Occasionally, one or more particles will be left behind with the disabled particle. This is occurring because the extremely high mass of the disabled particle creates a huge force on its neighbors that is then capped by $F_{\mathrm{max}}$. By default the Force_Maximum slider is set to one. Since the goal force is only 0.25, the neighbor(s) of the disabled particle cannot break free. Perhaps this is the behavior that we want. If not, we know what to do. The goal force must be increased sufficiently to break the bond between the disabled particle and its neighbors. Because $F_{\mathrm{max}} = 1$ the goal force must be increased to be somewhat greater than one. This explains the purpose of the Abandon Goal Force monitor. This monitor gives the amount of force required to fully abandon a disabled particle.[7]

What about obstacles? We will introduce them to the simulation via mouse clicks. NetLogo has a nice feature for handling mouse events, as shown in Listing 3.14, which is in the `run-and-monitor` procedure. This code always monitors the mouse. If the mouse is clicked, the code checks to see if the mouse pointer is within the black graphics pane. If so, the code sees how many obstacles have already been created. If none, the command `create-obstacles` creates one obstacle, sets it at the mouse location, makes it very heavy, and sets the color to green. If obstacles have already been created, a new one is cloned from a prior one, but is placed at the current mouse location. The `wait 0.2` command provides a slight pause so that one mouse click does not generate multiple obstacles in the same location.

The additional code needed in `ap-particles` to compute the repulsive force exerted by obstacles is given in Listing 3.15. First, it examines all of the obstacles. Then it checks to see if any obstacles are within sensor range of the particle. The Obstacle_Force slider determines both the sensor range (size of the obstacle) and the repulsive obstacle force law. If the slider is set at value `obstacleF` then if `r <= obstacleF` the force felt by a particle is equal to `obstacleF - r`, where `r` is the distance from the particle to the obstacle. For example, the default value for the Obstacle_Force slider is 10. Hence, when the particle and obstacle are nine units apart, the force felt by the particle is one.

---

[7] If you disable several particles, and they are close to each other, you may require an even higher goal force. But, because this is physics, it is easy to compute how high the goal force should be!

**Listing 3.14: The additional mouse event code needed to place obstacles where we want in the graphics pane**

```
if (mouse-down? and mouse-inside?) [
   ifelse ((count obstacles) = 0)
      [create-obstacles 1
          [setxy mouse-xcor mouse-ycor set vx 0 set vy 0
           set size (2 * obstacleF) set shape "circle"
           set mass heavy set color green]]
      [ask one-of obstacles
          [hatch 1
              [setxy mouse-xcor mouse-ycor set vx 0 set vy 0]]]
   wait 0.2        ; Pause so don't get a bunch of obstacles at once
]
```

**Listing 3.15: The additional code needed in `ap-particles` to respond to the repulsive obstacle forces**

```
   set hood [who] of obstacles          ; Get obstacle ids
   foreach hood [
      set deltax (([xcor] of obstacle ?) - xcor)
      set deltay (([ycor] of obstacle ?) - ycor)
      set r sqrt (deltax * deltax + deltay * deltay)
      if (r <= obstacleF) [
         set F (obstacleF - r)              ; Linear force law
         set Fx (Fx - (F * (deltax / r)))  ; Repulsive force
         set Fy (Fy - (F * (deltay / r)))
      ]
   ]
```

It is important to point out that goals and obstacles do not sense each other. Hence they do not react to each other. But there is an interaction between the parameters for the goal force and the obstacle force. The obstacle force must be stronger than the goal force. If not, the particles will move through the obstacles if we are using a simulation, or run into the obstacles if we are using actual hardware.

Let's show two examples, one with Obstacle_Force set to 10 and the other with it set to 50. Again, click Setup Agents and Move Agents. Allow the formation to become stable. Then slowly click in the graphics pane between the particles and the goal. This will create obstacles. When an obstacle is created

a green disk will appear and the #Obstacles monitor will increase by one. For
our example we created twelve obstacles. Then click Toggle Goal. When the
obstacle size is set to 10 and the obstacles are narrowly spaced, the formation
very slowly maneuvers around the obstacles by twisting and rotating the formation as minimally as possible. This occurs because the obstacles are not
sensed until they are close, and their effect is relatively small. Figure 3.25
shows the result as the formation navigates the obstacles (left) and as the
formation reaches the goal (right). In this case, all the particles made it to
the goal, although this does not always occur.

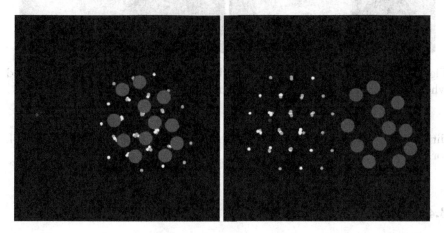

Fig. 3.25: The triangular lattice encounters an obstacle field (left picture),
and reaches the goal (right picture), when the obstacle size is 10

When the obstacle size is set to 50 the problem becomes different. In this
case, the virtual force wall thrown up by the obstacles is much larger. For the
example shown in Fig. 3.26, the whole formation rotated en masse around the
northern end of the wall. Note that the code says nothing about behaving
in this manner—it is an emergent property of the system and reflects the
easiest way for the formation to respond to the simultaneous repulsive and
attractive forces from the obstacles and goal.

It is instructional to set up your own obstacle courses and watch the behavior of the particles. Some obstacle courses can be traversed successfully and
some cannot.[8] This is still an open research problem in robotics. Our physicomimetics technique solves problems where the obstacle density is three
times higher than the problems used with the biomimetic techniques [93].
This will be explained more in Chap. 14, using a variation on another force
law referred to as the "Lennard-Jones" potential. This force law produces
formations that are more fluid-like than those shown in this chapter. The ad-

---

[8] Additionally, you can change the size of the obstacles while the simulation is running by
moving the Obstacle_Force slider.

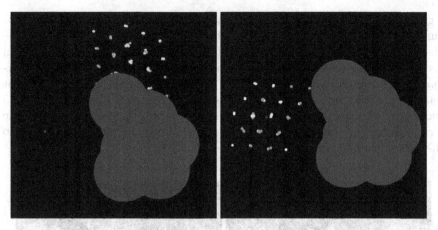

Fig. 3.26: A picture of the triangular lattice encountering an obstacle field when the obstacle size is 50, early in the process (left), and late (right)

ditional simulation "formation_lj_goal_obs.nlogo" is provided with this book for further experimentation.

## 3.4 Hardware Requirements

One of the most useful aspects of the simulations shown in this chapter is that they provide insight into the hardware requirements for actual robots. In order to make things interesting, assume the robots are outside and in an uncontrolled environment. Now, let's consider the various requirements.

First, the robots must be able to sense a goal location. This is not too difficult. The most obvious solution is for the goal to be a global positioning system (GPS) coordinate. GPS chips are now quite inexpensive, and if each robot has a GPS chip it can calculate the direction and range to the goal.

Second, the robots must avoid obstacles. There are a couple of ways to do this. Acoustic sensors are one possibility. They emit high frequency sounds which are reflected from objects. The time taken for the echo to return gives a distance to the obstacle. Infrared (IR) sensors are another possibility. These sensors emit an infrared beam, which is reflected back to the sensor by obstacles. One class of infrared sensors uses the intensity of the returned light as an estimate of distance. More sophisticated IR sensors use a method called "triangulation" that measure angles to determine the distance. We have used these latter sensors extensively in our research.

The sensors mentioned above are widely available and inexpensive. However, the most important requirement is that each robot obtain the bearing and range to its neighbors.[9] For this information researchers often rely on off-the-shelf cameras. Sometimes these cameras are mounted on the ceiling of an indoor environment. Then a central processor uses a vision algorithm to locate each robot. These locations are then broadcast to each robot. But, this is not a distributed solution. If the camera, the central processor, or the broadcaster fails, the system cannot run. Furthermore, this will not work in outdoor environments. An alternative is to mount an inexpensive camera on each robot. But cameras generally do not have a 360° field of view. Hence multiple cameras are needed, or conical and spherical mirrors may be used to provide a 360° field of view [62]. Regardless, vision algorithms are still required to locate the robots, and to distinguish them from rocks, trees, and other objects. Since vision algorithms are reasonably complex, this is difficult to accomplish when robots have inexpensive processors.

The solution we have taken is to design our own sensor for locating robots, based on our requirements. Imagine what happens when lightening strikes nearby, but you don't actually see the lightening bolt. You only notice a flash of light. After the flash you hear the thunder. By measuring the time between the flash and the thunder you obtain a distance $d$ to the lightening bolt. At this point you know that the lightening bolt struck somewhere along a circle of radius $d$ from yourself. Now, suppose two other people in different locations make the same measurement. Two more circles can be drawn, one each around the other two people. If you share your information you can calculate the intersection of the three circles, which is a point. That point is the location of the lightening bolt.

This technique is called "trilateration." Trilateration locates points based on distance computations. Triangulation (which is often confused with trilateration) locates points based on angle computations, as mentioned above. We built a specialized piece of trilateration hardware for each robot. Of course, we don't use lightening. However, each robot can simultaneously emit a radio frequency (RF) pulse and an acoustic pulse. When nearby robots detect the RF pulse they start a clock. There are three acoustic receivers on each robot. When each receiver hears the acoustic pulse, a distance is calculated. As with the lightening example given above, the three distances are sufficient to allow each nearby robot to calculate the range and bearing to the robot that just emitted the RF and acoustic pulse. Since each robot takes turns emitting, all robots can quickly calculate where their neighbors are.

For the triangular formations shown in this chapter, the trilateration technique described above is sufficient. However, what happens when square or honeycomb formations are required? Recall that two types of particles (robots) are needed. In this case each robot needs to know the type of each neighbor. How can this information be shared? The answer is quite simple. With the framework described above, the RF pulse contained no information. It merely served as a "gun" to start the clocks on the neighboring robots.

---

[9] GPS is not sufficiently accurate for this application.

But it is very simple to send information with the RF pulse. In the situation with square lattices only one bit is required (a 0 for one type of robot and a 1 for the other). However, we are not limited to sending one bit. Information packets can be sent along the RF, allowing for richer communication and more complex algorithms and behavior. In essence our trilateration technique merges two requirements: (1) the need to locate neighbors, and (2) the need to share information. Far more details on this technology are given in Chap. 10.

## 3.5 Summary

This chapter has provided the first examples of physicomimetics, as applied to the self-organization and movement of formations (Chap. 17 provides an alternative physicomimetics approach to flocking). A standard biomimetics flocking simulation was used to highlight some of the similarities and differences between the two approaches. One similarity is the concept that sensors have a limited range and that most information obtained by an agent is local. A second similarity is a more positive view of noise than we are normally taught. In fact, noise can be beneficial. In biomimetics noise enables creatures to explore their environment. In physicomimetics noise helps the system escape from suboptimal configurations.

A difference between biomimetics and physicomimetics is that we emphasize the concept of ground truth more strongly. We want to be able to specify a priori what we want our system to do, to measure how well it is doing, and to confirm that it is programmed properly. Although this is not something we can do perfectly yet, we've had excellent success in designing our systems from first principles. Another difference is that we want to be able to predict how the values of parameters affect behavior. Again, without being able to explain the full dynamics of these complex systems, we have shown that it is possible to use rather simple concepts from physics to predict and control important aspects of the dynamics.

One reason for our success thus far is that we have faithfully followed the basic laws of physics (i.e., Newton's second and third laws). However, to extend physicomimetics to more difficult problems, we will need to generalize our physics to some degree. This will be accomplished in the next chapter in three ways. First, we will start to incorporate minimal amounts of global knowledge. Second, we will mildly violate Newton's third law, and allow forces to be unequal and not opposite. Finally, we will explore physics that is not $F = ma$ and also show one example where physicomimetics and biomimetics are merged. *Physicomimetics is not merely the application of Newton's second law. It is a way of thinking about complex systems and how to design the behaviors that we want to accomplish.*

# Chapter 4
# Pushing the Envelope

William M. Spears

## 4.1 Perfect Formations

Chapter 3 examined physicomimetic techniques for the self-organization of formations, and for the movement of those formations towards a goal while avoiding obstacles. Standard Newtonian physics was followed faithfully, and each robot performed computations in its own frame of reference, using only information that was locally available. In this chapter we will "push the envelope" in four ways. First, we will occasionally permit more communication between robots, although the communication will still remain local. Second, we will use limited forms of global information for some applications. This will come in the form of a digital compass, which provides a common frame of reference for the robots. Third, we will also allow the simulations to carefully violate some standard physics assumptions. Fourth, we will introduce new forms of physics. As we will see, these changes allow us to expand the physicomimetics behavioral repertoire.

For example, perfect formations were not required in Chap. 3. But, what if we really wanted perfect formations? How can this be achieved? This section will show one technique. Our technique assumes that each particle has a preferred location in the formation. The preferred locations are shown in Fig. 4.1 (left), for an 18-particle triangular formation. The integer labels on the particles are the who identifiers provided by the NetLogo simulation. Although the ordering of the particles may appear to be arbitrary, it is in fact carefully chosen so that we can perform two transformations that are required to achieve perfect formations.

First, notice that the triangular lattice is made of rings, as in Fig. 4.1 (right). The first transformation is to convert the particle identification numbers to a "ring number" and "index" (numeric order) within the ring. The

William M. Spears
Swarmotics LLC, Laramie, Wyoming, USA, e-mail: wspears@swarmotics.com

W.M. Spears, D.F. Spears (eds.), *Physicomimetics*,
DOI 10.1007/978-3-642-22804-9_4,
© Springer-Verlag Berlin Heidelberg 2011

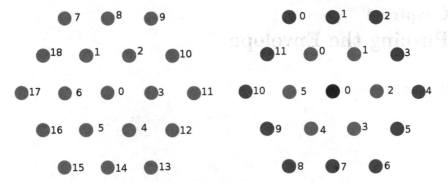

Fig. 4.1: The particle identification numbers in the triangular lattice are shown on the left. The right figure shows that the identification numbers can be converted to a ring number and index in the ring

rings in the lattice structure are shown by differently shaded particles. The index of each particle in a ring is shown by the integer label near each particle. The inner ring is composed of one particle. The next ring is composed of six particles. The outer ring is composed of twelve particles. The starting index (0) for each ring is always at the top left of the ring.

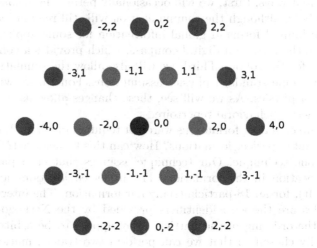

Fig. 4.2: The $(m, n)$ attributes for each particle

The second (more complex) transformation converts the ring number and index into two other attributes for each particle that we call the $(m, n)$

attributes.[1] The result is shown in Fig. 4.2. The $(m, n)$ attributes are not coordinates in the Cartesian coordinate system. However, they do serve to indicate where a particle should be with respect to another particle. If the $m$ of particle A is less than the $m$ of particle B, then A should be to the left of B. Similarly, if the $n$ of particle A is less than the $n$ of particle B, then A should be below B. But note that "left" and "below" indicate that the particles share a common frame of reference, and hence a piece of global knowledge. This knowledge can be as simple as "which way is north," which can be easily provided by an inexpensive digital compass that is mounted on each robot. Furthermore, in order for the robots to compare their $(m, n)$ attributes, these attributes must be communicated to each other. However, this is not difficult. As we saw in Sect. 3.4 of the last chapter, our trilateration localization module is capable of sending information along a radio frequency (RF) module as robot localization is occurring. In this case, two bytes of information would suffice, one for each attribute. This would allow for 256 values each for $m$ and $n$, which is certainly sufficient for the near future of swarm robotics.

In Chap. 3 we showed square lattices in addition to triangular lattices. Is it possible to extend the concept of $(m, n)$ attributes to square lattices? The answer is yes, and as a consequence each particle has two sets of $(m, n)$ attributes. One set is for triangular lattices and one set is for square lattices. The particular set is chosen when the user decides which formation he/she wants to see. The code for these transformations is shown in procedure init-m-n of the NetLogo simulation "formation_perfect.nlogo".

Figure 4.3 shows the graphical user interface for "formation_perfect.nlogo". The interface looks the same as those shown in Chap. 3. There are sliders to control the number of particles, the constants associated with the force law, the constants associated with the $F = ma$ simulation, and the desired separation. Three monitors and one graph show the quantities associated with linear and angular momenta. A Toggle Formation button allows you to switch between square and triangular formations. Finally, you are allowed to keep adding particles to the system with the One is Born button.

Go ahead and run the simulation. First, click Setup Agents and then click Move Agents. The square lattice formation is the default formation for this simulation. You should quickly see a result very similar to that shown in Fig. 4.3. The dynamics are fascinating. Particles revolve around each other for a while until the constraints imposed by the $(m, n)$ attributes are satisfied. Then the lattice "relaxes" into a perfect formation. Once the square formation is created, click Toggle Formation. Now a triangular formation is created from the square lattice. Again the dynamics are fascinating. You can keep switching between formations. Each time the whole formation is torn apart and rebuilt. An example of the square to triangular lattice transformation is shown in Fig. 4.4.

---

[1] The attribute $m$ should not be confused with the particle mass.

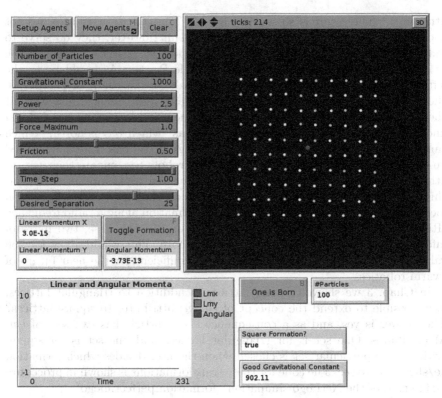

Fig. 4.3: A picture of the graphical user interface for the self-organization of perfect lattices using the "split Newtonian" force law. This picture shows a square lattice after formation

Click the One is Born button to add more particles to the formation. Each time you click the button, one of the existing particles is cloned and given new appropriate attributes. What you will see is that the new particle will ripple through the formation until the desired location is reached. The process is incredibly robust. You can also continue to switch back and forth between square and triangular lattices as you do this. In addition, you can slowly modify the Desired_Separation slider to expand and contract the formation. The Good Gravitational Constant monitor displays an estimate of the best value for the Gravitational Constant slider, based on Eq. 3.4.

All that remains is to describe how the extra attributes modify the force law. Once again we use the split Newtonian force law. The particles are still attracted to each other if they are greater than $D$ distance apart (the desired separation distance). The change to the force law occurs when the particles are less than $D$ apart. Normally these particles would repel each other. However, for perfect formations we make some exceptions, based on

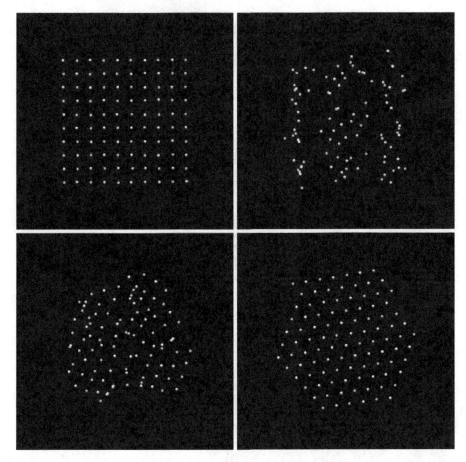

Fig. 4.4: Snapshots showing the transformation from square to triangular lattice

the values of the $(m, n)$ attributes. If the particles are placed such that the ordering implied by the attributes is incorrect, they will once again be attracted to each other. By doing so the particles will "dance" around each other until the ordering implied by the $(m, n)$ attributes is met. This can be seen in Listing 4.1, which shows a list of conditions under which the attribute conditions are not met, resulting in attraction between particles.

However, there are two special cases when the force is repulsive. These can only occur when the desired formation is the triangular lattice, as noted by the code and comments in Listing 4.1. The reason for these cases is exemplified in Fig. 4.5, which shows the second of the two cases. The $m$ attribute of the light gray particle is less than the $m$ attribute of the dark particle. Hence the light gray particle should be reasonably far to the left of the dark particle. Instead it is almost directly over the dark particle. In this situation the particles are

Listing 4.1: The portion of turtle code that calculates the split Newtonian force law for perfect formations

```
if (r < view * D) [
  set F (G * mass * ([mass] of turtle ?) / (r ^ p))
  if (F > FMAX) [set F FMAX]

  ifelse (r > D) [attract 1 1] [
  ifelse (((deltax < 0) and (pm < km) and
          (deltay < 0) and (pn < kn)) or
          ((deltax > 0) and (pm > km) and
          (deltay > 0) and (pn > kn))) [attract 2 2] [
  ifelse (((deltax < 0) and (pm < km)) or
          ((deltax > 0) and (pm > km))) [attract 2 2] [
  ifelse (((deltay < 0) and (pn < kn)) or
          ((deltay > 0) and (pn > kn))) [attract 2 2] [
  ifelse (((pn = kn) and (abs(deltay) > cy * D)) or
          ((pm = km) and (abs(deltax) > cx * D))) [attract 2 2] [
  ifelse ((not square?) and (pn != kn) and  ; <-- Special Case
          (abs(deltay) < cy * D)) [repulse 0 3] [
  ifelse ((not square?) and (pm != km) and  ; <-- Special Case
          (abs(deltax) < cx * D)) [repulse 3 0]
    [repulse 1 1]]]]]]]
]
```

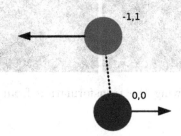

Fig. 4.5: An example of how a force law can break Newton's third law

repelled from each other in the direction of the arrows shown in Fig. 4.5. Note that the two forces that are depicted are "equal and opposite." In Chap. 1 we noted that Newton's third law states that the reaction between two bodies is equal and opposite. So, it looks as if Newton's third law is obeyed.

However, if you were observant, you may have noticed some unusual behavior in the Linear and Angular Momenta graph. When square lattices self-organize, these values stay close to zero, as they should. However, when triangular lattices are self-organizing,angular momentum oscillates. If you did not notice this behavior, toggle between the two types of formation. You will notice that linear momentum holds for both square and triangular lattices. However, angular momentum holds only for square lattices. What is happening? Clearly some aspect of Newton's third law is violated, but only during the self-organizing of triangular lattices.

The answer lies in the two special cases mentioned above for the triangular lattices. Although the forces shown in Fig. 4.5 are equal and opposite, they are *not* directly along the dashed line joining the two particles. A more precise description of Newton's third law comes from page 79 of Halliday and Resnick [85]:

> In other words, if body A exerts a force on body B, body B exerts an equal and oppositely directed force on body A; and furthermore the forces lie along the line joining the two bodies.

Hence, this simulation violates a basic assumption of Newtonian physics. If the "equal and opposite" assumption is violated, then both linear and angular momenta are unlikely to be conserved. If we break the assumption that forces must be directed along the line connecting two particles, then the conservation of angular momentum can be violated. The square lattice portion of the code does not violate any assumptions, and despite the added constraints from the $(m, n)$ attributes, both linear and angular momenta remain extremely close to zero. This indicates that the code for square lattices is accurately simulating $\boldsymbol{F} = m\boldsymbol{a}$ physics. However, the simulation breaks standard Newtonian physics for the self-organization of triangular lattices.

Why do we do this? Simply put, the triangular lattice self-organizes more quickly when we break Newton's third law in this manner. Introducing angular momentum into the system allows particles to revolve around each other more easily. This allows the additional constraints imposed by the $(m, n)$ attributes to be met more easily. We made a conscious design decision that performance was more important than strict adherence to physics in this case. And that is why we call our technique "physicomimetics." Physics inspires us, but we are allowed to modify physics to suit our needs.

As opposed to the earlier chapters, the simulations in this chapter are more cutting edge, and are ripe for further improvements and enhancements. For example, now that we have shown that the creation of perfect formations is possible, are there better ways to accomplish this? We leave the reader with some ideas for improvement. First, can the creation of perfect triangular formations be accomplished as well when Newton's third law is not broken? Second, could identification numbers be better assigned after the particles have been initially placed, but before they start to move? Third, instead of moving particles so that their $(m, n)$ constraints are met, could the attributes themselves be swapped via communication to minimize the amount of particle motion?

## 4.2 Chain Formations

Our next example was inspired by an application given to us by the military. Suppose that an environment is composed of long maze-like structures. One example is a sewer system or ventilation system. Another example is a network of caves in a mountain. The goal is for robots to self-organize into a chain that stretches from a point just outside the maze to a location of interest inside the maze, while maintaining communication connectivity. As an even more complex problem, how can a chain formation self-organize from an initially unorganized cluster of robots, given no environmental information whatsoever?

This section provides one solution to this task. Recall how we have created triangular lattices by assuming that forces are "circularly symmetric," which means that the force is the same regardless of direction. Figure 4.6 provides a concise summary. The left figure shows green-filled circles surrounded by red circles. The green circles are particles and the red circles represent the desired separation $D$ between particles. This formation is accomplished by using the force law shown on the right. There is a light red repulsive zone around the green particle, which extends to radius $D$. The blue zone is attractive and extends to the limit of the sensor range, which is $1.5D$. Hence the green particle is repelled from the red particle, but attracted to the yellow.

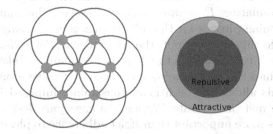

Fig. 4.6: An overview of how a circularly symmetric force creates triangular lattices

A circle is actually just a particular form of an ellipse. The major axis of an ellipse is the longest diameter of an ellipse, while the minor axis is the shortest diameter. In a circle the major axis is the same as the minor axis. If we want chain formations, it may be possible to create them using force laws that are truly elliptical rather than circular. This is because ellipses have an orientation, as do chains. Consider the left picture in Fig. 4.7. This shows how a chain can be considered to be a set of elliptical links, just as with a real physical chain. In fact, we are deliberately breaking the "circular symmetry" assumption that we have made thus far throughout this book.

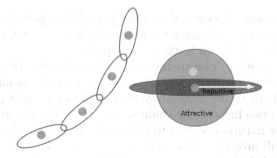

Fig. 4.7: An illustration of an elliptical force law

How does this change the force law? The right picture in Fig. 4.7 shows the modification. Because orientation is required, we use the heading (white arrow) to denote orientation. The blue circle still represents the area of attraction within the sensor range. But now the light red repulsive zone is elliptical in shape. The major axis is aligned with the heading of the robot and it is cut off at the sensor range (hence no force is felt at the light red regions outside the blue circle). As a consequence a particle is repelled from particles that are in front of it or behind it. In this situation the green particle is attracted to the yellow particle and repelled from the red particle. With this force law the distinction between the desired separation $D$ and the sensor range vanishes—they are the same.

But will this change yield a chain formation? As a example, consider the three pictures shown in Fig. 4.8. The left picture shows two particles, green and yellow. The blue circles represent the sensor ranges of both particles. The green particle is headed east, while the yellow particle is headed north. Both particles sense each other, since they are within sensor range.

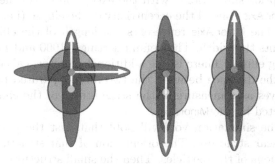

Fig. 4.8: An illustration of two particles with the elliptical force law

Because of the alignment of the ellipses, the green particle is attracted to the yellow particle. So its heading changes to south and it tries to get closer to the yellow particle. However, the yellow particle is repelled from the green particle. Its heading also changes to south as it tries to get farther from the green particle. This is shown in the middle picture.

At this point both particles are repelled from each other. The green particle changes its heading to north while the yellow particle maintains a south heading. They both try to move away from each other (right picture). Note the similarity to two links of the chain shown in the left picture of Fig. 4.7.

Of course, the natural question to ask is whether this works with multiple particles. We will investigate this with a NetLogo simulation. Before we do that, however, it is important to point out that the middle picture of Fig. 4.8 demonstrates an important feature of this new force law. Because it is no longer circularly symmetric, particles do *not* necessarily experience equal and opposite forces. In the middle picture both particles experience a force to the south! Since this violates Newton's third law, this has an important consequence, namely, that linear and angular momenta will not be conserved. Given that, what should we expect from our simulation? First, if angular momentum is violated, this should create chains that are curved. But that isn't an issue, because we don't want our chains to be straight lines—we want them to be able to bend according to the environment. A violation of linear momentum is more important, because the center of mass will move. However, this is not catastrophic. It suffices to have some nodes in the chain remain close to the initial cluster. In terms of practical use, we are assuming (as we have throughout this book) that robots are initially deployed from a central location. Since the applications we mentioned include the surveillance of sewers, tunnels, and caves, it is imperative that some robots remain near their original location, so that the eventual communication chain can be easily monitored.

Let's run the NetLogo simulation "formation_chain.nlogo". The picture of the graphical user interface is shown in Fig. 4.9. The simulation is extremely similar to our prior simulations, with the exception of two new sliders. Major_Axis and Minor_Axis control the eccentricity of the ellipse (i.e., the deviation from a circle). The Major_Axis represents the length of the ellipse, while Minor_Axis represents the width. The default settings (1000 and 1) create a very narrow and long ellipse. There is one additional monitor, called Max Distance, which reports the distance between the two most separated particles in the chain. This serves as a measure of the straightness of the chain, and as we will see, is affected by the Minor_Axis.

If you run the simulation you will note that first the particles collapse into a small linear structure. The orientation of that structure depends on the initial positions of the particles. Then the small structure slowly expands into a chain. Figure 4.10 shows two examples. As with our prior simulations, the small red dots represent the center of mass of the system. Not surprisingly, the center of mass moves. The large filled yellow circle represents the original center of mass of the system. As we stated above, it suffices to have

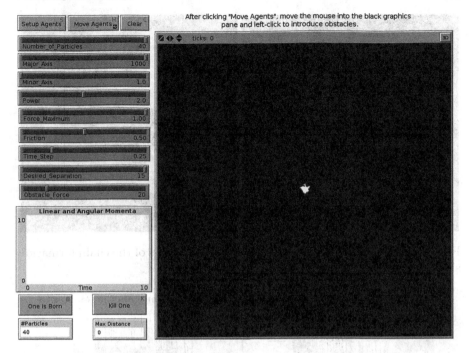

Fig. 4.9: The picture of the graphical user interface for the chain formation simulation in NetLogo

some robots remain near that original center of mass (deployment area). Our experiments indicate that this is true extremely often.

Note that both chains shown in Fig. 4.10 have a mild curve. The chain can be considered to be a two-armed structure with the center near the yellow circle. Figure 4.10 (left) shows a structure where both arms are curving clockwise. The right picture shows a structure where both arms are curving counterclockwise. Figure 4.11 shows the angular momentum graphs for the two chains. Note that the left graph indicates that angular momentum is generally negative, while the right graphs indicates positive angular momentum. This is in agreement with the chains shown in Fig. 4.10 because, by convention, a negative value for the angular momentum indicates a clockwise rotation, while a positive value indicates a counterclockwise rotation.

Now let's examine the impact that the Minor_Axis has on the formation of the chains. A larger Minor_Axis means that the ellipse is wider. This means that less of the sensed area is attractive. Attraction serves to enforce the straightness of the chain. Hence, a smaller attractive area allows greater curvature in the chain. We tested this by examining the maximum separation Max Distance, averaged over 50 independent runs, as the Minor_Axis is varied from one to five. The Major_Axis was kept at the default value of 1,000. Table 4.1 shows

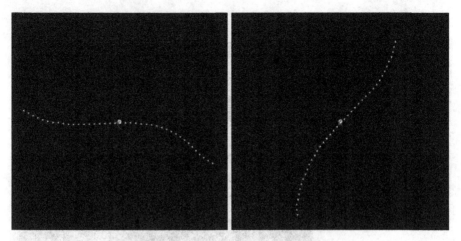

Fig. 4.10: Two chains formed using the default settings of the chain formation simulation

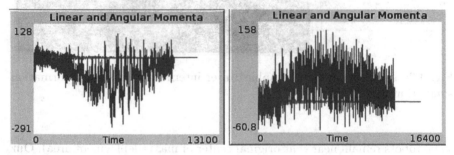

Fig. 4.11: The angular momentum graphs of the two chains shown in Fig. 4.10

Table 4.1: The relationship between the minor axis of the ellipse and the straightness of the chain

| Minor_Axis | Average Maximum Distance |
| --- | --- |
| 1 | 561 |
| 2 | 498 |
| 3 | 427 |
| 4 | 379 |
| 5 | 350 |

the results, which agree with our understanding of the system. There is a very clear correlation between the Minor_Axis and the average curvature of the emerging chain. As an example, Fig. 4.12 shows the two chains that are generated using a minor axis of one (left) and five (right). For this figure the random seed was set at one upon initialization, using random-seed 1 to

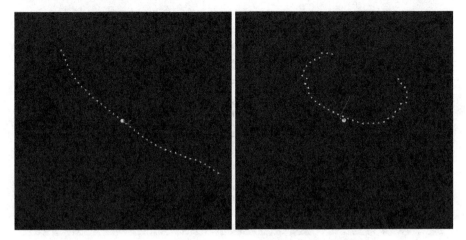

Fig. 4.12: Two chains generated with a minor axis of one (left) and five (right), using the same random seed at initialization

ensure that both runs started with precisely the same initial placement of particles. The right chain curves much more, due to the higher value of the minor axis. Yet even the right chain remains close to the original center of mass.

How precisely is the code implemented for chain formations? The change to the force law is surprisingly simple, as shown in Listing 4.2. As mentioned before, for this application the desired separation $D$ is the same as the sensor range. Hence, the code first checks to see if a neighbor is within sensor range. If it is, the force is computed using the split Newtonian force law. Then the bearing to the neighbor is computed. The bearing of the neighbor, combined with the heading of the particle, allows the variable dradius to be computed. This variable depends on the values a2 and b2, which are controlled by the Major_Axis and Minor_Axis sliders.[2] The dradius variable determines whether the neighboring particle is within the ellipse. If it is, the particle is repelled. Otherwise it is attracted.

There are several other features included with this simulation. You can add obstacles with a left mouse click and change the distance that they are detectable by moving the Obstacle_Force slider. Figure 4.13 shows the results of the self-organizing of the chain after several large obstacles were placed in the environment. Note that the chain formation acted as it should, by finding a route through the obstacles. Also, some robots remain near the original center of mass shown in yellow, as we desire. Finally, note that the chain displays the emergent behavior of bending minimally while within the obstacle course. Within the obstacle course the chain looks very much like a stiff wire that is minimizing its potential energy by bending as little as

---

[2] The full mathematical derivation is mildly complex and is shown in detail in Chap. 12.

Listing 4.2: The modified force law for the chain formation simulation

```
if (r < D) [                              ; Within sensor range
   set F (G * mass * ([mass] of turtle ?) / (r ^ p))
   if (F > FMAX) [set F FMAX]
      set theta (atan deltay deltax)   ; Bearing to neighbor
      set c1 (sin (heading + theta))
      set s1 (cos (heading + theta))
      set dradius (sqrt (1.0 / ((c1 * c1 / a2) +
                                (s1 * s1 / b2))))
   ifelse (r > dradius)                   ; Outside the ellipse?
     [set Fx (Fx + F * (deltax / r)) ; Attractive force
      set Fy (Fy + F * (deltay / r))]
     [set Fx (Fx - F * (deltax / r)) ; Repulsive force
      set Fy (Fy - F * (deltay / r))]
]
```

possible. This is an example of the principle of least effort, as discussed in Chap. 3.

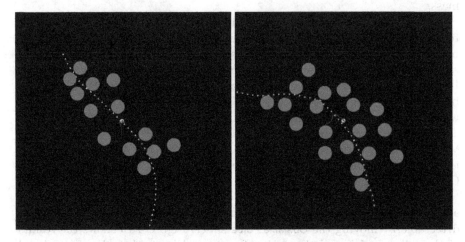

Fig. 4.13: The chain formation bends through obstacle courses

Finally, you can add and delete particles at any time. Adding particles works well and the algorithm is quite scalable. Figure 4.14 shows an example with 100 particles. Unfortunately, deleting particles does not work well after the chain has finished forming. The chain does not exhibit robustness in the face of agent failure, because it is unable to close the gaps that are formed. Similarly you have probably noticed that the final formation always leaves

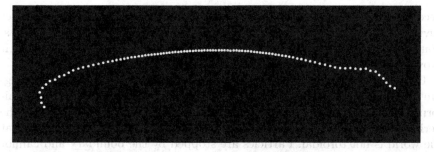

Fig. 4.14: One hundred particles self-organize into a chain formation

the particles just out of sensor range of each other. Both problems can be solved by assuming that the sensor range used in the simulation is actually shorter than the *real* sensor range. Chapter 12 provides details on how to deal with these issues, while generalizing the application of chain formations to maze environments. It also provides details on a hardware implementation and shows the results with five robots.

## 4.3 Uniform Coverage

Roughly 10 years ago Douglas Gage (of the Space and Naval Warfare Systems Command) proposed the following problem to me. How can a robot provide uniform coverage of an arbitrary environment? In addition, the robot behavior should not be predictable and should not neglect the periphery of the environment. There are numerous applications of uniform coverage, including surveillance, demining, cleaning and search.

Gage proposed an algorithm that is similar in spirit to physicomimetics. When a robot encounters the perimeter of the wall, it makes a random turn based on the concept of "diffuse reflection." The reader is probably more familiar with "specular reflection," where the angle of reflection of a ray of light equals the angle of incidence. With diffuse reflection a ray of light is reflected at many angles rather than one. Stated broadly, this means that all angles of reflection are equally likely, which can be used as a physics-based model for the behavior of the robot. Unfortunately this approach has some limits. First, it is claimed that "the searcher *must* be able to determine its position within the search area, and the distance to the opposite boundary in all directions." Second, it is stated that the algorithm does not generalize to nonconvex environments [73].

This section shows an alternative approach that combines biomimetics with physicomimetics. There are numerous advantages to this approach. First, the behavior is not predictable and does not neglect the periphery.

Second, the robot only needs to sense what is directly in front of it. No other environmental information is required. Third, it scales to multiple robots. Fourth, it works in arbitrary environments (including concave). Finally, it provides coverage that is highly uniform, even with a single robot.

Let's demonstrate the biomimetics portion of the approach first. Figure 4.15 shows the graphical user interface of the NetLogo simulation "uniform.nlogo". The first thing to notice is the yellow wall boundary around the perimeter of the particle environment. This boundary is used to ensure that the world is not toroidal. Particles are stopped by the boundary and cannot pass through it. The second obvious feature is that the environment has been broken into nine square "cells" of equal size. Our goal is to make sure that all cells receive roughly the same number of visits from a particle (or particles). The cells are colored according to their type—the red cell is the interior cell, the blue cells are the corner cells that have two walls, and the green cells are exterior cells that have one wall. The reason for this distinction will become clear shortly. Each cell has a counter associated with it so that it can record the number of times a particle has visited that cell.

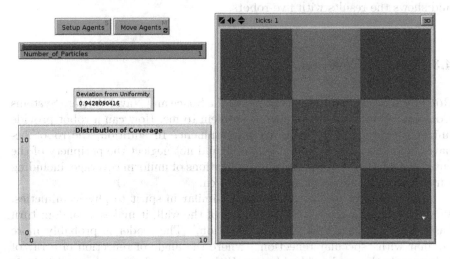

Fig. 4.15: The graphical user interface for the biomimetic portion of the uniform coverage simulation, after initialization

As you can see, this is a very simple simulation. There is one slider to control the initial number of particles. There is one monitor that indicates the deviation from uniformity of the coverage (this will be explained below). Finally, there is one histogram that shows the coverage of each cell.

Listing 4.3 gives the code that controls the behavior of the particle. A particle looks to see if the wall is ahead of it (straight ahead, right and ahead, or left and ahead). If the wall is in the way it makes a random turn from 0°

Listing 4.3: The code that controls the behavior of a particle

```
to go-particles
  let tries 1
  while [(tries < 10) and
         (([pcolor] of patch-ahead 1 = yellow) or
          ([pcolor] of patch-left-and-ahead 45 2 = yellow) or
          ([pcolor] of patch-right-and-ahead 45 2 = yellow) or
          (any? other turtles in-cone 3 30))]
    [set heading random 360 set tries tries + 1]
  if (tries < 10) [fd 1]
  monitor
end
```

to 359°. If the wall is still in the way it tries again. Similarly, a particle will turn if other particles are in front of it. To prevent infinite loops the particle makes at most 10 turns. If it succeeds in turning so that the wall or other particles are not in the way, it moves forward one step. Otherwise it does not move. Finally, the call to the monitor procedure increments the counter of the cell that the particle is currently visiting.

It is important to note that this algorithm differs in an important way from all previous algorithms in this book. The previous algorithms were deterministic. Given the initial conditions, we could determine the future position of every particle precisely. Even the flocking simulation was deterministic, although it may not have looked that way. In contrast, the uniform coverage algorithm is "stochastic," meaning that the future position of every particle is computed from both deterministic and random components. So, we will not be examining specific behaviors of particles, but their behavior in the aggregate (if there is more than one particle) and over the long term when the system is in "steady state" (or "equilibrium").

How does this differ from the "diffuse reflection" described by Gage? In order for a robot to perform diffuse reflection properly it needs to know the orientation of the wall with respect to itself. Only then can it generate a random turn from 0° to 180°, because that angle must be in terms of the orientation of the wall. This is not a trivial issue, since it involves the use of multiple or more sophisticated sensors. The code shown in Listing 4.3 is actually much easier to implement. As mentioned in Sect. 3.4, obstacles can be detected with acoustic or infrared (IR) sensors. These sensors do not indicate the orientation of an obstacle, only that the obstacle is in view. In fact, in this simulation the sensors do not even have to distinguish between a wall or a particle, since both are obstacles. The behavior of the particle is actually similar to that given in the NetLogo simulation for termites [267]. If termites need to move but are blocked, they make random turns from 0° to 359° until the way is clear. Hence we view this as a biomimetic approach.

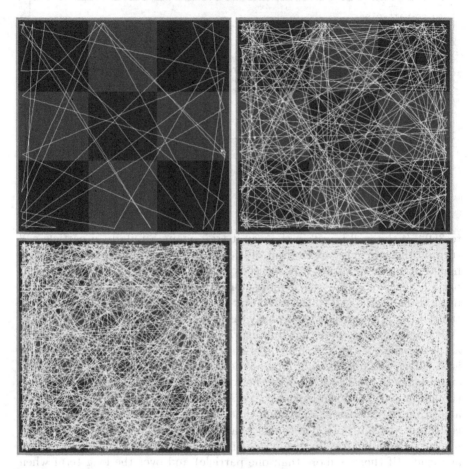

Fig. 4.16: The coverage of the environment by one particle, at 2,000, 12,500, 37,000 and 89,000 time steps

Let's run the simulation. Set the Number_of_Particles to one, and click Setup Agents and then Move Agents. You will see the particle move around the environment, bouncing off the walls. If you click the view updates button in the simulation (to the right of the speed slider at the top) you can turn off the graphics, which speeds up the simulation considerably. The first thing to examine is whether the one particle explores the whole environment, including the periphery. Although this is not the default behavior of the simulation, we used the "pen down" pd command so that the particle would draw a white path. We ran the simulation for a while and then clicked the view updates button again to turn on the graphics. Figure 4.16 shows the result, at different time steps. This shows that the particle does not neglect the periphery, which is one of our requirements. Since the particle will turn when the wall is one unit away you will see a very small amount of the cells between the white region and the boundary.

It is clear that the particle is hitting every portion of the environment, but is it a uniform coverage? The histogram in Fig. 4.17 shows the results. The height of each bar of the histogram represents the frequency of visitation for a cell in the environment. The left bar represents the bottom left cell. The second bar represents the bottom center cell. Finally, the last right bar represents the top right cell. You will note that the bars are given the same color as their corresponding cells. This helps us to make two observations. First, the coverage is not very uniform (ideally each bar would have height $1/9 \approx 0.1111$). Second, there are three levels of visitation frequency. The corner cells, shown with blue bars, are visited most often. Then the four green cells are visited less often. The central red cell is visited the least.

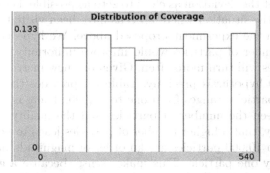

Fig. 4.17: The uniformity of the coverage by one particle

Why is this happening? Note that cells with more walls are visited more often. A reasonable hypothesis for this behavior is simply that the walls prevent the particle from moving from one cell to another. Hence the particle spends more time in cells with more walls. Put another way, suppose the particle is crossing the red cell. Then it will simply continue along a straight path to another cell. There is no opportunity for that particle to stop, turn, and spend more time in the red cell.

But that is with only one particle. Suppose we increase the number of particles. Then particles will see each other and make random turns. Will this help achieve better uniformity? We can test this with the NetLogo simulation. Before we do so, however, let's quantify what we mean by uniformity, and create a metric to precisely measure uniformity. The key is to note that in the case of perfect uniformity, each cell is visited $1/9^{\text{th}}$ of the time. This is what we would like to achieve. In reality, however, the fractions will not be $1/9$. Then it is necessary to measure how far off the fractions are from the ideal. This can be done with a Euclidean distance metric, as we now show.

Consider two probability distributions $P$ and $Q$.[3] The deviation from uniformity metric is

$$\sqrt{\sum_{i=1}^{C}(P_i - Q_i)^2}, \tag{4.1}$$

where $C$ is the number of cells. Let $Q$ be the distribution $(1/C, 1/C, \ldots, 1/C)$, which is the uniform distribution. The highest value for the metric is obtained at the beginning of the simulation, when only one cell has been visited. In this case, $P = (0, \ldots, 0, 1, 0, \ldots, 0)$. Then the distance is simply $\sqrt{(C-1)/C} = \sqrt{8/9} \approx 0.9428$, because $C = 9$. Note that after you click Setup Agents you see the value 0.9428090416 in the Deviation From Uniformity monitor (as shown in Fig. 4.15), which is correct. The Deviation From Uniformity value then decreases as the simulation is run. Because the simulation is stochastic, you need to run it for a long time to obtain a good estimation of the steady-state uniformity. The goal is to get the deviation as close to zero as possible, because a value of zero represents the situation where all cells have been visited equally often.

Now let's run the experiment proposed above. We hypothesized that increasing the number of particles would improve uniformity, due to the fact that the particles will turn more often. Given our new metric for uniformity, we can test this hypothesis precisely. Table 4.2 provides the results, where the number of particles ranges from one to 10,000. There is a very clear relationship between the number of particles and the quality of the uniform coverage, namely, that a higher number of particles leads to better coverage. The results with 10,000 particles are an order of magnitude better than the results with only one particle. This makes sense, because a gas would certainly fill the area uniformly, due to collisions between the enormous number of molecules.

Table 4.2: The relationship between the deviation from uniformity of the coverage and the number of particles in the simulation

| Number_of_Particles | Deviation from Uniformity |
|---|---|
| 1 | 0.027 |
| 10 | 0.026 |
| 100 | 0.016 |
| 1000 | 0.004 |
| 10000 | 0.003 |

---

[3] A probability distribution is a set of numbers from 0.0 to 1.0 that sum to 1.0. For example, $(1/2, 1/2)$ is the probability distribution of the outcome of a fair coin flip. See Sect. 15.2.2 for more details.

It is quite clear that the additional random turns, created by the presence of a greater number of particles, provide superior results. Unfortunately, it is unlikely that real-world systems will be composed of 1,000 or 10,000 robots. Instead they will probably only have a few. Is there a way to improve the performance? The answer is yes, and the key to success relies on the idea that although we won't have 10,000 robots, we can pretend there are as many *virtual* robots in the environment as we want. These virtual robots cause the actual robots to turn, providing better uniform coverage.

But how can we detect a virtual robot? In actuality, we don't have to. Instead we rely on the concept of "mean free path" from physics. Between particle collisions, the particle will travel a distance referred to as the mean free path. A robot can remember how far it has moved without turning. Once the mean free path has been reached, it can make a random turn. All that remains is to find a good value for the mean free path. But this involves a tradeoff. Extremely small mean free paths (in the nanometer range) will yield excellent uniform coverage, because this is like a gas at normal atmospheric pressure. However, this also means that the robot will take a long time to travel from one part of the environment to another. This would be a disadvantage for robots that are performing surveillance in a region. Hence a compromise is needed.

What we really want is the largest mean free path that provides high quality uniform coverage. This is analogous to a vacuum, where the mean free path is at the centimeter level. It is reasonable to define "high quality" as the deviation from uniformity provided by our 10,000 particle experiment above, which yielded a value of 0.003. If we can meet or surpass this level of quality with a small number of particles, we will consider the algorithm to be successful.

It can be shown theoretically that a good value for the mean free path is approximately 0.65 times the cell size. This is derived in detail in Chap. 11. Our environment is $84 \times 84$ in size. But it has a wall of width one, and the robots try to stay one unit away from the wall. This yields an environment of size $80 \times 80$. Hence, each of the nine cells has size $26\frac{2}{3} \times 26\frac{2}{3}$. Then the theory indicates that the mean free path should not be greater than $17\frac{1}{3}$.

We can test the theory via the NetLogo simulation "uniform_mfpl.nlogo". The graphical user interface is shown in Fig. 4.18. There is one additional slider, which controls the mean free path. The additional monitor shows the theoretically derived upper bound on the best mean free path.

The modifications to the code to implement the mean free path are shown in Listing 4.4. As can be seen, they are extremely simple. The particle monitors the number of steps that it has taken. This is identical to the length of the path that has been traveled, because the particle always moves forward one unit when it moves. If a turn is made, the number of steps is reset to zero. However, if the number of steps is equal to the mean free path a random turn is made, despite the lack of obstacles in the way. This mimics the presence of a virtual particle.

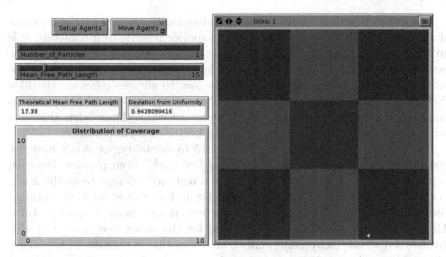

Fig. 4.18: The graphical user interface for the complete uniform coverage simulation, after initialization

Listing 4.4: The code that controls the behavior of a particle

```
to go-particles
   let tries 1
   set steps steps + 1
   while [(tries < 10) and
         (([pcolor] of patch-ahead 1 = yellow) or
          ([pcolor] of patch-left-and-ahead 45 2 = yellow) or
          ([pcolor] of patch-right-and-ahead 45 2 = yellow) or
          (any? other turtles in-cone 3 30) or
          (steps = Mean_Free_path_Length))]
      [set heading random 360 set steps 0 set tries tries + 1]
   if (tries < 10) [fd 1]
   monitor
end
```

To see how the mean free path affects the quality of the uniform coverage, two experiments were made. For both experiments, the mean free path was allowed to vary from one to 120.[4] In the first experiment, one particle was used, and the simulation was run for 200,000,000 time steps for each value of the mean free path. In the second experiment, 100 particles were used

---

[4] There is no reason to go beyond 120, since the long possible path in the environment is $80\sqrt{2} \approx 113$.

and the simulation was run for 2,000,000 time steps. Note that because the number of particles increased by a factor of 100, the number of time steps was decreased by a factor of 100 to provide a fair comparison between the two experiments.

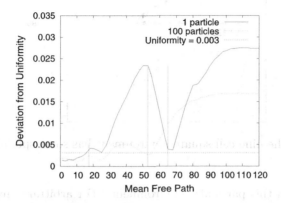

Fig. 4.19: The deviation from uniformity as the mean free path changes

The results are shown in Fig. 4.19. The mean free path is shown on the horizontal axis, while the deviation from uniformity is shown on the vertical axis. As one can see, very high mean free paths do not perform adequately. But small mean free paths work very well, providing a deviation lower than our goal of 0.003. Recall that our theory, which assumed one particle, indicated that an upper bound on the best mean free path is $17\frac{1}{3}$. This is in excellent agreement with Fig. 4.19, where a short vertical line is drawn in black. Good uniformity is achieved when the mean free path is approximately 15 or less.

In general, as expected, 100 particles usually perform better than one. However, there is an unusual area with mean free paths between 53 and 65 where the uniformity of one particle dramatically improves (the region between the two long vertical black lines). Why is this happening? The explanation is shown in Fig. 4.20. Suppose a particle has rebounded from the left wall and is traveling to the right. If we want the particle to spend more time in the center cell, the best place to introduce a random turn is at the right edge of the center cell. This does not guarantee that the particle will turn back to the center cell, but there is a 50% probability that it will do so. The shortest distance from the left wall to the right edge of the center cell is $53\frac{1}{3}$. Similarly, what if a particle is traveling down from the upper left corner? Again, we would like the particle to make a random turn at the right edge of the center cell. This distance is $66\frac{2}{3}$. This corresponds extremely well with Fig. 4.19. This indicates that, due to the structure of this particular environment, there is a range of good mean free paths that is higher than we would expect. Unfortunately this result is not general, because it stems from

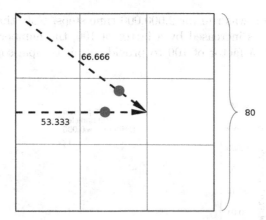

Fig. 4.20: The nine cell square environment has some special qualities

the structure of this particular environment.[5] For arbitrary environments it is still safest to use a mean free path that is approximately 0.65 times the cell size.

In summary, this new algorithm is a combination of biomimetics and physicomimetics. The physicomimetic portion is the concept of the mean free path. This concept is used in kinetic theory, radiography, acoustics, optics particle physics and nuclear physics [264]. Note that this is our first physicomimetics algorithm that is not based on $F = ma$ and illustrates our central tenet that *the imitation of physics principles and concepts is a powerful tool for problem solving.* Although we focus on swarms throughout this book, note that the use of physics reasoning was useful even for one particle.

Chapter 11 provides a more detailed explanation of the material in this section, including the derivation of the mean free path, and simulation results on a suite of structured environments. In addition, the chapter provides details of a hardware implementation on a robot, including analyses of the results.

## 4.4 Couette Flow

As shown in the last section, the concept of mean free path greatly improved the performance of a simple biomimetics algorithm for uniform coverage. Mean free path is a core concept in a branch of physics referred to as "kinetic theory." In this section we focus on a more complete application of kinetic theory [75]. This is an excellent alternative physicomimetics technique for

---

[5] One can also see an improvement with 100 particles with mean free paths between 40 and 55, for similar reasons.

swarm robotics that is not $F = ma$. As with the last section, this new technique is stochastic, not deterministic.

In kinetic theory, particles are treated as possessing no potential energy. The system consists entirely of kinetic energy. In fact, kinetic theory does not typically deal with forces at all. Instead, increases in particle velocity are modeled as being caused by collisions and/or temperature increase. This is very different from the $F = ma$ physics systems that we have examined.

This section focuses on a specific application of kinetic theory referred to as two-dimensional "Couette flow" [75]. Fluid moves through a corridor environment that has two walls. The fluid is composed of small particles, e.g., molecules, that locally interact with each other via collisions, providing "viscosity." Viscosity is a fluid's resistance to flow. Examples of fluids with high viscosity are honey and molasses. There are two sources of energy in the system. The walls have a temperature $T$, which imparts stochastic thermal energy to the system. The thermal energy is a form of kinetic energy that drives the particles. In addition, the walls can move, providing a particular direction for the particles to flow. The two walls can be configured in different ways—one can move while the other is stationary, or both can move in the same or opposite direction.

The task that we are interested in is related to uniform coverage. The goal is to "sweep" a large group of robots through a long bounded region. This is especially useful for demining, searching for survivors after a disaster, and for robot sentries. The type of sweep is determined by the wall velocities. Chapter 7 examines the situation where both walls move in the same direction.

This section will explore a variation of Couette flow where the two walls move in opposite directions. This creates an interesting sweep where one-half of the corridor is swept in one direction while the other half is swept in the other direction. The graphical user interface for "kt.nlogo" is shown in Fig. 4.21, after running for 20 time steps. The corridor is vertical, with a red wall on the right and a yellow wall on the left. The velocity of the walls is controlled with the Wall_Velocity slider. The red wall moves upward with velocity Wall_Velocity while the yellow wall moves downward with velocity Wall_Velocity.

In addition to controlling the wall velocity you can control the temperature $T$ of the walls. Higher temperatures mean that particles bounce with greater velocity from the walls. Two monitors provide information on the $x$- and $y$-components of linear momentum. Since the particles are not moving initially, the linear momentum should remain close to zero for three reasons: (1) the thermal noise from the wall is equal and opposite, (2) the walls move in equal and opposite directions, and (3) the collisions between particles are carefully designed to obey the conservation of linear momentum.

There is also a histogram that shows the distribution of the velocities of the particles in various columns of the corridor. The corridor is broken into 13 vertical columns, and the histogram shows 13 bars that indicate the

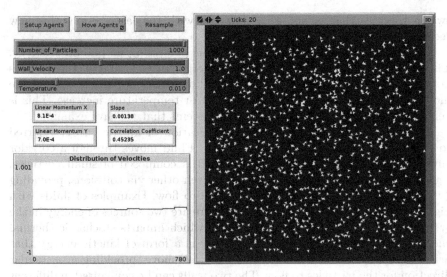

Fig. 4.21: The graphical user interface for the kinetic theory simulation, after
20 time steps

average velocity of the particles in each column. The leftmost bar gives the
average velocity of the leftmost column, and the rightmost bar gives the
average velocity of the rightmost column. Because there are an odd number
of columns, the central bar reflects the average velocity of the center column.
As expected, the histogram in Fig. 4.21 shows that there is almost no motion
anywhere in the corridor when the simulation is first started.

Now run the simulation by clicking Setup Agents and Move Agents. What
you will see is that the leftmost and rightmost histogram bars are the first
to move. This is because the particles at the left and right edges of the
corridor are subject to thermal and kinetic energy from the walls. As these
particles move, they collide with other particles. This spreads the kinetic
energy towards the center of the corridor. Figure 4.22 shows the simulation
after 15,032 time steps. If you watch the simulation you will see that the
particles on the left side of the corridor are generally moving down, while the
particles on the right side are generally moving up. In fact, the histogram
shows this very clearly (Fig. 4.22).

A very complete implementation of kinetic theory is presented in Chap. 7.
The implementation presented here does not incorporate all aspects, but it
contains the fundamental elements. The core code is shown in Listing 4.5.
Each particle sees if there are any particles nearby. If so, it randomly chooses
one with which to interact. This interaction is in the form of a virtual collision,
even if the two particles are not necessarily on a collision course. The concept
of virtual collisions is not what we would expect from standard Newtonian

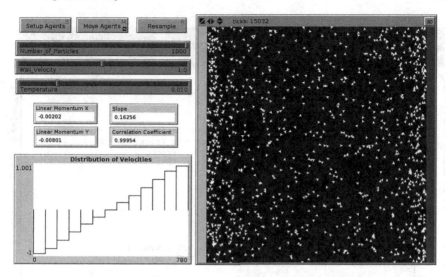

Fig. 4.22: The graphical user interface for the kinetic theory simulation, after 15,032 time steps

physics, but is central to the kinetic theory framework. Then the particle computes the relative speed of the interacting particle, as well as the center of mass velocity. Finally, a post-collision angle theta is uniformly randomly chosen and new velocity vectors are computed for the two particles. This calculation preserves linear momentum and the relative speed.

At this point the particle sees if either wall is adjacent. If so it receives thermal (kinetic) energy from the wall. This results in a new velocity for the particle. The vertical component vy is simply the wall velocity with some additional random Gaussian ("normal") noise. The amount of noise increases with higher temperature. The horizontal component vx is more complex, being drawn from a Rayleigh distribution. We will examine this in more detail below, but for now it suffices to state that higher temperatures yield larger values of vx on average.

This simulation is sufficiently accurate that it is able to illustrate two important concepts in kinetic theory. The first concept is illustrated by the histogram in Fig. 4.22. Note that the histogram appears to have a linear slope. It can be shown theoretically that if the two walls move at different speeds then the velocity distribution across the histogram *must* be linear [75]. This allows us to test the accuracy of our simulation. To do so we calculate the theoretical slope of the linear relationship. Recall that slope is defined to be "rise over run." The rise is two times the wall velocity. The run is equal to 12, since the histogram bars are numbered from zero to 12. Hence the slope should be 2 × Wall_Velocity / 12. With the default value of one for

Listing 4.5: The code that controls the behavior of a particle in kinetic theory

```
to go-particles
  let friends other turtles in-radius 2
  if (any? friends) [
    let friend [who] of one-of friends
    let rel_speed sqrt (((vx - ([vx] of turtle friend)) ^ 2 +
                         ((vy - ([vy] of turtle friend)) ^ 2)))
    let cm_vel_x 0.5 * (vx + ([vx] of turtle friend))
    let cm_vel_y 0.5 * (vy + ([vy] of turtle friend))
    let theta (random 360)
    let costh (cos theta)
    let sinth (sin theta)
    let vrel_x (rel_speed * sinth)
    let vrel_y (rel_speed * costh)
    set vx (cm_vel_x + 0.5 * vrel_x)
    set vy (cm_vel_y + 0.5 * vrel_y)
    ask turtle friend [set vx (cm_vel_x - 0.5 * vrel_x)
                       set vy (cm_vel_y - 0.5 * vrel_y)]
  ]
  if (any? patches in-radius 1 with [pcolor = red])
    [set vx (- (sqrt (2 * Temperature)) *
               (sqrt (- ln (random-float 1.0))))
     set vy (((random-normal 0.0 1.0) * (sqrt Temperature)) +
             Wall_Velocity)]
  if (any? patches in-radius 1 with [pcolor = yellow])
    [set vx ((sqrt (2 * Temperature)) *
             (sqrt (- ln (random-float 1.0))))
     set vy (((random-normal 0.0 1.0) * (sqrt Temperature)) -
             Wall_Velocity)]
  if ((vx != 0) or (vy != 0)) [set heading atan vx vy]
end
```

the Wall_Velocity, the default slope should be $1/6 \approx 0.1666$. The Slope monitor is always updating the slope of the histogram, and as you continue running the simulation it should approach a value very close to the theoretically derived value. As an additional check the simulation computes the Correlation Coefficient, which measures the quality of the linear relationship. A value of $+1$ represents a perfect positive correlation, while a value of $-1$ represents a perfect negative correlation (the slope will be negative). A value of zero means there is no correlation at all in the data. Values above 0.95 indicate excellent positive correlation, and you will see the simulation reach values in the range of 0.999.

It is important to note that the code computes the histogram, slope, and correlation coefficient using all of the data accumulated since the simulation started running. Hence it can be biased by the initial behavior of the system. Since we are most interested in the steady-state (long-term) behavior of the system, it is often useful to erase the data at some point and let the computations start again. This is the purpose of the Resample button. Once you click this button the old statistical data is erased and you should wait a while for the histogram, slope and correlation coefficient to stabilize as the simulation accumulates the latest data.

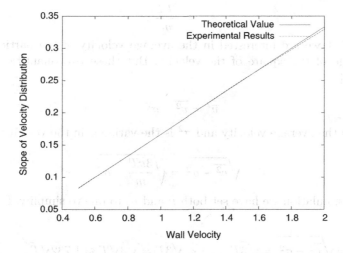

Fig. 4.23: The agreement of the slope of the velocity distribution with theory is almost perfect

Let's reexamine how well the simulation agrees with theory, by letting the wall velocity equal 0.5, 1.0, 1.5 and 2.0. For each value of wall velocity the simulation was run for 10,000 time steps. The Resample button was automatically activated at 5,000 time steps to delete the early data. Then the slope and correlation coefficient were reported at the end of the simulation. Figure 4.23 shows the results. The correlation coefficient was always greater than 0.999. The simulation agrees almost perfectly with the theoretically derived slope, showing only a minor deviation when the wall velocity is 2.0. This indicates that the simulation is in excellent agreement with theory.

A second concept from kinetic theory states that when the velocity of the walls is zero, the average velocity of the particles is linearly related to the square root of the temperature [75]. The formal proof of this concept is beyond the scope of this chapter, but we will summarize the important equations and test the accuracy of the simulation against the theory.

One of the fundamental laws of physics is that the temperature of a gas is related to the average kinetic energy of the molecules as follows [85]:

$$\frac{3}{2}kT = \frac{1}{2}m\overline{v^2} \ . \tag{4.2}$$

The constant $m$ is the mass of a single molecule in the gas, $T$ is the temperature and $k$ is the "Boltzmann constant" (which is not to be confused with the spring constant that was used earlier in this book). The notation $\overline{v^2}$ refers to the average of the square of the velocities of the particles in the gas. We can reorder the terms to say

$$\overline{v^2} = \frac{3kT}{m} . \tag{4.3}$$

Note that we are interested in the average velocity of the particles, not the average of the square of the velocity. But these two quantities are related [243]:

$$\overline{v}^2 = \overline{v^2} - \sigma^2 ,$$

where $\overline{v}$ is the average velocity and $\sigma^2$ is the variance in the velocity. Hence,

$$\overline{v} = \sqrt{\overline{v^2} - \sigma^2} = \sqrt{\frac{3kT}{m} - \sigma^2} .$$

In our simulation we have set both $k$ and $m$ to one to simplify. Hence:

$$\overline{v} = \sqrt{\overline{v^2} - \sigma^2} = \sqrt{3T - \sigma^2} < \sqrt{3T} = \sqrt{3}\sqrt{T} \approx 1.732\sqrt{T} . \tag{4.4}$$

Recall that we want to show that the average velocity of the particles is linearly related to the square root of the temperature. We have almost accomplished this, showing that an upper bound on the average velocity is a constant times the square root of the temperature, which is a linear relationship. A more detailed derivation takes the variance into account, yielding [75]:

$$\overline{v} = \frac{2\sqrt{2}}{\sqrt{\pi}} \sqrt{\frac{kT}{m}} \approx 1.5958\sqrt{T} .$$

Now, all of the derivations above have been in three dimensions. The final step is to recalculate the derivation for two dimensions, which yields [231]

$$\overline{v} = \frac{1}{4} \sqrt{\frac{8\pi kT}{m}} \approx 1.2533\sqrt{T} . \tag{4.5}$$

There is another way to see the relationship. If we examine the code again we note that the thermal noise is determined by the statement set vx sqrt (2 * Temperature) * sqrt (- ln (random-float 1.0)). This is the inverse transform technique for generating the Rayleigh distribution given by

$$P(v) = \frac{v}{s^2} \mathrm{e}^{-v^2/2s^2} ,$$

where $s = \sqrt{kT/m} = \sqrt{T}$. The mean of this distribution gives us the average velocity of the particles that bounce off the walls, and is $\bar{v} = s\sqrt{\pi/2} \approx 1.2533s = 1.2533\sqrt{T}$, which agrees with Eq. 4.5.

This equation can be tested against the simulation by adding a monitor for the average velocity of the particles (this is very simple in NetLogo and the interested reader is encouraged to do this). We let the square root of the temperature range from 0.01 to 0.20, in increments of 0.01. For each value of the square root of the temperature the simulation was run for 100,000 time steps, with an automatic **Resample** performed at 50,000 steps. Then the average velocity was reported at the end of each experiment.

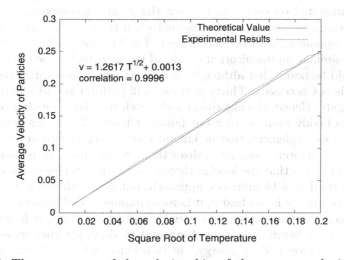

Fig. 4.24: The agreement of the relationship of the average velocity of the particles with the square root of the temperature is almost perfect

Figure 4.24 shows the results. The solid line is the theoretical relationship shown in Eq. 4.5. The dashed curve is drawn using the results from the NetLogo experiments just described. The agreement is excellent. A linear least squares fit to the experimental data yields

$$\bar{v} = 1.2617\sqrt{T} + 0.0013 \qquad (4.6)$$

with correlation 0.9996. These are excellent results. Note that the empirically derived slope of 1.2617 is extremely close to the theoretically derived slope of 1.2533. Also, the "$y$-intercept" is 0.0013, which is very close to the theoretical value of 0.0 (i.e., if the temperature is zero, there can be no motion). What this all means is that a reasonably simple kinetic theory simulation can accurately reflect theoretically derived behavior. Hence, when it is necessary to achieve the desired behavior, the user can set the Wall_Velocity and Temperature parameters properly, without guessing.

Of course, ultimately we want to implement kinetic theory on actual robots. There are two issues that arise. First, robots must be able to compute the orientation of Couette walls with respect to themselves, in order to calculate their new velocity vector. This can be done quite easily if the robot has a ring of IR or acoustic range sensors. The second issue arises due to the computation of the center of mass velocity. This requires knowing the velocity of yourself and the interacting (colliding) neighbor. The simplest way to accomplish this uses the same solution provided in Sect. 4.1 of this chapter. A common frame of reference can be established with an inexpensive digital compass mounted on each robot. Then the $x$- and $y$-components of velocity are transmitted along the RF channel when the trilateration localization module is operating, as described in Sect. 3.4. This merges localization with the very minimal communication overhead required by kinetic theory.

It should be noted that although we ran our simulations with 1,000 particles, this is not necessary. Thirty particles will perform almost identically, in the aggregate. Hence, the algorithm scales well to a large number of robots and is also highly immune to robot failure. Chapter 7 provides details on a more rigorous implementation of kinetic theory, taking into account robot capabilities. Chapter 7 also generalizes the sweeping problem by adding obstacles, and shows that the kinetic theory approach provides sweep coverage almost as good as a biomimetics approach, but in one-third the time. Also, since kinetic theory is stochastic, it is more immune to adversarial deception (i.e., someone trying to avoid the robotic sentries). Finally, it is important to note that the biomimetic algorithm works by dropping pheromones in the environment (a form of "stigmergy," as mentioned in Chap. 2), which can be difficult to accomplish in practice. Kinetic theory does not need to change the environment to perform the task.

## 4.5 Summary

All of the techniques shown in this chapter have several desirable qualities. First, they rely on information from local neighbors. As a result the techniques are very robust in the face of robot failure while also scaling well to large numbers of robots. When more information is needed, it is extremely minimal, requiring the use of a digital compass and the communication of a few bytes of information. Almost no information about the environment is needed. For uniform coverage the robot only needs to be able to detect obstacles (other robots or walls), but does not need to be able to distinguish between the different types of obstacle. For the sweeping task, the robot merely needs to know the global direction of a Couette wall. It then assumes

that *any* wall that is oriented in that direction is a Couette wall, even if it is
an obstacle [231].

This chapter has shown how physicomimetics can be extended in several
different ways. First, if we are using an $F = ma$ simulation, the standard
assumptions made by Newton's third law can be broken, if done carefully.
This allows physicomimetics to be applied in more situations, and to more
tasks. Second, we have shown how physicomimetics can be synergistically
combined with biomimetics to create a superior paradigm. Finally, we have
shown that physicomimetics is more general than $F = ma$. Kinetic theory is
yet another plausible paradigm for swarms.

# Part II
# Robotic Swarm Applications

# Part II
## Robotic Swarm Applications

# Chapter 5
# Local Oriented Potential Fields: Self Deployment and Coordination of an Assembling Swarm of Robots

Andrea Bravi, Paolo Corradi, Florian Schlachter, and Arianna Menciassi

## 5.1 Introduction

The creation of multi-robot organisms, defined as *"large-scale swarms of robots that can physically dock with each other and symbiotically share energy and computational resources within a single artificial life form,"* is one of the main recent challenges in robotics [139]. The interest in this field arises from the idea of merging the advantages given by distributed and centralized systems. The former are characterized by multiple modules capable of simpler actions. This design makes distributed systems robust (i.e., if one robot is destroyed, the functionality of the swarm is not affected), scalable (i.e., capable of adapting to different numbers of robots) and capable of parallelizing tasks. On the contrary, centralized systems are characterized by a single complex entity. This makes them capable of solving complicated tasks; however, they lack the advantages of a distributed approach. Multi-robot organisms share both of the advantages. When in the assembled phase (i.e., when the robots are docked with each other), they can solve complex tasks just as centralized systems. On the other hand, when unassembled, the organisms are nothing other than a collection of robots. From this idea the need for proper deployment techniques arose, where these techniques are capable of facilitating the assembly phase of the robots.

This chapter focuses on the pre-assembly deployment and coordination phase, which consists of preparatory swarm movements with the objective of reaching a multi-robot configuration most beneficial for subsequent self-assembly. The problems of deploying and coordinating a group of robots are not new in the literature [11]. For instance, triangular and square networks of robots were created by *Embedded Graph Grammars* [223]. In this technique

Andrea Bravi

Dynamical Analysis Laboratory, University of Ottawa, Ottawa, Ontario, Canada, e-mail: a.bravi@uottawa.ca

W.M. Spears, D.F. Spears (eds.), *Physicomimetics*,
DOI 10.1007/978-3-642-22804-9_5,
© Springer-Verlag Berlin Heidelberg 2011

each robot has a label, which corresponds to a specific control law; the labels change dynamically depending on the relative positions of the robots. They change according to certain rules of interaction, and this eventually leads to a specific deployment strategy. In other words, adaptive curves or regular formations can be obtained from predefined sets of geometric rules [49, 245, 290] or through the use of virtual hormones [218].

Despite several prior approaches in the literature, research on how to deploy a group of robots with the aim of enabling their assembly into three-dimensional structures is relatively new. The reason is that, to date, few robotic platforms are designed to behave as a swarm of modular robots capable of docking. Two exceptions are Replicator [139] and Swarm-bots [171]. The ability to dock introduces the possibility of creating structures that can be used to solve multiple tasks, like overcoming obstacles too challenging for a single robot, sharing energy (e.g., battery power) among the robots to prolong the life of the swarm, sharing computational power (e.g., communication between CPUs) with the aim of increasing the intelligence of the swarm, and many others [139, 171].

In addition to the problem of deployment, there is the problem of swarm coordination. Indeed, coordination is often required to achieve deployment as well as other features that do not belong to the individual robot, but to the robotic system as a whole. A popular solution to the coordination problem involves the combination of a set of basic actions to achieve more complex ones. This approach finds inspiration from nature. Synchronized firefly flashing and pattern formation of slime molds are just two examples of a variety of intriguing phenomena naturally occurring through the distributed application of simple rules [25]. There are also some examples in the robotics domain: (1) the foraging of a swarm composed of two types of robots, namely, foot-bots and eye-bots, respectively, which operate and guide the foraging [55], (2) the construction of architectures through the use of virtual pheromones, inspired by the study of ants' behavior (in the field of computer science called "swarm intelligence") [159], and (3) the flocking of a swarm of robots [289]. One researcher who stressed this idea and created a model capable of reproducing several different coordination properties is Matarić. In her 1995 work [154], Matarić extended a combination of basic actions to a combination of basic coordination behaviors, thereby creating complex coordination behaviors like flocking and herding from several basic coordination behaviors (e.g., aggregation and dispersion).

The present chapter addresses the problem of deploying a multi-robot system with the final aim of facilitating the assembly phase, and analyzes the problem of deployment coordination to solve task-oriented issues. To achieve this goal, we based our solution on the physicomimetics paradigm and the basis behaviors provided by Matarić in a framework that we will refer to as *behavioral coordination* [154]. Our solution has three underlying motivations behind it: (1) the need to create a system capable of self-organization and automatic adaptation to unexpected events, which are characteristics of phy-

sicomimetics, (2) the need to take advantage of the idea that many simple
coordination behaviors can lead to complex behaviors, which is a feature of
the behavioral coordination framework, and (3) the goal of creating a system characterized by simplicity. The advantage of creating a simple solution
is that it is easier to analyze, yet even simplicity can lead to complex and
interesting properties.

To achieve complex deployment and coordination properties, we extended
the above-mentioned paradigms and created a new technique that makes use
of distributed *potential fields* with a common, shared coordinate system; it
is called the *local oriented potential fields (LOPF)* approach. Like physicomimetics, the potential fields approach assumes the presence of virtual forces to
which agents react. Unlike physicomimetics, with potential fields the forces
are based on global knowledge. This global knowledge is useful in preparing
the agents for assembly, e.g., for achieving angular alignment between agents.
Although physicomimetics could be used with local reference frames and coordinate transformations (for example, see Chap. 10), the use of a global
coordinate system with potential fields simplifies the alignment process.

With the use of simple LOPFs, four deployment configurations (square,
line, star, tree) and the angular alignment behavior were created. The proposed ideas were then demonstrated with NetLogo simulations ("lopf.nlogo").
Properties of the proposed approach will be discussed toward the end of this
chapter.

In summary, this chapter presents an approach that combines physicomimetics, behavioral coordination, and local oriented potential fields, with the
objective of preparing a swarm of robots for assembly into a three-dimensional
multi-robot organism that can enable docking and energy sharing. The swarm
configurations and methods for achieving them will be described next.

## 5.2 Methods

In this section we present the basis of our approach. Assuming knowledge of
the physicomimetics background already presented in earlier chapters of this
book, we start with a short overview of the behavioral coordination paradigm.
Then we introduce local oriented potential fields (LOPF).

### 5.2.1 Behavioral Coordination

Flocking, schooling, and herding represent examples of behaviors that arise
from global coordination of a distributed system. Mataric et al. showed that
coordination within a multi-agent system can be achieved by combining mul-

tiple basis coordination behaviors [154]. In this article, five typologies of basis behaviors were defined:

- *Safe wandering:* the ability to move about while avoiding collisions with obstacles and other agents.
- *Following:* the ability to move behind another agent, retracing its path and maintaining a line or queue.
- *Dispersion:* the ability to spread out in order to establish and maintain some minimum inter-agent distance.
- *Aggregation:* the ability to gather together in order to establish and maintain some desired inter-agent distance.
- *Homing:* the ability to move toward a particular region or location of the arena.

By activating different combinations of these behaviors, different complex behaviors arise. For instance, flocking can be achieved by combining the effect of homing, dispersion, aggregation and safe wandering, while a foraging behavior is achievable as a combination of homing, dispersion, safe wandering and following. The power of this approach is that it is extendable (i.e., other coordination behaviors can be easily added to the model), and with a binary representation (e.g., 1 = activated, 0 = deactivated) it allows the system user to easily explore the effects of online learning. In fact, a group of robots could learn to activate or deactivate one or more of the basis behaviors depending on the environment in which they are embedded and the task they need to solve.

## 5.2.2 Local Oriented Potential Fields

The local oriented potential fields (LOPF) method introduced in this chapter is complementary to the use of physicomimetics. The basic idea is to use the LOPFs to regulate the angle between the robots (which is needed for assembly), and the physicomimetics forces to regulate their distance. The LOPF approach is based on the use of a common coordinate system. Essentially, each robot possesses a potential field whose characteristics are described within a reference system centered on the robot itself. The potential field affects the neighbors of the robot through the creation of a virtual force.

This force will depend on the absolute value and the direction of the distance vector between the robot and its neighbor. Considering a multi-robot system, each robot in the deployment will be affected by the potential fields of its neighbors. The resultant force on that robot will be the sum of all the single forces given by the potential fields, and it will be added to the standard interaction forces defined by physicomimetics (Fig. 5.1).

The reference system is represented by the vector **j**. Its direction represents the global knowledge needed by every robot when using LOPFs. Because our

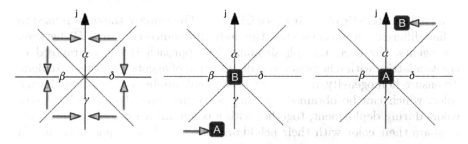

Fig. 5.1: In this example, the reference systems are divided into eight quadrants, each one characterized by a linear potential field. On the left, the direction and the intensity of four representative forces are reported. Those forces are constant within each quadrant because they arise from linear potential fields. The diagram in the middle represents the force that robot B exerts on robot A. On the right the opposite force is depicted

final objective is to apply the presented framework to real robotic platforms, it is important to outline the hardware requirements. To implement LOPFs in a robotic system, the following are needed: (1) an omnidirectional system of proximity sensors to compute the distances and angles between the robots, and (2) either a wireless position system, a compass sensor or reference landmarks in the environment to implement the global coordinate system.

In the next section, some applications of these methodologies will be shown. The LOPFs we present are based only on linear potential fields; this decision simplified the experiments without preventing the emergence of complex behaviors.

## 5.3 Applications

We now analyze how the methods just described can be applied in the context of preparing a swarm of robots for effective self-assembly. The results shown in this section were done in simulation.

### 5.3.1 Local Oriented Potential Fields for Deployment

The following are the different deployments (configurations) that have been achieved by using the local oriented potential fields paradigm:

**Square deployment**: In the physicomimetics framework a method exists for creating a square deployment by using the concept of virtual *color*, which is a binary characteristic of a robot (see Chap. 3). The value of the color is used to define different interactions (i.e., the desired distance is $D$ or $\sqrt{D}$) between the agents. However, this physicomimetics approach is not guaranteed to create a lattice with a homogeneous distribution of agents with the two colors. To ensure homogeneity, it seems we need equal numbers of agents with each color, which can be obtained by using a routine that dynamically assigns colors during deployment, together with a communication system for agents to share their color with their neighbors. See the video "square.avi" for an example.

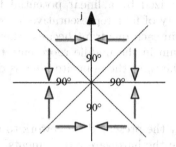

Fig. 5.2: LOPF needed to achieve a square deployment. The quadrant is divided into eight equal slices, affected by the same constant force with a different direction that depends on the position within the reference system

The use of a simple local oriented potential field represented in Fig. 5.2 leads to the same result, but without a communication system, a dynamic routine, or a specific number or agents.

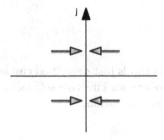

Fig. 5.3: LOPF needed to achieve a linear deployment. The quadrant is divided into two equal slices

**Linear deployment**: A linear deployment is achievable with the same LOPF as the square lattice, while avoiding two sets of parallel forces (Fig. 5.3). See "linear.avi" for an example.

**Star deployment**: The star deployment represents the family of all possible deployments with one agent in the middle (namely, the *leader*) and a set of agents in lines surrounding the leader (namely, the arms of the star). All of the agents know who the leader is through visual identification (via any identifier on the robot) or global communication (for instance, while broadcasting the call to the other robots to start the deployment formation). The star deployment is composed by combining two rules. The first rule creates the star; the second one extends the arms of the star to achieve the linear deployment. Because two different rules are applied, there must be a differentiation between robots. This differentiation is given by the leader, who is given the property that it will not be affected by any LOPF. See "star.avi" for an example.

An additional condition needs to be satisfied to aid this formation during the simulation. Because the physicomimetic interactions lead to a hexagonal deployment [235] (see "hexagonal.avi"), robots that are joining the deployment need only be affected by the LOPFs of the robots already deployed. This prevents the LOPF and physicomimetic forces from competing for agents' angles in the formation. In this way, the LOPFs become the only ones affecting the angles between the robots, and the physicomimetic forces are the only ones affecting the distances between the robots. It is worthwhile noting that this last constraint is not mandatory (i.e., the same effect could be achieved by increasing the absolute value of the LOPF forces); however, we noticed in our simulations that the solution we chose increased the stability of the deployment.

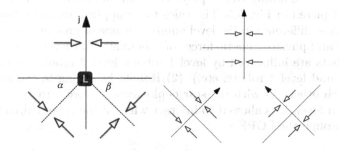

Fig. 5.4: The two LOPFs needed to achieve the star pattern. On the left, we see the LOPF exerted by the leader on its neighbors (basic star), and on the right we see the LOPFs exerted by all the other robots in the pattern (linear)

In Fig. 5.4, the LOPF for achieving a star with three arms is depicted. Any number of arms is achievable, as long as the physical dimensions of the agents are properly taken into account. Note that the dynamic change of the orientation of the **j** axes of the linear LOPFs (Fig. 5.4, right) gives actuative properties to the pattern; in other words, it is like creating a tweezer.

**Tree deployment**: The tree deployment corresponds to the dispositions of robots in a triangle-like deployment. The tip of the pattern is the leader (and again, the leader is not affected by any other robots). As before, all robots know who the leader is through local (e.g., with visible markings) or global communication (for instance by broadcasting the formation pattern to the other agents). To allow a proper formation, another bit of complexity is added that assigns each robot to a different *level*, depending on its distance and angular position respect to the leader. See "tree.avi" for an example.

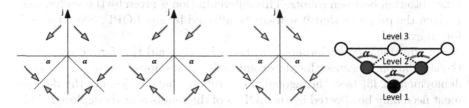

Fig. 5.5: From left to right: the LOPF on the leader and the robots in the middle of the deployment (the tip of the deployment); the LOPF of the robots facing the west edge of the formation (considering **j** to point north); the LOPF of the robots facing the east edge of the formation; an example of a deployment of six robots, with a relative subdivision into levels. The angles $\alpha$ regulate the width of the triangle

The LOPF promoting tree deployment, and an example of level subdivision, are depicted in Fig. 5.5. The rules to be applied are the following. (1) Robots whose difference in the level numbers is one or greater, are subject to the LOPF and physicomimetic forces of robots at lower levels (and therefore level 2 robots are influenced by level 1 robots, level 3 robots are influenced by level 2 and level 1 robots, etc.). (2) Robots belonging to the same level repulse each other. As with the star deployment, before a robot is assigned to a certain level, it is allowed to interact with all the robots already in the pattern through the LOPF.

## 5.3.2 Achieving Behavioral Coordination

Many activities can be involved in the coordination of a swarm. Example activities include surrounding an object to lift it, distributing homogeneously in a field to patrol it, and many others. To achieve the goals of these activities, it is possible to combine physicomimetics, LOPF and behavioral coordination. In doing so, we decided to add a basis behavior to the ones proposed by Matarić, namely, a behavior that we call *angularity*. We define angularity as angular alignment, i.e., the capability of two robots to remain at a fixed angle to each other. This property is well known in nature, where a clear example is Barnacle Geese (*Branta Leucopsis*), the migratory birds that fly in "V" formations. Without angularity, this behavior would not be achievable by the behavioral coordination paradigm.

The relationships between the different methodologies combined in our approach can be outlined as follows:

- *Safe wandering:* If the attractive aggregation force of the physicomimetics paradigm is turned off, the repulsive force between agents allows this behavior to occur. Each agent will avoid other agents or other elements in the environment within a certain radius $D$ (i.e., the desired radius of physicomimetics).
- *Following:* As shown by Matarić, this behavior is achievable by maintaining the position of an agent within a certain angle in relation to the followed agent. This can be achieved by making agents interact with the physicomimetics paradigm. The distance between the two agents will be the radius $D$.
- *Dispersion:* This can be achieved by extending the safe wandering behavior to the maximum repulsion radius.
- *Aggregation:* This behavior is achieved by using the attractive force of physicomimetics, without switching its sign when the agents get closer than the distance $D$.
- *Homing:* The direction towards home can be modeled as a potential field that leads the robots to the destination.
- *Angularity:* Angularity is the natural behavior emerging from the use of local oriented potential fields.

## 5.4 Experimental Results

These ideas were tested using NetLogo, a multi-agent simulation system described in previous chapters of this book. The simulated agents possessed an omnidirectional proximity sensor system, an omnidirectional movement system, and a coordinate system. Thus, the agents were able to turn in any direction, to sense the presence of other agents in any direction within a certain radius (obviously larger than the radius $D$ of the physicomimetics paradigm), and to know their position within a reference system. Gaussian noise was added both to the value of the proximity sensors and to the coordinate system coordinates in order to obtain more realistic simulations. In all of the simulations the robots started from a random position in the arena and were drawn to the center of the formation through a homing behavior.

Some snapshots from simulations of the square deployment are shown in Fig. 5.6, of the linear deployment in Fig. 5.7, the star deployment in Fig. 5.8, and the tree deployment in Fig. 5.9. A challenge we faced was that of dividing

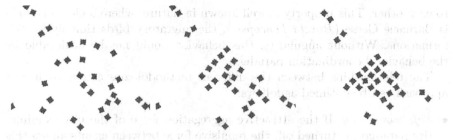

Fig. 5.6: Frames during the square deployment. During the simulation, a homing force was used to make the agents converge to the middle of the arena

Fig. 5.7: Frames during the linear deployment. During the simulation, a homing force was used to make the agents converge to the middle of the arena. Note that during the deployment the pattern adapted to the increasing number of agents trying to join the formation, which shows the scalability of this approach (and this was true for all the deployments and the patterns simulated). A limited number of robots was used to allow the deployment to fit the size of the arena

the robots into levels for the star and tree deployments. Because we did not want to simulate the use of local communication (since it would add hardware and software complexity to a real robotic system), we assigned the levels relying just on the distances and the angles between the robots and the leader. For both the star and tree deployments, we assigned a level every time a robot was placed within the proper angle (as designed in the deployment) and at a given multiple of the distance $D$ between two robots imposed by the physicomimetics paradigm. An example of this with the tree deployment is shown in Fig. 5.10. For the star deployment the same approach was used, but the assignment was limited to only two levels (where the robots closest to the leader are red, and the others inside the deployment are orange). The downside of this approach is that a considerable effort was needed for finding good thresholds (e.g., because the positions of the robots inside the

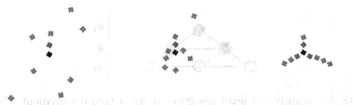

Fig. 5.8: A three arm star was simulated. The leader is the black agent. The homing behavior makes the agents point to the barycenter of the formation, calculated with respect to the position of the leader. While the agents join the pattern, their color changes according to their level. In red, there are the neighbors of the leader (which are affected only by its LOPF), and in orange are the other agents properly deployed inside the pattern. During the simulation, we allowed the possibility of changing the orientation of the linear LOPFs of the orange agents, which would have the effect of creating moving arms (as shown below). The use of the red level is not mandatory, i.e., it was only introduced to better show the movements of the arms (see below)

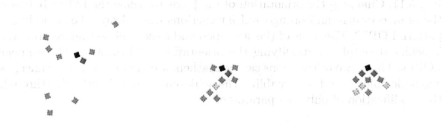

Fig. 5.9: The leader is the black agent. The homing behavior makes the agents point to the barycenter of the formation, calculated with respect to the position of the leader. When the agents join the pattern, their color changes according to their level (where red agents are the neighbors of the leader, and orange and yellow depict the other agents, depending on their position with respect to the leader). Again, the red robots interact only with the LOPF of the leader, the orange robots only with the LOPFs of the red robots, and the yellow robots only with the LOPFs of the orange robots

deployment always change due to their interactions with their neighbors and the effects of noise).

In addition to the different deployments, we also simulated the dynamic properties achievable by the star and tree deployments—by modifying the orientation of their LOPFs. In the case of the star deployment (Fig. 5.11), two of the three arms get close together, thereby simulating a gripping hand. In the case of the tree deployment (Fig. 5.12), the shape changes, making the formation more compact. Those examples represent clear ways to use angularity

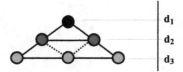

Fig. 5.10: An example of level assignment for a tree deployment with seven robots and three levels. The black circle represents the leader of the formation (level 0), and any robot that is positioned at the proper angle and within the range $[d_1,d_2]$ is assigned to level 1; the ones within the range $[d_2,d_3]$ are assigned to level 2. Clearly, this approach can be extended to any number of levels

Fig. 5.11: Changing the orientation of the $j$ axis to define the LOPF between the orange–orange and orange–red interactions (i.e., always the same linear pattern LOPF). The arms of the star open and close. This actuative property is achieved simply by modifying the orientations of two of the three linear LOPFs. Those two orientations can be mathematically linked, for instance, as a generic angle $\gamma$ and $-\gamma$, enabling the achievement of this behavior through the modification of only one parameter

Fig. 5.12: The shape of the triangle changes, modifying the angle $\alpha$ defining the LOPFs between the agents (Fig. 5.5)

behavior. In addition to these deployments, we also simulated the aggregation behavior (Fig. 5.13), the dispersion behavior (Fig. 5.14) and the homing behavior (Fig. 5.15), all translated into the physicomimetics paradigm.

In our simulations we faced a common problem encountered when performing self-organizing deployments, namely, the presence of holes in the formation [235]. A hole arises when a robot is not correctly deployed—either because of a bad initial position, or because the other robots self-organized in such a way as to prevent the robot from filling that space (and this second condition is itself dependent on the initial positions of the robots). Com-

Fig. 5.13: An example of the aggregation behavior. The radius $D$ of physico-mimetics was changed during the simulation, and the agents adapted their distance to the new value

Fig. 5.14: Starting close together, the agents disperse because of the repulsive force of the physicomimetics paradigm

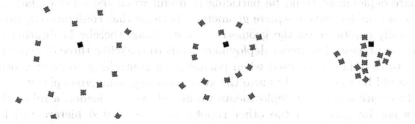

Fig. 5.15: An example of the homing behavior. The agent in black is the leader, and the home where the robots are pointing is the middle of the arena

mon solutions to this problem are the addition of noise into the movements of the robots, the inclusion of additional rules governing the movements of the robots while approaching the deployment, and the choice of a preferred starting position for the robots. The first one requires a delicate tuning of the level of noise (or the configuration might not be stable), while the second one adds complexity to the software. The third solution represents a deployment problem itself, and can be implemented by choosing a more easily achievable formation as the starting position. We noticed that the hexagonal deployment arising from physicomimetics interactions is, in general, a good starting point because of its isotropy and compactness. The square deployment is another possibility, even if less compact than the hexagonal one (because each robot is surrounded by four robots, instead of six).

## 5.5 Discussion

The presented framework combines the use of physicomimetics and basis coordination behaviors, together with the proposed local oriented potential fields. In the previous section, experimental results were provided in the form of figures showing the results of different deployment configurations. The interactions between the three approaches led to the achievement of several properties for the robot swarm. Now we discuss these properties.

### 5.5.1 Deployment Properties

Each of the presented deployments has its own advantageous features. The square deployment could be particularly useful when the robotic platforms to be assembled have a square geometry—because that configuration would naturally capitalize on the geometry of the robots, thereby facilitating the docking process. The linear deployment leads to snakelike three-dimensional structures, and can be used when patrolling or controlling an area; another use could be to coherently move the swarm through dangerous places.

To create additional deployments we introduced the leader, a robot that does not interact with the other robots, and we created hierarchical levels within the deployment. A small increase in the complexity of the model yielded several advantages. A star deployment enables the protection of a special element of interest placed in the center—the leader, and because the leader is not subject to any forces, it could either be a robot or an object placed in the arena. Furthermore, in the two-dimensional assembled phase (i.e., when all the robots are touching the ground) the star could be used to create a hand-organism able to move objects, while in the three-dimensional assembled phase it could be the starting configuration for spiderlike forms. The tree deployment allows the swarm to move in a compact way, and to create solid structures that could be used in the two-dimensional assembled phase (as a battering ram) and in the three-dimensional assembled phase (as the torso of an organism). Note that in both of the last examples the presence of the leader allows the deployment to be easily displaced—because moving the leader results in the movement of the whole formation (e.g., an external force affecting the leader would cause the whole formation to move, while still maintaining the formation).

## 5.5.2 Sensorial and Actuative Properties

The deployment of robots into a certain pattern, and thus into a geometric disposition in space, leads to different features that can be classified as *sensorial* and *actuative*. The sensorial properties describe how the pattern influences the information coming from the sensors. In particular, a fixed geometry enables the robots to gain knowledge of the properties underlying that particular deployment. An example of a sensorial property is given by the *angle of attention*. Given the number of robots involved in a deployment and their disposition, it is possible to understand the directions the robot has to oversee to know what is happening around the formation. For instance, consider a robotic platform with a square geometry and four cameras on each side. If a group of robots of this type adopts a linear deployment, then half of the information can be discarded—because each robot has half of its view covered by its adjacent neighbors (with the exception of robots at the tips). A reduction in the information to be processed means lower exercise of the robots' CPUs, thus allowing the possibility of implementing more complex data processing algorithms or gaining a longer battery life.

The actuative properties determine the type of scenario and the type of actuation best suited for the deployment. The actuative properties could be involved in either the assembled or the unassembled phase. As we showed during the simulations, an example is the modification of parameters defining the LOPFs, which leads to distributed complex behaviors such as the handlike grasping of the star pattern.

## 5.5.3 Behavioral Coordination, Physicomimetics, and LOPFs

It is well known that complex systems are nonlinear, and thus the whole system can synergistically show more properties and behaviors than the ones achieved by the sum of its individual components. This phenomenon is well known in nature, where many different collective behaviors can be achieved by the interaction of simple rules [25]. Following this reasoning, we used the approach proposed by Matarić, defining through physicomimetics and LOPFs a set of basis behaviors that can be used together or singularly, thereby widening the possibilities for swarm coordination. For example, a swarm can surround an object via an aggregation behavior towards the object along with inter-agent dispersion. As another example, the homogeneous covering of a field can be achieved with an initial phase of swarm aggregation, followed by a dispersion phase where the radius of interaction is brought to its maximum (so that the robots deploy at the maximal distance, which is usually given by the range of detection of the proximity sensors).

During our analysis, the angularity behavior was added. This behavior, implemented through the use of LOPFs, enables the robots to deploy in geometric structures. An interesting feature is that by rotating the coordinate axis **j**, there can be an ordinate rotation of the deployment where the relative positions between the robots are preserved. This feature could be important when the objective is to make the swarm move an object [140].

We have developed a system showing self-organization and simplicity. The value of these two features taken together is given by the fact that they easily allow the exploration of more complex behaviors. The self-organization takes care of the possibility of unexpected situations that were not considered during the design of the behavior, and the simplicity allows the application of this approach in different contexts. For instance, the parameters defining the physicomimetics forces, the LOPFs and the coordination behaviors could be easily implemented in a genome that would allow exploring the possibilities of online and offline evolution and learning, as done in the DAEDALUS approach to physicomimetics in the context of navigating through an obstacle field while moving towards a goal (see Chap. 14). Furthermore, the concepts of heterogeneity and modularity could be extended to LOPFs and coordination behaviors, as has been done in physicomimetics [263], thus creating a unique framework characterized by new emergent properties.

## 5.6 Conclusions

A new framework consisting of self-organizing deployment and coordination was proposed. The use of physicomimetics, together with the introduction of local oriented potential fields, allowed a robot swarm with docking capability to aggregate and deploy in square, linear, star and tree deployments. The known geometry gives the robots sensorial and actuative properties that are important for the management of the stream of information from the sensors and the applicability of the robots to defined scenarios. The translation of the behavioral coordination paradigm into the domain of physicomimetics, together with the addition of the angularity behavior, increased the exploration of new behaviors, and predisposed the proposed solution to evolution and learning.

In conclusion, the combination of physicomimetics, behavioral coordination, and the proposed local oriented potential fields led to the creation of a multipurpose approach characterized by simplicity, self-organization and scalability, and that is capable of facilitating the assembly phase of a swarm of robots.

**Acknowledgements** This research was supported by *PerAda*, 7th Framework Programme of the European Commission (Researcher Exchange programme). The authors would like to thank the University of Stuttgart for the support provided during the development of the work. The presented work was developed as part of the project REPLICATOR (http://www.replicators.eu/), funded by *Cognitive Systems and Interaction and Robotics*.

# Chapter 6
# Physicomimetics for Distributed Control of Mobile Aquatic Sensor Networks in Bioluminescent Environments

Christer Karlsson, Charles Lee Frey, Dimitri V. Zarzhitsky, Diana F. Spears, and Edith A. Widder

## 6.1 Introduction

In this chapter, we address a significant real-world application, namely, exploring and monitoring the *bioluminescence* of our oceans. Bioluminescence occurs when biological organisms produce and then emit light. Dr. Edith Widder (a coauthor of this book chapter) was recently interviewed by *The Economist*, where she discussed the importance of detecting and monitoring bioluminescence—for both pollution detection and submarine stealth [97]. Our approach to addressing this problem is to design and eventually deploy a swarm of physicomimetic-controlled aquatic vehicles that behave as a distributed sensing network for observing and determining the bioluminescent density within a specified aquatic region.

With this objective in mind, we first define the coastal bioluminescence problem, along with relevant background material on the subject of bioluminescence. Then, two tasks are defined within this application. These are the tasks that form the focus of the remainder of the chapter. Next, a

Christer Karlsson
Department of Mathematical and Computer Sciences, Colorado School of Mines, Golden, Colorado, USA, e-mail: ckarlsso@mines.edu

Charles Lee Frey
Harbor Branch Oceanographic Institute, Ft. Pierce, Florida, USA, e-mail: cfrey@hboi.fau.edu

Dimitri V. Zarzhitsky
Jet Propulsion Laboratory, California Institute of Technology, Pasadena, California, USA, e-mail: Dimitri.Zarzhitsky@jpl.nasa.gov

Diana F. Spears
Swarmotics LLC, Laramie, Wyoming, USA, e-mail: dspears@swarmotics.com

Edith A. Widder
Ocean Research and Conservation Association, Ft. Pierce, Florida, USA, e-mail: ewidder@oceanrecon.org

W.M. Spears, D.F. Spears (eds.), *Physicomimetics*,
DOI 10.1007/978-3-642-22804-9_6,
© Springer-Verlag Berlin Heidelberg 2011

NetLogo bioluminescence simulation is presented, which includes simulated aquatic surface vehicles, called "drones," that need to achieve two objectives in parallel—tracking bioluminescence while navigating to a desired goal location, e.g., a port. Using this simulation, experiments are run that test relevant hypotheses about the swarm behavior on the two tasks. The results are presented and conclusions drawn, with lessons on applying swarms for this problem. Because NetLogo is a limited language in terms of its expressiveness, we also present a more sophisticated bioluminescence simulator, called the "multi-drone simulator" (MDS), that we have designed and constructed. The MDS does a much better job of simulating reality, and it serves as a transition to the actual hardware drones. The chapter ends with a description of our drones and the indoor and outdoor aquatic environments in which they have been tested.

## 6.2 Background on Bioluminescence

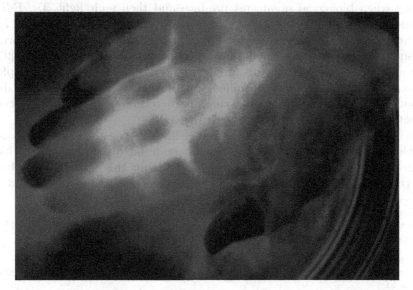

Fig. 6.1: Bioluminescent light from phytoplankton (single-celled algae)

Bioluminescence is a phenomenon whereby living organisms generate and emit light from their bodies. Bioluminescence is often confused with the phenomenon of *phosphorescence*. In phosphorescent materials, light energy is absorbed into the material and re-emitted after a period of time, in a process known as "photoluminescence." This is the predominant mechanism at work in "glow-in-the-dark" paints, for example. In contrast, bioluminescence is a "chemiluminescent" process whereby two or more chemicals are mixed together and a reaction occurs which emits light (see Fig. 6.1). This process is similar to that found in "glow-sticks." In order to start the chemiluminescent reaction, the stick is "snapped," which exposes the two chemical reactants to one another. In bioluminescence, the chemicals required to generate light are manufactured and regulated by a living organism. The chemical, called *luciferin*, reacts with oxygen, catalyzed by an enzyme called *luciferase*. This reaction emits a kind of "cold light," where the majority of the energy given off in the reaction is emitted in the form of light rather than heat. Although bioluminesce is found in a wide variety of colors, the two most common ones are blue and green. The understanding and use of bioluminescence has a wide variety of applications in defense, medicine, agriculture, environmental monitoring, and eco-friendly lighting technologies [97].

On land, bioluminescence is rare. The best known example is that of the common firefly. In the marine environment, however, especially the deep sea, bioluminescence is the rule. If one were to drag a net from 1000m deep to the surface in virtually any part the world's oceans, the vast majority of the creatures caught in the net would have some capability to generate their own light through bioluminescent means.

Fig. 6.2: Bioluminescent photophores on the underside of a lanternfish

Bioluminescent light can be generated inside or outside of cells. Some plants and animals produce their own luciferins and luciferases, while others harbor bioluminescent bacteria in their tissues in a symbiotic relationship, recruiting them to generate light on demand. In large organisms such as deep sea fishes, specialized light organs called *photophores* are used to emit and focus light, as well as to create complex illumination patterns (see Fig. 6.2).

Bioluminescence has many different uses for marine organisms. It can be used for defense, by either blinding a predator or by setting off a sort of "burglar alarm," which attracts larger predators to the scene. Some deep sea animals can even eject clouds of luminescent chemicals which act as "smoke-screens," allowing them to make a fast getaway. Bioluminescence can be used for camouflage, by counter-illuminating against a brightly-lit backdrop, thereby allowing a bioluminescent animal to obliterate its own shadow in the open ocean. It can also be used offensively, for attracting and illuminating prey, courting mates, or for communicating with others.

## 6.2.1 The Coastal Bioluminescence Application

In the coastal ocean, bioluminescence is also common, particularly in the form of microscopic phytoplankton known as *dinoflagellates*. These single-celled plants are ubiquitous in the ocean and form a large part of the foundations of many marine food webs. Some species of dinoflagellates are responsible for *harmful algal blooms (HABs)*, including "red tides." Most importantly for our interests, many dinoflagellate species are bioluminescent.

During daylight hours, many dinoflagellates "recharge" their luminescent chemicals via photosynthesis. At night, they enter a dark phase whereby mechanical stresses imparted to their cell membranes cause them to emit flashes of light. These stresses do not need to be fatal, as the luminescence occurs within the cell. In fact, many species of dinoflagellates can flash over and over again in a given evening. Flash dynamics vary among species, but as populations of dinoflagellates grow, the light generated after being disturbed by a swimming fish or passing boat can be extraordinary. In our own experience, surface bioluminescence while under sail in Florida at night can often be bright enough to enable reading a book.

Regions of coastal bioluminescence present unique challenges in military applications where stealth is desired. A warship, submarine, or special forces team moving through a bioluminescent "hot spot" can easily generate enough light to be seen from high points on a nearby shoreline or via aircraft or satellite. For example, aviators in World War II used to spot unlit ships by tracing their bioluminescent wakes. In fact, in a training exercise off the coast of Japan in 1954, Jim Lovell (later commander of Apollo 13) used this technique to find his aircraft carrier after his cockpit instruments had failed. It makes sense, therefore, that a large portion of the research in this area is supported by the US Navy through the Office of Naval Research (ONR) and the Naval Oceanographic Office (NAVO). Of primary concern to the Navy are (1) quantifying "bioluminescence potential," a measure of the amount of bioluminescent light that could be generated from a particular volume of water if disturbed, (2) forecasting bioluminescence potential, and (3) minimizing the bioluminescence signature during military operations.

Several sensors, called *bathyphotometers (BPs)*, have been developed over the years to quantify, understand, and help forecast marine bioluminescence [262]. For our purposes, we are interested in the application of this sensor technology along with a physics-based swarming framework to the development of a distributed marine surveillance network capable of detecting and predicting the bioluminescence potential of a given area. Our solution to this *biological plume tracing (BPT)* problem relies on a swarm of simple *autonomous surface vehicles (ASVs)*, i.e., drones equipped with minimal onboard sensor, communication, and processing power. By limiting the amount of resources committed to a single drone, we are able to achieve a cost-effective scale in the number of sensing platforms that can be deployed, and thus improve the spatial and temporal resolution of our system [72].

## 6.2.2 Tasks Within This Application

The creation of a swarm of simple, low-cost ASVs using physics-based swarming principles opens the door to a myriad of applications for marine bioluminescence sensing. In our case, we are interested in two primary tasks, one environmental and one defense-related.

### 6.2.2.1 Monitoring for Harmful Algal Blooms

As previously mentioned, many dinoflagellate species which are responsible for harmful algal blooms (HABs) are also bioluminescent. In Florida and the Caribbean, one bioluminescent species of particular concern is *Pyrodinium bahamense*, shown in Fig. 6.3.

*P. bahamense* produces a neurotoxin called *saxitoxin*, which is responsible for *paralytic shellfish poisoning (PSP)*. This toxin is dangerous to marine organisms, causing massive invertebrate and fish mortality. It is also harmful to humans, causing nausea, vomiting, tingling in the extremities, difficulty breathing, and other symptoms. In extreme cases, PSP can be fatal. In normal concentrations, dinoflagellates like *P. bahamense* are nontoxic. However, in large blooms they enter a toxic phase. Detecting the presence and concentration (density) of such organisms is critical to understanding and predicting HABs, and bioluminescence can be used as a metric to do so.

Therefore, our first task is HAB detection. This task functions much like a maximization problem; the goal is to navigate the drone swarm through a particular field of interest, while finding the bioluminescent maximum. If the HAB is caused by a point-source (e.g., nutrient-rich pollution outfall that causes the bloom), the problem becomes that of a plume-tracing task,

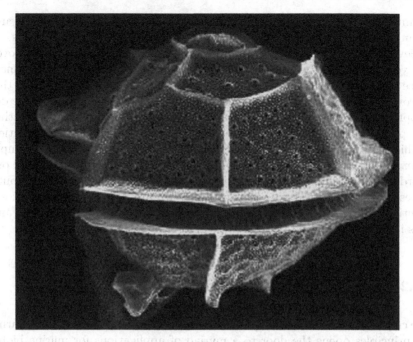

Fig. 6.3: The bioluminescent single-celled alga *Pyrodinium bahamense*, courtesy of MicrobeWiki, Kenyon College

which consists of searching the gradient of bioluminescence for a point of
global maximum flux. If the HAB is nonpoint-source in nature, the problem
becomes that of mapping the luminescent field and using the swarm to locate
the changing maximum (i.e., the developing HAB).

### 6.2.2.2 Navigating a US Navy Vessel on a Covert Coastal Operation

In our second task, which is a military defense scenario, the Navy wishes
to deploy a covert special operations team from open water to a particular
location on a hostile shoreline. The team is distributed among boats, divers,
submarines, and other vehicles, all of which will create large bioluminescent
wakes as they disturb the water. In order to deploy the team without being
detected via a light signature, i.e., to keep the team stealthy, it is critical to
determine the best path solution through the "bioluminescent minefield."

This task functions like a minimization problem. The swarm of autonomous surface vehicles is dispatched prior to the insertion operation, to
act as "scouts." It is assumed that the drones are small enough and move
slowly enough to create minimal bioluminescent signatures. Measurement
of bioluminescence potential is accomplished using closed-chamber bathy-

photometers, keeping the stimulated light obscured from view. As the ASV swarm passes over the field of interest, it searches for the path of minimal bioluminescence from the location of the (currently offshore) team to the goal site. The result is then relayed to the team members, who quickly follow the minimum path to shore.

### 6.2.2.3 Prior Research on the Two Tasks

There has been a great deal of prior research on the task of observing and monitoring HABs, including some research using multiple *autonomous underwater vehicles* (AUVs) and/or ASVs, e.g., [57, 137, 224, 225]. This prior work has made important advances in studying HABs. However, it has some limitations that we are working toward addressing. With physicomimetics implemented on ground vehicles, inter-vehicle coordination is done entirely onboard, and the swarm is online adaptive, reactive in real-time, and robust in the face of vehicle failures and unanticipated environmental conditions. These are features we plan to exploit as we transition to fleets of ASVs and AUVs in ocean environments for monitoring HABs.

Moline et al. are the only research group we have found working on the covert Navy operations task [163, 162]. Their approach consists of a single AUV to characterize ocean conditions as being more or less favorable to bioluminescent flashing, and this characterization is done long before any Navy operations are expected to be conducted. In contrast, our approach consists of using a swarm of multiple vehicles that collaborate to identify a path of minimal bioluminescence, and this path identification is done immediately preceding the deployment of a covert operations team.

This chapter also presents the first comparison between these two tasks, and draws experimental conclusions relevant to practitioners applying swarms of vehicles to the tasks.

## 6.3 NetLogo Bioluminescence Simulator

Now that we have defined our real-world application and the two tasks of interest within it, we would like to determine how swarms of drones perform on these tasks. This involves running experiments and varying parameters of both the environment and the aquatic drones. Unfortunately, it is both costly and time consuming to build and conduct huge numbers of experiments on full-sized drones in a real ocean environment. A computer simulation not only saves us money, but it also saves us time and allows us to perform many more experiments—especially since simulations can run both before and during the time the hardware is being built. Our hope is that both tasks (i.e., monitoring harmful algal blooms and navigating coastal bioluminescence) can be tackled successfully by a swarm of drones utilizing physicomimetics.

We proceed by testing this with a computer simulation. Following the precedents in previous chapters of this book, we select the NetLogo multi-agent programming language for our simulation. To accomplish the tasks, we will simulate a swarm of drones controlled with physicomimetics. We also need a way to simulate the bioluminescent dinoflagellates, so that we can create an environment for these drones to explore. Furthermore, we need to be able to program each drone with the task it is performing and the speed and direction in which it is to head at each time step. Additionally, we need a way to determine which simulation parameter settings lead to better or worse swarm performance on the tasks.

Our primary goal is to conduct initial testing to see if the two tasks can be solved using a swarm of drones. However a subsidiary goal of ours is to explore different parameter settings and discover theories that govern these settings. Although NetLogo has limitations as a programming language (e.g., because its vocabulary is high-level but restricted), it has proven to be sufficient for our experiments, described below. Before describing these experiments, we first describe the NetLogo simulator and how it has been implemented for our bioluminescence tasks.

## 6.3.1 Simulating Dinoflagellates Using NetLogo Patches

We need to create an environment that promotes the mobility and exploration capabilities of the drones. Our research focuses on dinoflagellates, in particular, for the drones to explore. Real-world marine organisms such as dinoflagellates occupy niches in the ocean. Their population distributions are based on a number of forcing factors, including ambient local ocean temperature, salinity, depth, tidal flushing, nutrient concentration and available sunlight [261]. They have some degree of self-propulsion, but are also largely at the whim of currents and tides. To detect where the dinoflagellates (called "dinos" for short in the simulator) are or have been, one can detect emissions of stimulated bioluminescence, which we model as "virtual chemicals." Higher concentrations of chemical indicate more dense dinoflagellate populations, and therefore higher bioluminescence potential. The vehicles that we intend to build will move on the surface of the water and will use sensors, such as bathyphotometers, to quantify bioluminescence output at the surface. Therefore, for simplicity, we simulate our drones as existing within a two-dimensional world. We are at least initially not interested in what forces act on and move the dinos below the surface. We further know that the dinos and their luminescent emissions will not provide any physical obstacle to the movement of the drones, i.e., any aggregation of plankton will be pushed to the side or below a drone that moves through it. This would of course

**Listing 6.1:** The code to create and randomly place 250 dinos

```
breed [dinos dino]    ; Introduce the ''dinoflagellates'' breed

patches-own [chemical]

to setup
    clear-all

    create-dinos 250 [set color yellow set shape ''circle''
                      set chemical 2 setxy (random world-height)
                                           (random world-width)]
    ask patches [set chemical 0]
end
```

cause a disturbance in the aggregation, but for simplicity we do not model this disturbance in our NetLogo simulator. But what we do need to model is the difference in the levels of concentration of the organisms. Hereafter, we focus exclusively on the chemical, which is considered to be directly proportional to the bioluminescence potential, and therefore to the concentration of dinoflagellates at any given location.

To model multiple different levels of chemical concentration, it is most appropriate in NetLogo to simply use the *patches* from which the world has been constructed. Every time we create a NetLogo world, it is built from a certain number of patches of a given size. Fortunately, NetLogo also provides us with tools to manipulate these patches. One of the simplest tools is the NetLogo command ask patches [set pcolor green], which asks all patches to change their color from the standard black to green. Recall that green is one of the two most common colors of bioluminescent light. Let us assume that we can take a measurement of every part of the world we are simulating, and that this measurement would tell us how much chemical material is at that specific location. If we record these measurements, then we would have a map of the simulated world. The good thing is that NetLogo has a tool that allows us to do this. Just like turtles-own [variable] defines the variables that all turtles can use, patches-own [variable] defines all the variables that all patches can use. However, we are not as interested in creating a map as we are in the ability to attach a value of the amount of chemical material to every patch. To accomplish this, we define patches-own [chemical], which ensures that each patch has a variable indicating its level of chemical density.

We now have a way to keep track of how much chemical (luminescence) there is at each patch. We next need a way to deposit the chemical within the

**Listing 6.2: The code to create a world with random chemical levels**

```
breed [dinos dino]    ; Introduce the ''dinoflagellates'' breed

patches-own [chemical]

to setup
   clear-all

   create-dinos 250 [set color yellow set shape ''circle''
                     set chemical 2 setxy (random world-height)
                                         (random world-width)]
   ask patches [set chemical 0]
end

to run-dinos
   repeat 100 [
      diffuse chemical 1
      ask patches [set chemical (chemical * 0.995)
                   set pcolor (scale-color green chemical 0 6)]
      ask dinos [dino-life]
      tick
   ]
end

; Moves dinoflagellate critters. Uses the original StarLogo
; slime mold model, which sufficed for dinoflagellates.
; http://education.mit.edu/starlogo/samples/slime.htm.
to dino-life
   turn-toward-max-chemical          ; Turn towards maximum
                                     ; chemical concentration
   rt random 40                      ; Make a random wiggle
   lt random 40
   fd 1                              ; Forward one step
   set chemical chemical + 2         ; Replenish chemical
end

; This portion of code is from the NetLogo code:
; http://ccl.northwestern.edu/netlogo/models/Slime
to turn-toward-max-chemical
   let ahead   [chemical] of patch-ahead 1
   let myright [chemical] of patch-right-and-ahead 45 1
   let myleft  [chemical] of patch-left-and-ahead 45 1
   ifelse ((myright >= ahead) and (myright >= myleft))
       [rt 45]
       [if (myleft >= ahead) [lt 45]]
end
```

Fig. 6.4: A screen shot of randomly placed yellow dinos

world in a somewhat realistic-looking manner, i.e., with some randomness. One way to do this would be to randomly create chemical "hot spots" and then push the chemical from patch to patch by means of fluid dynamics. This would be faithful to the world we are trying to simulate, but it would require a level of sophistication in the simulator that is quite challenging to implement in NetLogo. Let us instead adopt the idea of randomly created hot spots, but give these hot spots the ability to move and drop chemical in the patch they are currently occupying. The hot spots are in fact the dinoflagellates. This may not be the best representation of what goes on at the ocean's surface, but it will give us the opportunity to quickly create a random world for the drones to explore. To make it simple, let us create a breed and call it "dinos," i.e., `breed [dinos dino]`. Listing 6.1 shows our NetLogo code for setting up the world in the manner just described. Figure 6.4 shows the NetLogo world created by this code. Dinos are simulated by NetLogo turtles.

Recall that the randomly-located dinos need to deposit chemical material; the ASVs will then track this chemical. Because tracking point-particle dinos would be too difficult as a NetLogo task, we instead employ the

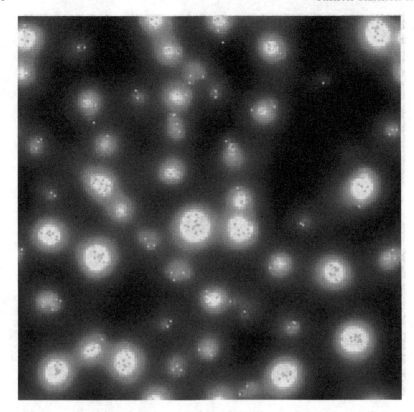

Fig. 6.5: A screen shot of a world with randomly located patches of diffused chemical

NetLogo "slime mold" model, in which the chemical is first diffused to surrounding patches and then the patches set their chemical level accordingly. This procedure begins with each dino turning toward the neighboring patch with the maximum chemical concentration. To avoid getting a perfect turn each time (i.e., to be more realistic), the dino wiggles a few degrees and then moves one step forward. It then drops some chemical in the patch, and the process is repeated. Listing 6.2 shows the NetLogo code for executing the procedure just described. The code `ask patches [set pcolor (scale-color green chemical 0 6)]` asks the patches to set their color shade proportionally to their level of concentration; lighter patches have higher levels of chemical density. The code `ask patches [set chemical (chemical * 0.995)]` simulates the fact that some of the chemical material sinks below the surface of the water. Figure 6.5 shows what the world looks like after chemical deposition, diffusion, and dispersion by the dinos.

There are now only two things left to create in our simulated world. We need to create and place the drones, and we need to place the goal. Both of these are generated as breeds, and they are separated as much as possible by initially placing them in opposite corners of the world. The goal stays stationary, and the drones move toward it. The drones are modeled as NetLogo turtles.

## 6.3.2 Simulating the Autonomous Surface Vehicles

In Chap. 3, we showed that physicomimetics can be used to create a formation of robots. This will keep our autonomous surface vehicles (aquatic drones) in formation, but we would also like to use physicomimetics to solve the task. Let us go back and review our tasks. We wanted to create a swarm that could find the route with the highest (for the monitoring task) or the lowest (for the stealthy navigation task) chemical concentration from one point to another. In physicomimetics this means we have a stationary goal B somewhere that attracts all the members of the swarm, thereby causing them to move from point A to point B. Furthermore, the chemical material in the world should either attract or repel each ASV. This sounds simple enough—all we have to do is to extend the physicomimetics algorithm to account for these new virtual forces—the goal force and the force pulling the agents toward higher (or lower, depending on the task) chemical concentration.

Before continuing, let us pause and consider what might happen when we introduce these new forces. If the new forces are too strong, they could rip the formation apart, and we have to ensure that this does not happen. If the goal force is weaker than the forces from the chemical material, then the ASVs may never reach the goal. But if the goal force is too strong, then our swarm ASVs would pay little or no attention to what path they take to reach the goal. We have to keep these issues in mind and address them when appropriate. In summary, we have two tasks—seek the maximum bioluminescence on the way to a goal, and seek the minimum bioluminescence on the way to a goal. In both cases, the forces must be appropriately balanced, as seen in earlier chapters of this book.

## 6.3.3 Seeking the Maximum or the Minimum

First, consider the task which involves seeking the maximum bioluminescence, for example to monitor a harmful algal bloom (HAB). Recall that our procedure has each dino move toward the highest known concentration during the creation of the world. In particular, we let the dino check the three patches (ahead, myright, and myleft) in front of it and turn toward the one with the highest concentration before depositing chemical. If our task is to seek out the extent of potentially harmful algal blooms composed of dinoflagel-

**Listing 6.3:** The code to move towards the maximum

```
if (chemical > ([chemical] of drone ?))
  [set Fx (Fx + (apoF * (deltax / r)))
   set Fy (Fy + (apoF * (deltay / r)))]
```

lates, then we need our swarm to move toward the highest concentration of the chemical deposited by these creatures. Unfortunately, our ASVs do not have the ability to explore what is ahead. Their sensors can only detect the area they are currently occupying. How would a group of human friends do this? They would probably spread out a little bit from each other, and all members of the group would take a sample at their location. They would then communicate their sample results to each other and everyone else would move toward the member that reported the highest result. They would then repeat the process at the new location.

We can simulate this process using the physicomimetics algorithm. Let each ASV share the chemical level at its own location with the other neighboring members in the swarm, and let the information regarding chemical level become a part of the force calculation. If we want the ASVs to follow the maximum concentration, then every ASV should be attracted to its neighbors that have a higher concentration, and be repelled by its neighbors that have a lower concentration than it has. The ASV performing the calculations then takes the vector sum (resultant) of these forces in order to move in the best direction to accomplish the maximization task. This solution (consisting of moving toward "better" neighbors and away from "worse" neighbors) is called *artificial physics optimization (APO)*, where "artificial physics" is an alternative name for "physicomimetics." We will see in Chap. 19 that APO is a very important algorithm for tackling optimization and tracking problems. Let us call the force corresponding to the APO algorithm the *APO force*. Right now let us just say it has to be a small part of the physicomimetics force between the two ASVs. In the code, the APO force is called apoF, and it is a global variable. We can now update our force calculations with the code in Listing 6.3. Clearly, at every time step the ASVs are attracted to those ASVs sensing a higher chemical concentration and repelled from those sensing lower concentrations, if the variable minimum? is set to false—which implies that we selected the maximization task (see Listing 6.4).

For seeking the minimum bioluminescence, we reverse the relationship by setting minimum? to true. In other words, the algorithm for finding minimum bioluminescence paths on the way to the goal is merely the dual of

Listing 6.4: The code to add forces towards the maximum or minimum

```
to ap-drones
    set Fx 0 set Fy 0
    set vx (1 - FR) * vx   ; Slow down according to friction
    set vy (1 - FR) * vy

    set hood [who] of other drones
    foreach hood [
        set deltax (([xcor] of drone ?) - xcor)
        set deltay (([ycor] of drone ?) - ycor)
        set r sqrt (deltax * deltax + deltay * deltay)

        ; The generalized split Newtonian law:
        if (r < 1.5 * D) [
            set F (G * mass * ([mass] of turtle ?) / (r ^ 2))
            ; Bounds check on force magnitude:
            if (F > FMAX) [set F FMAX]
            ifelse (r > D)
                [set Fx (Fx + F * (deltax / r)) ; Attractive force
                 set Fy (Fy + F * (deltay / r))]
                [set Fx (Fx - F * (deltax / r)) ; Repulsive force
                 set Fy (Fy - F * (deltay / r))]

            ; Move towards chemical minimum or maximum:
            ifelse ((minimum? and (chemical <
                        ([chemical] of drone ?))) or
                    ((not minimum?) and (chemical >
                        ([chemical] of drone ?)))
                [set Fx (Fx - (apoF * (deltax / r)))
                 set Fy (Fy - (apoF * (deltay / r)))]
                [set Fx (Fx + (apoF * (deltax / r)))
                 set Fy (Fy + (apoF * (deltay / r)))]
        ]
    ]

    set dvx DeltaT * (Fx / mass)
    set dvy DeltaT * (Fy / mass)
    set vx  (vx + dvx)
    set vy  (vy + dvy)

    set deltax DeltaT * vx
    set deltay DeltaT * vy
    if ((deltax != 0) or (deltay != 0))
        [set heading (atan deltax deltay)]
end
```

**Listing 6.5: The code for the goal force**

```
    ; Now include goal force, when particle has neighbors nearby
if (goal? and (ncount > 0)) [
    set hood [who] of goals        ; Get the ids of goals
    foreach hood [
        set deltax (([xcor] of goal ?) - xcor)
        set deltay (([ycor] of goal ?) - ycor)
        set r sqrt (deltax * deltax + deltay * deltay)
        if (r < 1) [set all_done? true]   ; Reached the goal
        set Fx (Fx + (goalF * (deltax / r)))
        set Fy (Fy + (goalF * (deltay / r)))
    ]
]
```

the maximization algorithm. APO is again used, but in this case each ASV is repelled by its neighbors sensing a higher concentration and attracted toward its neighbors sensing a lower concentration. The physicomimetics algorithm with APO for the maximization or minimization task is shown in Listing 6.4.

Finally, it is important to note that in order to succeed at either of the two tasks, three forces have to be properly balanced—the cohesion force that holds the formation together, the APO force apoF that attracts the ASVs toward higher chemical concentrations and repels them from regions of lower concentration (if the objective is maximization), and the goal force that attracts the formation to the goal. To balance these forces, we have taken a combined approach that entails both theoretical (Sect. 6.3.6) and experimental (Sect. 6.3.7) analyses.

## 6.3.4 Seeking the Goal

Goal-following is actually not that difficult to implement. Here, we introduce only a single goal. This goal is preprogrammed as the destination for each ASV, and we consider the mission to be completed as soon as a single ASV has reached (i.e., has gotten within the desired distance of) the goal. Because there is only a single goal, we can add the goal force after calculating the split Newtonian force and right before we calculate the velocity of the ASV. Let us call it goalF and make it a global variable. So far, we know it has to be strong enough to attract the formation toward the goal, but weak enough to not tear the formation apart. We will analyze this in detail later. For now, consider the code to accomplish this in Listing 6.5. The variable ncount

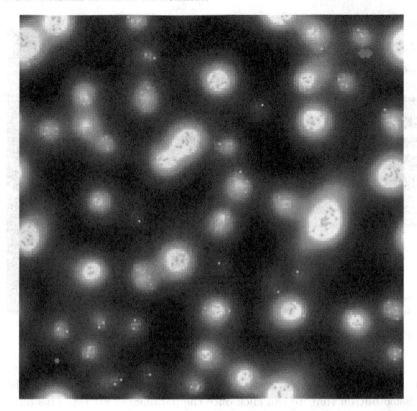

Fig. 6.6: A screen shot of a world with a goal and ASVs ready to begin the task

monitors the number of neighbors for each particle. Particles that do not have neighbors remain stationary and do not move toward the goal.

A snapshot of the ASVs about to tackle their task is shown in Fig. 6.6. The ASV drones in the upper right are orange, and the goal in the lower left is blue. The following section examines the graphical user interface and the system parameters in detail.

### 6.3.5 Graphical User Interface and Parameters

One major advantage of using NetLogo is that it is easy to create a graphical user interface (GUI) that not only allows us to set and change any desired parameters without having to recompile the code, but also display graphs and variable values. We will now show you how to take advantage of these capabilities.

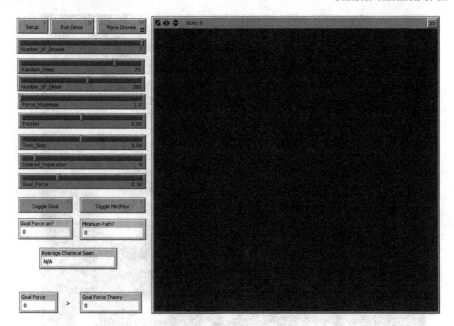

Fig. 6.7: A screen capture of the user interface and an empty world

The simulation begins with an empty world, as shown in Fig. 6.7. There is a **Setup** button that we can click once the parameters are set. This button initializes the simulation and resets it to the default settings. The button to begin the simulation is called **Move Drones**. There is also a third button, called **Run Dinos**, that runs the dinos for a fixed length of time to construct the chemical environment. We still have one other thing to consider before we start to consider any parameters. In the real world, even though the dinos have a mode of propulsion and the ability to move around, their speed is much slower than the speed of the drones. In our simulation, we model this in the following way. We make the assumption that the dino movement and the change in the chemical map are so small during the time it takes for the swarm to move from the deployment area to the goal, that we could consider the dino-chemical environment to be static.

Which system parameters need to be controllable with sliders? The answer to this question is based on both the user's needs and the experimental design requirements. For this chapter, our focus is on the latter. To mimic multiple different real-world scenarios for repeated experiments under varying conditions, we want to be able to randomly initialize the world, including the initial dino locations, at the beginning of each new experimental trial (also called a *run*). But at the same time we also want the capability of repeating experiments for possible reexamination as needed. The NetLogo random number generator will allow us to do both, and thus we create a slider with a minimum value of one and a maximum value of 100 for the Random_Seed. Like other pseudo-random generators, the NetLogo random number generator will generate the same sequence of numbers every time it is initialized with the same seed.

For our experiments, it would be useful to be able to methodically vary the number of dinos in order to have a more or less chemical-laden environment for the ASVs to explore, so we need a slider for this. Our slider is the Number_of_Dinos, and its values vary between zero and 500. We suspect that the size of the swarm will also have an impact on the results, and therefore we have created a Number_of_Drones slider that allows us to change the size of the swarm from one to seven drones. For many of the parameters in our simulation, the limits on the parameter values were largely driven by hardware and other practical constraints. The upper limit of seven drones, in particular, was motivated by our desire to create hexagonal formations, which are well-suited for our tasks. More than seven drones are used in the experiments below, however, and this is done by changing the code.

The swarm formation is controlled through several sliders. To establish the size of the drone formation, the Desired_Separation is used. To avoid overlapping drones, the minimum separation is set to two. Due to assumed limitations in drone sensing hardware, the maximum is set to ten. The Force_Maximum slider is a mechanism to ensure that the acceleration of the drones does not exceed the hardware capabilities. This ranges from one to five. Friction ranges from zero to one and serves as a form of inertia. With full friction (one), the formation stops at every time step and quickly reacts to the environmental forces. This setting is useful for rapid response times of the drones to their environment, and it is also useful for maximum cohesion of the formation. If the friction is lower than one, the formation has inertia, which produces a more sluggish (but also more realistic) drone response to new environmental information. The Time_Step slider establishes the time granularity of the simulation (from very fine to coarse). There is usually no need to adjust this particular slider.

To run the swarm correctly, we need to balance the three forces (cohesion, the APO force apoF, and the goal force goalF) correctly. This can be done theoretically, as shown in Sect. 6.3.6, although we also allow the user to modify goalF using the slider Goal_Force. Additionally, we added a Toggle Goal button that can switch the goal force on and off. This will allow us to make sure that the swarm has the optional capability of focusing solely on the objective of seeking the local maximum or minimum chemical concentration. We have also created a monitor Goal Force on? so that the user can quickly determine the value of this switch. Furthermore, we need to make sure that the balance between the goal force and the APO force is correct according to our theory. This can be done by comparing the value of this force in the monitor Goal Force in relation to our theoretical bound, whose value is displayed in the monitor Goal Force Theory—see Fig. 6.7.

We have two tasks: to seek the maximum or the minimum concentration on the way to the goal. We have created a button Toggle Min/Max that allows us to toggle between these, and a monitor Minimum Path? that displays the current value of this toggle.

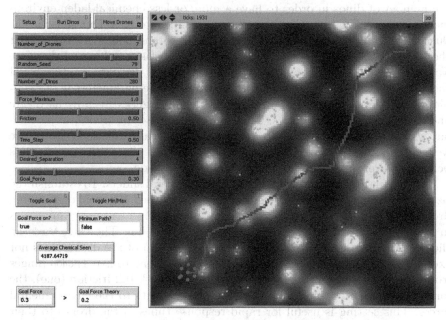

Fig. 6.8: A screen capture of the complete simulator

Finally, we need metrics to evaluate and compare the simulation performance between experiments with different parameter settings. NetLogo automatically provides us with one metric with which we can determine the time that elapsed from the beginning of a run (which begins at initialization and ends when the drones get to the goal or the time limit is reached), by counting the number of ticks performed by the simulator. The second metric will be the total amount of chemical measured at the *centroid* of the formation over the course of a run. What is the centroid, and why do we measure there? The centroid is the center of mass (assuming all masses are equal) of the formation. We mark this location with a red patch at every time step and assume that we have a sensor, such as a bathyphotometer, at this location. We do this because it will simulate the path that any ship (or covert special operations team) would take, given the information provided by the swarm's exploration. We add the reported chemical concentration at this location during each time step and accumulate the total amount of chemical measured over the duration of an entire run. Other performance metrics are defined in Sect. 6.3.7. Our NetLogo simulator is now complete and ready for action—see Fig. 6.8.

## 6.3.6 Theory for Setting Parameters

Now that we have studied the simulation, we are ready to see some theory about how to set the parameter values in order to achieve robot–robot formation cohesion, as well as movement to the goal without stagnation. To accomplish this balance-of-forces objective, it is necessary to derive bounds on the values of two parameters—the goal force goalF and the APO force apoF.

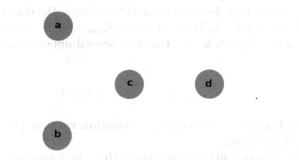

Fig. 6.9: An illustrative example for setting a bound on goalF

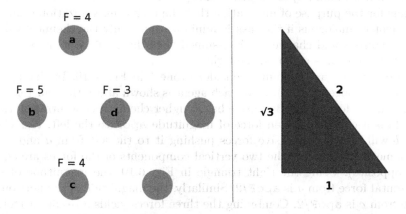

Fig. 6.10: An illustrative example for setting a bound on apoF

Assuming that the masses of all drones are set to one, the force felt by each pair is equal to $G/D^p$ (Eq. 3.3) when they are at the desired distance $D$ from each other. Recall from Chap. 3 that clustering at nodes can be an undesirable property. It is particularly undesirable when the agents are ASVs as in the application being addressed in this chapter. Using the phase transition theory that was developed for physicomimetics [236], shown in Eq. 3.4, we know that a good value for $G$ is approximately $0.9F_{\max}D^p/(2\sqrt{3})$ in order to avoid clustering.

With this theory, we can establish a bound on the goal force, goalF. Suppose there are four agents, $a$, $b$, $c$, and $d$, in the configuration visualized in Fig. 6.9. In this scenario, the agents $a$, $b$, and $c$ are stuck in place, whereas $d$ is not. Assume the goal is to the right. Then agent $d$ is experiencing a goal force that pulls it to the right. To keep $d$ from breaking away from the formation, we require that goalF $< G/D^p$ to ensure cohesion of the formation. By substitution with the value of $G$ established above, the desired inequality is goalF $< 0.9F_{\max}D^p/(2\sqrt{3}D^p)$. Because $F_{\max} = 1$ in our bioluminescence simulation, using algebra we get the final desired upper bound constraint on goalF:

$$\text{goalF} < \frac{0.9}{2\sqrt{3}} = 0.25981 \, . \tag{6.1}$$

This goalF bound ensures that the formation will not break apart, while also avoiding clustering.

The next issue to address is to ensure that the formation as a whole will not get stuck at a chemical patch, but will be able to continue to move toward the goal. For this, an upper bound on apoF has to be established. Consider Fig. 6.10, where $F$ is an abbreviation for the "fitness" (i.e., the chemical concentration) found by the agent at this position. Without loss of generality, assume for the purpose of illustration that the task is maximization (and the argument is analogous if the task is minimization—only the chemical values in the figure would change). It is assumed that the apoF force is constant, and that the goal is again on the right.

To derive the apoF bound, consider drone $b$ in Fig. 6.10. In this figure, the fitness (concentration), $F$, of each agent is shown above the four relevant agents in the formation. Because $b$ has a higher chemical concentration than $d$, it feels a chemical-driven force of magnitude apoF to the left. It is clear that $b$ will also feel repulsive forces pushing it to the left from $a$ and $c$ for the same reason (where the two vertical components of the forces are equal and opposite). Using the right triangle in Fig. 6.10, the magnitude of the horizontal force from $a$ is apoF/2. Similarly, the magnitude of the horizontal force from $c$ is apoF/2. Combining the three forces yields a resultant (total) force of apoF $+ 2$ apoF $/ 2 = 2$ apoF on $b$ toward the left. For the formation to continue moving toward the goal, which is on the right, it must be the case that $2$ apoF $<$ goalF, which implies that the desired upper bound constraint is $2$ apoF $< 0.25981$ or, equivalently:

$$\text{apoF} < 0.13 . \tag{6.2}$$

## 6.3.7 Experimental Performance on the Two Tasks

Given the two tasks defined above, namely, seeking the maximum and seeking the minimum bioluminescence on the way to a goal, we next evaluate the performance of the simulated ASVs on these tasks, where the ASVs use physicomimetics for vehicle coordination and control. Although the goal is optional in the simulation GUI, in all of these experiments it is included, and it appears in the lower left of the environment. In these experiments, the environment is initialized with chemical deposited by the dinoflagellates first. Then, the drones are initialized in a fixed starting location in the upper right of the environment, with random headings and zero initial velocity. The chemical gradient and goal forces then influence the subsequent motion of the agents. During experiments, the performance is monitored over the course of each run, where a run is defined as beginning with the agents at their starting location and terminating when either at least one drone is within the desired radius (two units) of the goal ("success") or a time limit is reached ("failure"). Note that although the software allows the possibility that the drones may never get within the desired radius of the goal before time runs out, in *all* of our experiments the "success" criterion was in fact achieved. This was fortunate because it made the comparisons more straightforward.

### 6.3.7.1 Experimental Design

To design experiments, it is necessary to define the independent and dependent variables. (Recall that the independent variable is typically plotted on the $x$-axis of a graph, and the dependent variable on the $y$-axis.) The independent variables are alternatively called the *experimental parameters*. These are the parameters whose values are varied methodically over the course of the experiments. The dependent variables are alternatively called the *performance metrics*. They are the metrics used to evaluate how well or poorly the simulated agents are doing the task.

These experiments assume two experimental parameters:

- *Agents*. This is the independent variable $N$, which is the number of agents. We use the values 3, 6, 9, 12, 15, and 18. These particular values were selected based on experiences with the simulation. They provided a good range of values for diverse simulation behavior. Note, however, that our algorithm and simulation scale to much larger groups of agents than 18. The reason for stopping at 18 drones is that it is the largest number we expect to have in our real-world testing (given the expense of drones).

- *Ratio.* This is the ratio goalF/apoF. We fix the value of goalF to be 0.40.[1] We choose four different values of apoF: 0.10, 0.02, 0.013, and 0.01. Therefore, the four values of the ratio are, respectively, 4.0, 20.0, 30.77, and 40.0. These particular values were selected because the lowest ratio allows the most meandering of the drones (within the bounds allowable by the theory), whereas the highest ratio results in agents heading straight for the goal.

To further ensure cohesion, Friction is set to 1.0. As a result of our parameter settings, on every run the swarm stays in formation and succeeds in getting to the goal.

Why is the ratio chosen as one of our experimental parameters? The reason is that it measures the tradeoff between *exploration* and *exploitation*. This is a classical tradeoff in the artificial intelligence literature [199]. The basic idea is that the better the agents exploit their environment to achieve a goal as quickly as possible, the more efficient and effective they will be at getting to this goal. Nevertheless, by behaving in a highly goal-directed manner, the agents sacrifice the ability to explore their environment for other purposes besides the goal. On the other hand, the more thoroughly the agents explore their environment, the more they will learn about the environment, but the slower they will be at getting to the goal. This tradeoff plays a key role in our two tasks. In particular, drones heading straight for the goal will minimize their time to get there, but will be suboptimal in their ability to find bioluminescence maxima/minima. On the other hand, drones whose goal force is lower in relation to the chemical density gradient force will be more adept at finding maxima/minima, but do so at the cost of taking longer to get to the goal. Which is better? There is no absolute answer to this question. It is a preference that is best decided by the practitioner who is in charge of deploying the drones for a particular task. Therefore, instead of deciding what is a "good" or "better" performance a priori, we display the results of scores for the ratio metric and allow the practitioner to select the "best" value of this ratio for his/her specific problem requirements.

In summary, there are two tasks—maximization and minimization (where goal-seeking is also assumed to be part of these tasks), and two experimental parameters—agents and ratio. Using all combinations of these leads to four *experiments*:

- *Experiment 1:* Max-Agents,
- *Experiment 2:* Min-Agents,
- *Experiment 3:* Max-Ratio, and
- *Experiment 4:* Min-Ratio.

---

[1] Because ASVs in water are extremely unlikely to get stuck in one location, we can safely exceed the theoretical upper bound in Eq. 6.1.

For clarity, we use the term *configuration* to describe any run with a fixed value of one of the independent variables. For example, suppose we are executing experiment 1 with maximization as the task and the number of agents is fixed at a value of 15. Then this is a particular configuration with the value 15 for the number of agents. Note that for experiments 1 and 2 there are six possible configurations, and for experiments 3 and 4 there are four possible configurations.

We have now defined the independent variables, and we have yet to define the dependent variables. However, before doing this we need to consider another aspect of the simulation that should be varied in order to evaluate performance, namely, the density of chemical in the environment. This is a parameter that is *not* under user control, but it is important to see how the ASVs will perform under these variable and uncontrollable conditions. Three different environmental densities have been chosen, based on our experiences with the simulation. The average density is measured by finding the sum of chemical density in all patches, and then dividing by the number of patches. Each of these average density levels is called an *environment class*:

- *Low Density.* The average density is around 2.0 for environments in this class. There are 290 dinoflagellates.
- *Medium Density.* The average density is around 2.7 for environments in this class. There are 400 dinoflagellates.
- *High Density.* The average density is around 3.5 for environments in this class. There are 510 dinoflagellates.

These different density classes provide variety for the experimental conditions, but for the purpose of collecting statistics on the performance it is important to also vary a random element of the simulation. We have chosen to do this by selecting different random seeds; each choice of random seed creates a unique specific environment. For each random seed, one run is executed. A reasonable sample size in statistics is 30 (e.g., see [71]). We far exceeded this norm, with 200 random seeds for each environment class. With 200 runs, it is possible to determine the *mean* and *standard deviation* *(std)* of performance, where the mean is the average over the 200 runs, and the standard deviation is the square root of the average of the squares of the differences of the 200 values from the mean.

Now that it is clear what aspects of the simulation are methodically varied, it is time to define the dependent variables, or performance metrics, whose values result from the methodical perturbations of the parameters just defined. Five different metrics are selected based on their relevance to the two tasks, and are evaluated over each run:

- *Mean Total Chemical.* This is the total amount of chemical seen over the entire run by the centroid of the multi-agent formation, averaged over the 200 runs for a particular experiment, configuration, and environment class. Section 6.3.5 explains why the centroid is used.

- *STD Chemical.* This is the standard deviation of the total chemical seen by the centroid over an entire run, calculated over the 200 runs.
- *Mean Peak Chemical.* This is the maximum chemical concentration ever found by the centroid over the entire run, averaged over the 200 experimental runs.
- *Mean Total Time.* The mean total time is the total time for the formation to travel from the initial location to the goal location for each run, averaged over the 200 runs.
- *STD Time.* This is the standard deviation of the total time, calculated over the 200 runs.

Other performance metrics were considered besides these, including normalized variants of the above metrics. For example, we also considered the mean total chemical divided by the mean total time and the average environmental density in the world. Although this is a potentially useful metric, it was rejected because it involves extra processing (arithmetic operations) on the raw data. We decided to stick with the raw data instead for two reasons. First, raw data can show some very striking and pronounced patterns that may be obscured by normalization. These patterns can lead to important observations and further analysis, ultimately leading to a better understanding of the system behavior. For an illustration of this process, see Hypotheses 3, 3.1, 3.1.1 and 3.1.2, below. Second, the purpose of extra processing is to infer what the practitioner will care about when deploying the swarm in the field. But it is a big assumption that we can make such an inference. For different variations of the task, different performance metrics may be desirable. Rather than anticipate, it seems more straightforward to present the raw data (metrics) and allow the practitioner to process it based on his/her needs at the time.

Note that for proper experimental methodology, we vary one parameter at a time while all the rest of the parameters have their values held constant. This methodology facilitates our understanding of the experimental results. For example, if more than one variable were varied at time, then it would be difficult to tease apart which variation caused the effects that are seen.

Finally, for fairness of comparison, every configuration is tested with the same suite of 200 environments. For example, when comparing six versus nine agents, the six-agent formation is tested on the same environments (same environment class, same random seed) as the nine-agent formation.

### 6.3.7.2 Experimental Hypotheses

Scientific experimental methodology involves formulating *hypotheses* (also called "conjectures"), running experiments, making observations, generalizing, and then drawing conclusions. Interestingly, although experimental hypotheses often precede the experiments, this is not a necessity. In fact, scientists often intermingle the hypothesis, experiment, and observation processes. After all of these processes are completed, general conclusions may be drawn. It is important to point out, however, that experimental conclusions are generalizations based on all of the experimental data seen, but they are never absolutely definitive (i.e., final, conclusive). To be definitive, they would have to be based on all possible data in the universe. Few of us scientists can gather all possible data. This is an important point. Contrary to the standard public view, *science does not prove anything with certainty*. It can *disprove* hypotheses, however. *The remaining hypotheses are our best understanding of how the universe works*. This scientific method is highly effective in most practical situations.

Scientists typically formulate many hypotheses when observing the data. Some hypotheses are refuted. Others remain as informal hypotheses (which never get published). The most important become formal hypotheses (which appear in published papers or book chapters). In this section, we present two formal hypotheses that were based on initial investigative runs of the NetLogo bioluminescence simulator by the authors of this chapter. These hypotheses were postulated based on watching the GUI prior to running any formal experiments. Note that in fact more hypotheses were formulated than these two (in particular, many informal hypotheses were formulated), but these two were selected as being more interesting than the others. For example, one informal hypothesis states that "As the chemical concentration increases (e.g., when going from a low density to a medium or high density environment), the mean total chemical and mean peak chemical should increase." This conjecture was not included in the final set of formal hypotheses because it is too intuitively obvious. Nevertheless, this and the other informal hypotheses were tested in the experiments (just to be sure!). Because they are highly intuitive and are confirmed by the data as expected, they are not discussed further in this chapter. Note, on the other hand, that if any of them had been refuted by the data, then their refutation would be a very interesting surprise and in that case the refutation would have been elevated to the status of a formal hypothesis.

The following formal hypotheses were postulated based on initial observations, prior to running the formal experiments:

**Hypothesis 1 (about maximizing and total chemical):** *When the task is maximization, it is preferable to have more agents, assuming the maximization task has a higher priority than getting to the goal quickly (as reflected by a low ratio value).*

**Hypothesis 2 (about minimizing and total chemical):** *When the task is minimization, there is a task-specific optimal number of agents that is somewhere between the minimum value (3 in our experiments) and the maximum value (18 in our experiments). This assumes that the minimization task has a higher priority than getting to the goal quickly (as reflected by a low ratio value).*

The rationale underlying Hypothesis 1 is that when more ASVs are in the formation, there appears to be (based on watching the simulation) a greater potential for exploration to find local chemical density maxima. In other words, more circuitous paths are taken by the formation along the way to the goal when the number of agents is increased. With more agents, speed seems to be sacrificed for more chemical.

The minimization objective is to have a low value of mean total chemical. The rationale for Hypothesis 2 is based on our initial observations in the simulation that when formations are composed of fewer agents, the formations are better able to "squeeze" through narrow passageways of low chemical, and to do this with little meandering (i.e., turning away from a direct path to the goal). But when the number of agents is really small, it becomes too difficult to get a representative sample of the environment. These observations lead to the expectation that the curve of the number of agents versus the mean total chemical should be a U-shaped parabola, where the lowest point on the curve corresponds to the ideal number of agents for this task.

### 6.3.7.3 Experimental Results and More Hypotheses

After the experimental design and formulation of hypotheses are complete, the formal experiments can be run. We have executed our experiments with the NetLogo simulator described above, but with the addition of a "wrapper" that fully automates the execution. This wrapper loops through each experiment, configuration, environment class, and environment to run the simulator. Recall that when one parameter is varied, all other parameters must have their values fixed. This raises the question of what value of the ratio should be selected for the Max-Agents and Min-Agents experiments, and what value of agents should be selected for the Max-Ratio and Min-Ratio experiments? If we use all possible values of these variables in all possible combinations, we would have to run a *factorial experiment design*, which can be extremely time-consuming, but often unnecessary. Instead of a factorial experiment design, let us select only the value of 4.0 for the ratio in the Max-Agents and Min-Agents experiments, and only the value of six for the number of agents in the Max-Ratio and Min-Ratio experiments. These values are selected because they seem to yield some of the more interesting behaviors. Despite restricting our experiments, we still get a huge amount of data, which has resulted in a large number of graphs. Therefore it will not be possible to show all graphs in this book chapter. We instead select representative graphs to illustrate our conclusions.

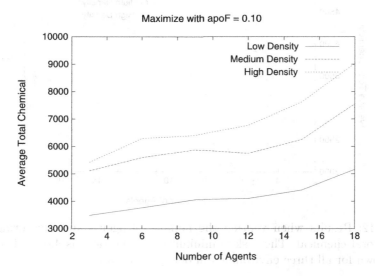

Fig. 6.11: Results when varying the number of agents and measuring the mean total chemical. The task is maximization, the ratio is 4.0, and results are shown for all three environmental density classes

Note that our hypotheses (see above) are about general trends in the data. Because of this, we consider a hypothesis to be "confirmed" if the general trend in the data satisfies the hypothesis, even if it is not the case that every single point on every graph perfectly satisfies the hypothesis. Our rationale is that although we would like to get smooth graphs, despite 200 runs per point on the graphs there may still be some irregularities in the curves due to unusual environments generated by the random number generator. Unless they are pronounced spikes, we will ignore them.

The experiments have been run, and all of the data collected and processed. We now present the experimental results in the form of graphs and analyses. We begin with Hypothesis 1, which states that for the experiment Max-Agents, more agents implies more exploration, thereby leading to a higher value of mean total chemical. Example results are shown in Fig. 6.11. The results confirm Hypothesis 1 because all three curves, corresponding to the three environment classes, in general have positive slopes as the number of agents increases along the horizontal axis.

Next, consider Hypothesis 2, which states that for the experiment Min-Agents, the graph of number of agents versus mean total chemical is expected to be U-shaped. Figure 6.12 presents an illustrative example. Again, the gen-

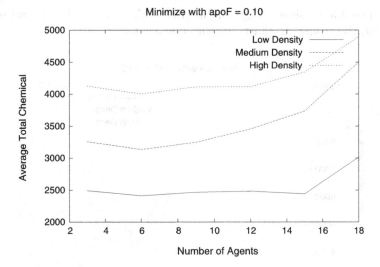

Fig. 6.12: Results when varying the number of agents and measuring the mean total chemical. The task is minimization, the ratio is 4.0, and results are shown for all three environmental density classes

eral trend matches the expectation. Performance differences along each curve are small with fewer than 15 agents, but six agents appears to be optimal for all three environment classes. It is also interesting to compare Figs. 6.11 and 6.12. To do this, the reader should observe that the vertical axes of both graphs are scaled to fit the data, and because of the big difference in mean total chemical values between the two graphs, these graphs have very different scales from each other. Taking the scales into account, clearly the swarms accumulate a considerably higher mean total chemical when performing the maximization task than when performing the minimization task, as expected.

Recall that hypotheses can be formulated before or after the experiments have been run. We have shown two examples that were formulated before, and now we will see one formulated afterward. After viewing a subset of the graphs displaying the mean total time, we observe that when the number of agents increases, so does the mean total time. Because this is not immediately intuitive, we therefore posit it as a formal hypothesis:

**Hypothesis 3 (about total time):** *As the number of agents increases, the mean total time is expected to increase.*

To test this hypothesis, we examine *all* graphs showing the mean total time on the $y$-axis (as the performance metric). Interestingly, this hypothesis is confirmed by all of the graphs. Although there is not room to show all of the graphs, Fig. 6.13 presents an illustrative example of both tasks, where

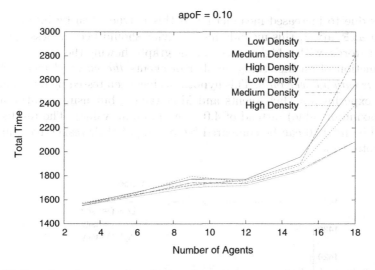

Fig. 6.13: Results when varying the number of agents and measuring the mean total time. Results are shown for both tasks, the ratio is 4.0, and results are shown for all three environmental density classes

there are three curves per task. The reason for this result is not immediately apparent. We have postulated and tested two possible reasons. The first is that formation inertia increases as the number of agents increases. Although this is true, it does not explain the behavior seen in the data (which we conclude based on some informal experiments with the simulator). The second possible reason is that with more agents, the formation will meander more, rather than heading straight to the goal. This second possibility is confirmed by watching the simulation. (The reader is encouraged to do this.) Note, however, that all you need is one simple counterexample to refute a hypothesis. On the other hand, to confirm it you need far more evidence. To absolutely confirm it, you would need all possible data that could ever be generated by perturbing the simulation. But scientists normally consider a hypothesis to be confirmed if a follow-up formal experiment is conducted that yields confirming results. This is the approach adopted here. We can express this as a subsidiary hypothesis to Hypothesis 3, so let us call it Hypothesis 3.1:

**Hypothesis 3.1 (about the reason for Hypothesis 3):** *As the number of agents increases, the mean total time increases because with more agents the formation takes a more circuitous path to the goal.*

Hypotheses 3 and 3.1 are based on our observations with a ratio value of 4.0, which is the lowest value possible. Hypothesis 3.1 states that the reason the agents take longer to get to the goal when there are more of them is due to increased meandering. If this is true, then by increasing the ratio goalF/apoF, this meandering behavior should become less apparent. In other words, with a higher ratio, in the graph showing the mean total time as a function of an increasing number of agents *the curve should be flatter than it is with a lower ratio*. This hypothesis has been tested by rerunning the formal experiments Max-Agents and Min-Agents, but using a ratio of 40.0 (the maximum value) instead of 4.0 (the minimum value). The results with ratios 4.0 and 40.0 can be compared by looking at their respective graphs of mean total time.

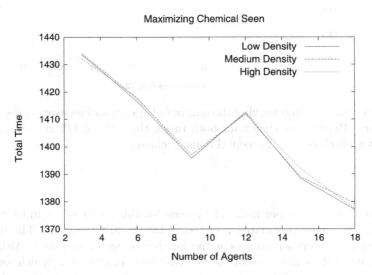

Fig. 6.14: Results when varying the number of agents and measuring the mean total time. Results are shown for the maximization task, the ratio is 40.0, and results are shown for all three environmental density classes

The results for the maximization task and ratio 40.0 are shown in Fig. 6.14. We do not show the graph for the minimization task because it is nearly identical to Fig. 6.14. When examining Fig. 6.14, three surprises emerge. First, Hypothesis 3 is refuted. Based on the new evidence, we refine Hypothesis 3 to state that:

**Hypothesis 3′ (about total time):** *If the ratio is low (e.g., has a value of 4.0), then as the number of agents increases, the mean total time is expected to increase due to additional meandering.*

This revised Hypothesis 3′ has been confirmed by all of our experimental data.

The second surprise is that instead of a flatter slope, the curves in Fig. 6.14 actually have *negative* slopes. Third, there is an unexpected increase in the mean total time when the number of agents is 12 in particular. The appearance of new surprises implies the need for further investigation to explain these behaviors. Two new hypotheses are therefore formulated, after watching the simulation:

**Hypothesis 3.1.1 (about the reason for the negative slopes in Fig. 6.14):** *The curve slopes are negative instead of flatter (and positive) because when the swarm goes relatively straight to the goal, with little meandering, a bigger formation will hit the goal sooner.*

A bigger formation will get to the goal more quickly because some of the perimeter agents (i.e., those on the part of the perimeter facing the goal) in a larger formation are initially closer to the goal than those in a smaller formation, and the formation heads right to the goal when the ratio is high. Furthermore, the first time any agent in the formation is close enough to the goal, the task is considered to succeed and the run terminates.

**Hypothesis 3.1.2 (about the reason for the significant increase in time with 12 agents in Fig. 6.14):** *There will be a peak in the curve showing time versus the number of agents when the swarm is composed of 12 agents, and this peak is due to the fact that the initial 12-agent configuration is farther from the goal than the other configurations (because the formation is a bit more concave on the side pointing toward the goal).* (The initial agent positions are deterministically set.)

Both of these hypotheses are most easily tested by running the swarms with no chemical in the environment, in which case the formation will head straight to the goal. These experiments have been executed, and the hypotheses confirmed on every run. There is no need to calculate the mean because without any environmental influences the behavior is deterministic. For example, the swarm always gets to the goal in 1295 time steps with nine agents, 1313 time steps with 12 agents, and 1292 time steps with 15 agents, and so on.[2]

This section has shown how the experimental results from hypothesis testing can be graphed and analyzed. Finally, note that none of our formal hypotheses are about the mean peak chemical, the three environment classes and their effects on the performance metric values, the Max-Ratio or Min-Ratio experiments, or the standard deviations. The reason is that although we believed that these were good variables to monitor and check, they did

---

[2] The astute reader may observe that this peak at 12 agents is not apparent in the graph with a ratio value of 4.0 (Fig. 6.13). This is because when the ratio value is low, exploration of the environment takes a higher precedence, and the initial distance from the swarm to the goal has virtually no effect on the performance.

not lead to any counterintuitive or especially interesting results. The one exception is that we found the need to vary the ratio from 4.0 to 40.0 in order to better understand the results of the Min-Agents experiment.

### 6.3.7.4 Experimental Conclusions

Hypotheses have been formulated, experiments run, and the results shown and analyzed. The following general conclusions may now be drawn based on our experimental results:

1. If the primary concern of the practitioner is maximization, then set the ratio to a smaller value and use a larger swarm—because in this case if the swarm is larger then the mean total chemical found by the swarm will be greater due to increased meandering.
2. If the primary concern of the practitioner is minimization, then set the ratio to a small value. In this case, there is a task-specific optimal swarm size for finding a path of minimal bioluminescence between the starting location and the goal location. If the swarm size is too small, then the agents will not get enough samples to find a good path. If, on the other hand, the swarm size is too large, then it will be unable to "squeeze" through narrow passageways of minimal bioluminescence. For this task, the practitioner is encouraged to model the task and try swarms of moderate size to find the optimal swarm size.
3. If the primary concern of the practitioner is for the swarm to get to the goal as quickly as possible, then it is important to use a large value for the ratio along with a larger swarm. The following two subsidiary conclusions provide more refined intuitions for this conclusion:
   a. If the ratio goalF/apoF is small, then larger swarms will take longer to get to the goal than smaller swarms. This is because they meander more.
   b. If the ratio goalF/apoF is large, then increasing the swarm size will result in getting to the goal faster, unless the initial swarm configuration is concave in the direction of the goal. This is due to the fact that the perimeter agents in the swarm facing the goal are closer to the goal initially, and they head fairly directly toward the goal with a larger ratio value.[3]

To summarize this section, we first presented two important real-world bioluminescence tasks that have motivated our research. Next, we described our NetLogo bioluminescence simulator and demonstrated its utility for deriving practical experimental results and conclusions. Unfortunately, the

---

[3] On the other hand, concavity may not be important if the formation rotates while traveling to the goal.

NetLogo simulator is inadequate for transitioning to real hardware ASV platforms. For this, we need a more realistic simulator—to be described next.

## 6.4 Multi-drone Simulator (MDS)
## for the Bioluminescence Application

The NetLogo simulation just described enables the reader to understand and appreciate the bioluminescence application at an intuitive level. However, NetLogo is limited in its ability to provide a complex, realistic model of the application. For this purpose, we have developed and implemented a *multi-drone simulator (MDS)*. The MDS provides a critical transition between the NetLogo simulator and the actual hardware drones (see Sect. 6.5, below).

### 6.4.1 Motivation

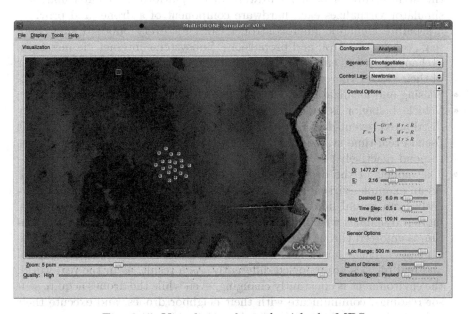

Fig. 6.15: Visualizing the task with the MDS

Our multi-drone simulator, shown in Fig. 6.15, is a comprehensive software tool that provides a configurable test environment for the bioluminescence plume-navigation problem. It is considerably more realistic than the NetLogo simulation of this problem, and it provides a high-fidelity emulation of key hardware robotic drone modules, such as communication, localization, and control.

### 6.4.1.1 A Realistic Model of the Actual Hardware Drones

The MDS software can be used to test, monitor, and evaluate a swarm of small autonomous drone vehicles as they actively engage in a bioluminescence-related task. During a mission, each simulated drone measures the amount of local bioluminescence, broadcasts that information to its neighbors, and then computes two distinct physicomimetic goal forces. The first is along the gradient of bioluminescence, and the second is attractive in the direction of a specified goal location. A vector sum is taken of three virtual forces, namely, the two goal forces just mentioned, and the inter-vehicle cohesion force. This vector sum translates into the velocity vector for the next move by the drone.

So far, this sounds like the NetLogo simulator. However, the MDS has sophisticated features that go beyond NetLogo—including the following:

- The MDS software architecture is structured in a manner that mimics the architecture of actual hardware drone platforms. In particular, the simulator virtualizes each hardware component of a drone, and provides a separate emulated microcontroller clock for every onboard subsystem (e.g., sensors, control, and propulsion). Subsystems interact via messages, exchanged over an emulated system data bus. Drones themselves may differ in their clocks, thereby enabling realistic asynchronous drone movement.
- Sensing and localization modules include optional added noise.
- Vehicle behavior is modeled accurately based on a decentralized, fully distributed, event-driven, multi-threaded model.[4] The operation of this pseudo real-time system is driven by periodic events to which the drones respond reactively.
- The models of the drones and their environment are strictly separated. As an example, virtual physicomimetics friction is treated entirely differently from actual fluid friction resulting from the impact of the drones' motion dynamics within the simulated marine environment. This separation is enforced through the use of different spatio-temporal scales for the simulated vehicle operations versus the simulated external world. Artificial and real physics are modeled differently. As a result, as in the real world, the environment is constantly changing, even while the drones acquire sensor readings, communicate with their neighbor drones, and execute their individual physicomimetic navigation strategies.
- Simulated hardware has precise and accurate physical dimensions and units. For instance, a simulated drone that has a mass of 1 kg and an active propulsion unit producing 1 Newton of thrust will experience an

---

[4] In computer science, a *thread* is a sequence of instructions, or a subroutine, that executes in parallel with other threads.

acceleration of $1\,\text{m/s}^2$ after one second (if we assume a frictionless environment).

- The MDS allows users to geo-reference the search area, so that all drone positions during the mission can be matched to real geographic locations.

The outcome of this realistic model is that the MDS provides us with an excellent intermediary step between the simple NetLogo model and the real-world aquatic drones. Without the MDS as an intermediary step, it would be far more difficult for us to develop and implement the hardware prototype vehicles. The MDS serves as our blueprint for hardware design.

### 6.4.1.2 A Parameterized Model

In addition to being a more drone-faithful model than the NetLogo simulator, the MDS provides us with the highly useful capability of running large numbers of parameterized experiments quickly. This is too difficult to do with actual hardware drones. With the MDS, we can vary any or all of the following for running methodical parameterized experiments to characterize the performance (and these choices are user-selectable, via the MDS user interface):

- Number of drones
- Localization range limit
- Communication range limit
- Type and degree of noise in the sensor and localization components
- Physicomimetics control law, including a choice between Hooke's law, the split Newtonian law, and the Lennard-Jones law, and the parameters within these laws
- Desired inter-vehicle distance $D$
- Size of time step

The MDS user interface includes other options as well, mostly for ease of monitoring and experimentation.

## 6.4.2 MDS Architecture

The MDS is implemented in C++ using the Open Source edition of the cross-platform Qt framework [169]. The most important feature of the simulated drone software model is the multi-threading aspect—every subsystem of the drone, as well as the drone itself, is implemented in a separate Qt thread. This approach ensures that the subsystems can operate independently of each other, mirroring the hardware configuration of a real drone platform. Multi-threading also provides a good method for emulating the physically distinct hardware modules that will be installed on the drone, and this approach is therefore useful in determining the proper data synchronization approaches to ensure data coherence among the different modules.

### 6.4.2.1 Functionality and Timing in the MDS

There is a single, user-controlled parameter called the Time Step that specifies how frequently the drone physicomimetics Control module recomputes the virtual force acting on the vehicle. This Time Step is used as the base period from which all other MDS sampling rates are derived, using predetermined scaling factors. For instance, in the current version it is assumed that the drone *localization* sensor (i.e., the sensor that locates and reports the positions of nearby neighbor drones) has a time update frequency of Time Step/2, so that on average, for every control decision, there will be two updates of the locations of the neighbor drones.

The unique pseudo real-time aspect of the MDS comes from the fact that the virtual Time Step parameter maps directly to the real "wall clock" time of the simulation. In other words, if the MDS is executing in "1X" speed mode, then one "second" of Time Step corresponds to one second of real time. This approach to modeling the relationship between simulated and real time steps is based on our decision to virtualize the drone hardware. The MDS provides an implementation of a separate, microcontroller-like clock for each simulated component, using the system clock of the host computer to emulate the virtual hardware timing. Furthermore, the MDS can emulate different clocks operating at different frequencies for multiple drones.

Figure 6.16 shows the MDS functional diagram. Note the clear separation between the drones and their environment. This architecture helps to ensure the locality of the MDS data, allowing us to create initial prototype simulations of hardware interfaces within the MDS. Also note the extensive multi-threading of the drone subsystems. Each Qt thread contains a unique timer object that emulates a microcontroller clock—effectively, every drone thread executes within its own "virtual" hardware.

Another use of the Time Step parameter occurs in the very fine-grained discretization of the environment. As was described in Sect. 6.4.1.1, one of our main motivations in designing the MDS was to separate the virtual physics model used for drone control from the simulated physics model used to predict the drones' interaction with the environment. To achieve this design goal, we chose to model different time scales for the drone subsystems and the environment simulation model. In particular, inside the Environment model we further discretized the Time Step, which is the period between successive executions of the physicomimetics controller, so that the "real" physical parameters of the simulated drone, such as its true velocity, orientation and location in the simulated environment are updated much more frequently than the output of the drone controller module. (Currently, the environment

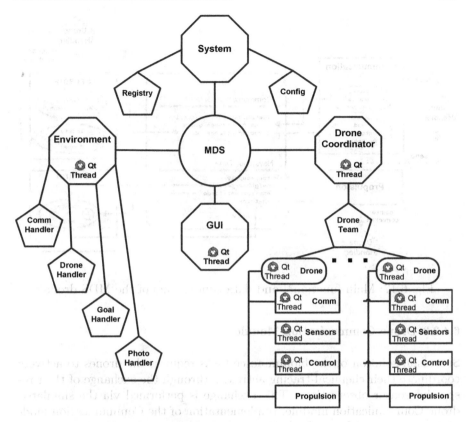

Fig. 6.16: MDS functional block diagram

update frequency is set to Time Step/25.) In effect, this mechanism allows for a more realistic, quasi-continuous motion of the simulated drones.

### 6.4.2.2 The Drone Architecture

The basic structure of the MDS drone is depicted in Fig. 6.17. The "external" modules shown with dotted outlines in Fig. 6.17 provide the drone with information about the simulated environment.

Recall from Sect. 6.4.1.1 that one of our MDS design goals is to use the MDS software as a blueprint for the final hardware design. Of particular value is the understanding of the data flow between all of the different drone subsystems. By modeling this aspect in software, we have acquired valuable insights into the required layout of the actual hardware data buffers to handle drone-to-drone communication, storage of sensor output, and other mission-specific tasks.

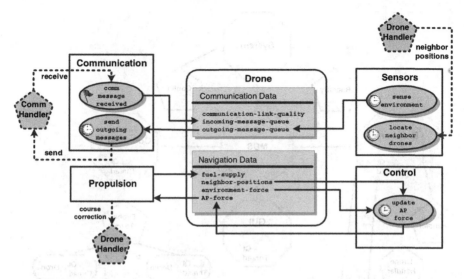

Fig. 6.17: Main functional and data components of the MDS drone

### 6.4.2.3 Drone Communication Module

Successful solution of the bioluminesce tasks requires the drones to actively coordinate their chemical-tracing activities through the exchange of their respective sensor observations. This exchange is performed via the simulated drone Communication module. Implementation of the Communication module is consistent with standard networking protocols, such as the Transmission Control Protocol and the Internet Protocol (TCP/IP). Because the drones will be equipped with wireless networking devices, communication is also subject to physical distance limitations, specified by the user and managed by the Environment model. To ensure that the MDS communication framework is compatible with the standard networking protocols, each drone is assigned a unique 32-bit *Internet Protocol (IP) address*, and every communication message contains a unique 16-bit identifier that is used by the drone communication message router to assign a proper storage location for each message within the incoming message buffer. These messages are time-indexed, and are typically processed by the individual subsystems that know how to handle these specific types of messages. For instance, messages containing chemical sensor measurements are created and processed by the chemical-sensor subsystem.

### 6.4.2.4 Drone Localization Module

The Localization module plays a key role in the MDS; it allows neighboring drones to estimate each other's positions. It is possible to use a GPS device for this purpose, although the lack of precision available to the civilian version of GPS locators (with the typical $\pm 3$ m position error) is likely to be insufficient for the small-scale drone prototypes currently being used in this project. Instead, local-coordinate positioning methods like trilateration (see Chap. 10) or *multilateration* (in three dimensions) are better alternatives, with superior spatial resolution accuracy at the expense of a limited range, which is typically a function of the transmitter power [238]. Such localization approaches yield the positions of nearby drones in the local coordinate system of the vehicle performing the localization. Staying true to the objective of faithful hardware emulation, the MDS drones also utilize a local-coordinate localization method, with the appropriate coordinate transformations performed between the drones' true geo-referenced locations and the local representations of the neighbors' coordinates [39]. In addition, the user may also specify the characteristics of the noise error model—in order to see the impact that reduced localization accuracy has on the physicomimetics controller.

### 6.4.2.5 Drone Environment Sensor Module

In this section we describe the Environment Sensor module, which is used by the drone to sense environmental variables, such as the chemical density in the case of the bioluminescence application. Specifically, the MDS drone Environment Sensor module queries the Environment model for the pertinent reading of a desired quantity (such as chemical density). The drone then computes a normalized vector pointing in the direction of interest (which is a vector sum of the formation and environment forces) to calculate the final control force acting on the drone.

The MDS GUI, shown in Fig. 6.15, provides a range control for the Max Environment Force variable to allow the user to vary the relative magnitudes of the formation and environment forces. Recall that if the environment force is too weak, the formation may never reach the goal; if it is too strong, the formation will break apart, and the drones may move out of the communication or localization range. Once forces are balanced, the task can be easily accomplished. For example, the video "mds.avi" shows 20 drones following a bioluminescent plume while moving toward the goal (shown in red). Figure 6.18 shows four snapshots from the video.

### 6.4.2.6 Drone Control Module

The Control module computes the virtual force vector acting on each physicomimetics drone. This vector is then interpreted by the drone Propulsion module (described in Sect. 6.4.2.7), which navigates the drone. The Control module requires a list of neighbor (i.e., within sensing range) drones' positions in order to estimate the distances to neighbors. The basic physicomimetics

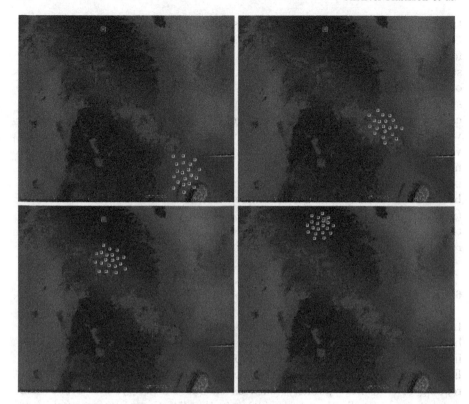

Fig. 6.18: Snapshots of 20 drones following a plume while moving towards a red goal

control principle for hexagonal formations is to repel the drones that are too close, and to attract the drones that are too far [240].

The current version of the MDS implements three distinct force laws for drone formation control: (1) Hooke's spring law, (2) the Lennard-Jones law, and (3) the split Newtonian law. The choice of control law can be altered at run time. As was mentioned in Sect. 6.4.2.1, the crucial **TIME STEP** parameter controls the frequency at which the drone controller updates the virtual navigation force. More frequent updates result in more accurate vehicle motion, allowing the drones to achieve more stable formations (with less error in the separation between the drones). Rapid updates to the control vector will also reduce collisions within a tight or a fast-moving drone swarm formation.

### 6.4.2.7 Drone Propulsion Module

The transition between the virtual forces output by the MDS Control module, and the simulated thrust that ultimately results in the displacement of the simulated drone, occurs inside the Propulsion module, as shown in Fig. 6.17. The MDS implements a model of a dual propeller thruster, as in our hardware drones. (For more about the thrusters in our hardware prototype drones, see Sect. 6.5.) The MDS Propulsion model captures the salient features of the drones' motion, such as the ability to apply thrust in any direction. The Propulsion module also keeps track of a generalized "fuel" resource consumed by the drone, so that the realization of larger control force vectors requires larger amounts of fuel.

### 6.4.3 Using the MDS

Figure 6.15 shows an example of the user interface, which includes a visualization of the drones performing the maximization task, as well as sliders and menus for selecting the values of key parameters. There is also a logging feature, which enables the user to archive system simulation data, and to display real-time performance analyses for the purpose of assisting in the selection and tuning of the physicomimetics parameters. As many as four different performance metrics can be graphed at any one time, for the purpose of multi-objective comparisons for design decisions. The MDS also includes parameterized noise on the sensors and localization. This provides the option of testing the robustness and overall system performance in the presence of noise.

In summary, the MDS provides the key transitional functionality that we require to move from NetLogo to actual hardware drone ASV platforms.

## 6.5 Hardware Drone ASV Platforms

This section concludes our chapter by describing the final transition—to the actual hardware ASV drone platforms, as well as the aquatic test tank and ponds in which the drones have been tested.

### 6.5.1 Hardware Objectives

Eventually our goal is to develop a series of low-cost, robust autonomous underwater vehicles (AUVs), purpose-built for three-dimensional underwater physics-based swarming tasks in the open ocean. As a first step towards that goal, we have endeavored to build a set of simple, low-cost autonomous surface vehicles (ASVs), called "drones" (**d**istributed **r**oving **o**cean **e**xplorers). The drones are intended to demonstrate proof-of-concept for two-dimensional physics-based swarming on the surface of an aquatic environment, and to serve as basic platforms for transitioning our swarm algorithms from simulation into the field.

There are many good reasons to approach such problems with simple surface vehicles initially. First, we keep per-agent costs low and benefit from economies of scale. Second, we keep the problem complexity lower by operating in a two-dimensional environment. Third, we keep our field operations simple, allowing for ease of deployment, observation, and testing in tanks, pools, ponds, and other shallow, calm open water bodies. Fourth, we gain the advantage of access to navigation, localization, control, and communications technologies which utilize radio frequencies (RF). This includes GPS, hobbyist R/C, WiFi, Bluetooth, etc. Radio frequencies do not propagate well through seawater, which means that such technologies become unavailable to us, and the most elementary navigation, localization, and communication tasks become major challenges. In order to focus primarily on demonstrating and refining our swarm algorithms, we have intentionally chosen to utilize surface vehicles for these very reasons.

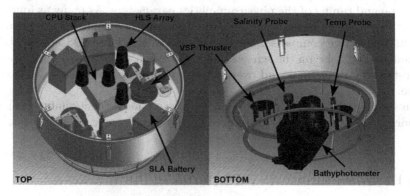

Fig. 6.19: Annotated three-dimensional model of the drone vehicle concept

In order to optimize our drone design for physics-based swarming, and to make it as faithful to our simulations as possible, we wanted to give it the ability to move like a two-dimensional particle. That is, our drone vehicles have no true forward orientation. They are essentially round particles and can move in an X–Y fashion. They possess an additional degree of freedom that our simulated particles do not, that of rotation about their vertical axis (yaw). This is an advantage, but in principle, not necessary. An image of the drone design is shown in Fig. 6.19.

## 6.5.2 Drone Specifications

Each drone is constructed from a round PVC plastic hull approximately 18″ in diameter and 6″ deep, making them portable and lightweight. The vehicle's power is derived from a set of conventional, sealed lead-acid batteries. These batteries are inexpensive, simple to recharge, and much safer in wet environments than other chemistries such as lithium ion.

Fig. 6.20: Drone vehicle prototype

The drone's propulsion is generated by two Voith–Schneider Propellers (VSPs). The VSPs operate as a sort of "water helicopter," allowing the drone to slide over the surface like a particle in the X–Y plane of our simulations. Each thruster consists of a rotor with five perpendicular blades protruding into the water. The VSP rotor is set in motion at a constant rotational velocity by an electric motor. Two control linkages are used to vary the pitch angles of the blades as they cycle through each revolution of the rotor. This allows a thrust vector to be generated in any direction, 360° around the plane perpendicular to the blades. The direction of the linkages (cyclic pitch) determines the direction of thrust, while the magnitude of the displacement of the linkages (collective pitch) along with the rotational speed of the rotor determines the magnitude of the thrust. When the linkages are centered, the net thrust is zero. In theory, only one VSP is needed to move the drone like a particle in the X–Y plane. However, we use a second counter-rotating VSP to counteract the torque induced by the thruster's rotation.

The drone's main "brain" is an ARM-9 processor on a PC/104 form-factor motherboard, running the Linux operating system. For navigation and geo-referencing, we provide a simple low-cost magnetic compass and GPS. As our goal is to investigate the relative movement and formation of the drone swarm, precise geo-referenced navigation is not essential at this point.

For localization of neighboring drones in the swarm, we have developed an RF-based Hyperbolic Localization System (HLS) for multilateration that utilizes an array of four antennas to detect and localize successive radio "pings" from its neighboring drones, based on the differences between the time-of-arrival of the ping at each antenna. This method, co-developed by HBOI and Textron Defense Systems, is influenced by previous work at the University of Wyoming, which utilizes differences between acoustic and RF pings to localize neighbors, as described in detail in Chap. 10.

Fig. 6.21: Drone vehicle in Florida pond

Finally, each drone is outfitted with a set of simple sensors to measure local water temperature, salinity, and bioluminescence. An image of an actual drone prototype is shown in Fig. 6.20.

## 6.5.3 Testing in Florida's Waters

A fleet of seven drone prototype vehicles has been constructed. The vehicles are still under development, but preliminary testing under remote control has been conducted successfully in a large test tank at the Harbor Branch Oceanographic Institute in Florida. The drones have also been tested in ponds near the Florida coast, e.g., see Fig. 6.21. In the future, we intend to continue our work with these vehicles, transitioning our bioluminescence swarm tracing algorithms from theory into practice.

## 6.6 Chapter Summary

In conclusion, this chapter began with the statement of an important real-world problem, and then proposed a physicomimetic solution to the problem that would involve a mobile sensing network consisting of physicomimetic-driven aquatic vehicles called "drones." The vehicles and their bioluminescent environment were first modeled in NetLogo, experiments were run, and lessons were learned. Then the vehicles and environment were modeled again, using a more realistic simulator called the "MDS." It was shown that the MDS provided a level of realism that facilitated the final transition to hardware. Finally, the actual hardware drones were described, along with the aquatic environments in which they operate. This chapter includes the full gamut to be explored by researchers—an application, simulations, empirical experiments, theoretical results that guide the selection of parameter values, and even the hardware platforms that will ultimately perform the tasks in real-world scenarios. In other words, it demonstrates the richness that can be experienced during a scientific journey.

**Acknowledgements** We are grateful to the Office of Naval Research (ONR) for partially supporting this project.

# Chapter 7
# Gas-Mimetic Swarms for Surveillance and Obstacle Avoidance

Wesley Kerr

## 7.1 Introduction

This chapter addresses a challenging coverage task that requires surveillance of a region about which there is no prior knowledge and for which movement is needed to achieve maximum coverage. We assume that there are too few robots with minimal sensing capabilities to continuously monitor the region. Our solution involves developing a novel control algorithm for a swarm of simulated robots to successfully achieve this task. Furthermore, characteristics of aggregate swarm behavior are demonstrated to be predictable using a theoretical physics framework.

The approach described in this chapter is based on the *kinetic theory* of gases, which is a subdiscipline of physics that models particles (atoms or molecules) with kinetic energy and collisions, rather than with potential energy and forces. Chapter 4 provides a brief introduction to kinetic theory (KT) and our agent-modeling approach that is based on KT, and it also provides a scaled-down and simplified version of the software that is used in this project for our KT version of physicomimetics. The swarm control algorithm presented in this chapter is a substantial extension of the algorithm described in Chap. 4 that has been modified to work on realistic simulated robots with limited sensing capabilities and without shared global knowledge. Furthermore, we extend the experiments here to more realistic situations, e.g., by including environmental obstacles, and we compare our KT algorithm experimentally with other algorithms on the task.

This chapter begins with a description of the coverage task. Then, we present our novel approach to solving this problem, founded on physics-based principles of modeling gaseous fluids. Next, physics-based theoretical laws are

Wesley Kerr
Department of Computer Science, University of Arizona, Tucson, Arizona, USA, e-mail: wkerr@email.arizona.edu

W.M. Spears, D.F. Spears (eds.), *Physicomimetics*,
DOI 10.1007/978-3-642-22804-9_7,

derived, which make predictions about aggregate stochastic swarm behavior. In other words, despite the fact that individual swarm members have unpredictable stochastic behaviors, using physics theory the swarm behavior is predictable in the aggregate. These physics laws are tested in simplified settings, and shown to be predictive of simulated swarm behavior. As mentioned in many chapters of this book, swarm predictability is a major concern of many swarm practitioners, and we have endeavored to ensure that it holds for all of our physicomimetic systems. Finally, experimental comparisons between our approach and state-of-the-art alternatives are run on the full simulated task. Our physics-based approach performs competitively, while making fewer assumptions than the alternatives.

## 7.2 The Coverage Task

The surveillance and coverage task being addressed is to sweep a large group of mobile robots through a long bounded region. This could be a swath of land, a corridor in a building, a city sector, or an underground passageway or tunnel. Chapter 12 also addresses the problem of traversing tunnels and other underground passageways. The distinction is that in Chap. 12, the goal is to maximize connectivity between swarm members in a long single-file chain formation.

This chapter focuses instead on maximizing the coverage of both the length *and* the width of the corridor. The goal of the robot swarm is to perform a search, thereby providing maximum coverage of the region. Chapter 11 has a similar goal and employs a similar algorithm to achieve it. The algorithms are similar because they are both stochastic and both are modeled on gas behavior, including the use of the *mean free path*. The difference is that the uniform coverage algorithm is designed to fill arbitrary-shaped regions, whereas the KT algorithm is designed for surveillance of a long corridor. This search might be for enemy mines (e.g., demining), survivors of a collapsed building or, alternatively, the robots might act as sentries by patrolling the area for intruders. All of these different examples require the robots to search the entire region to guarantee coverage.

For practical reasons, the robots in the swarm are assumed to be simple and inexpensive, and they have a limited sensing range for detecting other robots or objects. The robots have no GPS or global knowledge, other than that they can sense a global direction to move, e.g., with light sensors. Robots need to avoid obstacles of any size, possibly the size of large buildings. This poses a challenge because with limited sensors robots on one side of large obstacles cannot visually/explicitly communicate with robots on the other side. Chapter 14 of this book discusses the *obstructed perception* issues that arise when dealing with large obstacles. Because the obstacles considered in this chapter could be the size of large buildings, the issue of obstructed perception is a serious concern for us to address.

It is assumed that the robots need to keep moving because there are not enough of them to simultaneously view the entire length of the region. All robots begin at the corridor entrance, and move to the opposite end of the corridor (considered the "goal direction"). This constitutes a single *sweep*. A sweep terminates when all robots have reached the goal end of the corridor or a time limit is reached. The primary objective of the robots is to maximize the coverage in one sweep; a secondary objective is to minimize the sweep time.

Due to our two objectives, we are concerned with two primary forms of coverage: *spatial coverage* and *temporal coverage*. Two forms of spatial coverage are considered here: *longitudinal* (in the goal direction) and *lateral* (orthogonal to the goal direction). Longitudinal coverage can be achieved by moving the swarm as a whole in the goal direction, and lateral coverage is achieved by a uniform distribution of the robots between the side walls of the corridor. In particular, as robots move toward the goal they increase longitudinal coverage, and as they move or disperse orthogonal to the goal direction they increase lateral coverage. Temporal coverage is determined by the time spent performing a sweep of the corridor. A longer sweep time implies worse temporal coverage. Better temporal coverage can be achieved by increasing the average robot speed, for example.

Note that it is difficult to optimize both temporal and spatial coverage simultaneously, i.e., there is a tradeoff that must be balanced. Optimum spatial coverage takes time, whereas quick sweeps will necessarily miss areas. Poor spatial or temporal coverage is potentially problematic because in this case it is easier for an intruder to remain undetected.

## 7.3 Physics-Based Swarm System Design

Although swarms of insects and flocks of birds are adequate for solving our coverage task, we prefer a physicomimetic over a biomimetic approach. Our rationale for this decision is that although biomimetics is effective for swarm design, it produces systems that are difficult to analyze, as discussed in Chap. 3. Furthermore, many of the biomimetic approaches to coverage rely on the emission and subsequent detection of environmental markers for explicit communication. For example, one biomimetic method (with which we compare in the experiments section, below) is the Ant robot approach of Koenig and Liu [127]. The Ant algorithm is theoretically guaranteed to yield complete regional coverage, but to achieve such coverage, information must be communicated by leaving traces in the environment. This is a practical concern for three reasons: it requires additional sensors and effectors for detecting and emitting traces, it eliminates stealth, and it allows for deception by an adversary.

Our approach does *not* require environmental traces, any prior information about the environment to be surveilled, multi-robot formations, global information (other than the ability to detect the longitudinal direction of the corridor and to recognize that the far end has been reached), a sufficient number of robots or a sufficient sensing range to cover the region statically, or internal maps. Rather than a biomimetic, network-based [10, 158], or decomposition approach [191, 197], our approach was inspired by natural physics-based systems.

In particular, we have designed a swarm of robots to model a fluid composed of particles. Each robot in the swarm adopts the role of a particle in the fluid. Liquids, gases and plasmas are all considered fluids, but in this chapter we focus exclusively on gases. We decided to model a gas because: gases are easily deformed, they are capable of coalescing after separating to go around objects, and they fill volumes. These properties are best understood with a concrete example. Consider releasing a gas that is slightly heavier than the ambient air from a container at the top of a room. The room is filled with many different objects, and as the gas slowly descends towards the floor, it will naturally separate to go around obstacles. Also, it will diffuse—by either smooth or turbulent flow—to cover areas underneath obstacles. Over time, the gas will distribute itself throughout the room. This, in a nutshell, is our solution to the problems of maximum coverage and obstructed perception.

This example highlights the properties of a gas that we would like our robots to mimic. These properties make a physics-based gas model approach well-suited to coverage tasks. However, there is also another compelling reason for adopting a physics-based approach. Physics-based algorithms are easily amenable to physics-based analyses. Whenever a swarm solution is applied to a task, the question arises as to whether the emergent swarm behavior will be predictable, unpredictable or even chaotic. Unfortunately, the vast majority of swarm researchers are unable to answer this question. This is highly troublesome because *predictable aggregate swarm behavior is critical for swarm acceptability*. Therefore, we have turned to particle physics approaches, which have a long history of theoretical analyses and accurate multi-particle predictions.

Our research was also inspired by Jantz and Doty [107], who derived and applied a swarm analysis approach based on kinetic theory (KT). Kinetic theory is a stochastic model of gases that assumes a large number of small particles in constant, random motion. To test their analytical approach, they developed simple multi-agent systems in simulation and on real robots. Their KT-based formulas were quite predictive of robot collision frequency, collision cross section, and the rate of multi-robot effusion through a hole in a wall. Jantz and Doty applied kinetic theory to swarm analysis, rather than system design. Other than our earlier work, e.g., [121, 122, 231], the chapter presented here is the first of which we are aware that has explored kinetic theory primarily for swarm design.

We designed and implemented a physics-based approach based on kinetic theory (KT). The next section describes the details of this approach. KT is considered to be a *physicomimetic*, rather than a biomimetic, approach. Because the swarm is modeled with a physics-based approach, we are able to control, predict and guarantee that the behavior of the swarm will remain within desirable bounds.

## 7.4 Kinetic Theory

There are two traditional methods available for designing an algorithm to model a gas: an *Eulerian approach*, which models the gas from the perspective of a finite volume fixed in space through which the gas flows (which includes *computational fluid dynamics*—see Chap. 8), and a Lagrangian approach (which includes kinetic theory), in which the frame of reference moves with the gas volume [5]. Chapters 8 and 9 adopt the former Eulerian approach. This is because they address the task of localizing the source of a chemical emission, and the robot swarm moves through the chemical gas while the gas moves *around* the robots. On the other hand, in this chapter the robots *are* the gas particles. Because we are constructing a model from the perspective of the robots, we adopt the latter Lagrangian approach.

Before we consider how KT can be used to model robots, we first need to understand how KT is used to model real-world gases. When modeling a real gas, the number of particles (e.g., molecules) is problematic, i.e., in a dilute gas at standard temperature and pressure there are $2.687 \times 10^{19}$ molecules in a cubic centimeter. Because of this, we are unable to model a gas deterministically. Therefore, a typical solution is to employ a stochastic model that calculates and updates the probabilities of the positions and velocities of the particles, also known as a *Monte Carlo simulation*. This is the basis of KT [75]. One advantage of this model is that it enables us to make predictions about the aggregate behavior of the stochastic system, such as the average behavior of the ensemble. The second advantage is that with real robots, we can implement KT with probabilistic robot actions, thereby avoiding predictability of the *individual* robot, e.g., for stealth. It is important to point out that although KT is based on having large numbers of particles (e.g., $10^{19}$) we will demonstrate that the theoretical analysis is quite accurate with only hundreds of robots, and the actual task performance is quite adequate with as few as 10 robots.

In KT, particles are treated as possessing no potential energy, i.e., they are modeled as an ideal gas, and collisions with other particles are modeled as purely elastic collisions that maintain conservation of momentum. The system consists entirely of kinetic energy. KT typically does not deal with forces, which are central to Newtonian physics. Instead, with KT, increased particle movement is driven by collisions and/or a temperature increase.[1]

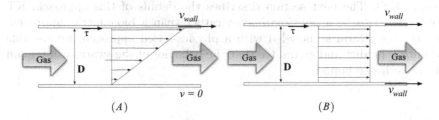

Fig. 7.1: (A) Schematic for a one-sided Couette flow. (B) Schematic for a two-sided Couette flow

Now that we understand KT for real gases, we next address the issue of using KT to model and control robot swarms. Our KT simulation algorithm is a variant of the particle simulations described in [75] and discussed earlier in Chap. 4. We substantially modified the algorithms in [75] to tailor them to simulated robots utilizing only local information.

An ideal model for our task of coverage of a long corridor is the standard model of gas flowing in a *Couette* container [75]. Figure 7.1, from [75], depicts one-sided Couette flow, where a gas moves through an environment between two walls—one wall moving with velocity $V_{wall}$, and the other stationary (and the static external environment is the frame of reference). In this Couette, fluid is assumed to move in the positive $y$-direction (i.e., longitudinally toward the goal end of the Couette corridor), and the positive $x$-direction goes from the stationary wall to the moving wall (i.e., laterally across the Couette corridor).[2] Because the fluid is Newtonian and has viscosity, there is a linear velocity distribution across the system. Fluid deformation occurs because of the *shear stress*, $\tau$, and the wall velocity is transferred (via kinetic energy) because of molecular friction on the particles that strike the wall. (The shear stress is a force parallel to the Couette wall.) On the other hand, the particles that strike either wall will transfer kinetic energy to that wall. This does not cause the wall to change velocity, since in a Couette the walls are assumed to have infinite length and depth and therefore infinite mass. We chose a Couette flow in order to introduce kinetic energy into the system and to give the particles a direction to move.

Our two-dimensional simulated world models a modified (two-sided) Couette flow in which both Couette walls are moving in the same direction with the same speed (see Fig. 7.1). We invented this variant as a means of propelling all robots in a desired general direction, i.e., the large-scale fluid mo-

---

[1] This is *virtual* system heat, rather than an actual heat, designed to increase kinetic energy and thereby increase particle motion.

[2] Note that the diagrams in Fig. 7.1 are rotated 90°.

tion is approximately that of the walls. We will revisit this point in more detail in Sect. 7.4.1.

Robots' velocities are randomly initialized (where initial velocities are directly proportional to a system temperature). These robot velocities remain constant, unless robot–robot or robot–wall collisions occur. (Note that with actual robots, all collisions would be virtual, i.e., robot–robot collisions occur when the robots get too close to each other. The velocities of the walls are also virtual.) The system updates the world in discrete time steps, $\Delta t$. We choose these time steps to occur on the order of the mean collision time for any given robot. Finer resolution time steps produce inefficiency in the simulation, while coarser resolution time steps produce inaccuracies.

Each robot can be described by a position vector, $\boldsymbol{p}$, and a velocity vector, $\boldsymbol{V}$. Because our simulation is two-dimensional, $\boldsymbol{p}$ has $x$- and $y$-components, i.e., $\boldsymbol{p} = \langle x, y \rangle$. At each time step, the position of every robot is reset based on how far it could move in the given time step and its current velocity:

$$\boldsymbol{p} \leftarrow \boldsymbol{p} + \boldsymbol{V}\Delta t .$$

Our KT robot algorithm is distributed, i.e., it allows each robot to act independently. Listing 7.1 shows the main pseudocode for this algorithm. Robots are modeled as holonomic objects with a size of $6''$, speed, and heading. At every time step, $t$, each robot needs to determine an angle, $\theta$, to turn and a distance, $\delta$, to move. At the beginning of the time step, the robot determines its proximity to the goal end of the corridor (e.g., based on light intensity). If its proximity to the goal exceeds a threshold, then one sweep has been completed; if one sweep is the robot's objective, then it halts. If the robot is not at the goal, it will continue moving. In the case of no collisions, the robot maintains its current heading and speed.

However, if there are any imminent collisions, then they affect the robot's turn and distance to move (i.e., speed). There are two types of virtual collisions modeled in the simulator: robot–wall and robot–robot collisions. The former are processed first. For imminent collisions of the robot with a wall, the robot gets a new velocity vector. If the robot–wall collision is "Couette," the (virtual) wall speed, $V_{\text{wall}}$, is added to the $y$-component of the velocity, which is in the goal direction. (Note that $V_{\text{wall}}$ is the wall speed, which is the magnitude of the velocity vector $\boldsymbol{V}_{\text{wall}}$, i.e., $V_{\text{wall}} = |\boldsymbol{V}_{\text{wall}}|$.) It is important to note that due to local sensing constraints, a robot is unable to distinguish between a Couette wall and any obstacle wall that is parallel to it. In other words, when applying this formula, obstacle walls can also increase the velocity of the robots toward the goal. Hence we treat all such walls as Couette walls. Further details are given in Sect. 7.4.1.

Next, robot–robot collisions are processed. The probability of a virtual collision with another robot is based on its proximity to that other robot, and it is independent of the angle between the two robots' velocity vectors. In other words, the two robots may not actually be on a collision course.

**Listing 7.1: KT pseudocode**

```
float distance, turn, vx, vy;
float v_mp   = 1.11777; (T=0.0030) // Most probable velocity
float stdev  = 0.79038;

boolean[] col;
int[] timeCounter;

void move()
    vx = distance; vy = 0;
    incrementCollisionCounters();

    // Read Goal Sensor (Light Sensor)}
    float bearing = sensor.getBearing();

    // Read Robot Localization Sensor
    int[]    robotIds = localization.getRobotIds();
    float[]  robot_d = localization.getDistances();
    float[]  robot_theta = localization.getBearings();

    // Read Sonar Sensors
    float[] sonar_d = sonar.getDistances();
    float[] sonar_theta = sonar.getBearings();

    if ( !robotDirectlyInFront() )
        wall-collisions(bearing, sonar_theta, sonar_d);

    robot-collisions(robotIds, robot_theta, robot_d);
    distance = max(sqrt(vx*vx + vy*vy), 6);
    turn = atan2(vy, vx);
```

The concept of collisions based on proximity is central to the kinetic theory framework. The new velocity vectors that result from this virtual collision are based on a center of mass vector, coupled with a random component. Further details are given in Sect. 7.4.2.

To prevent consecutive virtual collisions with the same robot/wall, collisions between the same entities are only processed again after a predetermined number of time steps have occurred. Because this process could in fact produce robots on actual collision courses, lower-level algorithms ensure that real physical collisions do not occur. The final robot velocity is the vector sum of all velocity vectors. Due to realistic assumptions concerning the maximum speed of a robot, the magnitude of this vector can be no greater than $V_{\max} = 6''$ per second. This KT algorithm halts when a sufficient percentage of the robots have completed a full sweep of the corridor. Figure 7.2 shows an example of a gas dynamically moving down an obstacle-laden corridor. The corridors used in our experiments are $1000'' \times 5000''$.

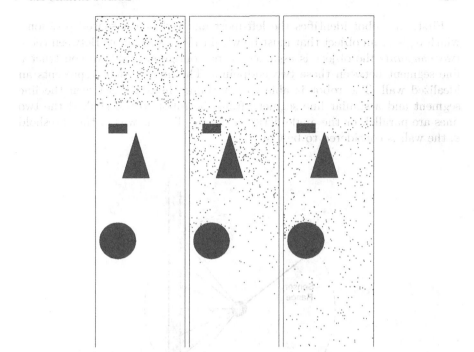

Fig. 7.2: KT-driven robots perform a sweep downward toward the goal. The three snapshots show the swarm at three different moments in time

## 7.4.1 Robot–Wall Collisions

As stated earlier, because robots cannot distinguish Couette walls from obstacle walls that are parallel to them, they assume that *any* wall parallel to the Couette walls (which is also parallel to the goal direction) is moving at speed $V_{wall}$. Therefore, the first portion of the wall collision algorithm is the determination of the angle of the wall with which the robot has virtually collided. If the wall is determined to be approximately parallel to the goal direction, i.e., within $\varepsilon$ degrees, the robot assumes that this wall is a Couette wall and therefore in motion. The user-defined threshold, $\varepsilon$, is based on the type of obstacles expected in the environment.

We model each robot as having a ring of 24 sonar sensors, positioned every 15° around the robot. These sonars are used for wall sensing and have a detection range of 60″. For illustration, Fig. 7.3 shows a potential collision of a robot with a wall. To determine if a wall is Couette, the robot has to determine the orientation of the wall with respect to the goal direction. For this calculation, the robot is only concerned with the sonar sensors in the direction that the robot is facing (between −45° to 45°).

First, the robot identifies the left-most and right-most sensed positions, which register an object that must be within $15''$ of the robot. Between these two *endpoints* the object is assumed to be contiguous, and we construct a line segment between these two endpoints. The line segment represents an idealized wall. The robot is able to determine the angle between the line segment and a similar line segment from the robot to the goal. If the two lines are parallel, or the angle between the two lines is within the threshold $\varepsilon$, the wall is considered to be Couette.

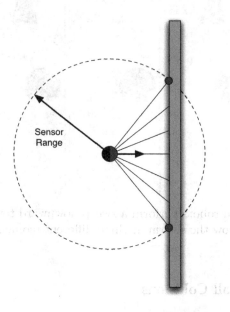

Fig. 7.3: The sonar information used to determine the orientation of a wall. The sonar at $0°$ represents the robot's current heading

Although this process is error prone, it performs well in simulation. If a collision is imminent, then the $x$-component (which is perpendicular to Couette walls) of the robot's new velocity is drawn randomly from a *Rayleigh distribution* (see Chap. 4)[3], and $V_{\text{wall}}$ is added to the $y$-component (which is parallel to Couette walls) of the velocity. The expression "drawn randomly from a Rayleigh distribution" means that the new value of $V_x$ for the $x$-component of the velocity is set to be equal to $v_{\text{mp}}\sqrt{-\ln(U(0,1))}$, where $v_{\text{mp}}$ is the most probable velocity. $U(0,1)$ is a uniformly generated random number between 0 and 1, $v_{\text{mp}} = \sqrt{2kT/m}$, $T$ is the system temperature, $m$ is the particle mass, and $k$ is *Boltzmann's constant* which equals $1.38 \times 10^{-23}$ J/K. The new value of $V_y$ for the $y$-component of the velocity is drawn randomly from a *Gaussian* (bell-shaped) distribution of possible values whose *mean*

---

[3] For three dimensions, Garcia [75] uses a *biased Maxwellian distribution*.

Listing 7.2: KT wall processing pseudocode

```
void wall-collisions(goalBearing, bearings, sonar)
  // If there is not an immediate concern for collision
  //    sonar[6] is the distance at 90 degrees
  //    sonar[18] is the distance at 270 degrees
  if (sonar[0] > SAFE and sonar[6] > 2 and sonar[18] > 2)
    return;

  // is the wall parallel with the goal?
  if (angle(sonar,goalBearing) < epsilon) parallel = true;

  // randomly reset the velocity for the wall collision
  ideal_vx = sqrt(-log(U(0,1))) * v_mp;
  ideal_vy = gaussian() * stdev;

  d    = sqrt(ideal_vx*ideal_vx + ideal_vy*ideal_vy);
  xi   = atan2(ideal_vy, ideal_vx);
  beta = goalBearing - 90 + xi;

  vx = d * cos(beta);
  vy = d * sin(beta)

  // Add in the wall velocity if it is a Couette wall
  if (parallel)
    vx += wall * cos(goalBearing);
    vy += wall * sin(goalBearing);
```

(average) is zero and whose *standard deviation* (i.e., the amount of deviation from the mean) is $kT/m$. Note that the temperature, $T$, plays an important role; increasing $T$ increases the kinetic energy and hence the speed of the particles. The result of the wall collision algorithm is a new velocity vector for the robot in its own local coordinate system. The pseudocode for the wall processing module of the KT algorithm is shown in Listing 7.2.

## 7.4.2 Robot–Robot Collisions

Next, a robot must process (virtual) collisions with other robots. This involves robot localization. Chapter 10 contains the details for a novel robot localization technique utilizing trilateration. We sketch the idea here. Each robot carries an RF transceiver and three acoustic transducers. A robot simultaneously emits an RF and acoustic pulse so that other neighboring robots can localize the emitting robot. The RF is also used for communicating sensor information. By using our trilaterative localization algorithm, a robot can sense the distance and bearing to all other neighboring robots, in its own local coordinate system. To aid in processing robot–robot collisions, two robots, $A$ and $B$, exchange bearing information. Communication is minimal, often just single float values. Exactly what is transmitted is discussed below.

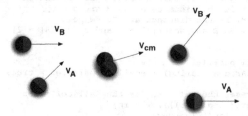

Fig. 7.4: How a robot–robot collision is processed. If two robots are too close, a collision is assumed (left). A center of mass vector is computed, conserving linear momentum (middle). Then the robots receive new stochastically determined velocity vectors (right)

Let us consider the robot–robot collision algorithm in detail. If the distance between two robots $A$ and $B$ is less than 15″ (the "collision zone"), then it is assumed that there will be a robot–robot collision (Fig. 7.4, left). In this case, two steps are performed. In the first step, each robot computes the same *center of mass velocity* as follows:

$$V_{\text{cm}} = \frac{1}{2}\left(V_A + V_B\right) .$$

The center of mass velocity represents the velocity of the two particles if they actually collided (Fig. 7.4, middle). Because conservation of linear momentum holds, this center of mass velocity remains unchanged by the collision, and it forms the basis for generating the new velocity vectors.

The second step involves computing new velocity vectors for each robot after the virtual collision. To do this, each robot first computes the *relative speed* of the other robot with respect to itself, e.g., $V_r = |V_A - V_B| = |V_B - V_A|$. In order to compute the relative speed for robot $A$ it needs to know the bearing from robot $B$ to robot $A$. When the robots separate from their collision, not only should linear momentum be conserved, but so should the relative speeds. The robots exchange bearing information with each other and are able to calculate the other's velocity in their own local coordinate system.

Because kinetic theory is stochastic and does not compute precise collision angles like Newtonian physics does, a post-collision angle, $\theta$, is drawn randomly from a *uniform distribution* (whose values are selected uniformly from zero to $2\pi$). The new relative velocity vector is $V'_r = < V_r \cos(\theta), V_r \sin(\theta) >$. Note that this transformation preserves the relative speed. Finally, at the end of the second step, robots $A$ and $B$ split the difference of the new relative velocity, centered around the center of mass velocity (Fig. 7.4, right). Again, the robots must synchronize in order to select the same $\theta$ and to determine which gets the negative fraction and which gets the positive:

$$V'_A = V_{cm} + \tfrac{1}{2}V'_r \,,$$

$$V'_B = V_{cm} - \tfrac{1}{2}V'_r \,.$$

This collision process is repeated for all robots within $A$'s collision zone. The final resultant velocity of each robot is a vector sum of all the new velocities from every collision. Again, we assume that the robots have low-level routines to prevent *actual* collisions.

Listing 7.3 contains the pseudocode for the robot–robot collision processing module. Each robot is assumed to know its own speed and maintains a new velocity vector that is updated when processing collisions. The pseudocode presumes that only one of the robots calculates the new velocities and relays the information to the other robots. This is just one of several options. An alternative has the robots exchange the randomly selected collision angle, since the remainder of the algorithm is deterministic. This reduces communication while preserving the integrity of the KT algorithm.

To summarize this section on kinetic theory, it is important to point out that all of these collisions (i.e., robot–wall, robot–obstacle and robot–robot) serve to generate a very dynamic swarm system that is in constant motion both laterally and longitudinally in the corridor. Furthermore, note that these omnidirectional post-collision particle trajectories remediate the "obstructed perception" problem. Sometimes (e.g., see later in this chapter) the dominant bulk swarm movement may favor one direction over another, but because the post-collision velocities are stochastic there will always be some movement in all directions through the corridor. This is precisely what we desire for effective surveillance.

## 7.5 Theoretical Predictions

One of the key benefits of using a physics-based multi-robot system is that extensive theoretical (formal) analysis tools already exist for making predictions and guarantees about the behavior of the system. Furthermore, such analyses have the added benefit that their results can be used for setting system parameters in order to achieve desired multi-robot behavior. The advantages of this are enormous—one can transition directly from theory to a successful robot demo, without all the usual parameter tweaking. To demonstrate the feasibility of applying physics-based analysis techniques to physics-based systems, we make predictions that support some of our claims regarding the suitability of gas models for our coverage task.

Listing 7.3: KT robot collision processing pseudocode

```
void robot-collisions(ids, bearings, distances)
    for (id in ids)
        if ( distanceToRobot(id) > SAFE) continue;
        if ( alreadyCollided(id) ) continue;

        // Compute id's velocity in our coordinate system.
        bearingToMe = getReverseBearing(id);
        thetar = 180 + bearings[id] - bearingToMe;

        c_speed = getSpeed(id);
        c_vx = c_speed * cos(thetar);
        c_vy = c_speed * sin(thetar);

        // The robot moves along its x-axis
        relative_s = sqrt((c_vx - speed) * (c_vx - speed) + (c_vy * c_vy))

        center_vx = 0.5 * (c_vx + speed);
        center_vy = 0.5 * (c_vy);

        // randomly choose a collision angle
        cos_theta = 1.0 - 2.0 * U(0,1);
        sin_theta = sqrt(1 - cos_theta * cos_theta);

        vx1 = center_vx + 0.5 * (relative_s * cos_theta);
        vy1 = center_vy + 0.5 * (relative_s * sin_theta);

        vx2 = center_vx - 0.5 * (relative_s * cos_theta);
        vy2 = center_vy - 0.5 * (relative_s * sin_theta);

        // choose the assignment of v1 and v2 so that it
        // doesn't result in an actual collision
        vx += vx1; vy += vy1;

        // Relay the information to the other robot.
        id_speed = sqrt(vx2 * vx2 + vy2 * vy2);
        id_angle = atan2(vy2, vx2);
        send(id, id_speed, id_angle);
```

Recall that our objectives are to sweep a corridor and avoid obstacles while maximizing coverage. We utilize our previous definitions of spatial coverage: *longitudinal* (in the goal direction) and *lateral* (orthogonal to the goal direction). Longitudinal coverage can be achieved by movement of the swarm in the goal direction; lateral coverage can be achieved by a uniform spatial distribution of the robots between the side walls. The objective of the coverage task is to maximize both longitudinal and lateral coverage in the minimum possible time, i.e., to achieve maximally time-efficient coverage as well.

To measure how well the robots achieve the task objective, we examine the following three metrics:

1. The degree to which the **spatial distribution** of the robots matches a uniform distribution. This is a measure of lateral coverage of the corridor and provides confirmation of the equilibrium behavior of KT.
2. The **average speed** of robots (averaged over all robots in the corridor). This is a measure of all three types of coverage: lateral, longitudinal, and temporal. Velocity is a vector with speed and direction. The type of coverage depends on the direction. To control the average swarm speed, one can directly vary the value of the system temperature, $T$. Therefore, our experiment explores the relationship between $T$ and average speed.
3. The **velocity distribution** of all robots in the corridor. This is a measure of longitudinal spatial coverage, as well as temporal coverage. For example, consider the one-sided Couette in Fig. 7.1 again, and in particular focus on the line representing the velocity distribution. The slope of that line is inversely proportional to the longitudinal spatial coverage (and the temporal coverage). In other words, for a given Couette diameter (width), $W$, if you increase the wall speed, $V_{wall}$, then the slope will be reduced and the longitudinal and temporal coverage will increase. Below, we run an experiment to examine the relationship between the wall speed, $V_{wall}$, and the velocity distribution in one-sided and two-sided Couettes. The intention is to enable the system designer to select a wall speed for optimizing this swarm velocity distribution.

The error between the theoretical predictions and the simulation results, denoted *relative error*, is defined as:

$$\frac{|\, theoretical - simulation \,|}{theoretical} \times 100 \,.$$

The theory presented in this chapter assumes simplified two-dimensional environments with no obstacles. To develop a theory for the full task simulation would require extensive theoretical physics analyses. That is beyond the scope of this chapter, but will be addressed in the future.

Although the theory assumes no obstacles, we varied the obstacle density to range from 0% to 50%, in order to determine the sensitivity of the theory to the obstacle density. Surprisingly, some of the theory holds well, despite the presence of obstacles.

There are two complementary goals for running these experiments. The first goal is to determine how predictive the theory is (see [231] for the derivations of all of our predictive theoretical formulas). The second goal is to determine the relationship between parameter settings and system behavior. If a system designer understands this relationship, he/she can more easily set parameters to achieve optimal performance. Finally, and most importantly, the reason why these two goals are complementary is that if the theory is predictive of the system simulation behavior, then future system designers no longer need to run the simulation to determine the optimal parameter settings—graphs generated from theory alone will suffice. This can greatly reduce the time required for system design and optimization.

We ran three sets of experiments, in accordance with the metrics defined above. For each experiment, one parameter was perturbed and eight different values of the affected parameter were chosen. For each parameter value, 20 different runs through the simulator were executed, each with different random initial robot positions and velocities. The average simulation results and relative error (over the 20 runs) were computed and graphed.

## 7.6 Experiment 1: Spatial Distribution

The first experiment examines the lateral spatial coverage provided by KT. Robots are placed in a square container in an initially tight Gaussian distribution and allowed to diffuse to an equilibrium. During this experiment, there is no goal direction.

The purpose of this experiment is to confirm the theoretically expected behavior of a KT system in equilibrium, and verify the correctness of our implementation. The KT gas model predicts that, upon reaching equilibrium, the particles will be spatially uniformly distributed. To confirm this prediction, we divided the square container into a two-dimensional grid of cells. Theory predicts that there should be (on average) $N/C$ robots per cell, where $N$ is the total number of robots and $C$ is the total number of grid cells. We ran the simulation with six obstacle densities, ranging from 0% to 50%, to determine the sensitivity of the spatial distribution to obstacle density.

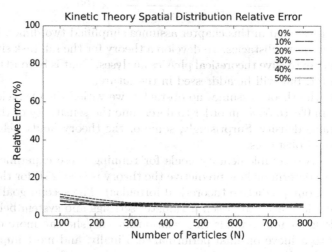

Fig. 7.5: Relative error for the KT spatial distribution

Figure 7.5 shows the results for the KT simulation. We varied the number of particles from 100 to 800, and for each setting of the number of particles we performed 20 simulation runs to generate an average relative error score. Note that despite the introduction of as much as a 50% obstacle coverage, we can still predict the spatial distribution with a relative error of less than 10%, which is very low. Furthermore, the error is even lower once the number of robots is about 300. This is a surprising result considering that 300 robots is far less than the $10^{19}$ particles typically assumed by traditional kinetic theory.

## 7.7 Experiment 2: Average Speed

For the second experiment, we examine the average speed of the robots in the system. We assume a Couette environment, but with no goal force or wall movement, and therefore no externally-directed bulk transport of the swarm. However, the robots will increase in speed if there is an increase in the system temperature, which causes an increase in kinetic energy. For example, the system temperature in the simulation directly affects the initial agent speeds and the new agent velocities after collisions, both of which are all drawn from a distribution that is a function of Boltzmann's constant, the temperature, and the agent mass.

The objective of this experiment is to examine the relationship between the temperature, $T$, and the average speed of the robots. The average robot speed serves as a measure of how well the system will be able to achieve complete coverage—because higher speed translates to greater lateral and/or longitudinal coverage, depending on the velocity direction. This experiment also serves to verify that our simulation code has been implemented correctly. Note that not all applications will require maximum coverage; therefore, we want to study the general question of precisely how specific choices of speed affect coverage.

Our predictive formula for the two-dimensional simulation is (see [231] for the complete derivation):

$$\langle V \rangle = \frac{1}{4}\sqrt{\frac{8\pi kT}{m}},$$

where $k$ is Boltzmann's constant ($1.38 \times 10^{-23}$ J/K), $m$ is the robot mass (assumed to be one), and $T$ is the initial system temperature (which is a user-defined system parameter analogous to real temperature).

This theoretical formula is compared with the actual average speed, $\langle V \rangle$, of the robots in the simulation, after the system has converged to the equilibrium state assumed to have occurred after a 1,000 time steps. There are 300 robots in the simulation. Because temperature affects speed, temperature is varied from 50° Kelvin to 400° Kelvin. We ran the simulation with six obstacle densities ranging from 0% to 50%, in order to determine the sensitivity of the average speed to obstacle density. For each setting of the

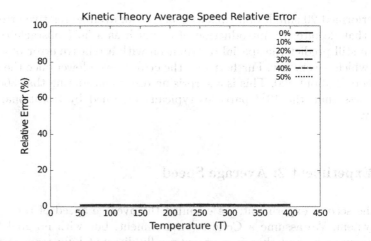

Fig. 7.6: Relative error for the KT average speed

parameters we performed 20 simulation runs additionally varying the initial starting locations and starting velocities.

The results are striking, as can be seen in Fig. 7.6. The difference between the theoretical predictions of the average speed and the simulated average speed results in less than a 2% error, which is very low considering that as much as a 50% obstacle coverage has been introduced. Finally, note that we can use our theory to not only predict swarm behavior, but also to *control* it. Specifically, by setting $T$, a system designer can easily achieve the desired average speed.

Of course, the theory can be used in the reverse direction. In this case, we knew what average robot speed we would like based on the movement capabilities of our robots. Then we solved for the temperature setting that would achieve the desired speed. The code in the robot simulator used $T = 0.0030°$ Kelvin (Listing 7.1), based on the robot capabilities.

## 7.8 Experiment 3: Velocity Distribution

The third experiment concerns longitudinal coverage and sweep speed via movement. Kinetic theory predicts what the velocity distribution will be for robots in a Couette. This theoretical prediction is compared with simulated behavior. Again, fluid flow is in the $y$-direction, which is toward the goal. The $x$-direction is lateral, across the corridor. This experiment examines the relationship between the wall speed, $V_{\text{wall}}$, and the velocity distribution of the robots in the system.

We examine a subtask in which a traditional one-sided Couette flow drives the bulk swarm movement. Our predictive formula is (see [231] for the complete derivation):

$$V_y = \frac{x}{W} V_{\text{wall}} ,$$

where $V_{\text{wall}}$ is the speed of the Couette wall, $x$ is the lateral distance from the stationary wall, and $W$ is the Couette width. In other words, the theory predicts a linear velocity distribution.

We set up an experiment to measure the relative error between theory and simulation, consisting of 300 robots. The corridor is divided into eight discrete longitudinal subcorridors. Theory predicts the average speed in the goal direction along each of the subcorridors. This experiment measures the $y$-component of velocity averaged across the robots within a subcorridor. We computed the relative error for each subcorridor. The overall relative error is averaged across all subcorridors for eight different wall speeds and six different obstacle percentages. This experiment was repeated 20 times for each different wall speed and obstacle percentage. The results are plotted in Fig. 7.7.

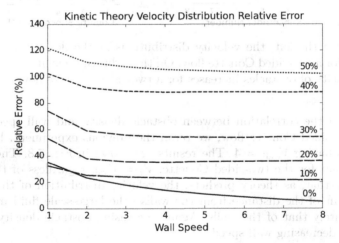

Fig. 7.7: Relative error for the KT velocity distribution

Although the error is reasonably low for 0% obstacles and high wall speeds, the error increases dramatically as obstacles are added. So, why is there a discrepancy between the theory and experimental results? The reason is that theory predicts a certain linear velocity distribution, but assumes that there are no obstacles. For simplicity, the theory assumes that robots never move backward (back up the corridor). In the simulator, on the other hand, robots *do* move backward, regardless of whether or not there are obstacles—because the simulation has a random component. In fact, as obstacles are introduced into the simulated world, the frequency of backward moving robots increases substantially.

To examine more closely the effect that obstacles have, Fig. 7.8 shows the velocity distributions themselves (where the wall speed $V_{wall} = 4$). Even with no obstacles, the maximum robot speed does not quite reach 4.0 (and we would expect 3.75 in the subcorridor nearest to the wall). This is caused by the backward moves. What is interesting is that the velocity distributions remain linear up to a reasonable obstacle density (30%), while the slope decreases as obstacles are added. Adding obstacles is roughly equivalent, therefore, to reducing the wall speed $V_{wall}$!

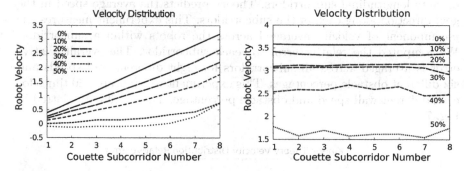

Fig. 7.8: On the left, the velocity distributions as the density of obstacles increases for a one-sided Couette flow. On the right, the velocity distributions as the density of obstacles increases for a two-sided Couette flow

To see if the correlation between obstacle density and wall speed holds in the two-sided Couette flow, we reran the previous experiment, but with *both* walls having $V_{wall} = 4$. The results are shown in Fig. 7.8. The theory predicts that for the two-sided Couette, $V_y = V_{wall}$ regardless of the value of $x$. Note that, as theory predicts, the velocity distribution of the flow is independent of the distance from the walls—the large-scale fluid motion is approximately that of the walls. Again, increasing obstacle density is very similar to decreasing wall speed.

In conclusion, without obstacles, the theory becomes highly predictive as the wall velocity increases. Furthermore, this very predictive theoretical formula can also be used to achieve a desired swarm velocity distribution, i.e., to *control* the swarm—simply set the value of $V_{wall}$, the virtual wall speed, to achieve the desired distribution, using the formula. On the other hand, with an increasing number of obstacles, the predictability of the theory is increasingly reduced. However, we have shown that (up to quite reasonable densities) the increase in the number of obstacles is roughly proportional to a reduction in wall speed, $V_{wall}$. Thus, there is a quite reasonable expectation that this effect can be captured theoretically. We intend to explore this in our future work.

## 7.9 Summary of Theoretical/Empirical Comparisons

With respect to the average speed and the spatial distribution, the empirical results are extremely close to the theoretically derived values. Although the theory was not derived with obstacles in mind, obstacles have almost no effect on the results. The errors are generally less than 10%. Furthermore, this level of theoretical accuracy is achieved with only hundreds of robots, which is quite small from a kinetic theory perspective.

Our results for the velocity distribution are acceptable for no obstacles, generally giving errors less than 20%. As the obstacle density increases, so does the error. However, we have shown that an increase in obstacle density changes the slope of the linear velocity distribution. This is roughly equivalent to a commensurate reduction in wall speed.

In summary, we can conclude that when the actual scenario closely coincides with the theoretical assumptions (e.g., few obstacles), the theory is highly predictive. Also, we have provided an important insight into the nature of the effect that obstacle density has on the system. The most important conclusion to be drawn from this is that in the future we can largely design KT systems using theory, rather than computationally intensive simulations, for the selection of parameter settings. A subsidiary conclusion is that we have verified the correctness of our swarm code, which is something quite straightforward for a physicomimetic approach but much more difficult for alternative approaches.

## 7.10 Performance Evaluation Algorithms

The next several sections describe the performance evaluations of our KT controller on the full coverage task described in Sect. 7.2. In order to provide a comparison, we have implemented two swarm algorithms. The "Default" controller is our baseline and represents the minimal behavior necessary for solving the task. The other controller is the biomimetic controller "Ant." Ant provides an upper bound on spatial performance because it is known to provide complete spatial coverage.

### 7.10.1 Default Controller

The first algorithm we compare against is a baseline controller. The Default controller attempts to exploit known information about searching "shadow" areas—regions just downstream (i.e., toward the goal direction) of obstacles. The robots used by Default are nearly the same as those used by KT. The primary difference is that Default robots have only eight possible moves (for algorithm simplicity), in 45° increments, as shown in Fig. 7.9*A*. The controller weights each move as more or less desirable based on the information known to the robot. The weights provide a bias to maintain the same direction, move toward the goal, and exploit shadow areas. We designed Default in this fashion because we believe that the shadow areas represent the hardest areas to cover.

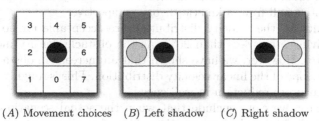

(*A*) Movement choices     (*B*) Left shadow     (*C*) Right shadow

Fig. 7.9: (*A*) The Default controller choices for moves. (*B* and *C*) Shadow preferences for the Default controller. The center dual-shaded (black and gray) circle is the current position of the robot. The dark gray squares represent walls or obstacles. The light gray circle represents the preferred choice of move to make

The Default algorithm stochastically determines the next move by first sensing the environment. The robot assigns a weight to each available move, and then decides to continue along the same path 80% of the time if the move is available. It then divides the remaining weight into three parts. Two of the three parts go to shadow areas (see Fig. 7.9) if they are available. The remaining one-third is again split two ways. Half of this weight goes to the goal direction and the other half is divided evenly between all the open moves. The sum of these weights is equal to one. The robot's next move is a weighted random choice.

## 7.10.2 Ant Controller

To experimentally compare our KT algorithm against what appeared to be the best performing and most appropriate alternative found in the literature, we duplicated the Ant algorithm designed by Koenig, Szymanski, and Liu [128]. This algorithm relies on graph theoretical proofs to guarantee complete coverage. It assumes *stigmergy*, i.e., robots that can detect and leave *pheromones* (odors) in the environment. It is important to note that the Ant robot described here has an explicit advantage over our other algorithms, because it is able to interact with its environment.

The Ant algorithm determines its next move by investigating the areas surrounding itself in four directions: north, east, south, and west. After each direction has been examined, the direction with the minimum pheromone value is chosen as the direction to move. The robot then updates its current position before repeating the process [128].

## 7.11 Performance Evaluation: Experimental Setup

To better understand the performance of our algorithms, we ran two sets of combinatorial experiments, varying the obstacle percentage in one set of experiments and varying the number of robots in the second set of experiments. For all of the combinatorial experiments, we compared all algorithms discussed above. To avoid unfairness in the comparison, the magnitude of the velocity of all robots is fixed to the same value, for all algorithms. First, let us formalize the problem for each of the experiments, and then define how the parameters were varied.

**Measurements:** The simulated world was divided into $50 \times 250$ cells. We applied the following performance metrics:

1. $w$, the **sweep time**, which measures the number of simulation time steps for *all* robots to get to the goal location. These experiments assume only one sweep. Sweep time is a measure of temporal coverage.
2. $p_c$, the **total spatial coverage**, which is the fraction of all cells that have been visited at least once by at least one robot, over one sweep.
3. $p_s$, the **shadow coverage**. This is the same as $p_c$, except that we take this measure over a small region downstream (with respect to the goal location) of the obstacles (see Fig. 7.10). The shadow region is the hardest to cover.

**Experiment 1:**

- $\Psi$, a bounded two-dimensional spatial corridor whose length is much greater than its width ($1000'' \times 5000''$).
- $\Gamma$, a swarm of 25 simulated robots.
- $\Lambda$, a set of randomly placed obstacles, varying from 0% to 40% coverage.

We varied the percentage of obstacles, $\Lambda$, to determine how this affects coverage. We generated five corridors for each obstacle percentage. Each individual experiment consisted of 10 trials through each of the five corridors, with random initial locations and orientations of the robots. The mean performance with 95% confidence intervals for each metric was graphed. Since each algorithm is not guaranteed to have all robots finish the obstacle course, we limited the total number of time steps to 150,000.

**Experiment 2:**

- $\Psi$, a bounded two-dimensional spatial corridor whose length is much greater than its width ($1000'' \times 5000''$).

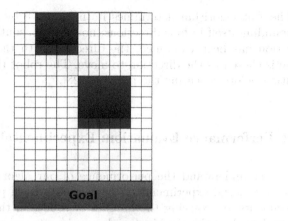

Fig. 7.10: Measurements taken by the simulation. Obstacles are dark gray. Free space is white. The light gray squares depict the shadow region

- $\Gamma$, a swarm of $N$ simulated robots, varying from five to thirty robots.
- $\Lambda$, a set of randomly placed obstacles at 30% coverage.

In the second set of combinatorial experiments we varied the number of robots in order to see how coverage is affected when robots fail or are not available. Similarly to experiment 1, we generated five corridors for each different number of robots, $N$. Each individual experiment consisted of 10 runs through each of the five corridors, with random initial locations and orientations of the robots. Again, we limited the total time to 150,000 steps.

## 7.12 Performance Evaluation: Experimental Results

For ease of presentation, we organize and present our results categorized according to the performance metric, instead of the combinatorial experiment. This allows us to look at different variations of parameter settings for each algorithm and how these settings affect specific metrics.

### 7.12.1 Temporal Coverage ($w$) Results

The first form of coverage the experiments measured was temporal coverage, $w$, also known as the sweep time. The results can be seen in Fig. 7.11.

**KT:** The KT algorithm was able to complete the task consistently, and in few time steps. As obstacles were added to the corridor, KT began to slow down. Although navigation became much tougher, KT's temporal coverage

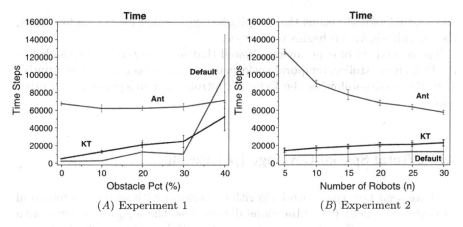

Fig. 7.11: Temporal coverage

performance degraded gradually. This can be explained as an emergent behavior of the KT system. Areas of the corridor that require backtracking in order to successfully reach the goal have plagued prior solutions to the coverage task. These areas, referred to as obstacle "wells," or box canyons, are areas surrounded by three walls where the only direct path to the goal is blocked. KT has an inherent behavior (i.e., occasional stochastic backward moves) that allows it to escape these wells rather quickly.

For the second set of experiments, robots were added. As robots were added to the corridors, KT was able to continue to finish in roughly the same time. This is intuitively reasonable—with such a small number of robots, interactions are quite rare. The robots are almost independent of one another.

**Ant:** The Ant algorithm extensively explored all areas. This caused Ant to take much more time than KT. As obstacles were added to the world, they did not appear to affect Ant. This is understandable, since obstacles just cover additional cells and Ant did not need to visit those cells.

In the second experiment, we varied the number of robots. In this case, the time required for Ant to complete one sweep decreased. Ant is the only algorithm studied for which we observed this phenomenon, but it is easily explained. The algorithm visits those surrounding cells that have been visited the least. As more robots are added, the surrounding cells are given pheromones more quickly, causing the individual robots to be able to explore faster. Robots interact with each other through the environment. Communication via stigmergy allows the robots in the Ant algorithm to avoid multiple visits to the same cell.

**Default:** Our baseline controller actually performed the best on temporal coverage, up to a point. Once the obstacle percentage passed 30%, the Default controller quickly dropped in performance. At 40% obstacle coverage, the corridors have many obstacle wells and are difficult to navigate. By making purely stochastic decisions that emphasize exploration in the shadow areas, the Default algorithm begins to degrade.

The second set of experiments showed that as more robots were added to the Default controller, temporal performance was unaffected. This is intuitive since the robots following the Default controller act independently of other robots.

## 7.12.2 Total Spatial Coverage ($p_c$) Results

We have seen how the control algorithms perform with respect to temporal coverage. The next two subsections discuss how the algorithms fared with respect to two different measurements of spatial coverage. The first measurement is the total spatial coverage. 100% coverage may not be achieved, because the robots are shut down once they pass the goal line. This means that there are areas that may not be explored. These spaces could be explored on a second sweep, but for our experiments we are only concerned with a single sweep. The results for $p_c$ can be seen in Fig. 7.12.

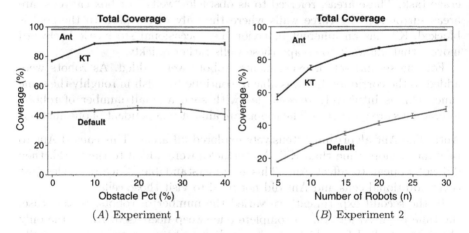

Fig. 7.12: Total spatial coverage

**KT:** Inspecting Fig. 7.12, it can be observed that as the obstacle percent increased, KT's performance improved by several percentage points, slowly tapering off as the corridors reached 40% obstacle coverage. As the obstacle percentage increased, the number of collisions with obstacles also increased. This slowed the robots' longitudinal movement (in the goal direction) and increased the lateral movement (orthogonal to the goal). This increase in the lateral movement increased spatial coverage. The slowing in the longitudinal direction can also be seen in Fig. 7.8. Recall that we already explained this relationship between obstacles and longitudinal velocity in Sect. 7.8, where we presented the velocity distribution experiment.

The second graph shows that as more robots were added, KT's coverage increased. This is as expected, and we can see that with only 30 robots, KT was able to achieve roughly 90% coverage, which is close to the success of the Ant algorithm. This was even accomplished without leaving pheromones in the environment and was done in roughly one-third the time.

**Ant:** The Ant algorithm is known to provide a complete solution to the task problem. Therefore, it is no surprise to see that for both sets of experiments Ant achieved maximum coverage.

**Default:** Our final algorithm was supposed to show the baseline performance. The Default controller attempts to combine greedy exploitation of shadow regions with exploration, in a hand-crafted manner. This algorithm performed surprisingly well, covering roughly 45% of the world, regardless of the number of obstacles.

The second set of experiments provided no new insights on Default. The results improved monotonically as more robots were added. Since the robots' decision making ability is independent of other robots in Random, this is not surprising.

## 7.12.3 Shadow Coverage ($p_s$) Results

The final measurement is shadow coverage. This is also a measure of how well the controller is able to explore. If the controller always goes directly to the goal, it will achieve a 0% shadow coverage. If the controller is optimal at exploration, then it will achieve a 100% shadow coverage. Note that greater shadow exploration affects not only shadow coverage, but it also affects total corridor coverage.

**KT:** The KT algorithm performed very well on shadow coverage, which is especially important for effective task achievement, as can be seen in Fig. 7.13. Regardless of the number of obstacles, KT was able to obtain roughly 80% shadow coverage. This implies that KT has a very good balance between exploitation and exploration. Good performance on the shadow coverage task is an emergent property of the KT system. There is nothing explicit in the KT algorithm for addressing shadow coverage.

As robots were added to the system, the performance of KT also improved. An implication of this result is that if robots begin to fail, then the performance of KT will also degrade gradually, which is a desirable form of system robustness.

**Ant:** The Ant controller performed exactly as designed and achieved 100% coverage behind obstacles, regardless of the number of robots or the number of obstacles in the environment.

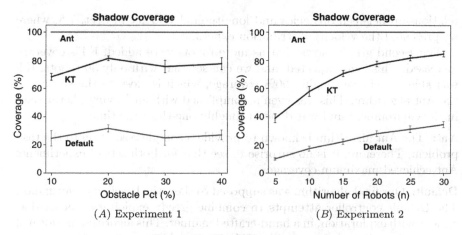

Fig. 7.13: Shadow coverage

**Default:** The Default controller provided us with a usable baseline. With random search, we were able to achieve roughly 20% shadow coverage. Clearly, the other algorithms outperform random search.

Our baseline algorithm's performance was improved when extra robots were added to the environment. This was another expected result.

## 7.13 Performance Evaluation: Conclusions

Several important results have been presented. First, the experimental results showed that Ant is superior in terms of spatial coverage. This superiority comes with numerous tradeoffs, however. First, Ant is much slower in terms of sweep time. Second, pheromones are required. Finally, the movements of the individual Ant robots are quite predictable. The latter two features imply that the Ant robots are susceptible to adversarial deception, lack of stealth, and interception.

Second, we showed the applicability of KT to the coverage task. KT was able to provide excellent coverage *without* leaving pheromones in the environment. Furthermore, if the coverage task requires multiple sweeps, KT clearly would have the advantage, since the sweep time is much less. Additionally, the lack of predictability of individual robots using KT makes it hard for an adversary to anticipate future movements.

## 7.14 Summary and Future Work

In this chapter we discussed a physics-based algorithm for controlling swarms of autonomous robots that is designed for a challenging coverage task. Our algorithm requires limited sensors and communication between robots, and only a small amount of global knowledge. The algorithm was inspired by particle motion in a gas, which is a highly effective mechanism for expanding to cover complex environments in the natural world. Like the natural gases from which it was inspired, our KT algorithm is efficient, effective and robust at maximizing coverage. It also has the advantages of stealth, very good noise tolerance [231] and predictability in the aggregate. Furthermore, because our KT controller is physics-based, behavioral assurances about the swarm are easily derived from well-known theoretical laws, e.g., [75]. These theoretical laws allow system designers to avoid costly trial-and-error because system parameters for optimal performance can be set using the theory alone.

When experimentally compared against alternative algorithms designed for coverage tasks, KT has proven to be competitive. It provides excellent coverage in little time and is exceptionally robust. It does not provide quite as much spatial coverage as the Ant algorithm, which is the best known competitor to date, but it provides much better temporal coverage and also does not have the explicit pheromone trail requirements of Ant. Therefore, KT is better-suited than Ant for situations where inexpensive robot platforms, stealth and/or speed of coverage are required.

Work will continue on developing the theoretical analysis and will move toward implementing KT on actual robotic platforms. Chapter 10 shows that all of the sensors necessary for KT to run on actual robots are available. Finally, further parallel theoretical and simulation developments will lead to more efficient, effective, and predictable physicomimetic swarms for practical tasks.

# Chapter 8
# A Multi-robot Chemical Source Localization Strategy Based on Fluid Physics: Theoretical Principles

Diana F. Spears, David R. Thayer, and Dimitri V. Zarzhitsky

## 8.1 Introduction

The *chemical source localization*, or *chemical plume tracing* (CPT), task is a search and localization problem in which one must find locations where potentially harmful substances are being released into the environment. The ability to quickly locate the source of such a chemical emission is fundamental to numerous manufacturing and military activities. In light of the current international concerns with security and the possibility of a chemical terrorist attack, many private and government agencies, including the Japan Defense Agency (JDA), the US Department of Homeland Security (DHS), the US Defense Advanced Research Projects Agency (DARPA), the Danish Environmental Protection Agency (EPA), the UK Health Protection Agency (HPA), and local emergency response agencies throughout the world have expressed interest in updating techniques used to track hazardous plumes and in improving the search strategies used to locate an active chemical toxin emitter [17, 87, 220]. Minimization of human exposure to the toxin is, of course, an objective. In addition to airborne chemicals, there is also an interest in locating deep sea hydrothermal vents by tracing their chemical plumes [63]. As a result of the variety of chemical plume problems to be tackled, there is a growing interest among both researchers and engineers in designing robot-based approaches to solving this problem.

Diana F. Spears
Swarmotics LLC, Laramie, Wyoming, USA, e-mail: dspears@swarmotics.com

David R. Thayer
Physics and Astronomy Department, University of Wyoming, Laramie, Wyoming, USA, e-mail: drthayer@uwyo.edu

Dimitri V. Zarzhitsky
Jet Propulsion Laboratory, California Institute of Technology, Pasadena, California, USA, e-mail: Dimitri.Zarzhitsky@jpl.nasa.gov

W.M. Spears, D.F. Spears (eds.), *Physicomimetics*,
DOI 10.1007/978-3-642-22804-9_8,
© Springer-Verlag Berlin Heidelberg 2011

Prior research on the topic of CPT has been predominantly biomimetic in nature. The purpose of this chapter is to present a physicomimetic alternative, and to compare it with the leading biomimetic approaches to the problem. Our physicomimetic CPT algorithm, called *fluxotaxis*, is based on principles of *fluid dynamics*, which is defined as the study of the properties and behavior of moving fluids. The name *"flux*otaxis" was chosen because robots that apply this algorithm follow the chemical mass *flux* within the fluid. Flux will be defined shortly, but for now you only need to know that it is a fluid property. A "taxis" is a locomotor response toward or away from an external stimulus; in this case, the stimulus is the flux. Using fluxotaxis, robots can stay in formation while localizing the source of a chemical emission. This is ideal for solving the CPT task. In particular, using virtual cohesion forces, the robots can achieve lattice (grid) formations. These lattice formations are ideal for computing derivatives, which are central to most CPT navigation strategies. Other virtual forces attract the lattice toward desirable regions in the environment, such as chemical mass flux. This physicomimetic approach to multi-robot CPT exploits fundamental principles of fluid dynamics to drive a swarm lattice to a chemical source emitter. In summary, the robots are driven by the resultant vector of two virtual forces—an inter-robot cohesion force, and an environmental source localization force. The outcome is that the robots form a mobile computer that stays in a regular grid formation, while using the grid for computing spatial derivatives that drive it in the correct direction toward the chemical source emitter. This approach is extremely well-suited to large swarms of robots; in particular, the greater the number of robots, the greater the spatial extent explored by this distributed differential computer, and the more successful the approach becomes.

This book chapter is organized as follows. Section 8.2 provides the necessary background material for reading this chapter. It presents the fluid dynamics basics for the reader who is unfamiliar with this topic, describes the CPT task and its subtasks, and explains how swarms are applicable to fluid-based problems such as CPT. Section 8.3 then presents prior research on the CPT problem, with a focus on biomimetic approaches, which have dominated the literature. After this background, Sect. 8.4 presents our fluxotaxis algorithm, along with its motivation and theoretical underpinnings. The previous background on fluid dynamics enables the reader to understand the fluxotaxis algorithm at this point in the chapter. The remainder of the chapter presents intuitive theoretical comparisons between the leading biomimetic CPT algorithms and fluxotaxis, followed by a summary and concluding remarks.

## 8.2 Background

This section provides background material on fluid dynamics, the CPT task, and the application of swarms of robots to this task.

### 8.2.1 Brief Overview of the Chemical Source Localization Task

The objective of the chemical plume-tracing task is rapid localization of an emitter that is ejecting the chemical. Plume *tracing* is different from plume *mapping*—the objective of the latter is to acquire a map of the plume [228]. In CPT, complete mapping is typically unnecessary, i.e., only the source emitter must be found. The underlying motivation for CPT is typically to minimize human exposure to the chemical toxin; therefore, the use of autonomous robots is especially appropriate. It is difficult to predict ahead of time the conditions that the plume-tracing robots will encounter at the site of the contamination—it may be an urban setting after a deliberate terror act, or an industrial facility after a chemical accident (see Fig. 8.1). Therefore, it is not surprising that both government and commercial customers have expressed interest in obtaining an effective CPT solution [24, 36, 37, 99].

Fig. 8.1: Envisioned CPT scenario of an intelligent robotic swarm responding to a chemical accident at an industrial facility. The autonomous drones establish an *ad hoc* communication and sensor network (bold lines) to collaboratively search the area for the source of the toxin

The evolution of chemical plumes occurs in three dimensions. However, in most cases, due to practical constraints, chemical plume-tracing activities must be carried out in a narrow, two-dimensional horizontal plane. (An example is ground-based robotic vehicles tracing an airborne chemical plume.) This can obscure the perception of the robots (also called "agents") performing the task, thereby making it considerably more difficult. The CPT problem is even more challenging when the plume-tracing robots occlude each other, or perturb the surrounding flow due to their own movement. To simplify our presentation, we do not discuss these effects here. Furthermore, this chapter omits a discussion of robotic sensing issues, so as to avoid confounding the main presentation of the relevant fluid dynamics. Instead, we focus on the theoretical properties of fluid flow, and what chemical signals and signatures the robots should follow in order to track the plume toward the source, as well as the information necessary to correctly recognize the source emitter. Nevertheless, it should be noted that our theoretical results apply regardless of the nature of these specific robotic implementation issues.

## 8.2.2 Fluid Dynamics As a Physicomimetic Foundation for Solving the Task

The foundation for effective CPT algorithms can be naturally and effectively built upon the fundamentals of fluid dynamics, as we have discovered. This section shows the relationship between fluid physics and chemical plume tracing. Note that the term "fluid" refers to both gaseous and liquid states of matter. All of the theoretical results presented in this chapter are applicable to both gases and liquids. Therefore, our findings are applicable to the design of ground, marine, and/or aerial autonomous vehicles.

A physical description of a fluid medium consists of *flow-field variables*, i.e., characteristics of the fluid such as its density, velocity, specific weight and gravity, viscosity, temperature, pressure, and so on. Here, due to their relevance and importance to the goals of CPT, we focus on the scalar density $\rho$, which has the units of mass per unit volume, and the vector velocity $V$. Because our focus is on CPT, we use the term "density" and its symbol $\rho$ to denote the concentration of the chemical being traced, rather than the density of the ambient carrier fluid. For simplicity of presentation, unless stated otherwise this chapter assumes that the ambient fluids are compressible—because in reality *all* materials are to some extent compressible. Furthermore, relative to a plume-tracing vehicle in motion, the distinction between the velocities of the fluid-borne chemical and that of the chemical-bearing fluid is often well below the sensor resolution threshold. In our model of the chemical plume, we assume that the chemical and ambient fluid velocities are identical.

Consider an infinitesimal fluid element moving within the plume. The velocity of this element can be represented in three dimensions as the vector $V = u\hat{i} + v\hat{j} + w\hat{k}$, where $\hat{i}$, $\hat{j}$, and $\hat{k}$ are the orthonormal unit basis vectors along the $x$, $y$, and $z$ axes in $\mathbb{R}^3$, i.e., $V = <u, v, w>$. Note that $u = u(x, y, z, t)$, $v = v(x, y, z, t)$, $w = w(x, y, z, t)$, and $t$ denotes time. Because chemical flow is typically unsteady, $\partial V / \partial t \neq 0$ in the general case. The density and other characteristics of the fluid, i.e., the flow-field variables, are likewise functions of space and time.

Sensor measurements of these flow-field variables are collected by the robots. In particular, it is assumed that a swarm of mobile vehicles equipped with chemical and fluid flow sensors can sample the $\rho$ and $V$ of the plume to determine the current state of the contamination, and then use their onboard computers, along with the outputs of their neighbors, to collaboratively (in a distributed fashion) decide the most effective direction in which to move next. The calculations and collaborative decisions can be effectively based upon the physical rules that govern the flow of fluids, as we will demonstrate shortly.

From Chap. 2 of this book, the reader should be familiar with conservation equations of physics. Fluid flow is no exception; it too has conservation equations that it obeys. The three governing equations of fluid dynamics for modeling fluid flow are the *continuity, momentum,* and *energy* equations. The continuity equation states that mass is conserved, the momentum equation captures Newton's second law $F = ma$, and the energy equation states that energy is conserved. To model fluids on a computer, the relevant field of study is *computational fluid dynamics.* According to computational fluid dynamics, each one of the three governing equations can be expressed in at least four different forms, depending on one's perspective with respect to the fluid: (1) a finite control volume fixed in space with the fluid moving through (or past) it, (2) a finite control volume moving with the fluid, (3) an infinitesimal (differential) fluid element fixed in space with the fluid moving through (or past) it, or (4) an infinitesimal fluid element moving with the fluid [5].

The CPT problem is most naturally tackled by viewing the ambient fluid from perspective (1) or (3), namely, by taking the perspective of a robot that has fluid flowing past it. Because we focus on differential calculus for the most part in this chapter (see below), we select perspective (3), which implies writing the continuity equation as:

$$-\frac{\partial \rho}{\partial t} = \nabla \cdot (\rho V), \qquad (8.1)$$

where $\rho V$ is the *chemical mass flux* or *mass flux* for short (and simply called the "flux" above). One can see that this equation expresses the physical fact that the time rate of the decrease of mass inside the differential element (left hand side of the equation) is caused by the net mass flux flow (right hand side) out of the element.

To model fluid flow in computational fluid models, the governing equations are solved using numerical methods. This is because scientists have found them too difficult to solve in closed form, except in very simple cases of fluid flow. In the case of perspective (3), the numerical *method of finite differences* is typically employed [5]. This method approximates continuous derivatives on a computer using discrete differences.

For the CPT problem, a distinction must be made between the *forward solution* and the *inverse solution*. We have just been discussing the former, which consists of a simulation that models the chemical flow. The latter requires tracing within the fluid to find the source emitter, i.e., the latter is CPT. For the purposes of developing and testing new CPT approaches in simulation, the forward solution is required. Once the CPT algorithm has been refined in fluid-dynamic simulations, then the forward solution is no longer necessary, since the robots are tested with real laboratory-scale chemical plumes. However, note that the inverse solution (i.e., the CPT algorithm) is required for both simulation and laboratory testing.

A numerical model of the forward solution for a given fluid provides data at discrete points in space, i.e., at "grid points." As mentioned above in the introduction to this chapter, one can see an abstract relationship between grid points and swarm robots in a geometric formation, or lattice. This abstract correspondence motivates our view of a swarm of robots as a distributed computer that jointly senses the flow-field variables, shares them with their immediate neighbors, and then decides in which direction to move in order to locate the chemical emitter, thereby performing embodied computation.

Consider one very important variable to measure in the context of CPT— the *divergence of the velocity* of the fluid (and therefore of the chemical). Note that as a fluid flows, any element within that fluid has invariant mass, but its volume can change. The divergence of the velocity, $\nabla \cdot V$, is the time rate of change of the volume of a moving fluid element, per unit volume. This divergence is the rate at which a fluid expands, or diverges, from an infinitesimally small region. A vector field with positive divergence expands, and is called a *source*. A vector field with negative divergence contracts, and is called a *sink*. The magnitude of the divergence vector, $|\nabla \cdot V|$, is a measure of the *rate* of expansion or contraction. A vector field with zero divergence, when measured over the entire velocity vector field, is called *solenoidal*. The definition of solenoidal implies fluid incompressibility. This can be understood by considering the $\nabla \cdot V = 0$ constraint on the conservation of mass in the continuity equation (8.1). In this case, $(\partial/\partial t + \nabla \cdot V)\rho = 0$, so that the total time derivative is zero, i.e., $d\rho/dt = 0$. This indicates that the fluid density is constant when viewed along the trajectory of a moving fluid element.

Compressible fluid flow with sources and sinks is not only realistic, but is also desirable—because the sources and sinks can facilitate local navigation in the absence of global landmarks [47]. Furthermore, it is important to note that a mathematical source (in terms of divergence) that remains a source over time (i.e., is not transient) indicates a chemical emitter, whose identification is the key to successful CPT. This will be discussed in depth below.

First, however, for CPT we need to broaden the notion of divergence to include the chemical density, in addition to the wind velocity. The key notion we seek is that of the (chemical) mass flux, which is the product of the chemical density and the velocity, i.e., $\rho V$. We already encountered the mass flux in the continuity equation (8.1). Informally, this is "the stuff spewing out of the chemical emitter." In fact, what is particularly relevant for CPT is the *divergence of the mass flux*, i.e., $\nabla \cdot (\rho V)$, again from (8.1). The divergence of the mass flux is the time rate of change of mass per unit volume lost (if positive) or gained (if negative) at any spatial position. In other words, if this divergence is positive, it indicates a source of mass flux; if negative, it indicates a sink of mass flux.

To conclude our discussion of fluid basics, consider two modes of fluid transport: *diffusion* and *advection*. Diffusion, which consists of a slow spatial spread of the chemical in all directions, dominates in an indoor setting where the air is relatively stagnant, e.g., when the windows are closed and there is little disturbance in a room. It also dominates at smaller spatial scales, e.g., insects tend to be more sensitive to diffusive effects than larger animals [40]. In the absence of air flow, diffusion of chemical mass away from a chemical source will result in a Gaussian chemical density profile (see Fig. 8.3 for an example of such a profile). Advection, which is a directional flow of the fluid (for example, caused by wind or water currents) is a more macroscopic phenomenon than diffusion. Both disperse the chemical away from an emitter, but they have different effects on the chemical density profile that makes up a chemical plume. In this chapter, we examine the effects of both diffusion and advection processes on CPT.

## 8.2.3 The Chemical Source Localization Task and Its Subtasks

Computational fluid dynamics is typically used for modeling the forward solution of real fluid flows for the purposes of theoretical analysis and engineering design (for example, of airplane wings). Chemical plume tracing, on the other hand, seeks a solution to the inverse problem. Assuming that at least one source emitter is present in the environment (and for simplicity, herein we assume a single emitter, though generalization to multiple emitters is straightforward)[1], finding that source requires environmental information. Ideally, the CPT robots would have access to global knowledge, such as geo-referenced coordinates of the emitter, which can then be used to navigate directly to the source. But in most realistic situations a local navigation strategy is the best that the robots can do, and that is what is assumed in this chapter.

The chemical plume-tracing task consists of three subtasks: (1) detecting the toxic chemical, (2) tracing the chemical to its source emitter, and (3) identifying the source emitter. We consider each of these subtasks, in turn. However, after this section, the remainder of this chapter ignores the first subtask (because algorithms to solve this subtask are heuristic, rather than

---

[1] Multiple emitters induce a separation of the swarm into smaller sub-swarms [292].

being based on fluid principles) and focuses instead on the second and third tasks only.

### 8.2.3.1 The Chemical Detection Subtask: Exploration

Typically, as CPT begins, the robots are not initially in contact with the chemical. Therefore, their first job is to locate the chemical plume. Under the assumption that they have no background knowledge about where the emitter is, the best they can do is to engage in an undirected exploratory search. In the literature, this search typically takes the form of a zigzag or spiral pattern [86, 198]. It is called *casting* because the robots are essentially casting about the environment, looking for a detectable (i.e., above sensor threshold) amount of the chemical. Casting strategies are often motivated by observation of biological creatures, such as insects, birds, or crustaceans [78].

### 8.2.3.2 The Plume-Tracing Subtask: Following Local Signals

The objective of the plume-tracing subtask is to follow clues, which are typically local if robots are used, to increase one's proximity to the source emitter. We have identified five major classes of prior (not physicomimetic) algorithms for this subtask, and some of the algorithms are exemplified in [38]:

1. **Heuristic** strategies, which apply intuitive "rules of thumb," such as, "Follow the edge of the plume."
2. **Chemotaxis** strategies. Usually, these algorithms follow the gradient of the chemical.
3. **Anemotaxis** strategies. These algorithms direct the robots upwind.
4. **Hybrid** strategies. They typically combine both chemical and upwind strategies.
5. **Infotaxis** strategies. A wide variety of algorithms fall into this class. One popular version follows the frequency of chemical puffs (i.e., filamentary structures), which typically increases in the vicinity of the emitter. Others use stochastic approaches.

Two of the above-listed strategies are biomimetic, namely, chemotaxis and anemotaxis. This chapter focuses on the biomimetic–physicomimetic distinction, and therefore these two strategies are the only ones discussed hereafter. Our choice of chemotaxis and anemotaxis for comparisons is a good one for another reason—because these are the two most popular CPT strategies in the literature; therefore, they are the most relevant of all the CPT strategies with which to compare our fluxotaxis strategy.

The chemotaxis strategy of chemical gradient following is motivated primarily by the chemical diffusion process that is responsible for some of the dispersion of the chemical away from the emitter. Diffusion creates a Gaussian chemical profile, i.e., the chemical density is highest near the emitter and it is reduced exponentially as distance from the emitter increases. In the case of Gaussian diffusive chemical dispersion, chemical gradient following is the logical way to proceed when seeking the source emitter. For an implementation of chemotaxis, one robot with multiple spatially separated chemical sensors would suffice, but a much more efficient and effective approach uses a swarm of tens or even hundreds of robots [292].

Advection is a physical fluid process that motivates the anemotaxis strategy of traveling upwind. Because advection transports the chemical downwind, the robots expect to get closer to the origin of the chemical emission by moving upwind. Considering the practical aspects of robotic CPT, observe that anemotaxis does not require any sharing of information between robots. Therefore, it is a strategy that could be employed by either a single robot or a collective of independent robots.

### 8.2.3.3 The Emitter Identification Subtask: Recognition

The final subtask, once the robots have gotten close to a suspected emitter, is to try to determine whether or not the suspect is indeed an emitter. The robots can look for a characteristic emitter signature, or use heuristic methods.

## 8.2.4 Applying Swarms of Robots to the Task

This section addresses the issue of applying swarms of robots to CPT problems. We need to address the role that the swarm will adopt in the context of this task. To better understand this role, it will help us to revisit previous chapters of this book.

### 8.2.4.1 Lagrangian Versus Eulerian Approaches to Modeling Fluids

Recall that the governing equations of fluids can be expressed in four different forms, depending on one's perspective with respect to the fluid, as (1) a finite control volume fixed in space with the fluid moving through (or past) it, (2) a finite control volume moving with the fluid, (3) an infinitesimal fluid element fixed in space with the fluid moving through (or past) it, or (4) an infinitesimal fluid element moving with the fluid [5]. The first and third perspectives are called *Eulerian*, whereas the second and fourth are called *Lagrangian* [167]. Chapter 7 describes a Lagrangian approach to physicomimetic modeling of fluids, whereas Chap. 6 describes a Eulerian approach. In Chap. 7, the robots themselves behave as a fluid. The robots are the fluid particles, i.e., the molecules within the fluid. Using kinetic theory to model particle–particle and particle–object collisions, the robot particles adopt fluid-like movement as they travel through corridors or across environments. In contrast, Chap. 6 models both biological dinoflagellate creatures and artificial robotic drones as moving through water, which is the fluid of interest. At any time step, these "particles" have the ambient fluid medium flowing around/past them, thereby facilitating a Eulerian fluid perspective. In summary, in Chap. 7 we adopt the perspective of a particle (robot) that is part of the fluid and moves with it. On the other hand, in Chap. 6 we adopt the perspective of a robotic vehicle that moves through the fluid, not as part of it. In this case, the fluid is considered to be moving (flowing) around the vehicle.

Either perspective is valid, but we must always choose the perspective that best suits our application. The CPT application is most similar to the bioluminescence one, and therefore for this application we adopt a Eulerian approach.

### 8.2.4.2 Swarms Navigating Through Fluids to Localize a Chemical Source: A Eulerian Approach

Chapter 1 defines a "swarm." Here, we define a "robotic swarm" as a team of mobile autonomous vehicles that cooperatively solve a common problem. The swarm size and type may vary anywhere from a few simple, identical robots to thousands of heterogeneous hardware platforms, including ground-based, aquatic (autonomous surface/underwater vehicles or ASVs/AUVs) and/or aerial (unmanned aerial vehicles or UAVs). In our swarm design, each robotic vehicle serves as a dynamic node in a distributed sensor and computation network, so that the swarm as a whole functions as an adaptive sensing grid, constantly sharing various fluid measurements between adjacent vehicles (see Fig. 8.1). Meanwhile, the agents are embedded within a dynamically changing fluid environment, and their perspective is therefore clearly Eulerian.

Applying a collective or swarm of robots for the CPT task does not require global coordination or communication. Furthermore, as mentioned above, swarms greatly increase the spatial exploration, thereby resulting in superior performance. Even in the context of biomimetic strategies of blue crabs, there is convincing evidence for the CPT performance advantage of sampling the plume at multiple, spatially separated locations simultaneously [260]. Crabs, for instance, have evolved spatially separated arrays of sensors on their claws, legs and antennae for concurrent sampling of odor density at different spatial locations. Our prior research (e.g., [293, 295, 296]) shows that we have tackled the CPT task with robot swarms in mind from the outset. Back then, only a couple of other research groups were doing swarm-oriented CPT research, e.g., [41, 86]. However, more recently a swarm-based approach to CPT is becoming increasingly popular due to its effectiveness.

With our particular physicomimetic approach to swarm robotic CPT, namely, fluxotaxis, fluid velocity and chemical concentration are measured independently by each robot, and the collected sensor data is communicated between neighboring vehicles. Our physics-based fluxotaxis strategy requires the computation of (partial) derivatives. Derivative calculations are performed by the robots using the numerical finite-difference method of second-order accurate central differences [291]. Note that this short-range sharing of information between robots is sufficient for calculating local navigation gradients, so there is never a need for a "global broadcast" of these values. In other words, implementation of the theory we present here requires no leaders or central controllers or global information, which in turn increases the system's overall robustness, while lowering its cost. Furthermore, any local information is implicitly merged into the swarm via the formation.

We have demonstrated in a robot-faithful simulation, with swarms of hundreds of robots, that our physicomimetic CPT theory is directly applicable to a distributed swarm robotic paradigm. See the next chapter for convincing results that scale up the number of robots in the swarm methodically and, as a result, show increasingly better CPT performance. Evidently, swarms are an effective means to succeed at the CPT task, and we have therefore used them throughout our research on CPT.

## 8.3 Prior Research on Biomimetic Chemical Plume Tracing

This book is about physicomimetics, and most of the comparisons with alternative algorithms focus on biomimetics. We follow suit here, not only because this follows the precedent in the book, but also because biomimetic approaches have dominated the CPT literature. Most robotic approaches to CPT are modeled after natural biological organisms that trace olfactory scents. When the *Environmental Fluid Mechanics* journal published a special issue on the topic of CPT in [38], the papers in that issue laid an initial foundation for the field. The strategies presented in that special issue were based on observations of living organisms, e.g., insects, land animals, aerial and aquatic creatures. Such organisms trace olfactory clues, thereby enabling them to satisfy their feeding and/or reproductive requirements. That journal special issue set a precedent, and the majority of CPT research since that time has been biomimetic.

Therefore, before presenting our physicomimetic algorithm for CPT, we first describe prior related research on the two leading biomimetic CPT algorithms, namely, chemotaxis and anemotaxis.[2]

---

[2] For more general and comprehensive surveys of CPT approaches, see [130] and [232].

### 8.3.1 Chemotaxis

Chemotaxis is the best understood and most widely applied CPT approach. It consists of tracing the chemical signal. Recall the two modes of fluid transport, diffusion and advection. Diffusion typically dominates indoors or at smaller spatial scales, such as closer to the ground [40]. Chemotaxis is designed to be maximally effective in such diffusion-dominated environments. Insects that crawl on the ground, for example, typically apply chemotactic strategies.

Although there are various types of chemotaxis, the most common one follows the local gradient of the chemical concentration within a plume [132, 146]. Some of the earliest research on chemotaxis is by Sandini et al. at the University of Genoa [205]. Among the most extensive applications of chemotaxis are those of Lilienthal and colleagues. In some of their work they show chemotaxis success in an uncontrolled indoor environment [143]. They have also explored chemotaxis in ventilated corridors with weak chemical sources [144].

Feddema et al. apply a control theoretic approach to chemotaxis that allows for stability and convergence proofs with optimality guarantees [61]. On the other hand, it makes a strong assumption that the chemical plume density profile can be closely approximated by a quadratic surface.

While chemotaxis is very simple to perform, it frequently leads to locations of high concentration in the plume that are not the real source, e.g., a corner of a room [228]. Cui et al. have investigated an approach to solving this problem by using a swarm majority vote, along with a communication and routing protocol for distributing information to all members of a robotic collective [41]. However, that is a strong requirement, and they also make an even stronger assumption that each robot in the collective has a map of the environment.

### 8.3.2 Anemotaxis

Another common approach, usually called "anemotaxis" but sometimes alternatively called *odor gated rheotaxis*, has been proposed for the CPT task. An anemotaxis-driven robot focuses on the advection portion of the flow. Assuming the chemical density is above some minimum threshold, an anemotaxis-driven robot measures the direction of the fluid's velocity and navigates "upstream" within the plume. Hayes et al. have done seminal work in this area [86]. Li et al. model their anemotaxis strategy on the behavior of moths [142]. Grasso and Atema combine anemotaxis with casting, so that the robots alternate moving upstream and cross-stream in the wind [78]. The simulation results of Iacono and Reynolds show that the effectiveness of anemotaxis improves with increased wind speed [103]. More complex wind-driven strategies by Kazadi et al. may be found in [116]. They have success-fully explored a form of anemotaxis that combines wind velocity information with passive resistive polymer sensors and strategic robot placement, which enables effective anemotaxis. Ishida et al. gain performance improvements by coupling anemotaxis with vision capabilities [106].

One of the biggest advantages of anemotaxis is that it can be performed successfully with either a single robot, or with a group of independent robots. On the other hand, Grasso and Atema have found that two spatially separated wind sensors (which could be on two different robots) outperform a single wind sensor [78].

### 8.3.3 Hybrid Strategies

Recently, there has been an increasing trend toward the development of hy-brid CPT strategies that combine chemotaxis with some form of anemotaxis. For example, Ishida et al. use a simple hybrid strategy that consists of apply-ing anemotaxis when the chemical density $\rho$ is high, and otherwise applying chemotaxis [105].

Russell et al. employ a more complex hybrid approach to CPT for finding the source of a chemical emission in a maze [198]. Their single CPT robot employs a heuristic algorithm for traveling upwind, following the chemical and avoiding obstacles.

More unusual hybrids are the approaches adopted by Lytridis et al. [145] and Marquest et al. [147]. They combine chemotaxis with a biased random walk, and discover that this hybrid outperforms its components.

### 8.4 Physicomimetic Chemical Source Localization Algorithm

We have developed, implemented, and analyzed a physicomimetic algorithm for CPT, called "fluxotaxis," that is based on the physics of fluids. It is a theo-retically based, not *ad hoc*, hybrid of chemotaxis and anemotaxis. This section describes the motivation for our algorithm, its mathematical and physics-based foundation, and presents intuitions for why it has been demonstrated to be successful under a wide variety of common simulated flow conditions.

Let us begin with our motivation for exploring a physics-based alternative to biomimetic CPT. The chief difficulty with biomimetic algorithms is that in order to mimic natural biological strategies with robots, sophisticated sensors are needed. However, despite recent advances in manufacturing techniques, state-of-the-art chemical sensors remain crude and inefficient compared to their biological counterparts. Furthermore, it is difficult to achieve optimal or near-optimal performance in artificial CPT systems by relying on mimicry of biological systems alone. For example, Crimaldi et al. analyzed crustaceans' antenna sensor arrays, also called "olfactory appendages," and found that an antenna structure with multiplicity and mobility improves an organism's CPT performance [40]. But we can surpass this with cooperating mobile robots. In particular, the inherent flexibility afforded by the multiplicity and mobility of a swarm of robots far exceeds that of a single organism's antennae. Therefore, rather than mimic single biological organisms that use CPT for food or reproduction, we instead developed an entirely new approach to CPT that was inspired by swarms of robots performing the task. This physics-based approach, fluxotaxis, is built on our knowledge of fluid physics.

This section motivates and describes our fluxotaxis method, by specifying its underlying mathematical principles. We begin with the original motivation, which is the definition of a chemical source.

## 8.4.1 Emitter Identification Based on Fluid Dynamics

Fluxotaxis is a fluid dynamics-based source localization strategy. At the heart of this strategy is a theoretically founded formula for emitter identification. The rest of the fluxotaxis algorithm emerges naturally from this emitter identification process. Therefore, we begin our presentation of the fluxotaxis algorithm by stating its foundation—the formula for identifying a chemical emitter, i.e., although the emitter identification subtask comes last for the robots, we address it first in this exposition because it is foundational.

A key fluid dynamics concept that enables identification of a source of chemical flow is the divergence of the chemical mass flux. Recall from Sect. 8.2.2 that the divergence of mass flux, $\nabla \cdot (\rho V)$, is the time rate of change of mass per unit volume. The mass flux combines the chemical density and the velocity of the chemical flow in one term. A positive divergence of this quantity implies a mathematical source.

Formally, the divergence of the mass flux $\rho V$ at a point $p$ is defined as the limit of the net flow of this flux across a smooth boundary (e.g., the surface area $A_o$) of a volumetric region $W_o$, divided by the volume of $W_o$, as this volume shrinks to $p$. Mathematically, this is expressed as:

$$\nabla \cdot (\rho V)(p) = \lim_{W_o \to p} \oint_{A_o} \frac{(\rho V) \cdot \hat{n}}{|W_o|} \, dA \,, \tag{8.2}$$

where $\hat{n}$ is a unit vector pointed as an outward normal from the surface of the volume. Recall from calculus that $\oint$ denotes a surface integral in three dimensions. This equation defines the divergence of the mass flux as the surrounding surface area integral of the mass flux in the outward normal direction. It is the total amount of "stuff" spewing out from point $p$, where $p$ is a point source of mass flux.

Consider a practical method for a swarm of robots to perform an emitter identification test, using Eq. 8.2. The robots can surround a suspected emitter and calculate the above integral. This is a theoretically sound emitter identification procedure. In other words, assume that there is a robot lattice consisting of hundreds of robots surrounding the emitter, thus creating a full sensing grid. Also, assume that the chemical emitter within the control volume $W_o$ behaves as a "source" by ejecting chemical material in a puff or in a continuous fashion. Then a key diagnostic of the chemical plume emitter location is when we have $\oint_{A_o} (\rho V) \cdot \hat{n} \, dA > 0$, where $A_o$ is the surface area of the volumetric region $W_o$ that the robots are known to surround. As the size and density of the swarm surrounding the emitter increase toward infinity, the computation approaches that of Eq. 8.2. If $\oint_{A_o} (\rho V) \cdot \hat{n} \, dA > 0$, which is indicative of a source, then the robot lattice surrounding the emitter can locate it by going opposite to the direction of flux outflow. Of course, in a practical situation the swarm size will be limited and may even be relatively small. In this case, the theory is approximated; however the strategy is still well-founded because it is theoretically based.

Furthermore, in a practical sense, a source may be either transient or sustained. To differentiate a true source associated with the chemical emitter from a transient source that is merely a puff of chemical created by turbulence or other chemical distribution phenomena, we require a *sustained* positive value of the integral in order to identify a clear emitter signal. *In summary, the physical signature of a chemical emitter is $\oint_{A_o} (\rho V) \cdot \hat{n} \, dA > 0$ for a period of $\tau$ experimentally determined time units.*

To detect this signature (using a differential approximation defined below), and to perform CPT, the only additional sensors (beyond standard sensors, like IR for obstacle avoidance) required for robots are chemical sensors to detect the density $\rho$ and anemometer sensors to detect the velocity $V$. Sensors for discriminating *which* chemical is present are not discussed here. We assume they have already been applied to determine the chemical class. This chapter focuses instead on robots that need to determine the value of $\rho$, i.e., chemical density rather than type, for a known contaminant.

Note that the characteristics of an emitter can affect the characteristics of its signature. There are continuous emitters, as well as single-puff emitters, and regularly and irregularly pulsing emitters. In some of our discussions below, we address single-puff and continuous emitters—because real-world CPT problems typically contain a source that is a combination of these basic emission modes. Nevertheless, note that these are just examples; our theory is general and is not restricted to these two types of emitters. The size, shape, orientation, and height of an emitter can also affect the plume characteristics. The theoretical results in this chapter are deliberately general, but could be parameterized as needed to model particular emitter geometries and configurations.

## 8.4.2 Deriving Our Fluxotaxis Swarm Navigation Strategy from the Emitter Signature

Now that we understand how to identify a source emitter, we can derive a strategy to trace its signal. This section describes our fluxotaxis algorithm for solving the plume-tracing subtask. For simplicity, here (and throughout this chapter), we assume a single stationary source emitter in the region of interest. Extensions to multiple and/or mobile emitters are straightforward. In real-world situations, the entire region that will be explored by the robot swarm will be filled with transient fluid flow phenomena, which will be likely to include transient local sources and sinks of chemical mass flux, as well as solenoidal regions. How will the swarm navigate through this larger region, while moving toward the true "permanent" (during the CPT task) source emitter? The swarm navigation strategy that we have invented and designed, called "fluxotaxis," is one that avoids all sinks, traverses toward all sources (both transient and "permanent"), and uses casting to navigate through solenoidal regions. The motivation behind this navigation strategy is that the robots should avoid getting trapped in sinks or in solenoidal regions, and the number of transient sources increases as one approaches the true source emitter, so these sources should be followed. The presence of an increasing number of transient sources (and sinks) as one approaches the true emitter is due to a "shedding" effect from the true emitter, which releases transient sources and sinks due to the emitter-generated turbulence. This fluxotaxis navigation strategy is theoretically sound (see below) and has outperformed its leading biomimetic alternatives (see Chap. 9) in a wide variety of simulated flow and swarm conditions.

We have just learned that fluxotaxis essentially moves robots toward sources and away from sinks. Therefore, it must be able to identify sources and sinks. What do we know so far about how robots can identify a source? Recall that identification of the source emitter requires evaluating (approximating) an integral, as in Eq. 8.2. Unfortunately, although this works well for an emitter test, it is not what we want when navigating through unsteady fluid regions farther from the emitter—especially when we do not want the agents to have to "surround" something. In this latter case, we need a more local perspective for the agents. Integrals are suitable for global volumetric calculations, whereas derivatives are better suited for local point calculations. Therefore, we henceforth assume an infinitesimal analysis based on differential calculus that reasonably approximates the global volumetric (integral) perspective. Although in any given time step we have a finite, limited grid size (both in simulation and with the robots), we can compute an approximate solution by assuming that the divergence of the mass flux can be calculated at *every* point in space. Mathematically, we are stating that $\oint_{A_o} (\rho \boldsymbol{V}) \cdot \hat{\boldsymbol{n}} \, dA$ becomes equivalent to $\boldsymbol{\nabla} \cdot (\rho \boldsymbol{V})$ in the limit as we move from discrete grids to continuous space, as shown in Eq. 8.2. The advantage of adopting this local differential perspective is that it simplifies our analysis without loss of correctness. Furthermore, the emitter (source) test becomes $\nabla \cdot (\rho V) > 0$. With this new, local perspective, we can use the differential calculus definition of the divergence of the mass flux [102] (in three dimensions):

$$\nabla \cdot (\rho V) = u \frac{\partial \rho}{\partial x} + \rho \frac{\partial u}{\partial x} + v \frac{\partial \rho}{\partial y} + \rho \frac{\partial v}{\partial y} + w \frac{\partial \rho}{\partial z} + \rho \frac{\partial w}{\partial z} \ .$$

Note that this differential formula can be applied at any point in space, by an agent, to decide what direction to go next. It can be used to detect both sources *and* sinks. At what point(s) in a region will the value of this be maximized/minimized? If the agent is near a source emitter, then the closer the agent gets to this emitter, the greater the (positive) value of the divergence of the mass flux will be. The opposite holds as an agent goes toward the strongest "drain" within a sink. Therefore, the obvious differential signal for each agent to follow is the *gradient of the divergence of the mass flux* $(\overrightarrow{GDMF})$:

$$\overrightarrow{GDMF} = \nabla \left[ \nabla \cdot (\rho V) \right] = \nabla \left( u \frac{\partial \rho}{\partial x} + \rho \frac{\partial u}{\partial x} + v \frac{\partial \rho}{\partial y} + \rho \frac{\partial v}{\partial y} + w \frac{\partial \rho}{\partial z} + \rho \frac{\partial w}{\partial z} \right) \ .$$

This gradient points away from sinks and toward sources. (See more about this below.)

When using derivatives rather than integrals, each agent can make its navigation decision based on highly local information. In other words, each agent will calculate its environmental force based on the $\overrightarrow{GDMF}$ and then calculate its formation cohesion forces, based on the ranges and bearings to its neighboring agents; then it will take the vector sum of these. This vector sum determines the direction the agent will go. The desired behavior of a robot grid/formation moving toward the emitter will result.

By following the regions that behave like sources, fluxotaxis-driven robots quickly and easily find the true source emitter. In numerous computational fluid dynamics simulations we have found that our gradient-following fluxotaxis algorithm quickly and reliably locates the source emitter using these ambient chemical flux signals (see Chap. 9). Even in complex, simulated urban environments filled with many large buildings and narrow passageways, we have shown that fluxotaxis demonstrates rapid and reliable success, whereas the leading alternative chemotaxis and anemotaxis approaches typically flounder and fail [2].

At this point, the following theorems should be intuitively obvious. For their formal proofs under general, realistic conditions, see [232].[3] The theorems have been rigorously experimentally demonstrated under a wide range of fluid-laden environments, ranging from laminar to turbulent.

---

[3] Fluids are realistically modeled and so are the robots, with the exception that robots are modeled at an abstract level and therefore sensor and actuator noise are not included.

**Theorem 8.1. (Source Theorem)** *Fluxotaxis $\overrightarrow{GDMF}$-controlled robots will always move toward a chemical source.*

**Theorem 8.2. (Sink Theorem)** *Fluxotaxis $\overrightarrow{GDMF}$-controlled robots will always move away from a chemical sink.*

To conclude this section, we briefly mention some practical considerations for applying fluxotaxis and other CPT methods. For one, *any* CPT strategy will only be valid in regions where the value of $\rho$ and its gradients (if computed by the strategy) are above some minimally significant threshold. Below this minimum density (e.g., too far from the source emitter), *all* CPT algorithms will be poor predictors of the source location, and the robots should resort to casting. For example, in solenoidal regions in which the fluid is incompressible and the chemical density is invariant along velocity trajectories, heuristic approaches such as casting are needed.

Another consideration is the scope of our guarantees regarding the behavior of fluxotaxis. Our theoretical results state that $\overrightarrow{GDMF}$ locally points in the correct direction—toward sources and away from sinks. But we have *not* given any global optimality guarantees. This is in line with the physicomimetics philosophy; control theory gives global optimality guarantees, but provably optimal control theoretic swarm strategies have shown difficulty in scaling up to real-world problems. Physicomimetics, on the other hand, proves local properties of the swarm so that there are more assurances than those of biomimetics, but it is not as formally assured (or as restricted) as control-theoretic approaches.

The above two considerations identify the limitations of fluxotaxis (and other CPT algorithms). What are its strengths? Fluxotaxis navigates best in the presence of sources and sinks, which are common in realistic fluid flows, and it performs well in all other CPT fluid conditions in which we have tested it. It is robust in its avoidance of sinks, and exceptionally precise in homing in on sources. Despite the lack of global optimality guarantees, fluxotaxis works extremely well at finding the global optimum, i.e., the emitter, in practice when tested on a wide variety of simulated realistic and challenging conditions (see Chap. 9).

## 8.5 Theoretical Comparisons of Biomimetic and Physicomimetic Approaches to the Task

In the related work section of this chapter, above, we presented prior research on the leading biomimetic approaches, namely anemotaxis and chemotaxis. We continue to focus on these two biomimetic approaches, as well as our physicomimetic fluxotaxis approach, throughout the remainder of the chapter. In this section, we compare these two biomimetic algorithms with the physicomimetic fluxotaxis algorithm theoretically. To complement the theoretical comparisons in this chapter, Chap. 9 presents experimental comparisons between the two biomimetic and the physicomimetic fluxotaxis approaches.

## 8.5.1 Anemotaxis Versus Fluxotaxis

This section focuses on a comparison between the expected performance of anemotaxis and fluxotaxis. For a theoretical comparison, we first need to state the vectors that the robots will follow when executing these two algorithms.

Although there are several variations of anemotaxis, we focus on the most common version here. Mathematically, anemotaxis-driven robots follow an upwind vector, i.e., they move in the direction $-V$. Their speed (velocity magnitude) may be constant, or it could be proportional to $|V|$. Of course, anemotaxis is only employed if the chemical density is above-threshold, so it is a perfectly reasonable strategy for CPT.

Recall that the mathematical formula for the fluxotaxis $\overrightarrow{GDMF}$ in three dimensions is:

$$\nabla \left[ \nabla \cdot (\rho V) \right] = \nabla \left( u\frac{\partial \rho}{\partial x} + \rho\frac{\partial u}{\partial x} + v\frac{\partial \rho}{\partial y} + \rho\frac{\partial v}{\partial y} + w\frac{\partial \rho}{\partial z} + \rho\frac{\partial w}{\partial z} \right) .$$

This is the navigation vector followed by fluxotaxis.

The $\overrightarrow{GDMF}$ combines information about both velocity and chemical density, in a manner motivated by the theory of fluid dynamics. Mathematically, the divergence of the mass flux can be subdivided into two terms:

$$\nabla \cdot (\rho V) = \rho (\nabla \cdot V) + V \cdot (\nabla \rho) . \tag{8.3}$$

Therefore, the $\overrightarrow{GDMF}$ is a gradient $\nabla$ of the sum of two terms:

1. $\rho(\nabla \cdot V)$, which is the density times the divergence of the velocity field, and
2. $V \cdot (\nabla \rho)$, which is the flow velocity field in the direction of the density gradient.

When the chemical flow is divergent, the first term takes precedence in directing the robots, and it is analogous to anemotaxis. When the fluid velocity is constant, or the flow is stagnant, then $\nabla \cdot V$ is zero and the second term is the only relevant term. This second term is the flow velocity in the direction of the density gradient, and is analogous to chemotaxis. What is important here is that $\overrightarrow{GDMF}$ *theoretically* combines the essence of both chemotaxis and anemotaxis, and *the relevant term is activated automatically based on the environment.* No deliberate strategy selection or switching mechanism is required—$\overrightarrow{GDMF}$ automatically becomes whatever it is supposed to be in an environmentally-driven manner.

From this formalization of the two algorithms, we infer that $\overrightarrow{GDMF}$ can behave in either a chemotaxis-like manner or an anemotaxis-like manner, depending on the environmental conditions, whereas pure anemotaxis does not possess this hybrid behavior. Under what conditions is the contrasting behavior most apparent? We find that the differences are most striking when the swarm is near the chemical emitter, especially when there are wind currents near the emitter that move past the emitter. In this situation, the fluxotaxis $\overrightarrow{GDMF}$ observes that it is near a fluid source and homes right in on the emitter. Because the ambient winds that move past the emitter do not appear as sources, they are ignored by fluxotaxis. Anemotaxis-driven robots, on the other hand, are misled by the ambient wind currents, especially if these currents are stronger than the fluid flow coming from the emitter, or if the robots are closer to those misleading wind currents. In these cases, anemotaxis-driven robots will overshoot the emitter. We have observed this behavior many times in our CPT simulations, as shown in Chap. 9.

## 8.5.2 Chemotaxis Versus Fluxotaxis

We next compare the chemotaxis and fluxotaxis algorithms. Chemotaxis-driven robots will not travel upwind because they ignore the advective component of the fluid flow. On the other hand, as seen in the previous section, fluxotaxis becomes similar to either anemotaxis or chemotaxis, depending on the environmental conditions. Therefore, there is no point in comparing chemotaxis and fluxotaxis in an environment with strong advection—because it is intuitively obvious that fluxotaxis should perform better under such conditions. For example, we already know that fluxotaxis will outperform chemotaxis in the presence of fluid sinks. This is because chemotaxis-driven robots are attracted to local chemical maxima, such as sinks, and can get stuck there, whereas we know from our Sink Theorem that fluxotaxis-driven robots avoid all sinks and instead move toward sources.

Therefore, for our comparisons in this section, we deliberately select a situation that is maximally favorable for chemotaxis, namely, a chemical diffusion dominated flow. As mentioned above, insects and crustaceans that navigate in diffusion dominated regimes are known to employ the biomimetic chemotaxis strategy. Therefore, one would expect that in such a regime, chemotaxis would outperform fluxotaxis. Our experimental results have shown otherwise, however. In this section, we provide intuitions for why this is the case.

The scenario assumed in this section is void of sinks or sources other than one source—a chemical emitter. To emphasize diffusion, the wind/current flow speed is assumed to be temporally invariant, and the velocity field has a constant radial component. In other words, diffusion dominates and advection plays a minor, subsidiary role.

The diffusive chemical density profile, i.e., the forward solution, can be modeled using the *density diffusion equation*, simply called the *diffusion equation*. Because we focus more on the chemical and less on the advective flow, we further refine our notion of the chemical density, $\rho$. In particular, it is useful to consider a *particle number density* formulation, $n(r, t)$, sometimes written as simply $n$, at a three-dimensional spatial coordinate $r = x\hat{i} + y\hat{j} + z\hat{k}$, at time $t$. Assume that $r$ is a vector that points away from the source emitter. The particle number density $n$ is the chemical particle count per unit volume. Then the mass density can be split up into its components, i.e., $\rho = mn$, which is the mass $m$ of each chemical particle times the number density $n$. With this new notation, our prior expression for the mass flux, $\rho V$, can be rewritten as $mnV$. We can now express the three-dimensional density diffusion equation in terms of $n$ as

$$\frac{\partial n(r, t)}{\partial t} = \mathcal{D}\nabla^2 n(r, t) , \qquad (8.4)$$

where $\mathcal{D}$ is the *diffusion coefficient*, which is proportional to the square of the *mean free path* divided by the *collision time*.[4] The mean free path is the average distance traveled by particles in the fluid before they collide, and the collision time is the average time between collisions. This equation describes the evolution of the density of particles, $n(r, t)$, as a function of the spatial coordinate, $r$, and time, $t$. It uses the *Laplacian operator* $\nabla^2$, which is the divergence of the gradient of a function in Euclidean space. For the diffusion equation, we assume a temporal initial condition (at $t = 0$) where *all* the $N_p$ particles (chemical filaments) are contained at the origin ($r = 0$). The particles then gradually diffuse into the region surrounding the emitter.

We wish to solve the diffusion equation (8.4). The reason for doing this is that a solution to this equation will show us what the chemical density profile looks like, so that we can model the forward solution before examining the inverse CPT solution. Such a density profile will have the distance from the origin of the chemical emitter on the horizontal axis, and the density on the vertical axis. In other words, it will show us how the density varies as a function of one's distance from the emitter. This global perspective of the density in the region surrounding the emitter is not something that the robots perceive—they only sense local density. But it gives *us* a global perspective for analyzing and evaluating the performance of the agents and enables us to determine whether they will, as desired, move in the correct direction (toward the source emitter) at each time step.

Using a *Green's* function technique, we can solve the diffusion equation (8.4) for any type of emitter (see [232] for details). The Green's function

---

[4] The reader should be careful not to confuse the diffusion coefficient $\mathcal{D}$ with the desired inter-agent radius $D$ used in other chapters of this book.

for the diffusion equation can be used by integrating over the emitter source of particles to achieve the density solution for any localized particle emitter source. In the next two sections we present the solutions for the cases of a continuously-emitting constant-speed emitter and a single-puff emitter, respectively, and analyze the comparative performance of chemotaxis and fluxotaxis for these cases. First, however, we need to formally define chemotaxis.

Recall that fluxotaxis-driven robots follow the $\overrightarrow{GDMF}$ vector, which is defined as $\nabla \left[ \nabla \cdot (\rho V) \right]$. To identify the chemotaxis vector that robots follow, we assume the most popular version of chemotaxis in the literature. The vector that these chemotaxis-driven robots follow is the *density gradient*, or the $\overrightarrow{DG}$ for short. Formally, the $\overrightarrow{DG}$ is defined as $\nabla \rho$. Now we can consider two cases of emitters that result in diffusion dominated environments favorable to chemotaxis, and compare the performance of fluxotaxis and chemotaxis for these cases. Note that the following sections present the outline and intuitions for our theoretical results only—because we wish to avoid having the reader get overwhelmed by the mathematics. However, mathematically inclined readers will find the complete formal derivations in [232].

## 8.5.3 The Continuous Emitter Case

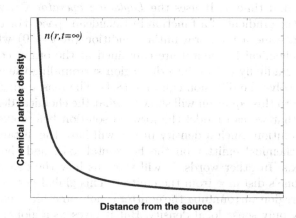

Fig. 8.2: Steady-state chemical concentration profile as a function of distance from the center of a continuous chemical emitter, expressed in terms of the particle density function $n(r, t)$

Here, we consider the case of a continuous emitter, i.e., one that continuously emits chemical particles. For this type of emitter, the solution to the diffusion equation (8.4), in the limiting case as $t \to \infty$, is

$$n(\boldsymbol{r}, \infty) = \frac{S_o}{4\pi \mathcal{D} r} \, , \tag{8.5}$$

which has an inverse radius decay, so that the number of particles drops off as the inverse of the radius from the source emitter. This limiting case solution for $t \to \infty$ is called the *steady-state solution*. The steady-state solution is used here for simplicity of exposition, though the conclusions presented in this section hold prior to the steady state as well. Figure 8.2 shows a graph of this inverse radius decay function, where the distance from the source emitter increases along the horizontal axis. Here, we are assuming that $S_o$ is the initial particle source rate (number of particles per unit time) from the source emitter, and $S(t) = S_o$ for all time, $t$, at the origin $r = 0$, i.e., a constant rate emitter. Also, the total number of particles $N_p$ increases with time, in particular, $N_p = tS_o$. Furthermore, this solution assumes a constant outward velocity profile, i.e., $\boldsymbol{V} = V_o \hat{\boldsymbol{r}}$, where $\hat{\boldsymbol{r}}$ is a unit vector pointing away from the emitter, and $V_o$ is the initial magnitude of the chemical velocity at $t = 0$. This solution uses $r$ as an abbreviation for the magnitude of the position vector, i.e., $r = |\boldsymbol{r}|$.

Given that we now have the steady-state solution (8.5) to the diffusion equation (8.4) for the case of a continuous emitter, next consider the question of whether the $\overrightarrow{GDMF}$ of fluxotaxis and the gradient of chemotaxis are good guides (predictors of the direction of the source) for CPT, i.e., "Do either or both of these formulas/algorithms point in the correct direction toward the source emitter?" in a continuous source situation dominated by diffusion. In this chapter, we interpret the $\overrightarrow{GDMF}$ and other CPT algorithm vectors as "predictors" of the correct direction to the emitter. Chapter 9 extends this interpretation to actual robot moves.

First, consider the $\overrightarrow{GDMF}$. After a series of mathematical operations (see [232] for specifics), we find that the $\overrightarrow{GDMF}$ is equal to

$$\overrightarrow{GDMF} = -\frac{mV_o S_o}{2\pi \mathcal{D} r^3} \hat{\boldsymbol{r}} \, , \tag{8.6}$$

where $r$ is the magnitude of the vector $\boldsymbol{r}$ and $\hat{\boldsymbol{r}}$ is a unit vector that points in the same direction as $\boldsymbol{r}$.

Recall that $\boldsymbol{r}$ is a vector that points away from the source emitter. This $\overrightarrow{GDMF}$ vector points in the direction

$$\hat{\boldsymbol{s}}_{\text{GDMF}} = \frac{\overrightarrow{GDMF}}{|\overrightarrow{GDMF}|}$$

$$= -\hat{\boldsymbol{r}} \, ,$$

where $\hat{\boldsymbol{s}}$ is a unit vector pointing in the direction denoted by the vector subscript. In other words, the $\overrightarrow{GDMF}$ points toward the origin (the source emitter) in a radially symmetric stationary flow. This is exactly as desired.

Next consider the density gradient, $\overrightarrow{DG}$ approach used by chemotaxis. From [232] we derive the fact that

$$\overrightarrow{DG} = -\frac{S_o}{4\pi\mathcal{D}r^2}\hat{\boldsymbol{r}} . \tag{8.7}$$

We can see that chemotaxis also predicts the correct source direction since

$$\hat{s}_{\text{DG}} = \frac{\overrightarrow{DG}}{|\overrightarrow{DG}|}$$

$$= -\hat{\boldsymbol{r}} .$$

The conclusion is that both the $\overrightarrow{GDMF}$ and the $\overrightarrow{DG}$ vectors point in the correct source direction, and therefore fluxotaxis is as good as chemotaxis for CPT domains where diffusion dominates the spread of the chemical away from a single continuous emitter. This is a very reassuring result. The intuition that lies behind this result is that, as shown in the section comparing fluxotaxis with anemotaxis, fluxotaxis contains elements of both anemo- and chemo-like behavior within it. The chemo-like behavior enables it to perform in a competitive manner to chemotaxis, even in situations most favorable to chemotaxis.

In addition to pointing toward the source emitter, both $\overrightarrow{GDMF}$ and $\overrightarrow{DG}$ have another advantage in the case of a continuous chemical source, i.e., they maximize their values at the emitter. Larger values of the navigation vector imply a stronger, more confident prediction of where the source is located. Thus the predictive capabilities of $\overrightarrow{GDMF}$ and $\overrightarrow{DG}$ are both maximized close to the source emitter. However, it should be noted from Eqs. 8.6 and 8.7 that as $r \to 0$, the magnitude of the fluxotaxis predictor, $|\overrightarrow{GDMF}| \propto r^{-3}$, becomes larger than that of the chemotaxis predictor, $|\overrightarrow{DG}| \propto r^{-2}$. (The notation $A \propto B$ means that $A$ is directly proportional to $B$.) Therefore, the $\overrightarrow{GDMF}$ makes better predictions than the $\overrightarrow{DG}$ as the robots get closer to the emitter.

## 8.5.4 The Single-puff Emitter Case

The solution to the three-dimensional diffusion equation (8.4) with a *delta function* initial condition and a single-puff emitter is the *Gaussian evolution function*:

$$n(\boldsymbol{r}, t) = \frac{N_p}{(4\pi\mathcal{D}t)^{3/2}}e^{-r^2/4\mathcal{D}t} . \tag{8.8}$$

A "delta function initial condition" means that all of the chemical begins at the emitter and nowhere else.

The interpretation of such diffusion of chemical concentration, where the density is initially peaked at the origin, $r = 0$, is that after a time $t$, the density has spread over a radial distance $r \approx \sqrt{4\mathcal{D}t}$ from the origin. Furthermore, the amplitude decreases by $1/e$, and the width of the density distribution increases as the square root of time, i.e., $\Delta r \propto \sqrt{t}$, as shown in Fig. 8.3. The origin $r = 0$ is at the center of the horizontal axis in this figure and it increases in either direction from the center. The vertical axis shows the density as a function of the distance from the origin. The notation $\Delta r$ represents the change in the value of $r$.

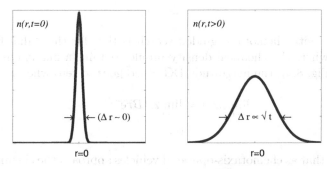

Fig. 8.3: Temporal growth of a Gaussian chemical density profile generated by a single-puff source, as a function of the distance from the emitter

Again, we assume a stationary outward flow velocity profile, $\boldsymbol{V} = V_o\hat{\boldsymbol{r}}$, where $V_o$ is the magnitude of the velocity at $t = 0$. For this single-puff emitter solution to the diffusion equation (8.4) with a Gaussian density profile, we wish to determine whether the $\overrightarrow{GDMF}$ of fluxotaxis is equally as effective as the density gradient ($\overrightarrow{DG}$) of chemotaxis, as it is in the case of a continuous emitter.

First, consider the $\overrightarrow{GDMF}$. To simplify the expression, let $A = N_p/(4\pi\mathcal{D}t)^{3/2}$ and $B = 1/(4\mathcal{D}t)$. Then after a sequence of mathematical operations [232], we find that

$$\overrightarrow{GDMF} = -2mV_oAB\left(3 + \frac{1}{Br^2} - 2Br^2\right)e^{-Br^2}\hat{\boldsymbol{r}} .$$

For plume-tracing robots inside the inflection point $r_i = 1/\sqrt{2B}$ (i.e., the region where CPT is predictive), the $\overrightarrow{GDMF}$ vector correctly predicts the source direction to be

$$\hat{s}_{\text{GDMF}} = \frac{\overrightarrow{GDMF}}{|\overrightarrow{GDMF}|}$$

$$= -\hat{\boldsymbol{r}} .$$

Next, consider a chemotaxis-driven robot, which tracks the density gradient $\overrightarrow{DG}$. For the same Gaussian density diffusion function, mathematical operations [232] result in:

$$\overrightarrow{DG} = -2ABre^{-Br^2}\hat{r} .$$

While this simpler scheme also predicts the correct source direction, i.e.,

$$\hat{s}_{DG} = \frac{\overrightarrow{DG}}{|\overrightarrow{DG}|}$$

$$= -\hat{r} ,$$

the problem with chemotaxis-guided search is that for the radial flow configuration, where the chemical density profile contains a maximum point at $r = 0$ (see Fig. 8.3), the magnitude DG of $\overrightarrow{DG}$ goes to zero when $r \to 0$, i.e.,

$$\lim_{r \to 0} |\overrightarrow{DG}| = \lim_{r \to 0} 2ABre^{-Br^2}$$

$$= 0 .$$

This means that as chemotaxis-operated vehicles approach the chemical emitter, their navigation vector $\overrightarrow{DG}$ disappears.

In contrast to this, the quality of the $\overrightarrow{GDMF}$ prediction is maximized at the emitter location. In particular,

$$\lim_{r \to r_i} |\overrightarrow{GDMF}| \to 8mV_oABe^{-1/2} > 0 , \text{ and} \tag{8.9}$$

$$\lim_{r \to 0} |\overrightarrow{GDMF}| \to \infty , \tag{8.10}$$

where $r_i$ denotes the inflection point. This result indicates that the $\overrightarrow{GDMF}$ vector is an effective and accurate source direction predictor for all of the relevant radial domains. Apparently, the magnitude of the $\overrightarrow{GDMF}$ is highest near the origin of the plume, since $|\overrightarrow{GDMF}| \to \text{MAX}$ as $r \to 0$. The implication is that the predictive capability of $\overrightarrow{GDMF}$ increases as fluxotaxis-controlled vehicles approach the emitter. In practice, an important additional benefit of the fluxotaxis approach is that the plume-tracing region where $\overrightarrow{GDMF}$ correctly finds the emitter *grows* in time as $\Delta r \propto \sqrt{t}$.

These results are surprising. Apparently, despite the fact that we have chosen environmental conditions that are expected to be most favorable for chemotaxis and we have set up chemotaxis as the "golden standard" for this chemical flow regime, fluxotaxis has proven to be superior to chemotaxis on such problems when the emitter generates a single puff of chemical emission.

Although both algorithms point in the correct source direction, as the robots get closer to the source emitter, chemotaxis loses its predictive ability, but the predictive ability of fluxotaxis improves.

What is the intuition behind this seemingly counterintuitive result? The intuition lies in the derivative calculations of $\overrightarrow{GDMF}$ versus $\overrightarrow{DG}$. The chemotaxis $\overrightarrow{DG}$, i.e., $\boldsymbol{\nabla}\rho$, uses only the first derivatives of $\rho$ as guidance, whereas the fluxotaxis $\overrightarrow{GDMF}$, i.e., $\boldsymbol{\nabla}\left[\boldsymbol{\nabla} \cdot (\rho\boldsymbol{V})\right]$, also uses the second derivative. In a manner analogous to improving approximations with a Taylor series by adding the higher-order derivative terms, we can likewise improve the predictions of our navigation formula by adding higher-order derivatives.

## 8.6 Summary and Conclusions

This chapter has presented a physicomimetic algorithm for solving the chemical source localization problem, and compared it with the leading (biomimetic) alternatives. This algorithm, called "fluxotaxis," is designed based on the fundamentals of fluid physics. It is also especially "swarm friendly" because it uses the swarm robots as computational units in a mobile differential computer leading toward the source emitter. The algorithm has the desirable proven properties of avoiding misleading sinks and zeroing in on potentially toxic sources of chemical emissions. Furthermore, fluxotaxis does not require global or fully distributed knowledge, maps, explicit swarm consensus, or any other extra mechanisms to work. Formation movement is driven by local neighbor-shared values, resulting in *implicit* consensus without explicit voting or other costly computational overhead.

The chapter concludes with analytical results and intuitions demonstrating that fluxotaxis has equal or better predictive capabilities than its leading competitors, which are biomimetic. Because of the importance of the CPT application, the next chapter explores fluxotaxis in the context of experimental simulations.

**Acknowledgements** We are grateful to Dr. Michael Pazzani, formerly with the National Science Foundation, for partially supporting this project.

# Chapter 9
# A Multi-robot Chemical Source Localization Strategy Based on Fluid Physics: Experimental Results

Dimitri V. Zarzhitsky

## 9.1 Introduction

Inspired by the theoretic analysis of our physics-based solution for the *chemical plume-tracing* (CPT) problem (see Chap. 8), we were eager to find out how the predictions regarding the use of chemical mass flux as an indicator of the location of a chemical emitter would scale to teams of mobile robots. Just as we did in the earlier chapters, we continue to refer to the group of mobile robots as a *swarm*, and for our baseline, we require only a small set of restricted capabilities in terms of each vehicle's onboard sensing, communication, and computation equipment. Even though the individual robots are limited in what they can do on their own, operating as a team they can solve a challenging search and localization task. Here, we will show that the principle of the "whole being greater than the sum of its parts" plays a critical role when designing and implementing a distributed CPT solution. Most of the material discussed in this chapter originated in the context of an empirical evaluation effort that compared our physics-based CPT strategy, called *fluxotaxis*, against two of its main leading biomimetic competitors: *chemotaxis* and *anemotaxis*. As explained at length in Chap. 8, the fluxotaxis CPT strategy takes into account the fundamental physical laws that govern fluid flow, using the virtual surface formed by the robots to measure and trace changes in the chemical mass flux when searching for the origin of the plume. In contrast, the two biomimetic approaches essentially emulate food and mate seeking behaviors observed in insects and crustaceans.

To ensure an accurate and fair comparison, we formulated new environmental and goal forces for the multi-robot versions of chemo- and anemotaxis CPT algorithms, and collected plume-tracing performance data from four

Dimitri V. Zarzhitsky
Jet Propulsion Laboratory, California Institute of Technology, Pasadena, California, USA,
e-mail: Dimitri.Zarzhitsky@jpl.nasa.gov

W.M. Spears, D.F. Spears (eds.), *Physicomimetics*,
DOI 10.1007/978-3-642-22804-9_9,

different experiments, split evenly into two separate studies, spanning an exceptionally wide range of parametric conditions and plume-tracing performance metrics, which we tailored specifically for swarm-oriented algorithms. We started out by focusing on the issues that affect collaborative control, and then looked at how task performance was affected by the number of collaborating robots. Each experiment evaluated one particular aspect of the CPT problem according to performance measures that capture the most salient characteristics of each plume-tracing technique. It is important to note that the physicomimetics design philosophy was applied consistently in each experiment. Physicomimetic vehicle formations acted as a distributed sensor network by monitoring fluid flow in real time. Each robot, working as a member of a cooperative team, used its low-power, embedded microcontroller to combine observations from the onboard sensors with the information shared by its neighbors to perform chemical detection and distributed analysis of the fluid environment.

Incidentally, as part of a related research effort, we carried out a number of laboratory experiments with real chemical plumes (see Sect. 9.2), but for the purposes of our present discussion, we focus on the use of computational methods for simulating both the chemical environment and the plume-tracing robots. This restriction is motivated by two practical constraints. First, the high costs associated with laboratory work using real chemicals and robots would prematurely limit us to just a handful of experiments, which in turn would prevent us from fully understanding the pros and cons of the swarm-based solution. The other reason we opted to rely on physically accurate numerical simulations is the unprecedented degree of control over the environment that this alternative provided. In fact, more than 40,000 distinct plume and swarm configurations were modeled in simulation, and we meticulously recorded and tracked each parameter of interest in order to understand how it affected the robots' performance. Many representative fluid conditions were created with the help of a software toolkit incorporating a computational fluid dynamics (CFD) solver. As explained in Sect. 9.3, different numerical models of the chemical flow served as the foundation for the comparative study of various plume-tracing methods under controlled conditions. Sections 9.4.4.4 and 9.4.5.4 describe two software implementations of fluxotaxis: the first is a faithful model of a laboratory-scale prototype of seven CPT robots, and the second can accommodate much larger teams of plume-tracing vehicles. The aim of the first study, presented in Sect. 9.4.4, is to demonstrate how changes in different parameters for the physicomimetic formation control algorithm affect a small contingent of seven vehicles during plume tracing. Section 9.6 documents the second study dedicated to understanding the effects of increasing swarm size on chemical emitter localization performance. Early experiments, summarized in Sect. 9.5.2, suggested that one important advantage of our physicomimetic approach is its natural integration of obstacle avoidance capabilities directly into CPT navigation routines. Further analysis in Sects. 9.6.1 and 9.6.2 confirmed that even a very large swarm can navigate around obstructions in its path in much the same way that a fluid flows around objects, with the virtual vehicle-to-vehicle formation forces mimicking the role of real molecular bonds.

Since the physicomimetic foundation of fluxotaxis makes it amenable to standard mathematical analysis techniques, we can explain why the fluxotaxis approach is inherently more effective at finding and identifying sources of chemical plumes than the biologically motivated CPT methods. Indeed, one common observation from all of the tests is that robots controlled by fluxotaxis were more likely to satisfy multiple objectives of a CPT mission, whereas chemo- and anemotaxis teams rarely excelled in more than one performance category. Keep in mind that the long-term goal of this research effort is a fully functioning swarm of autonomous robots capable of performing cooperative chemical source localization in both indoor and outdoor environments. It is our hope that this chapter will clearly illustrate many of the specific concepts and mechanisms we use to guarantee superior swarm performance, including the exploitation of the underlying fluid and other physical principles, their numerical implementation, and the practical techniques for navigating teams of robots through obstacle-dense environments. The main "take-away" point for the reader should be the experimentally proven fact that chemical plume tracing is best tackled with many robots. However, our analysis also reveals that a "many-robot" solution in and of itself does not necessarily increase performance—the swarm must be managed in an intelligent and robust manner in order to realize the full benefits of the multi-robot solution. Designing a scalable, distributed, fault-tolerant control architecture that scales efficiently on an arbitrarily large fleet of vehicles constitutes a formidable challenge. This chapter is a practical, hands-on illustration of how we accomplished this task using physicomimetics as our guide.

## 9.2 Motivation and Background

One of the main reasons we developed an interest in the chemical plume-tracing problem is its relevance to the state-of-the-art research in distributed robotics. Autonomous robotic solutions for handling emergency situations, such as a chemical spill at a manufacturing facility, a deliberate terror act, or a search-and-rescue effort after a natural disaster, can address the critical need of minimizing human exposure to dangerous toxins. Likewise, the potential to deploy and operate rugged machinery for long, uninterrupted periods of time, which in turn translates into a faster and more effective neutralization of the environmental threat, further motivates us to invest time and resources into researching and developing the required technological "building blocks," from which capable and robust systems can be built. In this context, the practical usefulness of such systems is often measured in terms of the amount of human supervision needed to oversee the mission, as well as the amount of logistical support needed to ensure the system's effectiveness. In the ideal case, the role of the human "operator" is actually more along the lines of an "observer," because once deployed, the robots should be able to self-configure, adapt to the changing environment, and perform their task with minimal dependence on the existing infrastructure for vehicle-to-vehicle communication, localization, and perception. Note that we have assumed right from the onset a distributed, multi-vehicle, "swarm" approach to solving the problem. It should be intuitive that for the type of large-scale, unexpected, incident–response situations we are considering, the multiplicity aspect is going to be important. After all, the bigger the fire, the more firefighters you will need to put out the blaze quickly!

It turns out that the same conclusion holds for the problem of locating sources of a chemical contamination, as we will demonstrate with experiments in the later portions of this chapter. But first, let us begin with a somewhat simplified example, just to motivate our approach and to introduce some common terminology that will make it easier to understand the more technical sections that follow. Assume that we have a small robot equipped with chemical sensors, and we want to program it to detect and find a leaking laboratory container filled with a volatile organic compound (such as ammonia, acetone, or alcohol). One of the simplest methods that comes to mind is that of chemical gradient following—just like you can usually identify a nearby kitchen in an otherwise unfamiliar building by the smell of food being cooked, the robot can locate the chemical emitter by sniffing the air and moving toward higher concentrations within the chemical plume emanating from the source. Of course the farther away you are from the source, the more difficult it becomes to find it, since the air tends to mix and disperse the signal, making it harder to identify the correct direction toward the origin. You could recruit a few friends to help you search, and by spreading out over a larger area, you might have better luck in finding the origin of the smell. To test exactly this hypothesis, we conducted a few experiments in a controlled indoor environment, where we opened a small jar with liquid ethanol and observed how well one robot (called the MAXELBOT) can track the resulting vapor cloud as opposed to three identical robots organized in a triangular formation (using the physicomimetic methods explained in the earlier chapters of this book). All of the robots had Figaro TGS 2620 chemical sensors for detecting the ethanol. A robot motion trace capture from one of the test runs is depicted in Fig. 9.1, and the summary of the outcomes from the entire experiment is given in Table 9.1. The conclusion here is unambiguous—the single robot simply does not perform as well as the multi-robot version on any of the performance categories that matter. Understanding why this happens is precisely the question that motivated the work described in the remainder of this chapter.

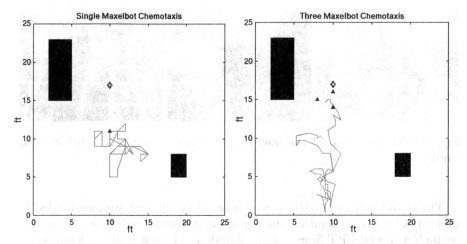

Fig. 9.1: Movement traces for the single robot (left) and a triangular lattice of robots (right) while tracing a concentration gradient of ethanol vapor. The starred diamond marks the location of the ethanol emitter, and bulky laboratory equipment is shown with black rectangles

Table 9.1: CPT performance of the swarm-based robot implementation outperforms the single robot configuration on each evaluation metric. The performance metrics are defined in Sect. 9.5.1.3

| PERFORMANCE METRIC | SINGLE ROBOT | THREE ROBOTS |
|---|---|---|
| Success Rate | 22.2% | 60.0% |
| Search Time | 630.0 sec | 415.0 sec |
| Emitter Proximity | 2.1 m | 1.6 m |

## 9.2.1 Chemical Plume

Although our first robot experiment with the real chemical plume showed that the robots can locate the source, the uneven, back-and-forth path they took (see Fig. 9.1), and the long time required for the search, made it very clear that we would need to make considerable and numerous improvements to this basic setup before we would be able to construct a practical CPT system. One of the first questions we had to answer concerns the way in which the moving robot affects the plume, because that determines where and how the chemical and air flow sensors should be mounted on the vehicle. We carried out several flow visualization studies using smoke as a tracer, as shown in Fig. 9.2, to get a better insight into the types of chemical flow structures the robots might encounter during a CPT mission.

Fig. 9.2: Visualization of the flow of air around a small CPT robot using smoke

The complex smoke patterns we observed meant that we also needed to understand how the chemical plume grows and changes with time. To do that, we built a special CPT flume—a small-scale wind tunnel that was optimized for conducting chemical studies over a flat region of enclosed space with controlled air velocity. In order to maximize the usefulness of this experimental testbed for our robotics work, we used the same sensing and processing components as would be used on the actual robots, so that any data we collected could be used as a baseline for predicting and analyzing robot behavior. Figure 9.3 shows the flume in operation, connected to a visualization workstation, a chemical sensor array installed inside the test chamber, and a small chemical diffuser positioned at the front air inlet. This hardware enabled us to collect a large set of chemical data that empirically captured the impact that the airflow conditions, the type of the chemical emitter, and the robots'

Fig. 9.3: A photo of the CPT flume in operation, showing a configuration with 32 chemical sensors affixed to a measurement scaffold inside the flume, a pressurized chemical emitter below the air inlet filter assembly, and a PC workstation displaying a real-time visualization of the plume

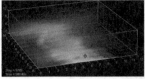

Fig. 9.4: False-color three-dimensional interpolation of changing plume density inside the CPT flume

positions all have on the development of the plume. To help visualize the chemical during the post-processing phase of the analysis, we used a false-color black and blue mapping for low chemical densities, green and yellow for medium density, and red for high density chemical regions (one example of this representation for an evolving plume is rendered in Fig. 9.4).

However, not long after we started working with the flume it became apparent that the amount of effort needed to collect this chemical data and the number of prerequisites for conducting experiments with real, large-scale plumes were all prohibitively high, and that the best we could hope for was to use our laboratory setup for verification and validation of a complete CPT solution, but the development, debugging, testing, and analysis of many individual steps and modules of the complete system all had to be performed "offline"—using numerical models and software simulations. But of course we did have the advantage of having access to the real data, so that we were able to make periodic cross-checks to ensure that the conclusions we made as the result of simulated CPT scenarios were still valid when applied to a real chemical environment.

## 9.2.2 CPT Algorithms

In order to solve the CPT problem with robots, they need to be programmed with some sort of CPT algorithm (terms such as "method," "approach," or "strategy" are used interchangeably in this context). Many of the algorithms in current use are based on olfactory systems of living organisms, e.g., insects such as moths, land animals, and aquatic creatures (especially crustaceans). In other words, most of the CPT strategies that have been tried and evaluated to date are *biomimetic*, i.e., they are designed to mimic biological systems. There are two general algorithms, called *chemotaxis* and *anemotaxis*, that are representative of a large proportion of biomimetic approaches, and we describe them very briefly in the next two sections. Since the goal of this book is to explain the principles of physicomimetic design, our main interest lies in physics-based systems and methods, such as the *fluxotaxis* CPT algorithm first described in Chap. 8. Here, we will not redevelop the fluid dynamics theory that forms the foundation for fluxotaxis, but rather simply restate the main points of Chap. 8 for easier reference.

### 9.2.2.1 Chemotaxis

*Chemotaxis* is the type of a greedy search strategy we used in our first robot CPT example, where robots followed a local gradient of the chemical concentration within a plume [132, 146]. In general, chemotaxis is very simple to perform, but it blindly favors regions with high concentration, which may not contain the chemical source.

### 9.2.2.2 Anemotaxis

*Anemotaxis* is essentially an upstream follower—an evolutionary survival strategy often exploited by predators while hunting [78, 86]. When implemented on robotic vehicles, the speed of the airflow is frequently taken into account, and we have adopted this modification in our simulations for consistency with the solutions in the literature [103]. Although anemotaxis is very effective for some specific flow conditions, it can also mislead the robots to a wind source that is not the chemical emitter [9].

## 9.2.3 Fluxotaxis for CPT

Chapter 8 explains how we used the knowledge of the physical principles that control the flow of fluids to construct a new physicomimetic CPT method. Here, we simply restate some of the key concepts and findings from Chap. 8, and introduce the basic mathematical notation that is needed to understand our analysis of the experimental results in this chapter. Throughout this work, we often discuss the chemical density (also called "concentration") $\rho$, and the fluid velocity $V = u\hat{i} + v\hat{j}$, where $\hat{i}$ and $\hat{j}$ are the usual unit vectors along the $x$-axis and $y$-axis in a two-dimensional plane. Because of the somewhat limited number of chemical flow scenarios likely to be encountered by a ground vehicle tracing an airborne plume, we make the simplifying assumption here that the chemical pollutant is fully dissolved in the carrier fluid, so that the $V$ term represents the combined chemical flow.

Formulation of the fluxotaxis algorithm revolves around the physical concept of *chemical mass flux*. Simply put, the mass flux is a coupling of the chemical density, $\rho$, and the fluid velocity, $V$. Its formula in two-dimensional space (after application of the calculus Product Rule) is

$$\nabla \cdot (\rho V) = V \cdot \nabla \rho + \rho \nabla \cdot V$$
$$= u\frac{\partial \rho}{\partial x} + \rho\frac{\partial u}{\partial x} + v\frac{\partial \rho}{\partial y} + \rho\frac{\partial v}{\partial y} . \tag{9.1}$$

If chemical mass divergence is positive, it describes a *source* of mass flux, such as a chemical emitter that is leaking a toxin into the air. Otherwise, if it is zero, it means that there are no detectable chemical changes in the flow, and when it is negative, it represents a flux *sink*, or a region of space which is undergoing a local (and usually temporary) buildup of the chemical without the emitter. The fluxotaxis CPT strategy causes the robots to trace the gradient of Eq. 9.1 to the source, i.e., the swarm follows the *gradient of the divergence of mass flux*, abbreviated $\overrightarrow{GDMF}$, where the gradient is simply a vector with the steepest increase. The mathematical formula for the $\overrightarrow{GDMF}$ in two dimensions is

$$\nabla\left[\nabla \cdot (\rho V)\right] = \nabla\left(u\frac{\partial \rho}{\partial x} + \rho\frac{\partial u}{\partial x} + v\frac{\partial \rho}{\partial y} + \rho\frac{\partial v}{\partial y}\right) . \tag{9.2}$$

Chapter 8 contains the proofs of the following two theorems, which are quite useful when interpreting results of our CPT simulations:[1]

- *Source Theorem:* Both fluxotaxis and chemotaxis will move robots toward a chemical source, i.e., a region with positive chemical mass flux divergence.
- *Sink Theorem:* Fluxotaxis will lead robots away from a chemical sink, i.e., a local density maximum that is not a true emitter. On the other hand, chemotaxis will fool the robots by misleading them right into the sink.

Because Eqs. 9.1 and 9.2 make extensive use of partial derivatives, such as $\partial/\partial x(u)$, it makes sense to look at a practical way of computing them on the robots. Table 9.2 specifies a *discrete* approximation for some of these derivatives using a mathematical method known as *finite differences*. This formulation assumes that we can measure values of a function, such as $u$— the component of fluid velocity along the $x$-coordinate dimension, at discrete points in space. Then, using a Taylor's series approximation [75], the value for different orders of the derivative, with varying degrees of accuracy, can be estimated. Note that the central difference technique in Table 9.2 has the lowest discretization error, on the order of $O[(\Delta x)^2]$, but it also uses values from data points that are more "spread out" around the point at which the derivative is being computed. In practice, this means that the underlying sensor grid which provides these values must be large, and have a geometric shape that helps maximize the information content of the sensor readings.

Of course, at this point it is easy to recognize the value of physicomimetic swarm formations (a few examples are pictured in Fig. 9.5) for performing distributed sensing and computation: the CPT algorithms can treat the robots

---

[1] These theorems are stated slightly differently for our purposes here, but they follow directly from the material in Chap. 8.

as sensor nodes for measuring both spatial and temporal flow characteristics at specific locations in space. Since the robot formations are dynamic, capable of adapting to a local geometry near obstacles, building corners, etc., and each vehicle is equipped with a sensor that can estimate the range and bearing to neighboring robots, each member of the team can continually monitor as well as influence the local topology of the virtual sensor grid. Furthermore, each vehicle can be thought of as a discrete vertex in a large, mobile sensor array, where the onboard processor can "solve" these differential equations by using the sensor data obtained from the nearby robots. In this way, the gradients and derivatives in the mathematical description of the chemical plume can be resolved and then translated into navigation decisions by the CPT algorithm. This is why obtaining an accurate model of the plume is important for algorithm development, and the next section provides an overview of the computational fluid dynamic approach to chemical plume generation.

## 9.3 Numerical Model of the Chemical Plume

In order to develop and evaluate new multi-robot CPT strategies, we needed a way to create many different types of chemical plumes. Using real chemical pollutants in a controlled environment was impractical, due to the high costs and the effort involved. Instead, we followed the fundamental paradigm shift that occurred in the field of aircraft design, where the traditional use of ex-

Table 9.2: Useful finite-difference approximations for partial derivatives of a continuous function $u(x, y)$ evaluated at discrete points $(i, j)$ in a sensor grid. See [247] for the derivation and additional details

| | |
|---|---|
| FORWARD DIFFERENCE | $\left.\dfrac{\partial u}{\partial x}\right|_{i,j} = \dfrac{u(i+1, j) - u(i, j)}{\Delta x} + \mathrm{O}(\Delta x)$ |
| | $\left.\dfrac{\partial^2 u}{\partial x^2}\right|_{i,j} = \dfrac{u(i, j) - 2u(i+1, j) + u(i+2, j)}{(\Delta x)^2} + \mathrm{O}(\Delta x)$ |
| BACKWARD DIFFERENCE | $\left.\dfrac{\partial u}{\partial x}\right|_{i,j} = \dfrac{u(i, j) - u(i-1, j)}{\Delta x} + \mathrm{O}(\Delta x)$ |
| | $\left.\dfrac{\partial^2 u}{\partial x^2}\right|_{i,j} = \dfrac{u(i, j) - 2u(i-1, j) + u(i-2, j)}{(\Delta x)^2} + \mathrm{O}(\Delta x)$ |
| CENTRAL DIFFERENCE | $\left.\dfrac{\partial u}{\partial x}\right|_{i,j} = \dfrac{u(i+1, j) - u(i-1, j)}{2\Delta x} + \mathrm{O}[(\Delta x)^2]$ |
| | $\left.\dfrac{\partial^2 u}{\partial x^2}\right|_{i,j} = \dfrac{u(i+1, j) - 2u(i, j) + u(i-1, j)}{(\Delta x)^2} + \mathrm{O}[(\Delta x)^2]$ |

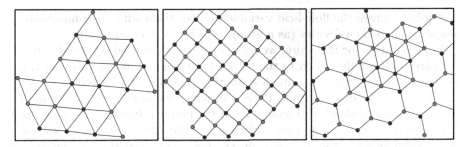

Fig. 9.5: Representative types of physicomimetic grid formations. In the hexagonal lattice on the right, note the non-uniform spacing between different nodes in the lattice. This ability enables the swarm to quickly increase sensor coverage of a particular area during a mission. Each vehicle computes the physicomimetic control forces independently, which means that all of the swarm's navigational decisions are fully decentralized and distributed

perimental wind-tunnel testing has been largely supplanted by computational methods, and adopted a strategic decision to concentrate on computational models of the plume. This section explains the most important aspects of modeling the properties and long-term behaviors of chemical plumes using computational fluid dynamics, with a particular emphasis on the plume tracing task.

Chapter 8 motivated a categorization of the CPT task as a combination of two conceptually distinct *forward* and *inverse* solutions. The "forward" solution consists of a time-accurate description of the chemical flow, while the "inverse" solution focuses on the plume-tracing algorithm for finding the source emitter. Numerical modeling supplies the majority of the data for an incremental design and refinement of the CPT strategy, which in turn reduces the number of validation experiments required with the actual robots and chemical plumes. This section focuses on the forward solution; Section 9.4 will address the inverse solution.

To accurately simulate the environment, we solve a system of equations that describe the physical flow of fluids. Because analytic (i.e., closed-form) solutions for the general case do not exist, computational techniques are needed to predict what the fluid will look like within a region of space at a given point in time. This fluid, which may be in a liquid or gaseous state, is described by several physical quantities, such as density, temperature, pressure, velocity, and so on. Collectively, these are called the *flow field* variables (see Chap. 8 for more information). The output of a numerical flow simulation consists of a set of time-indexed matrices containing flow field variables calculated at discrete positions in space, i.e., at "grid points." We construct a mapping between these grid points and the robots, motivated by our view of a team of cooperating robots as a distributed, adaptive, computational mesh that jointly senses the flow-field variables, shares them with their immediate neighbors, and then decides (as a group) in which direction to move.

In preparation for this work, we researched and evaluated several different software packages described in the CPT literature, and as the result of this investigation, we selected an implementation developed by Farrell et al. [59] at the University of California, Riverside, which we believe to be the most practical and well-developed among all of the plume solvers accessible to us. It is optimized for computational efficiency, while also serving as a realistic and faithful model of the environment, i.e., both its immediate and long-term distribution statistics match closely the observations of actual airborne chemical plumes. A notable feature of this computational fluid dynamics (CFD) application is its multi-scale aspect, including molecular diffusion of the chemical and the advective transport due to wind movement. Rather than a more conventional, continuous, time-averaged model, Farrell's framework models the plume as a collection of filament-based emissions of chemical "puffs." Air currents smaller than the mean distance between puff centers are modeled as a white noise process. In other words they "mix" the puff components, mimicking the effect of small-scale flow turbulence on the chemical plume. Air currents on the order of the puff size induce growth and distortion of the puffs' template shape, and are therefore modeled using differential equations. Fluid advection (obtained via a numerical solution) transports each puff as a whole, thus causing the ensemble of puffs to appear as a meandering sinuous plume, as visualized in Fig. 9.6.

Because of the open source license for Farrell's work, we were able to reimplement the original simulator to better fit our current research goals and to adapt it to newer, more advanced workstation hardware. Our reimplementation improves the computational efficiency of the code via memory

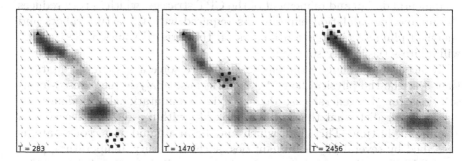

Fig. 9.6: CPT simulation with a seven-vehicle physicomimetic lattice (black rectangles), no obstacles, and a meandering plume originating in the top left corner (higher chemical concentration is shown with darker colors, and the arrows denote the airflow). The lattice starts out in the lower right corner and uses the fluxotaxis algorithm to successfully trace the plume to its source (the triangle)

optimization methods and multi-threading technologies. Our new fluid model is also more general than that in the original code because it allows for solid obstacles to be introduced into the flow. It is very important to emphasize that the model is accurate for a wide range of fluids, and is capable of resolving dynamics of both gaseous and liquid plumes. The next few sections provide a more detailed description of the plume simulation model.

### 9.3.1 Advection Due to Wind

Fluid velocity is generated using a simple procedure. First, random wind vectors are placed in the four corners of the rectangular mission area. Then, using linear interpolation between the four corners, new velocity vectors are computed along each of the boundary edges, such that at the end of this step, wind velocity is completely specified on the world boundaries (this is known as a *Dirichlet* condition). Once the boundary conditions are determined, we use a second-order accurate forward Euler in time, central difference in space numerical approximation algorithm [75] to calculate wind velocity across the interior points. In the current version all grid spacings are uniform, and wind velocity is obtained directly from the discretization equation.

### 9.3.2 Mixing Due to Random Velocity

During each step of the simulation, a random amount $\sigma$ (i.e., white noise) is added to the advective velocity vector to model the diffusion of the plume about its centerline (i.e., the mixing of the filaments within the plume due to the molecular Brownian motion). The implementation of time correlation, bandwidth, and gain in the wind vectors follows the exact specification and code examples provided in [59].

### 9.3.3 Growth Due to Diffusion

It is assumed that each puff of the chemical ejected by the emitter has the same "template shape," and for convenience and efficiency, it is treated as a sphere (and approximated by a disk in two-dimensional simulations) when the plume density is computed at each location. Thus the only parameter that affects the diffusion model is the size of each filament $k$, with the radius $r_k(t)$ at time $t$ specified by

$$r_k(t) = (r_o^{\frac{2}{3}} + \gamma t)^{\frac{3}{2}} , \tag{9.3}$$

where $r_o$ is the initial radius of a puff when it is first ejected by the emitter, $t$ is the age of the puff (i.e., the number of time steps since emission), and $\gamma$ is the volumetric diffusion rate.

### 9.3.4 Density Computation

The plume simulator only needs to keep track of the movement of the chemical puff filaments, driven by the wind and dispersed by intermolecular forces. However, since we need a density map of the entire chemical plume for the tracing experiments, the individual density contributions of each of the $M$ puffs to the overall volumetric chemical concentration $\rho(x, y, t)$ at point $(x, y)$ and time $t$ are determined using the following relation:

$$\rho(x, y, t) = \sum_{k=1}^{M} \rho_k(x, y, t),$$

where

$$\rho_k(x, y, t) = \frac{Q}{\sqrt{8\pi^3} r_k(t)^3} \exp\left(\frac{-S_k^2(t)}{r_k(t)^2}\right) \qquad (9.4)$$

and $Q$ is the amount of chemical contained in each puff, $S_k(t)$ is the distance between the $k$th puff center at time $t$ and the point $(x, y)$, for which the effect of puff $k$ with the radius $r_k(t)$, as defined in (9.3), on the density $\rho(x, y)$ is calculated. Farrell et al. [59] employed this update rule to reduce a spherical puff into a flattened, disk-like filament, assuming that the chemical has a Gaussian (i.e., normal) distribution within the puff. Once the plume concentration is thus calculated for the entire search area at each time step, we can start the actual mission with the CPT robots—a task that is explained in detail in the following section.

## 9.4 Swarms of Chemical Plume-Tracing Robots

The software framework we just described gives us the forward solution, i.e., it supplies us with large amounts of high resolution plume data. In this section we focus on the inverse solution, and explain how we develop and test new CPT strategies. A review of related literature (see Chap. 8), shows that even recently developed CPT strategies tend toward either non-cooperative or centralized emitter localization algorithms [146]. However, our goal is the development of a robust, fully distributed, multi-robot (i.e., swarm) solution to the CPT problem. To meet this primary objective, two other problems must be solved: the first is the cooperative control of the swarm, and the second is the extension/adaptation of the CPT strategies to a distributed, decentralized grid of mobile sensor nodes. The next few sections describe our methodology for addressing these challenges.

## 9.4.1 Physicomimetics Control Architecture

The physicomimetic design method provides all of the essential control technologies that make it possible for us to achieve CPT objectives using a mathematically consistent approach. Of course one of the core ideas of this book is that physicomimetics is a powerful swarm design tool, optimized for self-assembly and self-repair of robotic lattices, where dynamic grid-like vehicle formations are constructed via short-range (i.e., local) virtual physics forces.

A high-level description of our physicomimetic control algorithm implementation is consistent with the basic concepts explained in the earlier chapters. During each decision cycle (or time step), every team member observes its neighbors and the environment. Each unit then (independently) calculates the virtual forces imposed upon it by nearby objects, e.g., neighboring vehicles, obstacles, etc. After taking a vector sum of all virtual formation and goal forces acting on the agent, the onboard controller converts the net force into a velocity vector, which in turn determines the robot's next move. Once again, note that the physicomimetic swarm controller adheres closely to the physical model that governs the macroscopic behavior of real systems, which is why the equations of motion for our physicomimetic-controlled vehicle look identical to those commonly used in classical particle physics.

Here, we employ a control law based on the Lennard-Jones potential, which we generalize as

$$F_{\text{formation}} = \varepsilon \frac{\mathcal{F}(D)^{\alpha}}{r^{\beta}} - \kappa \frac{\mathcal{F}(D)^{\gamma}}{r^{\delta}} . \tag{9.5}$$

Assuming that all parameters are non-negative, the first term, $\varepsilon \mathcal{F}(D)^{\alpha}/r^{\beta}$ describes the attractive component of the formation force, and the second term, $\kappa \mathcal{F}(D)^{\gamma}/r^{\delta}$ specifies the repulsive component. As in previous chapters of this book, $D$ and $r$ are the desired and actual inter-robot distances, respectively. Exponential coefficients $\alpha$, $\beta$, $\gamma$, and $\delta$ determine the distance proportionality of the force power law, and in typical usage, $\alpha < \beta \leq \gamma < \delta$. Optional scalar coefficients $\varepsilon$ and $\kappa$ are used to linearly scale the attractive and repulsive components of the formation force, which in turn affect the cohesion of the swarm by altering the strength of vehicle-to-vehicle bonds. The function

$$\mathcal{F}(D) = \left[ \frac{\varepsilon D^{\delta - \beta}}{\kappa} \right]^{\frac{1}{\gamma - \alpha}}$$

is used to compute the Lennard-Jones separation parameter in (9.5) for a given desired distance between the vehicles $D$, since in the Lennard-Jones

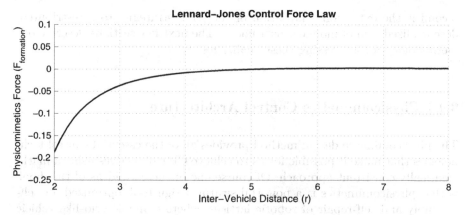

Fig. 9.7: Lennard-Jones physicomimetic force law with $D = 5$, $\varepsilon = \kappa = 1$ and $\alpha = 1, \beta = \gamma = 2, \delta = 3$. The $x$-axis is marked with units of distance and the $y$-axis with units of force

control law, the exact distance at which the formation force achieves equilibrium depends on all of the scalar and exponential parameters. Figure 9.7 shows a plot of (9.5), with $\varepsilon = \kappa = 1$ and $\alpha = 1, \beta = \gamma = 2, \delta = 3$, so that for this limited, simplified case $\mathcal{F}(D) = D$. The Lennard-Jones force is mostly repulsive, with a weak attractive component (in Fig. 9.7, $F_{\text{formation}}$ is greater than zero for $r > 5$, albeit by a small amount). Because the attractive component of the force is small, this control law is especially well-suited for constructing swarms with liquid-like virtual formation bonds, which is desirable for robots operating near obstacles or narrow passageways.

## 9.4.2 Simulated Environment

In order to faithfully emulate our physical vehicle platforms, the CPT robots in simulation are modeled as homogeneous disk-shaped entities of a fixed radius, so that each simulated vehicle occupies a small amount of space, which is an important prerequisite for accurate modeling of obstacle and collision avoidance. The robots' onboard sensor package consists of an anemometer, which can compute local wind velocity $V$, and a chemical concentration sensor, which measures and reports the value of $\rho$ when it exceeds a predetermined threshold. The simulator makes a simplifying assumption that the obstacle and chemical detectors are mounted at the center of the robot's circular body, so that keeping track of the current location of the vehicle is sufficient to determine which region of the environment is visible to the simulated sensors. To improve performance of memory management algorithms, we discretized the simulated environment into an array-like collection of cells.

Each of these cells contains the CFD solution for the chemical plume and the ambient flow, i.e., fluid velocity $V$ and chemical density $\rho$, within a specific region of the environment. A test configuration may also contain rectangular obstacles, which are impenetrable by both the fluid and the robots. For measurements of discretized quantities, such as the plume chemical density, the simulator maps the robots' real-valued coordinates into the corresponding two-dimensional environmental data array.

As mentioned briefly in the introduction, the long-term goal of this work is a fully distributed, swarm-based CPT system capable of finding chemical sources across a wide range of practical, real-world scenarios. We anticipate that early implementations of such a system will consist of ground-based robots and airborne chemical plumes, partly due to the more accessible and cost-effective logistics associated with the ground/air scenario; however, we should not forget that the theoretical foundation of this work is equally applicable to both two-dimensional and three-dimensional air, ground, and aquatic surface/subsurface platforms. As previously stated, a successful, swarm-based solution of the CPT problem requires advances in both the control and emitter localization algorithms, which is what motivated us to organize our experimental work into two major categories—the first one is focused on robot team control, and the second is more concerned with the plume tracing and emitter localization performance of the different CPT strategies, as explained below.

### 9.4.3 General Implementation Details

Most of the work we discussed so far originated as our response to the notable lack of cooperative, multi-robot solutions for the CPT problem. To address this unfortunate gap in research coverage, we carried out many numerical CPT simulations of different plume and swarm configurations in order to study the key aspects of cooperative CPT, such as distributed sensing, local information sharing, scalability, and robustness. To better manage such a large undertaking, we adopted a two part approach for our investigation. First, we studied the issues associated with physicomimetic control of CPT robots using a fixed number of vehicles. Once the control and communication technologies were designed, tested, and evaluated, we relaxed the fixed team size constraint, and explored scalability and the distributed aspects of swarm-based CPT as part of our second study.

All of the experimental results presented in Sects. 9.5 and 9.6 are based on a faithful software simulation of our laboratory plume-tracing robots and their chemical sensor payload [291, 294]. In simulation, swarm movement and vehicle formations are coordinated via strictly local interactions between agents, using the physicomimetics framework. We use a parameterized version of the Lennard-Jones formation force law (9.5), $F_{\mathrm{LJ}}$, instantiated as

$$F_{\text{LJ}} = \frac{D}{r^2} - \frac{D^{1.7}}{r^{2.7}} .$$

This equation gives the virtual formation force between two vehicles separated by a distance of $r$, for a lattice with the ideal vehicle-to-vehicle separation set to $D$. We selected these specific numeric values for the exponent coefficients based on the dimensions of our laboratory prototypes.

Chemical plume data for all of the experiments comes from the two-dimensional flow solver described in Sect. 9.3. We controlled the plume configuration in each of the four experiments according to a physically consistent model by changing the boundary conditions of the differential equations that describe the fluid. To ensure that these simulations have practical CPT relevance, each study is made up of a balanced mix of flow regimes, such as laminar, transitional, and turbulent. All source localization results we report here are based on a methodical combinatorial evaluation of each CPT strategy against identical, side-by-side pairings of flow conditions [295]. We performed several experiments for each environment and plume type with different lattice parameters, but the CPT objectives remained the same in each experiment—the lattice must first search the environment for the chemical plume (using the process called *casting*), and then determine the location of the single stationary source. In other words, each individual CPT run consists of a search for the chemical plume, followed by the trace to its source emitter. Because of some important implementation differences, we will explain the relevant details shortly.

But first, note that each CPT algorithm can be broken down into low-level lattice movement control routines and emitter localization functions. The low-level routines are responsible for moving each agent in a formation, executing collision avoidance, and reporting flow-field variable sensor readings when requested by the onboard CPT algorithms. Listing 9.1 shows the control decisions in every simulation step. Vehicle velocity is modified in response to the different constraints; thus the call to `ap_maintain_formation` will alter each robot's desired velocity according to the formation forces acting on the vehicle. The `agent_do_cpt_strategy` modifies the agents' velocities according to the current plume-tracing algorithm (implementation details are presented in the next few sections). Once the new velocity vector of each vehicle is computed, the final call to `move_agents_with_constraints` ensures that no agent goes out of the search area boundaries, that the agents' velocities are consistent with mechanical limits, and that there are no collisions between vehicles and obstacles. The robots employ a behavior subsumption algorithm to avoid colliding with obstacles and each other. In other words, the output of the CPT strategy is ignored when the collision avoidance behavior is triggered by the sensor module in the vicinity of obstacles or when other vehicles are detected closer than the safe separation radius.

Listing 9.1: Top-level control operations of the CPT simulator

```
ALGORITHM: CPT_simulation(lattice)
while ( TRUE )
    ap_maintain_formation(lattice)

    for agent in lattice
        agent_do_cpt_strategy(agent)
    end-for

    move_agents_with_constraints(lattice)
end-while
```

## 9.4.4 Seven-robot Lattice Implementation

To better understand the requirements of physicomimetic-based CPT, we
started out with a conceptually simple configuration of seven robots arranged
in a hexagonal formation (with one vehicle in the center of the formation).
The hexagonal lattice geometry was selected because it requires the least
amount of sensor information to construct and maintain [240], it represents a
fundamental structural element within larger physicomimetic swarms [241],
and it provides a computationally convenient surface for measuring the volumetric flow of chemical mass flux [247, 294]. The following few sections give
a more detailed description of the CPT strategies implemented as part of the
first seven-robot hexagonal lattice study.

### 9.4.4.1 Seven-robot Casting Algorithm

The sole purpose of the casting algorithm is to help the robots locate the
chemical plume. As is the case with most search algorithms, it is important to minimize search time by means of an aggressive exploration of the
environment. To help maximize spatial coverage, our implementation (see
Listing 9.2) combines the translational side-to-side, up-down motion of the
center vehicle with the periodic expansion and contraction of the outer ring
of the lattice, resulting in a pulsating motion. The center robot uses local
waypoints to navigate (i.e., the location of each new waypoint is computed
using the position data obtained by the vehicles on the lattice perimeter),
while the expansion and contraction of the lattice is implemented by changing the desired inter-vehicle separation $D$ on each robot. Each of the CPT
strategies (defined below) uses this casting method to find the plume at the

Listing 9.2: The casting algorithm implemented on a seven-robot lattice

```
ALGORITHM: cast
while ( TRUE )
   if ( lattice is expanding )
      if ( expansion factor is less than maximum )
         increment expansion factor
      else
         change lattice mode to contraction
      end-if
   else
      if ( expansion factor is greater than minimum )
         decrement expansion factor
      else
         change lattice mode to expansion
      end-if
   end-if

   radius = expansion_factor * D_o

   if ( horizontal advance is blocked )
      reverse horizontal direction
   end-if

   if ( vertical advance is blocked )
      reverse vertical direction
   end-if

   waypoint = direction_unit_vector * D_o
end-while
```

start of the mission, as well as during the tracing in cases where the robots unintentionally exit or otherwise lose track of the plume.

### 9.4.4.2 Seven-robot Chemotaxis Algorithm

The operational premise of the chemotaxis CPT algorithm comes from the intuitive fact that the concentration of the trace element increases in the vicinity of the chemical source. Therefore, one possible way to find the source is to follow the gradient of the chemical density. Since the gradient computation requires the chemical density reading at several spatially separated points, the chemotaxis algorithm relies extensively on the neighboring vehicles to share their chemical sensor information. The implementation of the algorithm in pseudocode is given in Listing 9.3.

**Listing 9.3:** The chemotaxis algorithm implemented on a seven-robot lattice

```
ALGORITHM: chemotaxis
while ( TRUE )
    ensure lattice radius and location are within limits

    if ( lattice is within plume )
        execute move_to_max_density
    else
        execute cast
    end-if
end-while


STRATEGY: move_to_max_density
    obtain the sensor reading of ρ across the lattice
    move to the location of the maximum ρ reading
```

#### 9.4.4.3 Seven-robot Anemotaxis Algorithm

The intuition behind the anemotaxis CPT strategy is to move the lattice upstream while keeping the vehicles inside the plume. This implementation of the anemotaxis algorithm (see Listing 9.4 for the pseudocode) is based on the examples in the literature; however, one notable improvement is the explicit averaging of the ambient wind direction, calculated by combining the information about wind conditions as reported by each robot in the lattice.

#### 9.4.4.4 Seven-robot Fluxotaxis Algorithm

Recall that our fluxotaxis approach for finding chemical emitters follows the gradient of the divergence of chemical mass flux ($\overrightarrow{GDMF}$) as a guide. As explained in Chap. 8, we originally postulated this methodology as a fluid physics based method for identifying chemical sources. But we also realized that the chemical flux conveniently combines the information about the chemical concentration and fluid velocity. We extended and adapted the basic $\overrightarrow{GDMF}$ flux method for traversing and searching an evolving chemical plume. The fluxotaxis algorithm presented here (see Listing 9.5) is one of two implemented versions.

This initial version of fluxotaxis emphasizes the interactions between the CPT algorithm and the physicomimetic control framework that manages the robot lattice, reflected prominently in the tight integration of the CPT actions

Listing 9.4: The anemotaxis algorithm implemented on a seven-robot lattice

```
ALGORITHM: anemotaxis
while ( TRUE )
    ensure lattice radius and location are within limits

    if ( lattice is within plume and wind sensors detect V )
        execute move_upstream
    else
        execute cast
    end-if
end-while

STRATEGY: move_upstream
    average the direction of V across the lattice
    move one time step along the -V direction at maximum speed
```

(e.g., acquiring plume sensor readings at different lattice radii, $D_i$) with the lattice formation control actions. Although we realize that this coupling does not support our overall goal of a scalable CPT algorithm for arbitrary-sized swarms, we created this version to study the impact of the lattice formation on the CPT algorithm's performance. The chem_region strategy in Listing 9.5 contains references to chemical *centroids*, $C_p$, which serve as waypoints for the lattice. The position of each centroid, $r_{C_p}$, is computed as a weighted sum based on the chemical density measurement reported by each robot. Mathematically, this computation is

$$r_{C_p} = \begin{bmatrix} x_{C_p} \\ y_{C_p} \end{bmatrix} = \frac{1}{N \sum_{i=1}^{N} \rho_i} \sum_{i=1}^{N} \rho_i \begin{bmatrix} x_i \\ y_i \end{bmatrix} ,$$

where $N = 7$ is the number of CPT robots in the lattice, $\rho_i$ is the output of the chemical detector on robot $i$, whose location is $(x_i, y_i)$.

When performing this flux computation, the robots distinguish between incoming and outgoing fluxes, as determined by whether the chemical flow is into or out of the lattice. This explicit separation of the two types of fluxes is necessary because in this study, the single, virtual, hexagonal "surface" constructed by the seven-robot lattice is the only available *control volume* suitable for calculating the chemical mass flux. In order to compute the gradient of the divergence of mass flux, the lattice expands its radius, making it possible to measure the surface flux across three virtual surfaces of increasing size. Because of the limited number of CPT agents, the gradient estimate based on the expanding-radius surface is consistent with the spatial charac-

**Listing 9.5: The fluxotaxis algorithm implemented on a seven-robot lattice**

```
ALGORITHM: fluxotaxis
while ( TRUE )
   ensure lattice radius and location are within limits

   if ( lattice is within plume )
      if more than 50% of total ρ is sensed by the center agent
         contract the lattice to minimal radius
      else
         execute chem_region
      end-if
   else
      execute cast
   end-if
end-while


STRATEGY: chem_region
   sense total lattice ρ over 3 different lattice radii

   compute ρ centroid Cp, where p ∈ RADIUS{inner,middle,outer}
   if ( ρ increases with each radial increase )
      move to the centroid of the centroids Cp
   else
      if ( outermost ρ is greater than innermost )
         move to the location of the Couter centroid
      else
         if ( ρ decreases with each increasing radius )
            execute flux_ring
         else
            execute cast
         end-if
      end-if


STRATEGY: flux_ring
   compute the maximum incoming flux, ρV, at 3 different lattice radii
   if ( maximum influx exceeds a flux threshold )
      move to the location of the maximum incoming flux, ρV
   else
      compute the maximum outgoing flux, ρV
      if ( maximum outflux exceeds flux threshold )
         move to the location of the maximum outgoing flux
      else
         execute cast
      end-if
   end-if
```

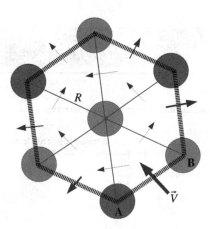

Fig. 9.8: A typical hexagonal lattice formation assumed in the first CPT study. The lattice radius, $D$, is a dynamic parameter that controls lattice expansions and contractions. The fluid passes through the virtual surfaces formed between the robots with velocity $V$. The chemical mass flux across a lattice edge, $\rho V$, can be estimated via an interpolation of the chemical density and flow velocity measurements obtained by the adjacent robots

teristics of our $\overrightarrow{GDMF}$ analysis, but it is not accurate with respect to time, since the plume continues to evolve and move while the lattice is undergoing the radial expansion. We address this limitation in our second study, when we extend the fluxotaxis algorithm to work with arbitrary-sized swarms.

In the flux_ring strategy, the robots measure the chemical concentration and the airflow across each of the outer edges of the lattice, first computing the maximum *incoming* flux, and if that metric is below a preset threshold, they attempt to find the location of the maximum *outflux*. Note that the incoming flux is indicative of the condition in which the plume "impinges" on the leading *edges* of the lattice. Thus the algorithm selects the location of maximum chemical influx in order to keep the robots inside the plume. Here, we use the term "lattice edge" as a label for the virtual connection between the outer edges of the hexagon formed by the vehicles, shown with bold black lines in Fig. 9.8. Fluxotaxis agents compute the chemical mass flux, $\rho V$, across each of the outer edges of the lattice using an average of the $\rho$ and $V$ values measured by each robot, and for the purpose of the flux computation, the distance between the adjacent vehicles is taken as the "surface area" of unit depth, matching the length of the virtual edge. If we denote the flux through the virtual edge between vehicle $A$ positioned at $(x_A, y_A)$ and vehicle $B$ at $(x_B, y_B)$ by $L_{\text{AB}}$, then the "origin" of this vector $(x_L, y_L)$ is taken as the midpoint of the $AB$ edge:

$$\begin{bmatrix} x_L \\ y_L \end{bmatrix} = \frac{1}{2} \begin{bmatrix} x_A + x_B \\ y_A + y_B \end{bmatrix} .$$

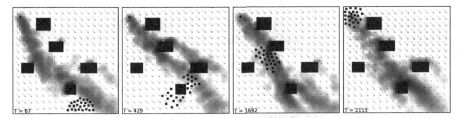

Fig. 9.9: CPT trace near obstacles (dark rectangles). Note how the swarm "breaks up" into two groups in order to negotiate the obstacle (second image from the left), and then "rejoins" due to formation and plume-tracing forces (second image from the right). After finding the emitter, the fluxotaxis-controlled swarm surrounds the source and continues to enclose it throughout the rest of the CPT mission (the rightmost image)

Any time the fluid flows *out* of the lattice, the averaged flux value is treated as the outflux by the fluxotaxis algorithm, and any time the fluid flows *into* the lattice, the flux calculation is treated as the influx. In the numerical portion of the code, this flux computation is performed using a standard technique of two-dimensional finite-volume discretization for control volumes [247].

### 9.4.5 Large Swarm Implementation

Our second CPT study removes the seven robot restriction. In fact, it makes no assumptions about how many vehicles are deployed, nor does it attempt to explicitly control the outermost shape of the virtual surface formed by the vehicles. Instead, it models a real-world situation, where each robot makes the navigational decisions based solely on the *local* information from its onboard sensors. This means that the number of neighbors that each vehicle has, the neighbors' positions and their sensor readings, as well as the behavior of a given CPT strategy all reflect dynamic swarm properties that update continuously during the simulation. Once again, we limit our attention to three different CPT algorithms and, just as we did in the previous study, we adapt the standard implementations of chemotaxis and anemotaxis from the literature, and extend the fluxotaxis method to an arbitrary-sized and shaped swarm.

During the mission, vehicles form many dynamically stable, embedded hexagonal formations as they move about the plume-tracing area. These virtual formation bonds often rearrange or break completely as the result of obstacle avoidance, and the movement of other neighboring vehicles, as shown in Fig. 9.9. The most important difference between this implementation of the swarm controller as compared to its counterpart in Sect. 9.4.4 is the lack

### Listing 9.6: Implementation of the swarm casting procedure

```
ALGORITHM: casting
    if ( horizontal advance is blocked )
        reverse horizontal direction
        broadcast new horizontal direction
    end-if
    if ( vertical advance is blocked )
        reverse vertical direction
        broadcast new vertical direction
    end-if

    velocity = direction_vector * time_step
```

of waypoint navigation. Instead, the vehicles' velocities depend on real-time changes in the local swarm formation topology, as well as the presence of obstacles or the chemical in the vicinity of the robots.

#### 9.4.5.1 Swarm Casting Algorithm

Regardless of the CPT strategy, an identical casting algorithm is used in each experiment to help the robots with plume localization. For this study, we implemented a modified casting procedure, detailed in Listing 9.6, that moves the robot along 45° diagonal paths. Near obstacles and search area boundaries, each vehicle executes a "bounce-like" reflection maneuver. To ensure the swarm has the ability to coordinate its movement during casting, we simulate an explicit communication message, for which we use a recursive local broadcast to synchronize each vehicle's vertical and horizontal movement bias, as shown in Listing 9.6. Each CPT agent starts out with the same casting state, and stops casting once the chemical plume is located. Each robot's casting state is distinct, so that a part of the swarm that has split up and moved out of the communication range can cast, while the remaining group can continue tracing the plume. Vehicles that are within the communication range will periodically synchronize their casting goals in order to reduce formation stresses.

#### 9.4.5.2 Swarm Chemotaxis Algorithm

The chemotaxis algorithm has a simple swarm-oriented implementation (see Listing 9.7): all vehicles broadcast their own chemical sensor measurements,

**Listing 9.7: Implementation of the swarm chemotaxis algorithm**

```
ALGORITHM: chemotaxis
    if ( neighbors are present )
        find the agent with the highest ρ reading
        compute the local gradient ∇ρ = ρmax − ρself
        if ( |∇ρ| > 0 )
            velocity = ∇ρ * time_step
        else
            execute casting
    else
        execute casting
    end-if
```

wait for their neighbors to do the same, and then compute a local chemical gradient, which acts as a goal force that propels the robots toward regions with high concentrations of the trace element. This algorithm implementation is a good model for a versatile sensor network that takes advantage of the automatic aggregation of the chemical signal which emerges directly from the physicomimetic control laws. Because the goal forces are balanced by the formation forces, the trajectory of each individual vehicle is an implicit function of *all* chemical density measurements collected by the swarm. This emergent behavior has two notable benefits: first, the sensor fusion method is implicit in the topology of the network—we never had to design one specifically for this problem, and second, the movement of each vehicle is minimized because the impact of the high-frequency transients (noise) in the chemical density signal is automatically filtered out due to the implicit averaging of the individual robots' sensor readings. The practical utility inherent in these swarm properties presents yet another compelling reason for integrating the physicomimetic approach into a variety of distributed sensing tasks.

### 9.4.5.3 Swarm Anemotaxis Algorithm

The anemotaxis CPT strategy implementation shown in Listing 9.8 benefits from the emergent collaboration in the swarm. Please note that this "baseline" anemotaxis implementation does not explicitly share any information between the neighboring agents, nor does it attempt to keep a history of the vehicle's locations or plume observations. But it is still able to achieve similar functionality due to its operation within the lattice formation. Once again, the formation acts as a distributed sensor fusion network, performing implicit averaging and filtering of the fluid flow observations throughout the swarm.

**Listing 9.8:** Implementation of the swarm anemotaxis algorithm

```
ALGORITHM: anemotaxis
   if ( ρself > 0 and |Vfluid| > 0 )
      velocity = -Vfluid * time_step
   else
      execute casting
   end-if
```

### 9.4.5.4 Swarm Fluxotaxis Algorithm

When introducing this second CPT study, we stated that a large-scale swarm implementation would require a fluxotaxis redesign that is somewhat different from the initial version that we employed in the first study with a seven robot hexagonal lattice. Before, we had explicit control over the lattice radius, and it was very straightforward to compute the boundaries of the virtual flux surface. However, for a large swarm, repeating the same steps leads to an unnecessary coupling between the agents that we wish to avoid. Again, physicomimetics inspired our new design, where all swarm control operations are based on local inter-vehicle forces. The second version of the fluxotaxis algorithm is also derived from the basic $\overrightarrow{GDMF}$ theory on the flow of chemical mass flux within a plume (see Chap. 8). Similarly to its earlier implementation, the revised fluxotaxis algorithm still requires that the vehicles be able to obtain point measurements of the chemical concentration, $\rho$, and ambient wind velocity, $V$. The vehicles then communicate this information to their local neighboring robots so that all of the CPT agents can calculate the chemical mass flux, $\rho V$. Revisiting the important concepts of *influx* and *outflux* first defined in Sect. 9.4.4.4, we must construct a set of virtual surfaces in order to compute the local surface fluxes that can be used to measure the volumetric divergence of chemical mass flux in the vicinity of the CPT robot. The basic operation behind this procedure is for each agent to compute the product $\rho_n \nu_n$, where $\rho_n$ is the chemical density measurement collected and reported to the agent by neighbor $n$, and $\nu_n$ is the component of wind velocity $V_n$ (observed by the neighbor) projected onto the *neighbor line* that joins the centers of the two agents. Figure 9.10 depicts this situation from the perspective of agent $A_0$ in the center of the diagram.

The virtual surface, displayed as a solid line through the circular body of a CPT vehicle in Fig. 9.10, across which the neighbor mass flux, $\rho_n \nu_n$, flows is defined as a planar region that is perpendicular to the (dotted) neighbor line between the robot pairs. The orientation of this virtual patch with respect to the $A_0$ agent determines the amount of chemical mass flux that is associated with the position of each $A_n$ neighbor vehicle. Depending on whether the flux is *toward* or *away* from $A_0$, the agent will record the corresponding *influx* or *outflux*. For the geometry and the airflow pattern shown in Fig. 9.10, from the viewpoint of robot $A_0$, the position of $A_1$ represents an outflux of $\rho_1 \nu_1$, and the position of $A_2$ contains an influx of $\rho_2 \nu_2$. The robot's onboard procedure for determining the magnitude and the classification of the chemical mass flux at a neighbor's location is displayed in Listing 9.9.

Because the local plume density measurement, $\rho_n$, is a scalar value, neighboring vehicles can simply exchange their individual chemical sensor readings. The software implementation assumes that the robots can also share vector values, such as anemometer readings of wind speed and direction. This too is a trivial data exchange if the swarm shares a common frame of reference. However, standard coordinate transformations can be applied in a direct manner for cases where only relative orientation between the neighbors is known (see Chap. 6 for another case where such a transformation is utilized). To obtain $\nu_n$, each robot must compute the dot product, $V_n \cdot \hat{n}$, of the airflow velocity, $V_n$, and the unit normal vector for the virtual surface patch, $\hat{n}$. This is a straightforward computation, because each virtual surface patch is defined in terms of the neighbor line, that is itself simply a translation vector between the centers of the two neighbors. Thus, the unit normal vector, $\hat{n}$, can be computed from the information about the relative positions of the vehicles, which is readily available from the onboard physicomimetic control software. Finally, the familiar geometric identity, $A \cdot B = |A||B|\cos(\theta)$ is applied to calculate the $V_n \cdot \hat{n}$ dot product.

The dot product method we just described allows fluxotaxis agents to distinguish between the outgoing and incoming chemical mass fluxes, and like our earlier fluxotaxis implementation from Sect. 9.4.4.4, this version of the fluxotaxis CPT algorithm first attempts to move the robots in the direction

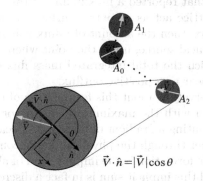

Fig. 9.10: Structure and position of the virtual surfaces (shown with a dark solid line across the circular body of agents $A_1$ and $A_2$) in the fluxotaxis implementation for an arbitrary-sized swarm

**Listing 9.9: Implementation of the swarm fluxotaxis algorithm**

```
ALGORITHM: fluxotaxis
  if ( more than one neighbor )
    for neighbor in neighbors
      execute neighbor_flux( neighbor )
    end-for
    if ( influx detected )
      compute bearing unit vector F_influx toward the neighbor with
      maximum influx
      velocity = F_influx * time_step
    else if ( outflux detected )
      compute bearing unit vector F_outflux toward the neighbor with
      maximum outflux
      velocity = F_outflux * time_step
    else
      execute chemotaxis
    end-if
  else
    execute anemotaxis
  end-if

STRATEGY: neighbor_flux
  retrieve ρ_neighbor and V_neighbor
  let u and v be the x- and y-components of wind velocity V_neighbor
  let dx and dy be the x- and y-components of the neighbor
  separation vector

  return ρ_neighbor * |V_neighbor| * cos[arctan(v/u) − arctan(dy/dx)]
```

of the neighbor who reported the maximum incoming flux, if one exists. Otherwise, if none of the neighbor fluxes are "influxes," the algorithm will search for a position that reported a maximum outward flux. We found that if we let the robot lattice act as a control surface in order to compute the divergence of mass flux, then the amount of influx will gradually decrease as we approach the chemical source, up to the point where the robots surround the emitter, after which the total integrated mass flux over the boundaries of the swarm will appear as a positive outflux.

An interesting observation about this formulation of the algorithm is that by selecting a location with the maximum incoming or outgoing flux, each robot is actually computing a gradient of mass flux. Because all of the robots interact with each other through the physicomimetic control mechanism, the trajectory of the whole formation is the implicit sum of all of these individual goal force vectors, and this implicit sum is in fact a discrete approximation of the chemical mass flux divergence $\overrightarrow{GDMF}$ in the vicinity of the robots. Thus, the preservation of individual formation bonds inside a physicomimetic swarm leads to an implicit computation of the "source" or a "sink" of the mass flux gradient. Effectively, when executing this modified version of fluxotaxis, the swarm computes a first-order approximation of the $\overrightarrow{GDMF}$ metric defined in Chap. 8 and again in Sect. 9.2.3. Due to the geometry of the resulting robot grid, this first-order approximation becomes increasingly closer to the second-order $\overrightarrow{GDMF}$ as the swarm size increases. In the limit, it becomes equal to the $\overrightarrow{GDMF}$ defined in Sect. 9.2.3. In other words, this version of fluxotaxis is particularly tailored for large swarms, although, as seen below, it even works quite well (and better than its competitors) with small swarms.

It is noteworthy that this version of fluxotaxis, when combined with the physicomimetics Lennard-Jones force law, facilitates fluid-like swarm movement around obstacles and through the environment. It is a truly fascinating phenomenon to watch, like a more viscous fluid such as syrup flowing upstream and smoothly around obstacles toward the source of an ink jet dispersing within a less viscous (than the syrup) fluid such as water.

Most importantly, although the explicit objective of fluxotaxis is to trace the chemical to the emitter, as an emergent property the flux-following behavior facilitates navigation, despite not being programmed to do so. In particular, the chemical flux itself provides a path around obstacles, which generate chemical sinks. This is an excellent illustration of the power of physicomimetics.

## 9.5 Seven-robot CPT Study

This first study uses a hexagonal lattice of seven robots. Here, we demonstrate how to use physicomimetics to navigate the robot formation toward the chemical source. Since the number of agents is fixed, the swarm control parameters are selected a priori to the deployment. The study consists of two experiments. The first one is based on obstacle-free environments and looks at the relationship between the lattice radius $D$ and the end-of-mission CPT performance. The second experiment uses the optimal control parameters found in the first experiment and introduces obstacles into the search area. In both experiments, chemotaxis, anemotaxis, and fluxotaxis algorithms are evaluated and compared using several different performance metrics.

### 9.5.1 Experiment in an Unobstructed Environment

#### 9.5.1.1 Purpose

The goal of this first CPT study is to understand the comparative performance of three different CPT algorithms as the *maximum lattice expansion factor* is methodically varied. Here, the maximum lattice expansion factor is defined as the largest separation distance between any two vehicles. The physicomimetic control framework has several parameters that determine key behaviors of the lattice, and we are interested in the effect that the lattice radius, $D$, has on the comparative CPT task performance of each strategy.

### 9.5.1.2 Setup and Methodology

The test configuration for this experiment covered a representative mix of laminar, transitional, and turbulent flow regimes, each containing a dynamic chemical-gas plume evolving over a $930\,m^2$ area. All CPT algorithms were pairwise compared over consistent plume conditions, meaning that all forward solution parameters, as well as the lattice initial state were matched for each CPT strategy run, i.e., we were extremely careful to ensure fair and consistent evaluation of each CPT algorithm. Note that the plume evolution and lattice movement are concurrent, i.e., the plume continues to develop during the tracing. Each CPT run lasted 3000 time steps, simulating approximately an hour of plume time. Lattice movement is determined using a set of waypoints spaced one meter apart. Since only the fluxotaxis algorithm includes a control output for the lattice radius, in order to keep the experimental conditions as similar as possible, anemotaxis- and chemotaxis-driven lattices were allowed to expand or contract their radii at random to increase exploration. The initial size of the lattice radius was fixed at 0.5 m, and the maximum expansion factor was set at 15, which means that the maximum lattice diameter was 7.5 m. We selected 35 plume configurations and 10 expansion factors, and then evaluated each combination of the plume and expansion factor in a separate CPT scenario. For each plume and expansion factor setting, we selected 200 random locations within the environment where the lattice started the tracing.

### 9.5.1.3 Performance Metrics

The evaluation metric consists of two related components: the *proximity* of the center vehicle to the true location of the chemical emitter, and a Boolean measure of emitter containment (i.e., whether the chemical source is inside the lattice) that we called a *CPT success*. Here, both metrics are normalized with respect to the "optimal" value, with 1.0 being the best, and are calculated at the conclusion of the trial. Note that the second metric indirectly measures the impact of the maximum lattice radius expansion factor: allowing a larger expansion radius increases the probability that the final emitter location will be inside the lattice, thus also increasing the likelihood of a CPT success.

### 9.5.1.4 Results

The simulation results are plotted in Fig. 9.11, showing the average performance of each CPT algorithm over all plumes with respect to the maximum lattice expansion factor. This experiment demonstrated that on the proximity metric, a higher expansion factor allows anemotaxis to beat chemotaxis. This is due to a characteristic oscillation in the anemotaxis lattice at a larger radius: the lattice moves upwind to get nearer to the emitter, but moves past it, and then exits out of the chemical plume. At this point it switches to casting, which causes the lattice to reverse its direction and move back into the plume, where the anemotaxis upwind-following behavior activates again, creating a cycle near the emitter. As mentioned earlier, due to the increased likelihood of the robots surrounding the emitter when the lattice is allowed to expand, the CPT success rate for all of the algorithms improves slightly as the expansion factor is increased. Performance averages obtained in this experiment are summarized in Table 9.3.

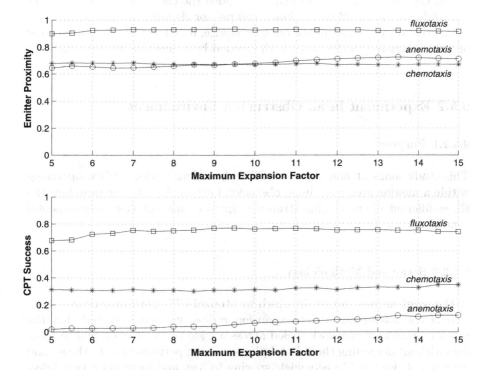

Fig. 9.11: Performance of the three CPT algorithms on a lattice of seven robots as a function of the varying expansion factors. Data is averaged over 35 plumes with 200 random starting locations. The $y$-axis is normalized with respect to dimensionless units

Table 9.3: CPT algorithm performance on a lattice of seven robots, showing mean ± standard deviation. Data is averaged over 35 plumes and all parameter variations. Higher values indicate better performance

| CPT Algorithm | Emitter Proximity (%) | Localization Success (%) |
|---|---|---|
| Anemotaxis | $68.43 \pm 0.29$ | $6.67 \pm 3.62$ |
| Chemotaxis | $67.45 \pm 0.49$ | $31.84 \pm 1.32$ |
| Fluxotaxis | $92.35 \pm 0.89$ | $74.60 \pm 2.50$ |

#### 9.5.1.5 Conclusions

The plots in Fig. 9.11 and the data in Table 9.3 support the use of the fluxotaxis strategy over the alternative algorithms—fluxotaxis shows a consistent improvement in being able to locate the emitter. The data shows that at the end of the mission, it located and surrounded the chemical source in 75% of the trials, as compared to its closest competitor chemotaxis, which only contained the source in 32% of the trials. Likewise, it achieved a proximity rating of 92%, compared to the 68% for the second-best anemotaxis technique.

## 9.5.2 Experiment in an Obstructed Environment

#### 9.5.2.1 Purpose

This study looks at how different CPT algorithms perform when operating within a mission area containing obstacles. Our aim is to understand how the three different plume-tracing strategies operate under diverse environmental conditions.

#### 9.5.2.2 Setup and Methodology

In the previous experiment, in which we studied CPT performance as a function of the maximum lattice expansion radius, we demonstrated that the physicomimetics framework scaled across a wide range of inter-vehicle spacings without degrading the robots' plume-tracing performance. In the second experiment, we fixed inter-vehicle spacing to 3 m, and focused on two different aspects of the mission: the number of obstacles and the starting position of the lattice. For each CPT trial, we generated 500 random vehicle starting locations, where the starting distance from the emitter varied from zero to 25 meters, which shortened the casting (i.e., plume search) phase of the mission, and allowed us to focus on the behavior of the CPT algorithms instead.

We evaluated the effect of obstacles on CPT algorithms by solving fluid dynamic equations in varied environments with the following configurations: no obstacles, one $3\,m \times 3\,m$ obstacle (1% obstacle coverage), four $1.5\,m \times 1.5\,m$ obstacles (1% coverage), two $3\,m \times 3\,m$ obstacles (2% coverage), and with eight $1.5\,m \times 1.5\,m$ obstacles (2% coverage), with a total of 150 different plume and obstacle configurations. CFD parameters were the same for each obstacle course, but the chemical plumes in each new environment varied due to the size and placement of the obstacles. As before, all of the experimental conditions were carefully controlled to ensure consistent and unbiased evaluation of each CPT method. Since chemical contamination (e.g., a toxic spill) typically precedes the start of CPT efforts, we "turned on" the chemical emitter for 3000 time steps (about an hour of real plume time) before activating the CPT robots. Each plume then evolved for an additional 7000 steps as the CPT robots searched for the source.

### 9.5.2.3 Performance Metrics

For this experiment we studied emitter localization characteristics of each CPT method as a function of the initial distance between the lattice and the chemical source. Therefore, this second portion of the seven-robot study required a new performance metric to indicate *how well* a given CPT strategy localizes the source. We called this a *stability metric* to differentiate it from the end-of-the-run CPT success measure used in the previous experiment. This criterion is a helpful evaluation tool because it measures consistency of the source localization solution obtained with each CPT algorithm. To measure this benchmark we again used a global observer function, which computed the fraction of simulation time the plume emitter was contained inside the lattice.

### 9.5.2.4 Results

Results of this test are displayed in Fig. 9.12, showing fluxotaxis with an average emitter containment rate of 49.1%, which is much higher than that of anemotaxis (8.4%) and chemotaxis (7.2%). Since anemotaxis always moves upwind in the plume, it often drives past the emitter, and then spends over 90% of its mission in a cycle of first moving upstream in the plume, and then switching to casting after a subsequent miss of the emitter. This explains why the anemotaxis performance is low even when it begins its tracing next to the emitter, and why the performance curve deteriorates with the increasing starting distance to the emitter. Chemotaxis does somewhat better when it starts out near the emitter, but its performance falls off rapidly, caused by the increase in the number of obstacle-induced local density maxima (i.e., "false" sources). However, even when chemotaxis begins its tracing within

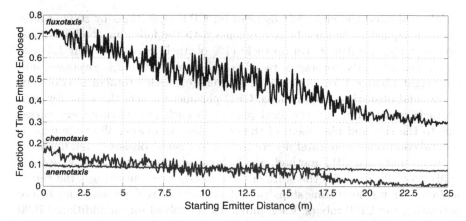

Fig. 9.12: CPT algorithm performance results averaged over 150 plume and obstacle configurations. The $x$-axis is the distance between the initial starting location of the lattice and the emitter, and the $y$-axis is the fraction of total CPT time (7000 simulation steps) when the vehicles successfully surrounded the emitter

three meters of the emitter, the average containment in this case is still only 14.3%, which means the lattice failed to find the source more than 85% of the time. The poor chemotaxis performance is caused by periodic turbulence and the resulting variance (e.g., wind gusts) in the flow, both of which give rise to a "shedding" effect, that manifests itself when large "chunks" of the plume in the vicinity of the emitter are "torn off" and carried away from the source.

### 9.5.2.5 Conclusions

Fluxotaxis consistently and significantly outperforms chemotaxis and anemotaxis algorithms on the emitter containment metric. This experimental outcome is consistent with the theoretic predictions of the *Source* and *Sink* theorems we introduced in Sect. 9.2.3. Turbulence is the reason why fluxotaxis does not achieve 100% emitter containment when it starts with the emitter already enclosed, because periodic wind gusts transport large portions of the ejected chemical away from the emitter, and the moving mass of chemical appears as a temporary pseudo-source. Future work on improving the fluxotaxis algorithm will address this detrimental impact of turbulence. In addition, we found that the CPT algorithms usually manage to navigate around obstacles well before active collision avoidance even becomes necessary. The obstacle avoidance problem is often simpler within a chemical plume, since the lattice follows the plume as the carrier fluid flows around the obstacles. This is an important and relevant observation, since CPT systems are designed for hazardous areas, such as debris-filled passageways.

## 9.6 CPT Study of a Large Decentralized Swarm

Section 9.5 described our first CPT simulation study, which used a fixed-size lattice of seven vehicles. The results were definitive—our fluxotaxis approach is an improvement over both the chemotaxis and anemotaxis CPT methods in terms of being able to find the emitter (what we previously called a *CPT success*), and in being able to consistently contain the chemical source within the bounds of the seven vehicle hexagonal formation (a property we called *emitter containment*). Of course one notable limitation of the previous implementation is its dependence on a particular lattice configuration—all of the performance metrics, as well as the CPT algorithms themselves, assumed that the "swarm" consists of exactly seven robots, and the vehicles had to maintain the hexagonal formation at all times in order for the experiments to be valid. Therefore, we conducted a follow-up CPT study to answer an important question: does fluxotaxis scale to an arbitrary-sized swarm? Experimental results we present in this section show that fluxotaxis retains its CPT advantage across a wide range of swarm sizes and environment conditions.

In this study, we rejected any a priori knowledge regarding the number of agents participating in the plume-tracing task, along with any restrictions on the swarm topology. This is a practical requirement, because if we consider the logistical concerns of operating a large number of robotic vehicles out in the field, we can no longer justify assumptions of the fixed swarm layout.

### 9.6.1 Experiment with Increasing Number of Obstacles

#### 9.6.1.1 Purpose

For the first experiment in this study, we measure the performance of modified, swarm-oriented, fully decentralized CPT algorithms on different plume environments with many different obstacles. As before, our motivation here is to demonstrate how the three different approaches to chemical source localization compare against each other across a range of plume conditions.

#### 9.6.1.2 Setup and Methodology

We evaluated emitter localization by variable-sized swarms driven by casting, chemotaxis, anemotaxis, and the fluxotaxis algorithms on a suite of 81 simulated plume scenarios with physically distinct flow configurations, each containing an airborne chemical plume evolving over a large $8{,}360\,\mathrm{m}^2$ area. As in the previous study, we picked a range of CFD boundary conditions that produced an even mix of different flow types within randomly created environments with: no obstacles, with 9, 18, 27, and 36 obstacles of size $1.5\,\mathrm{m}$ × $1.5\,\mathrm{m}$, and with 2, 4, 7, and 9 obstacles of size $3\,\mathrm{m}$ × $3\,\mathrm{m}$.

The trace chemical was ejected for 3000 simulation steps (about an hour of real plume time) before a swarm of CPT robots was first deployed, and the plume-tracing mission lasted for an additional 7000 steps (corresponding to a realistic two hour CPT time frame). The initial swarm starting location varied from a position precisely over the emitter to $60\,\mathrm{m}$ away from the emitter in one meter increments (compare this with the $25\,\mathrm{m}$ maximum we examined as part of the first study in Sect. 9.5.2). We varied the number of vehicles in the swarm from 7 to 70 robots, with a total of 10 different swarm sizes per plume, obstacle, and initial starting location combination. Thus, a total of 40,500 CPT evaluation runs were performed as part of this experiment. For every new evaluation run, we made sure that the plume and the search area configuration were the same for all CPT strategies, thus the observed differences in the swarm's performance are the direct result of the different navigation choices made by the individual plume-tracing strategies.

### 9.6.1.3 Performance Metrics

In the first CPT study, we took advantage of the fact that the hexagonal lattice of seven robots would surround the chemical source in a very predictable manner, and all of our CPT performance metrics were based on this assumption. However, for this study, we have very little a priori knowledge of how the swarm will approach the emitter. Because each CPT vehicle functions as a completely independent entity, the old performance metric of emitter proximity based on the "center" robot in the lattice no longer makes sense, since there is no "center" in a decentralized swarm. Likewise, the containment metric needs to be adapted for very large swarms, otherwise we cannot measure the CPT performance as a function of swarm size.

Therefore, we developed two new CPT performance metrics that evaluate the swarm aspect of our physicomimetic solution. The first metric, called the *arrival time*, is designed to evaluate the speed with which the swarm locates the source. The arrival time value is equal to the earliest time step of when a vehicle from the swarm first drives within the sensor range, $R_s$, of the chemical emitter, and lower values indicate better CPT performance. Note that we do not assume that the vehicle somehow *knows* that it is close to the emitter. Because in order to "succeed" on the arrival time metric the vehicle only needs to be near the emitter, we also evaluated the casting strategy as one of the CPT "algorithms." Because casting is essentially a random method, this comparison provides us with the baseline of the performance

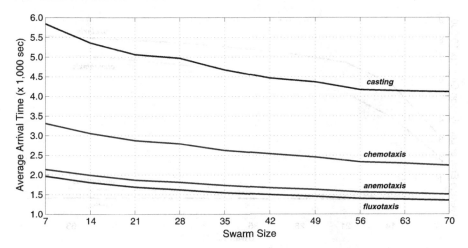

Fig. 9.13: Arrival time results for each CPT algorithm averaged over 81 plumes and 50 starting locations. Lower values indicate better performance

Table 9.4: Arrival time metric of each CPT algorithm versus the world obstacle coverage, averaged over 10 swarm sizes, 50 starting locations, and 81 plumes

| CPT ALGORITHM | OBSTACLE COVERAGE | | | | |
|---|---|---|---|---|---|
| | 0.0% | 0.25% | 0.50% | 0.75% | 1.0% |
| Fluxotaxis | 1508 | 1523 | 1615 | 1572 | 1610 |
| Anemotaxis | 1885 | 1760 | 1725 | 1700 | 1731 |
| Chemotaxis | 2451 | 2603 | 2887 | 2437 | 2778 |
| Casting | 5292 | 4848 | 4739 | 4498 | 4490 |

level that can be achieved via an uninformed search of the environment. Of course we expect all of the actual CPT strategies to exceed the performance of this random search of the environment.

The second metric, which we call the emitter *localization frequency* or *localization count* is a measure of *how many* vehicles drove within the sensor range, $R_s$, of the emitter. This is a cumulative metric—it simply sums up the number of vehicles located within the circle of radius $R_s$ for each time step in the simulation. We again use the score of our random casting strategy as the baseline for this characteristic, with the expectation that all three of our CPT algorithms will perform better than the random search.

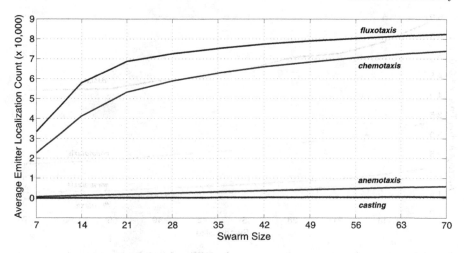

Fig. 9.14: Frequency of emitter localization by the swarm, averaged over 81 plumes and 50 starting locations. Larger values indicate higher performance

Table 9.5: Cumulative count of emitter detections by the swarm for each obstacle coverage category, averaged over 10 swarm sizes, 50 starting locations, and 81 plumes

| CPT ALGORITHM | OBSTACLE COVERAGE | | | | |
|---|---|---|---|---|---|
| | 0.0% | 0.25% | 0.50% | 0.75% | 1.0% |
| Fluxotaxis | 73253 | 72189 | 70085 | 70657 | 69369 |
| Chemotaxis | 62925 | 59398 | 55802 | 62649 | 56368 |
| Anemotaxis | 3604 | 3667 | 3479 | 3429 | 2939 |
| Casting | 368 | 308 | 330 | 593 | 271 |

#### 9.6.1.4 Results

Results for the arrival time performance metric (i.e., the elapsed time before the first vehicle detects the emitter) are plotted in Fig. 9.13, and are broken down by the type of obstacle course in Table 9.4. The number of times that the chemical emitter was successfully localized by swarm vehicles is shown in Fig. 9.14, with the performance breakdown based on obstacle coverage presented in Table 9.5.

Our first observation from the two figures is that only fluxotaxis performs well on *both* metrics: it combines the speed performance demonstrated by anemotaxis with the localization frequency advantage shown by chemotaxis. Examining the performance data in Fig. 9.14, we conclude that chemical plume tracing is inherently a swarm application. We base this claim on the improvement in the consistency of the emitter localization, as captured by the localization frequency metric, for the fluxotaxis and chemotaxis algorithms observed as the result of the increasing swarm size. However, the poor emitter localization performance of the random casting algorithm suggests that the robot team must be managed in order to gain performance benefits of the swarm approach—simply adding more vehicles to the group is not sufficient to improve task performance.

The large difference in performance between fluxotaxis and anemotaxis, seen in Fig. 9.14, can be attributed in part to the fact that the baseline implementation of the anemotaxis algorithm is not specifically designed to direct the vehicles toward the emitter. Instead, the upwind follower reaches the emitter in approximately the same time as does fluxotaxis (which can be inferred from Fig. 9.13), but the swarm continues to move upstream, and thus moves past the emitter. After losing the emitter, the anemotaxis swarm switches to the casting behavior in order to find the plume, which is why its localization performance is similar to that of the casting strategy. In the literature, several heuristic methods have been suggested to address this problem. However, we view the fluxotaxis approach of seeking out the chemical mass flux as a superior method, because our physics-based approach does not require additional heuristics to achieve CPT goals.

Data in Tables 9.4 and 9.5 provide the evidence for the robustness and scalability of the physicomimetics framework, which are manifested in the robots' ability to adapt in a dynamic environment. The force law parameters are the same in each experiment, but the control algorithm maintains its operational efficiency and oversees interactions within the swarm regardless of the number of robots or the size and number of obstacles.

### 9.6.1.5 Conclusions

The overall CPT performance of each algorithm in this experiment must be interpreted in the context of the type of plume information that is extracted and then acted upon by the given CPT strategy. In other words, the variations in the rates of success for each algorithm are due to the natural characteristics of each CPT approach. For instance, for the random casting strategy, the arrival time result improves with the increasing obstacle coverage because greater obstacle coverage implies less open area that the random searcher has to explore. On the other hand, performance of the chemotaxis strategy on both metrics is generally worse in the environments with more obstacles. This is a direct consequence of the fact that obstacles create local maxima in the density profile of the plume, because the chemical has a tendency to slow down and build up in the flow that impinges on the obstacles' surface. From the mathematical perspective, each obstacle induces a local chemical sink, and these sinks mislead the naïve chemical gradient follower. This behavior is the direct consequence of the chemotaxis algorithm's reliance on the first-order derivative of the chemical concentration, which does not contain enough information to distinguish a true chemical source from an obstacle-induced sink. The *Sink* Theorem we introduced in Sect. 9.2.3 provides a precise mathematical formulation for describing and detecting such regions with high chemical concentration that do not contain the emitter. The time spent by chemotaxis investigating these temporary "false emitters" is reflected in the algorithm's increased localization time, and it also shortens the time that the chemotaxis-controlled agents spent in the vicinity of the true chemical source. However, when operating near the actual emitter, chemotaxis is expected to perform well on the localization frequency metric, since the peaks in the chemical concentration landscape are in fact close to the real source (see the *Source* Theorem in Sect. 9.2.3). The fact that fluxotaxis consistently outperforms the other two CPT methods on *both* performance metrics is a straightforward validation of our argument in Sect. 9.4.4.4, and Eq. 9.2 shows how the fluxotaxis algorithm can automatically select the best plume characteristic (i.e., the density gradient or the flow direction) to follow in accordance with the changing environmental conditions.

## 9.6.2 Experiment with Increasing Swarm Size

### 9.6.2.1 Purpose

Results of the previous experiment showed that the CPT performance is affected by the size of the swarm to a much greater degree than it is influenced by the obstacle configuration. Therefore, in this final experiment we increased the number of CPT robots in the swarm to determine what performance gain can be realized with very large swarms.

### 9.6.2.2 Setup and Methodology

We simulated 10 different flow conditions, with the chemical emitter located inside a $8,360\,m^2$ region. As before, a choice of appropriate boundary conditions resulted in a diverse mixture of airflow patterns. Each plume-tracing area contained ten $1.5\,m \times 1.5\,m$ randomly placed obstacles. Similarly to our previous experiments, the chemical emitter activated 3000 simulations steps (an hour of real plume time) before the CPT swarm deployed. We advanced each plume for 7000 steps (corresponding to a two hour time frame), and recorded the emitter arrival time and localization statistics for each CPT algorithm.

All of the data in this experiment comes from matching evaluation runs for each CPT algorithm and the casting strategy on a large set of CPT scenarios, consisting of 15 swarm sizes (ranging from 10 to 150 vehicles) and 30 different starting locations per each swarm size. The initial position of

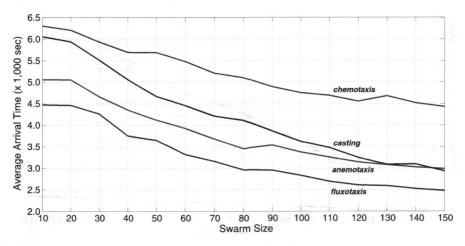

Fig. 9.15: Arrival time metric for each CPT algorithm over a range of swarm sizes (smaller values indicate faster localization). Arrival event occurs when any swarm vehicle makes the first sensor contact with the emitter (as determined by a global observer)

the swarm is selected at random, with the starting distances ranging from zero to 60 m away from the emitter. As was the case in all of the other experiments, we made sure that the chemical plume, the search environment, and the evaluation criteria were precisely the same before assessing algorithm performance.

### 9.6.2.3 Performance Metrics

Swarm performance in this experiment was evaluated using the same *arrival time* and *localization frequency* metrics first defined in Sect. 9.6.1.3.

### 9.6.2.4 Results

The performance of each CPT strategy as a function of the number of robots is given in Figs. 9.15 and 9.16. Table 9.6 lists the cumulative performance average for each CPT method. Results of this experiment confirm that an increase in the size of the swarm improves both the speed and accuracy of the source localization. Each algorithm we tested displayed improved performance on both evaluation metrics, and we want to point out that the fluxotaxis performance curves clearly show the algorithm's ability to satisfy CPT objectives in a stable and predictable manner. At the same time, note that chemotaxis is the most vulnerable of the CPT techniques, so that even

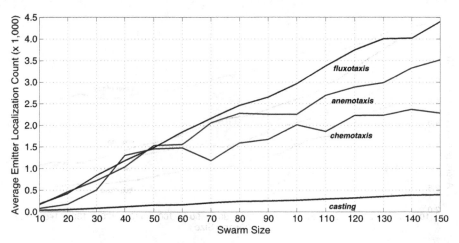

Fig. 9.16: Cumulative count of chemical emitter localizations by each CPT algorithm. The emitter is considered to have been localized when a swarm vehicle makes sensor contact with the emitter, as determined by a global evaluation algorithm

Table 9.6: Average performance of the three CPT algorithms over 30 starting locations, 15 swarm sizes, and 10 plumes

| CPT Algorithm | Arrival Time | Localization Frequency |
|---|---|---|
| Casting | 4224.4 | 224.1 |
| Chemotaxis | 5208.5 | 1496.5 |
| Anemotaxis | 3780.5 | 1983.6 |
| Fluxotaxis | 3249.9 | 2384.8 |

the uninformed (i.e., random) world exploration yields better arrival time results than the greedy gradient-following chemotaxis. The chemotaxis-driven swarm is frequently misled by the local concentration maxima around obstacles, thus lowering its performance on the localization time metric. The anemotaxis approach initially outperforms simple casting in terms of the arrival time, but this advantage decreases as the swarm size increases. This finding can be explained by the fact that instantaneous wind velocity is generally a poor indicator of the true direction of the plume's source, and as the fluid periodically changes the flow direction, the anemotaxis strategy is misled by the isolated pockets of a fragmented plume, resulting in a time-inefficient zigzag plume traversal pattern often observed in insect pheromone studies [77]. This important observation confirms our earlier assertion that just adding more sensors and robots into the swarm is not enough, and once again we conclude that mobile sensors must be continually repositioned according to an informed and theory-guided policy. Given the decentralized swarm architecture, each team member must provide the means to facilitate such perception optimizing ability. Our fluxotaxis algorithm provides a practical example of how such functionality can be engineered into a complex system.

### 9.6.2.5 Conclusions

At this point, it is instructive to reiterate the fact that the physicomimetic framework contains several mathematical tools that a swarm designer can use to predict and thus assure the robots' long-term performance on a given task. The local nature of vehicle interactions results in efficient scalability, and the reactive control is what allows the swarm to succeed regardless of the environment configuration and initial conditions. Because both the fluxotaxis CPT algorithm and the swarm control framework have a rigorous mathematical foundation, fluxotaxis consistently outperforms the biomimetic methods. This experiment further supports our earlier claims regarding the robustness and scalability of the physicomimetics approach. The virtual forces controlling the swarm formation are identical for all of the experiments, even as the number of robots is increased by an order of magnitude! And again, there is no explicit programming needed to achieve this level of performance—the pairwise formulation of the control law means that it does not depend on the number of vehicles in the formation.

Due to space constraints, this chapter does not directly address questions pertaining to error recovery, communication loss, vehicle breakdowns, and so on. But the reader should already be able to see that a scalable, reactive physicomimetics control paradigm opens up many potential avenues of dealing with these important real-world concerns in a structured and informed manner, with mathematically-robust analysis tools to guide future designs.

## 9.7 Lessons Learned

Our step-by-step construction of the fluxotaxis CPT algorithm from observations of fluid flow properties to a computationally practical implementation illustrates a robust, theory-guided approach for accomplishing complex tasks with autonomous robots by intelligently exploiting the underlying physical principles of the problem. Our simulation experiments demonstrate that fluxotaxis is able to compute correct navigation waypoints using local sensor observations in a way that is superior to the most popular biologically-inspired plume-tracing methods. In addition, all three team-oriented CPT strategies we evaluated showed a gain in performance due to the cooperation between neighboring vehicles. This is a unique emergent property of the swarm supported by the automatic sensor aggregation feature of the physicomimetics control framework. By sharing the information about local flow conditions between neighboring vehicles in the group, each team member is able to construct a more accurate view of the surrounding plume, which in turn improves the accuracy of the emitter search algorithm. However, since fluxotaxis is founded on insights from fluid mechanics, the physics-based foundation allows this new algorithm to be consistently more effective in achieving the CPT goals than what is possible with the biomimetic approaches.

The outcomes of our comprehensive study translate into a convincing argument that *CPT is inherently a swarm application*, which means that significant gains in performance, such as reductions in the required search time and increased consistency of the localization estimate are realized in a predictable manner as the size of the swarm is increased. We also showed that these improvements in the CPT performance require a scalable algorithm; in other words, the *CPT algorithm must manage the information flow in an efficient and intelligent manner*. Our experiments revealed that a class of single-robot oriented CPT algorithms like anemotaxis, which do not explicitly take advantage of the collaborative functionality of the swarm platform, can realize only small increases in the CPT performance when used on large-sized swarms. This finding further reinforces the motivation for our dedicated effort of designing a new CPT algorithm specifically for swarms—by making sure that our fluxotaxis approach utilizes all of the cooperative mechanisms offered by the swarm implementation, we automatically receive all the benefits of using the swarm platform for the chemical source localization problem.

The fact that our fluxotaxis algorithm realizes this increase in CPT performance in a fully emergent fashion speaks to the power and the flexibility of the physicomimetic design. First, we took advantage of the existing fluid dynamics understanding to construct the fluxotaxis algorithm, which allowed us to address significant gaps in the current state-of-the-art CPT research regarding emitter identification in obstacle-filled environments with unsteady, turbulent flows. Next we employed the physicomimetic swarm control methodology to build a massively parallel, distributed computer out of simple, inexpensive robots with limited onboard capabilities. The resulting sensor "mesh" is ideally suited for a variety of in situ analyses and monitoring activities that benefit from the robust self-organizing behavior of the grid-like robot formations. For the CPT problem, we identified the physical property of chemical mass flux flow as the crucial indicator of the location of the chemical source, and showed how a mathematically-derived technique is implemented on simulated mobile robots. Finally, we demonstrated that this methodical, step-by-step construction of all the key components of our implementation resulted in a distributed system with a predictable long-term behavior.

We also showed that the physics-based fluxotaxis plume-tracing strategy combines the strengths of the popular chemo- and anemotaxis approaches, and outperforms these two biomimetic methods in terms of search time and

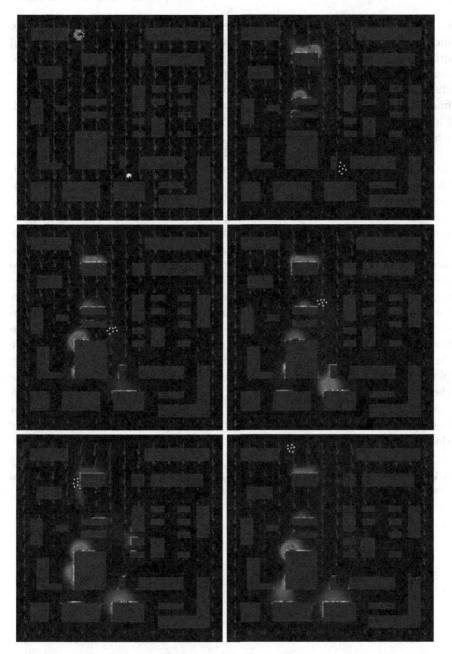

Fig. 9.17: Fluxotaxis succeeds at CPT in a challenging urban scenario that include buildings. Used with permission from Textron, Incorporated

localization accuracy. As a final measure of the accuracy and robustness of the fluxotaxis algorithm, we implemented and tested it on a highly challenging urban environment that includes buildings, providing a *much* higher obstacle density than has been discussed thus far in this chapter. Without making any changes to the algorithm we found fluxotaxis to be extremely consistent at finding the emitter, while chemotaxis and anemotaxis almost always failed. Figure 9.17 shows an example of fluxotaxis's success at meeting this challenge, where the emitter is at the top left of the environment.

In addition to showing the strengths of our approach, the experiments have also revealed several specific areas where additional work is needed. One such area is discussed in Sect. 9.5.2, where we noted that air turbulence creates transient concentrations of the chemical that may appear as temporary sources to the plume-tracing robots. Therefore, future revisions of the fluxotaxis method will include mitigation strategies for transient effects present in turbulent flows. Another planned improvement is a more accurate modeling of the chemical flow that takes the movement of the vehicles into account. In the current formulation of fluxotaxis, we assume that both the chemical concentration and airflow sensors can provide an accurate, instantaneous assessment of local plume conditions. However, in practice, issues such as sensor noise and flow occlusions will periodically violate this assumption, and will require incorporation of new mathematical equations into the fluxotaxis theory to account for these physical phenomena. Other challenging plume-tracing problems we plan to address in the near future include multiple chemical sources and mobile emitters. Given the gains in fluxotaxis performance realized through an increase of the vehicle fleet size, we believe that our physicomimetic, swarm-centric approach will be effective in addressing these extended CPT scenarios.

**Acknowledgements** We are grateful to Dr. Michael Pazzani, formerly with the National Science Foundation, for supporting this project.

# Part III
# Physicomimetics on Hardware Robots

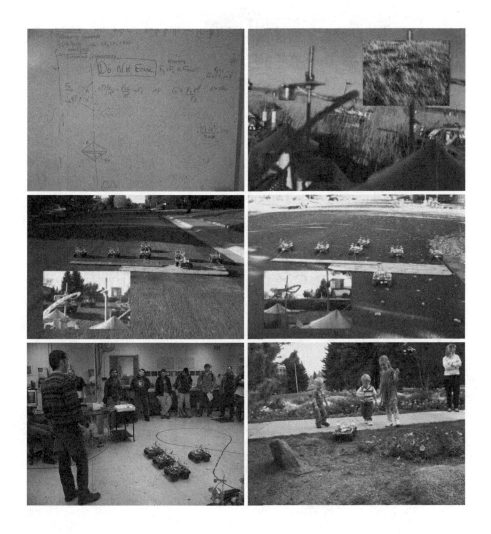

Part III
Physioanimatics on Hardware Robots

# Chapter 10
# What Is a Maxelbot?

Paul M. Maxim

## 10.1 Introduction

The previous chapters have focused on physics and computer science. This chapter brings us to the electrical computer engineering portion of our work, by presenting the theory behind our approach to trilateration-based localization. It continues by showing how we designed and built our trilateration localization module. Our MAXELBOT robot platform is presented in the end, together with other hardware modules that augment the capabilities of the robot platform.

## 10.2 Localization

In general, "localization" can be defined as the process of determining the place where someone or something is.[1] As stated earlier in this book, it is important that each robot in a team has the ability to obtain the bearing and range to its neighbors. From this perspective, localization can be categorized based on the frame of reference. A popular use of localization is to determine a position in a global coordinate system (i.e., "where am I?"). Examples of such localization vary from pinpointing a location on a map using celestial navigation to complex localization technologies like global navigation satellite systems (e.g., the Russian GLONASS, the upcoming European Union Galileo, and the United States Global Positioning System—GPS).

Paul M. Maxim
Wyoming Department of Transportation, Cheyenne, WY 82009, USA, e-mail: paul.maxim@dot.state.wy.us

[1] WordNet 3.0, Princeton University (http://wordnet.princeton.edu/)

W.M. Spears, D.F. Spears (eds.), *Physicomimetics*,
DOI 10.1007/978-3-642-22804-9_10,
© Springer-Verlag Berlin Heidelberg 2011

Localization can also be looked at from an egocentric point of view (i.e., "where are you?"). In this case, localization is used to determine the positions of neighbors in a local coordinate system. There are many examples of using such a technique that we observe in nature. When Canadian Geese (*Branta canadensis*) migrate in the winter, they form a V-shaped flying formation. The spiny lobsters (*Panulirus argus*) migrate in single-file lines [14]. In both examples the formation helps to conserve individual energy, and its shape is maintained due to the fact that each member in the formation is keeping the same relative position with respect to its neighbor ahead.

Similar to the latter localization category, our goal is to create a new "enabling technology" that allows a robot to localize other robots in a multi-robot team. Because we do not want to impose a global coordinate system on the robot team, this means that each robot must have its own local coordinate system. In contrast with the more traditional robotic localization that focuses on determining the location of a robot with respect to the coordinate system imposed by an environment [19], we focus on the complementary task of determining the locations of nearby robots, from an egocentric point of view.

We continue with a comparison between two well-known localization techniques: *triangulation* and *trilateration*. Section 10.4 introduces our trilateration approach to robust localization, which is fully distributed and assumes that each robot has its own local coordinate frame (i.e., no global information is required). Each robot determines its neighbors' range and bearing with respect to its own egocentric, local coordinate system. Details on how we built the hardware localization module based on trilateration as well as its evaluation are then presented. The robot platform developed by us and the synchronization protocol that allows a team of robots to locate each other are presented in Sects. 10.7 and 10.8. At the end of this chapter, Sect. 10.9 introduces additional hardware modules, some developed by us, that equip our robots.

## 10.3 Triangulation Versus Trilateration

Two methodologies for robot localization are triangulation and trilateration [19]. Both compute the location of a point (e.g., a robot) in a two-dimensional space. In triangulation, the locations of two "base points" are known, as well as the interior angles of a triangle whose vertices comprise the two base points and the object to be localized. The computations are performed using the law of sines. In two-dimensional trilateration, the locations of three base points are known as well as the distances from each of these three base points to the object to be localized. Looked at visually, two-dimensional trilateration involves finding the location where three circles intersect.

Thus, to locate a remote robot using two-dimensional trilateration, the sensing robot must know the locations of three points in its own coordinate system and be able to measure distances from these three points to the remote robot. The discussion presented here expands on the overview in Sect. 3.4.

## 10.4 Our Trilateration Approach

Our trilateration approach to localization is illustrated in Fig. 10.1. Assume two robots, shown as round-corner rectangles. An RF transceiver is in the center of each robot. Each robot has three acoustic transducers (also called *base points*), labeled **A**, **B**, and **C**. Note that the robot's local X–Y coordinate system is aligned with the L-shaped configuration of the three acoustic transceivers. This simplifies the math [238]. $X$ points to the front of the robot.

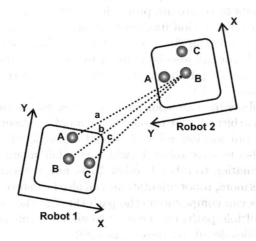

Fig. 10.1: Three base points in an X–Y coordinate system pattern

In Fig. 10.1, robot 2 simultaneously emits an RF pulse and an acoustic pulse from its transducer **B**. Robot 1 then measures the distances **a**, **b**, and **c**. Without loss of generality, assume that transceiver **B** of robot 1 is located at $(x_{1B}, y_{1B}) = (0,0)$ [88].[2] Let **A** be at $(0,d)$, **B** be at $(0,0)$, and **C** be at $(d,0)$, where $d$ is the distance between **A** and **B**, and between **B** and **C**. For robot 1 to determine the position of **B** on robot 2 within its own coordinate system, it needs to find the simultaneous solution of three nonlinear equations, the intersecting circles with centers located at **A**, **B** and **C** on robot 1 and respective radii of **a**, **b**, and **c**:

---

[2] Subscripts denote the robot number and the acoustic transducer. Thus transducer **A** on robot 1 is located at $(x_{1A}, y_{1A})$.

$$(x_{2B} - x_{1A})^2 + (y_{2B} - y_{1A})^2 = a^2 \,, \qquad (10.1)$$
$$(x_{2B} - x_{1B})^2 + (y_{2B} - y_{1B})^2 = b^2 \,, \qquad (10.2)$$
$$(x_{2B} - x_{1C})^2 + (y_{2B} - y_{1C})^2 = c^2 \,. \qquad (10.3)$$

The form of these equations allows for cancellation of the nonlinearity, and simple algebraic manipulation yields the following simultaneous linear equations in the unknowns:

$$\begin{bmatrix} x_{1C} \ y_{1C} \\ x_{1A} \ y_{1A} \end{bmatrix} \begin{bmatrix} x_{2B} \\ y_{2B} \end{bmatrix} = \begin{bmatrix} (b^2 + x_{1C}{}^2 + y_{1C}{}^2 - c^2)/2 \\ (b^2 + x_{1A}{}^2 + y_{1A}{}^2 - a^2)/2 \end{bmatrix} \,.$$

Given the L-shaped transducer configuration, we get [88]:

$$x_{2B} = \frac{b^2 - c^2 + d^2}{2d} \,, \qquad\qquad y_{2B} = \frac{b^2 - a^2 + d^2}{2d} \,.$$

An interesting benefit of these equations is that they can be simplified even further, if one wants to trilaterate purely in hardware [238].

Analysis of our trilateration framework indicates that, as expected, error is reduced by increasing the "baseline" distance $d$. Our robots have $d$ equal to 15.24 cm (6″). Error can also be reduced by increasing the clock speed of our trilateration module (although range will decrease correspondingly, due to counter size) [88].

By allowing robots to share coordinate systems, robots can communicate their information arbitrarily far throughout a robotic network. For example, suppose robot 2 can localize robot 3. Robot 1 can localize only robot 2. If robot 2 can also localize robot 1 (which is a fair assumption), then by passing this information to robot 1, robot 1 can now determine the position of robot 3. Furthermore, robot orientations can also be determined. Naturally, localization errors can compound as the path through the network increases in length, but multiple paths can be used to alleviate this problem to some degree. Heil provides details on these issues [88].

In addition to localization, our trilateration system can also be used for data exchange. Instead of emitting an RF pulse that contains no information but only performs synchronization, we can also append data to the RF pulse. Simple coordinate transformations allow robot 1 to convert the data from robot 2 (which is in the coordinate frame of robot 2) to its own coordinate frame.

Our localization system does not preclude the use of other technologies. Beacons, landmarks, vision systems, GPS [19], and pheromones are not necessary, but they can be added if desired. It is important to note that our trilateration approach is not restricted to any particular class of control algorithms—it is useful for behavior-based approaches [8], control-theoretic approaches [60, 65], motor schema algorithms [22], and physicomimetics [90, 291].

## 10.5 Trilateration Implementation

This section will present details on how we designed and implemented our trilateration-based localization module.

We will start by presenting the initial tests performed on an off-the-shelf ultrasonic range finder and how we developed and implemented our own (Sects. 10.5.1–10.5.3). Our improved trilateration module implementation is described in Sect. 10.5.4. Finally, Sect. 10.6 presents the performance results of our trilateration module.

### 10.5.1 Measuring Distance

Before a trilateration-based localization module can be built, we need a way to perform accurate distance measurements. Our distance measurement method exploits the fact that sound travels significantly slower than light, employing a *Difference in Time of Arrival* technique. For this, we need to have an accurate way of transmitting and receiving sound.

The following section presents the SRF04, an off-the-shelf ultrasonic range finder. Next, the development of our new range finder, based on the SRF04, is described. The main contributions we brought to the new range finder over the existing design are: (a) an increase in range, due to additional amplification and filtering of the incoming signal, (b) increased accuracy and (c) automatic transmit/receive switching.

### 10.5.2 SRF04 Ultrasonic Range Finder

An important aspect of the trilateration module is its ability to transmit and receive sound. The first component tested was the off-the-shelf SRF04 ultrasonic range finder produced by Devantech Ltd. (Fig. 10.2, top). Its performance, low current requirements (30 milliampere), and low cost (US $30 at the time of this writing) made it a good candidate. The SRF04 provides non-contact distance measurements and, according to its specifications [52], it can detect an object placed in front of it from about 3 centimeters (cm) up to 3 meters (m). The exact timing diagram of the SRF04 is shown in Fig. 10.2, bottom [52]. In order to start a measurement, a short 10 microsecond (µs) pulse needs to be supplied to the range finder, through the trigger input. When triggered, the SRF04 sends out an eight-cycle burst of ultrasound through its 40 kilohertz (kHz) transmit (Tx) transducer and then it raises the echo line. It then waits for the sound to bounce off a surface and return to the second ultrasonic transducer, which is tuned to receive (Rx) 40 kHz ultrasound. When the SRF04 detects that the signal has arrived, it lowers the echo line. The variable width pulse generated through the echo line corresponds to the distance to the object.

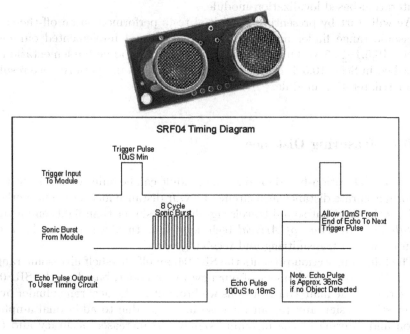

Fig. 10.2: Top—SRF04 ultrasonic range finder (courtesy of Acroname Robotics). Bottom—SRF04 timing diagram (courtesy of Devantech Ltd.)

To make sure that this sensor suits our needs, we modified this range finder by removing the Rx transducer and placing it away from the Tx transducer. This modification was necessary, since we are interested in measuring the direct distance between the Tx transducer and the Rx transducer. We used a twisted pair cable to connect the Rx transducer back to the SRF04. The purpose of this test was to measure the accuracy of the sensor at various distances. Both the Tx and Rx transducers were placed 30 cm above the floor. A reading was performed by the sensor every 10 milliseconds (ms), and for each of the distances, a set of 130 readings was performed. The distance interval varied from 10 cm to 200 cm with a 10 cm increment. As the results of this test show in Fig. 10.3, the modified SRF04 does very well with distances close to 1 m, having a tendency to report a slightly shorter distance for distances close to 2 m.

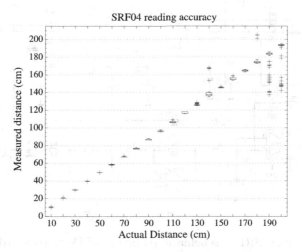

Fig. 10.3: SRF04 reading accuracy

### 10.5.3 Our Prototype XSRF

Although the performance of the SRF04 was encouraging, we decided that for our application, we needed more accuracy and range. The SRF04 schematic shown in Fig. 10.4 [52], which is available from the manufacturer, was the starting point for the design of our own ultrasonic range finder.

At the core of the SRF04 lies a Microchip PIC12C508 programmable microcontroller that performs the control functions. Our first prototype took shape on a breadboard (Fig. 10.5) and used a Microchip PIC12F675, or "PIC" for short, microcontroller from the same family (first integrated circuit, IC, from the left in Fig. 10.5). We used the same air ultrasonic ceramic transducers, 400ST/R160 (shown on the right side of Fig. 10.5). These transducers are an excellent choice since they have their frequency centered at 40.0±1.0 kHz, which means they will only receive or transmit acoustic signals at that frequency [183]. This is very useful for noise filtering. The Tx transducer was connected to the PIC microcontroller through a MAX232 chip (second IC from the left in Fig. 10.5), usually found in RS232 serial communication hardware. This chip helps boost the voltage of the transmitted signal, providing about 16 volts direct current (VDC) of drive, which in turn helps to generate a stronger acoustic signal. The transducer that receives the signal is connected to the input of two signal amplifiers (third IC from the left in

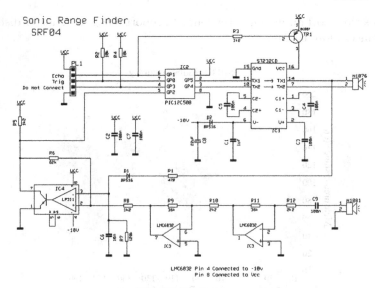

Fig. 10.4: SRF04 schematic (courtesy of Devantech Ltd.)

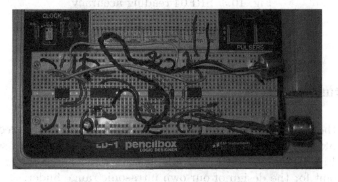

Fig. 10.5: XSRF prototype based on the SRF04 schematic

Fig. 10.5) that amplify the signal and then pass it to the comparator. The comparator (fourth IC from the left in Fig. 10.5) alerts the PIC microcontroller when the signal has been received.

To program the PIC12F675 microcontroller, we used the Integrated Development Environment (IDE) MPLab v.7.20, provided by Microchip Technology Inc., the manufacturer of this device. A compiler for the C language was available but at a high cost, and therefore we instead went with the freely available assembly language compiler, Microchip MPASM Toolsuite. To upload the compiled code to the microcontroller we used a PICStart Plus programmer available from the same manufacturer.

We modified an existing assembly code [35] to fit our microcontroller. Also, the echo mode is not useful for us, and it was removed. In a simple trilateration setup, where we have two localization modules, one localization module will only transmit, and the other module will only receive. This is why we modified the code to use pin 4 of the PIC to select between receiving and transmitting. When pin 4 is high, the PIC is in receive mode, which means there will be no signal generated on pins 2 and 3. This modification was necessary to eliminate the potential noise that this may introduce in the circuit. To account for the time spent generating the signal in transmit mode, we use a simple delay routine of 200 µs. This time interval reflects the exact amount of time it takes the PIC to generate the eight-cycle burst of ultrasound. When pin 4 is low, the PIC is in transmit mode and the 40 kHz signal will be generated.

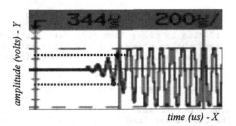

Fig. 10.6: Oscilloscope snapshot of a 40 kHz acoustic signal showing the two thresholds fine-dotted lines that decide when a signal was received

#### 10.5.3.1 Increasing the Accuracy

One important aspect of the range accuracy relates to the frequency of the acoustic signal. In dry air at 20° C, the speed of sound is roughly 343 m/s. At 40 kHz, the wavelength, or the distance between repeating units of the propagating wave, is $\lambda$ = (343 m/s) / 40 kHz = 8.58 millimeters (mm). To determine if an acoustic signal was received, the SRF04 listens for the incoming acoustic wavefronts. This incoming signal has an attack/decay envelope, which means it builds up to a peak then fades away. Depending on which wavefront is the first to be detected (which could be the first, second, or even third), the result can oscillate by 8.58 mm multiplied by the number of times the detection was missed (Fig. 10.6). One possibility to reduce this jitter is to use transducers that operate at frequencies higher than 40 kHz. However, these transducers are more expensive and we looked for other alternatives.

To help with the problem mentioned above, we made modifications at both software and hardware levels. At the software level, we modified the code such that the PIC will switch between the normal signal and its complement when generating the 40 kHz signal. This should improve the average range accuracy

Listing 10.1: PIC12F675—assembly code fragment

```
...
main:
  clrwdt              ; clear the watch dog timer
  btfss  trig         ; wait for the trigger signal from user to go high
  goto   main         ; keep looping

m2:
  clrwdt
  btfsc  trig         ; wait for the trigger signal from user to go low
  goto   m2

  btfss  nc           ; check the status of pin 4 (nc - noise control)
  goto   delay        ; receive mode, no signal generation

  comf   flag,f       ; complement the value stored in flag

  btfsc  flag,0       ; if flag is clear then skip next instruction
  call   burstA       ; send the complement of the ultrasonic-burst
  btfss  flag,0       ; if flag is set then skip the next instruction
  call   burst        ; send the ultra-sonic burst

  bsf    pulse        ; signal the user to start the timer

; Detect the incoming acoustic signal: wait for a transition on the
; ''echo'' line. The transducers and filters provide the selectivity.

m1:
  btfsc  echo         ; wait for the low signal coming from the comparator
  goto   m1           ; keep looping
  bcf    pulse        ; signal the user to stop the timer

  goto   main
...
```

by at most 4.25 mm. A fragment of the assembly code that was written for the PIC microcontroller is shown in Listing 10.1. In this listing we provide the first three functions that drive the PIC: (1) *main*, which is the main function that waits for an external signal to start the measurement process, (2) *m1*, which does the detection of the incoming acoustic signal and then signals the user that a signal was received, and (3) function *m2* that does the 40 kHz signal generation (alternating with its complement), if in transmit mode, as well as signaling the user that a signal was generated, if in receive mode.

The first thing we did to help improve the sensing accuracy at the hardware level was to add a second comparator to our prototype. The reason the second comparator would work is still related to the incoming acoustic wavefronts. These wavefronts, when changed to electrical signals, take the shape of an exponentially increasing sine wave (Fig. 10.6). This wave is amplified until it hits the upper and lower bounds of the onboard signal amplifiers. The original design only called for one comparator, and that comparator was on the negative side of this wave (the bottom fine-dotted line in Fig. 10.6). As soon as the wave drops below a certain threshold, the comparator sends a signal to the PIC microcontroller. If the signal does not quite exceed the threshold value, the comparator will only detect the signal one full wavelength later. With a second comparator, on the positive side (the top fine-dotted line in Fig. 10.6), the signal would be detected half a wavelength sooner.

Another hardware level improvement that we made was to add two potentiometers to work in conjunction with the two comparators on this circuit. One potentiometer is connected to the positive threshold and the other to the negative threshold. These potentiometers allow us to adjust the threshold voltage of the comparators. The threshold voltages control the sensitivity of the sensor to the input signal that sensor receives and they must be adjusted to optimize the sensor. If this threshold is set too low, the sensor will never recognize that a signal has been received. Ideally we would like to make this threshold as high (i.e., as close to zero volts), as possible. This would allow the sensor to receive the signal as soon as it is received with minimal delay.

However the problem with this approach is that no signal that comes into the sensor will be completely clean, that is, without noise. Noise will cause the sensor to receive a signal when it should not have received one (i.e., "a false alarm"). For these reasons we have to choose a threshold voltage that is somewhere between these two extremes. After a series of tests we found that a +2.0 VDC voltage for the positive threshold and a −2.0 VDC for the negative threshold provide the best response. The incoming signal is detected quickly and also the noise is reduced.

### 10.5.3.2 Sending Versus Receiving

Up to this point, our acoustic range finder has separate transducers to transmit and receive ultrasound. For each trilateration module we need three of these range finders to receive and one to send the acoustic signal. Although this setup is realistic, it would require redundant hardware. We decided to improve the current design in such a way that the same range finder would have the capability of transmitting and receiving. To automate this task, we added two input control lines to the setup. With these lines we are able to control a switch that would swap the transmitting and receiving lines automatically. This allows us to have one single transducer configuration.

The next step was to find a transducer that would perform well both transmitting and receiving. The acoustic transducers come in sets, a transmitter and a receiver. After a series of tests, we chose to use the receiving transducer for both transmitting and receiving.

For a switch, we selected the sophisticated MAX362, which proved to do an excellent job. This type of switch has isolated inputs and it is designed so that cross-talk is minimal.

**10.5.3.3 Signal Amplification and Filtering**

In Sect. 10.5.4.1 we present how we use parabolic cones in our localization module. While making use of the parabolic cones works very well for spreading out the signal, this comes at the cost of reducing the energy of the signal. The acoustic pulse will now be much weaker when it arrives at the receiving transceiver. At this point in time there are two stages of amplification on the XSRF, and since each amplifier has a gain of 18, we have a total gain of 324. With this much amplification, we are able to get a better signal, but the range is still not sufficient for our application. The maximum range where the readings are consistently accurate is about 2 feet (ft). For greater distances, the received signal becomes unstable or too weak to accurately detect the incoming signal. We would like to have a sensor range of at least 3 ft. To overcome this problem, we decided to add a third operational amplifier to further amplify the signal. This stage multiplies the existing gain by 18, bringing the total gain up to 5,832. This gain helps to amplify the signal, but unfortunately it also amplifies the noise, which in turn causes the sensor to produce unreliable data.

To fix this problem, we used a bandpass filter. A bandpass filter is a circuit that will only amplify a certain range of frequencies and ignore the rest. This filter works well for us since we know exactly what frequency we are interested in. All the other remaining frequencies are considered noise in the system. We designed this filter to have a gain of 18 with a pass band centered on 40 kHz [134]. All the frequencies in the 40 kHz range are amplified by this filter, and the frequencies out of this range are dampened. The fact that the ultrasonic transducers are tuned to transmit and receive 40 kHz, together with our bandpass filter, should give us a significant reduction of signal noise.

After integrating this bandpass filter in our sensor, we immediately noticed a much cleaner signal together with a much greater gain. The acoustic signal is also steadier and more accurate at greater distances. In the preliminary tests we performed measurements up to 4 ft and obtained excellent accuracy. This distance is one foot longer than our initial minimum requirements for this project. As an example, the video "2robots.avi" shows a radio-controlled MAXELBOT being followed by a second autonomous MAXELBOT using our trilateration hardware. The quality of the trilateration is reflected in how smoothly the second robot follows the R/C MAXELBOT.

Since it is based on the SRF04, we chose the name "XSRF" for our first prototype of the ultrasonic range finder, which stands for eXperimental ultraSonic Range Finder. The Electrical Engineering shop at the University of Wyoming was in charge of fabricating and populating the Printed Circuit Boards (PCB) for the XSRF. The PCB version of our acoustic sensor boards produce a cleaner signal and are able to sense at a longer distance than the bread-boarded prototype. Further testing revealed a useful range of 9–11 ft, with accuracy sufficient for numerous applications [155]. The video "3robots.avi" shows three fully autonomous robots using trilateration to maintain formation while moving outside. Figure 10.7 represents our modified schematic and the finalized board can be seen in Fig. 10.8.

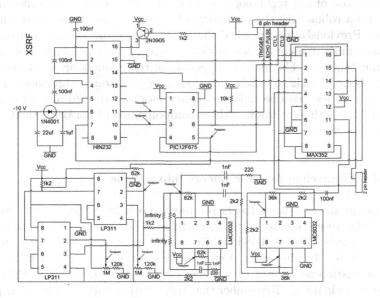

Fig. 10.7: XSRF schematic version 2.1

Fig. 10.8: The XSRF printed circuit board version 2.1 (reprinted from "Robotics Research Trends," 2008. Used with permission from Nova Science Publishers, Inc.)

## 10.5.4 Trilateration Module

The purpose of our technology is to create a plug-in hardware module that provides a robot with the capability of accurately localizing its neighboring robots. Previously, we described the XSRF board that enables us to perform distance measurements. This section presents how the localization module was put together. The main components of this module are (a) three XSRF sensor boards with (b) their corresponding ultrasonic transducers and (c) parabolic cones, (d) one radio frequency module, and (e) a main controller board that both synchronizes these components and performs the trilateration computations.

### 10.5.4.1 Channeling Acoustic Energy Into a Plane

Ultrasonic acoustic transducers produce a cone of energy along a line perpendicular to the surface of the transducer. The width of this main lobe, for the inexpensive 40 kHz 400ST/R160 transducers used in our implementation, is roughly 30° for distances around 36 inches (in) (Fig. 10.9, left[3]). To produce acoustic energy in a two-dimensional plane would require 12 acoustic transducers in a ring. Remember that trilateration requires three base points; hence we would require 36 transducers. This is expensive and is a large power drain. We adopted an alternative approach. Each base point is comprised of one acoustic transducer pointing downward. A parabolic cone [88] is positioned under the transducer, with its tip pointing up toward the transducer (Fig. 10.9, right). The parabolic cone acts like a lens. When the transducer is placed at the virtual "focal point," the cone "collects" acoustic energy in the horizontal plane, and focuses this energy to the receiving acoustic transducer. Similarly, a cone also functions in the reverse, reflecting transmitted acoustic energy into the horizontal plane.

The parabolic cones were specifically designed to have their focal point at one inch above their tip. After several tests we determined that a distance of 7/8 in from the face of the transducer to the tip of the cone corresponds to the optimal signal strength. We can safely assume that the distance from the face of the transducer to the inside oscillating membrane is about 1/8 in.

### 10.5.4.2 Radio Frequency Module and Manchester Encoding

In order for the trilateration module to produce accurate results, it is very important to synchronize the transmitter with the receiver. Since the trilateration module is designed to be installed on independently moving robots, we have to provide the trilateration module with wireless communication

---

[3] http://www.robot-electronics.co.uk/images/beam.gif

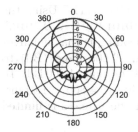

Fig. 10.9: Left—400ST-R160 beam pattern (courtesy of Pro-Wave Electronics Corp.). Right—parabolic cones setup (reprinted from "Robotics Research Trends," 2008. Used with permission from Nova Science Publishers, Inc.)

capabilities. The RF module that we use, AM-RTD-315 (Fig. 10.10), is manufactured by ABACOM Technologies, Inc. and operates at 315 MHz, which is part of a license-free band. Allowing for implementation of half-duplex two-way radio link, this transceiver is capable of transferring data at rates up to 10 kilobits per second (kbps) [1]. Other important features that made this module a good candidate are low power consumption and low cost (US $28.95 at the time of this writing).

This type of RF module requires a balanced signal to be sent through. This means we need to have almost the same number of 0s and 1s sent alternatingly over a short period of time. Sending unbalanced data will make it harder for the RF receiver to detect a change in the line from 0 to 1 or vice versa. Both hardware and software solutions are available to tackle this issue. Since one of our goals is to keep the hardware design simple, we decided to go with the software solution. The Manchester encoding algorithm [185] is widely used to encode the regular signal to a balanced signal. Instead of sending the binary data as a sequence of 0s and 1s, each bit is translated to a different format according to the following rules:

Fig. 10.10: AM RF transceiver module AM-RTD-315

- 0 encodes to 01
- 1 encodes to 10

As an example, let us consider the pattern of bits "0000_1111." Using the rules mentioned above, this pattern will encode to "0101_0101_1010_1010." One of the main disadvantages of the Manchester encoding is that it halves the transmission speed. Since currently we are using 9,600 baud (Bd) to communicate between modules the actual speed will be only 4,800 Bd. To send one byte of information it will take approximately $8 \div 4,800 = 1.67$ ms. Since the information that we plan to exchange over the wireless is limited to a small number of bytes, this speed is adequate.

### 10.5.4.3 Encoding

In order to make the implementation easier, we will use a slightly modified version of the Manchester encoding algorithm [185]. The process of encoding one byte is as follows:

- Extract the odd bits [7, 5, 3, 1] of the byte by using 0xAA as a bit mask (using the *and* operator).
- Extract the odd bits [7, 5, 3, 1] of the complemented byte by using the same bit mask and *shift* the result one bit to the right using 0 fill; this way we will obtain the complement of the odd bits positioned one bit to the right.
- Use the *or* operator between the results obtained above; the result will contain the odd bits encoded.
- Extract the even bits [6, 4, 2, 0], this time using 0x55 as a bit mask.
- Extract the even bits [6, 4, 2, 0] of the complemented byte and *shift* the result one bit to the left using 0 fill; we do not *shift* the result to the right because it would involve a *circular shift* which is more computationally expensive.
- Use the *or* logic operator between the two previous results to obtain the even bits encoded; these bits will be encoded slightly differently since we replace 0 with 10 and 1 with 01 but this does not affect the overall goal of having a balanced stream.

Next, we will show an example of how the encoding is implemented. If we consider encoding byte 0x0E or in binary "0000_1110" we will perform the following steps:

1) First, extract the odd bits: $0000\_1110 \wedge 1010\_1010 = 0000\_1010$.
2) Obtain the complemented odd bits, first computing the complemented byte $\neg 0000\_1110 = 1111\_0001$, then $1111\_0001 \wedge 1010\_1010 = 1010\_0000$.
3) Then shift the result (2) one bit to the right: $1010\_0000 \gg 1 = 0101\_0000$.
4) Now *or* the two results, (1) and (3): $0000\_1010 \vee 0101\_0000 = 0101\_1010$.

So far, we encoded only the odd bits; the encoding of the even bits follows.

5) Obtain the even bits from the byte to be sent $0000\_1110 \wedge 0101\_0101 = 0000\_0100$.

6) Obtain the complemented even bits, first computing the complemented byte $\neg0000\_1110 = 1111\_0001$, then $1111\_0001 \wedge 0101\_0101 = 0101\_0001$.

7) Then shift the result (6) one bit to the left: $0101\_0001 \ll 1 = 1010\_0010$.

8) Now *or* the two results, (5) and (7): $0000\_0100 \vee 1010\_0010 = 1010\_0110$.

Finally, the initial byte "0000_1110" is encoded to "0101_1010_1010_0110." By directly applying the Manchester encoding we would have obtained "0101_0101_1010_1001." As we can see, both encoded results are comparable as far as the balancing between 0s and 1s is concerned.

### 10.5.4.4 Decoding

Since we are using standard serial communication, our transmission is done with one-byte packets, so each byte of data is sent, after encoding, using two packets. At the other end, two packets need to be received before the decoding process can be performed. The decoding process takes an encoded 16-bit value as an input and returns a decoded 8-bit value. The following steps are performed:

- Extract the encoded odd bits, which represent the first 8 bits of the 16-bit value (shift 8 bits to the right using 0 fill).
- Apply the 0xAA bit mask to the encoded odd bits.
- Extract the encoded even bits from the 16-bit value, which represent the last 8 bits of the 16-bit value.
- Apply the 0x55 bit mask to the encoded even bits.
- Use the *or* operator between the two previous results to obtain the original decoded byte.

Using the same value that we encoded above, "0101_1010_1010_0110," the next example will show how the decoding process takes place:

1) Extract the odd bits: $0101\_1010\_1010\_0110 \Rightarrow 0101\_1010$.

2) Apply the bit mask: $0101\_1010 \wedge 1010\_1010 = 0000\_1010$.

3) Extract the even bits: $0101\_1010\_1010\_0110 \Rightarrow 1010\_0110$.

4) Apply the bit mask: $1010\_0110 \wedge 0101\_0101 = 0000\_0100$.

5) Apply the *or* operator between the two results, (2) and (4), to obtain the decoded byte: $0000\_1010 \vee 0000\_0100 = 0000\_1110$.

Because of the balanced nature of Manchester encoding, we can easily implement a method to detect bit errors. If the received data is not balanced, we can choose to have that data retransmitted or we can simply discard it.

### 10.5.4.5 The Minidragon

Before our trilateration module can be put together, there is one more part that needs to be carefully chosen: the main controller. Since we already had experience working with the Minidragon board, we decided to use it for this purpose. This board is manufactured by Wytec Co. [278] and is equipped with a 25 MHz MC9S12DG256 Freescale microcontroller [70]. Some of the features of this board that are important for our application are two separate serial communication ports, 16-bit timers, the $I^2C$ interface, and 256 kilobytes (kB) of non-volatile memory or EEPROM. The 16-bit timers are necessary to measure the time interval produced by the three XSRF boards. Finally, the Minidragon does not have an operating system running on it, which allows us to better control the synchronization between the microprocessors on the localization modules.

To program the Minidragon we are using the ICC12 IDE version 6.16A produced by ImageCraft Inc., which provides a C language compiler for the HCS12 series of Freescale microprocessors. Normally, the programs are uploaded to the 12 kB of volatile memory, which has obvious drawbacks (e.g., the program is wiped out when the power is cycled or lost). We are using a DragonBDM flash programmer available from Wytec Co. (Fig. 10.11) to upload the code directly to the FLASH memory [277].

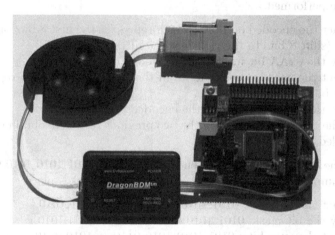

Fig. 10.11: DragonBDM flash programmer connected to the Minidragon (right side) and to a serial adapter (left side)

Given that we now have all of the necessary parts, the next step is to put everything together. One complete trilateration module is shown in Fig. 10.12. The module consists of three XSRF sensor boards (the three outlines at the bottom) each connected to an acoustic transducer that is pointed down to a parabolic cone, a radio frequency transceiver, and the main controller which is outlined in the center. Two XSRF boards are hardwired to receive and one XSRF board can be switched between transmitting and receiving.

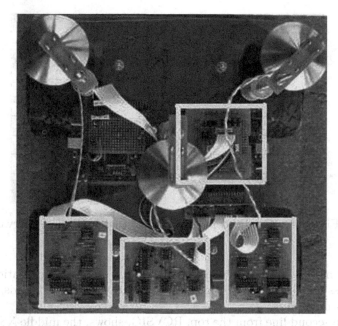

Fig. 10.12: Trilateration module top-down view (reprinted from "Robotics Research Trends," 2008. Used with permission from Nova Science Publishers, Inc.)

## 10.6 Trilateration Evaluation

At this point we are ready to use this trilateration hardware to perform the trilateration calculations. The trilateration system involves at least two modules. One module transmits the RF–acoustic pulse combination, while the other uses these pulses to compute (trilaterate) the Cartesian coordinates of the transmitting module. Hence, trilateration is a one-to-many protocol, allowing multiple modules to simultaneously trilaterate and determine the position of the transmitting module.

To better understand how the three distances necessary to perform trilateration are measured, let us examine Fig. 10.13. The oscilloscope is connected to the receiving module. The top line, SCITRG, is connected to the RF line, while the second line, RCVSIG, shows the amplitude of the incoming acoustic signal for the middle XSRF sensor (RECVB or B). The bottom line, TRIGGR, shows the trigger, which is shared by the three XSRF boards. After the RF signal is received (RF arrival time), the trigger line is raised, signaling the three XSRF boards, which in turn raise their own "echo" line (A, B, and C or RECVA, RECVB, and RECVC).

Note that there is a 250 μs delay between when the trigger and the echo line is raised. This delay is imposed by hardware limitations, and it represents

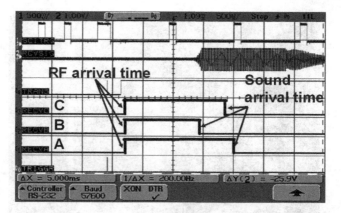

Fig. 10.13: Oscilloscope snapshot of the three time intervals

the amount of time it takes the controller on the receiving localization module to process the incoming RF signal and prepare the XSRF boards to receive the ultrasonic pulse.

As the second line from the top, RCVSIG, shows, the middle XSRF board is the first one to detect the acoustic signal; therefore its echo line (B) is the first one to be lowered. The other two XSRFs lower their echo lines shortly after. These three echo lines are connected to three input capture pins on the Minidragon board, which accurately times their widths, and converts them to distances.

### 10.6.1 Trilateration Accuracy Experiment

To test the accuracy of the trilateration-based localization module, we placed a robot on our lab floor, with transducer **B** at $(0'', 0'')$. Then we placed a transmitter along 24 grid points from $(-24'', -24'')$ to $(24'', 24'')$.

The average error over all grid points is very low—$0.6''$, with a minimum of $0.1''$ and a maximum of $1.2''$ (Fig. 10.14). This amount of error is acceptable for our application, given the size of the robots we plan to install it on. Even at distances of 9–11 ft the error was at most 11% [155].

### 10.7 Synchronization Protocol

The purpose of this trilateration-based localization module is to allow any robot equipped with it to determine the positions of all its neighboring robots, which are also equipped with this localization module. For this to be possi-

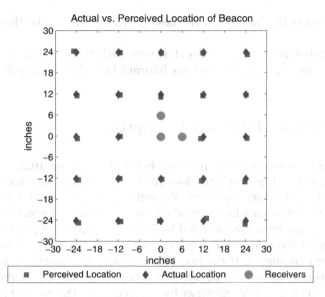

Fig. 10.14: Trilateration accuracy

ble, it is necessary that each robot take turns transmitting the RF–acoustic combination pulse (by broadcasting). In this section we present our synchronization protocol, which is similar to the token ring network protocol [104]. This protocol also allows for limited data transmission through the RF pulse, and currently carries information about the identity of the triggering robot and, depending on the application, information about the heading of the robot or the concentration of the traced chemical ($\rho$). (For more about the chemical tracing application, see Chaps. 8 and 9.) From now on, when we use the word "robot," we mean a robot that is equipped with our localization module.

## 10.7.1 High-Level Protocol Description

Our synchronization protocol is similar to a token ring network protocol. Each robot has a unique hardware encoded identification number (ID). The IDs start from zero and go up to $N-1$, where $N$ is the total number of robots. Both $N$ and the IDs must be known a priori. When a robot is broadcasting, it also sends its own ID. When the robots that are performing the localization (trilateration) receive the ID of the broadcasting robot, they increment this broadcast ID by one and then compare it with their own ID. The robot that matches the two IDs is considered to have the token and it broadcasts next. The other robots whose IDs did not match will continue to trilaterate. When the ID reaches the last robot, the next computed ID will be that of the first robot.

Each robot maintains a data structure with the coordinate information, as well as any additional broadcast information, of every neighboring robot.

## 10.7.2 Low-Level Protocol Description

The main loop running this protocol is listed in Listing 10.2. This loop has two main parts. The first part does the broadcasting, while the second part controls the trilateration process. We will start with the first half of the loop.

When the robots are first turned on, the main controller on the localization board reads the hardware encoded ID (own ID). Initially, the ID of the broadcasting robot (broadcast ID) is zero for all robots. The own ID and broadcast ID are next compared. If the two IDs are equal, the controller will start the broadcasting sequence. First, one XSRF board and the RF module are set to transmit, then a synchronization byte is sent over the RF link, followed by the heading of the robot provided by a digital compass (see Sect. 10.9.3), in this particular example. The own ID is the next byte to be transmitted and it can be thought of as being the RF pulse. Two blank bytes are sent next. These two bytes have the role of delaying the acoustic signal generation for a very short period. This period is necessary for the trilaterating robot to process the information that was already sent over the RF link and to prepare the localization board for receiving the acoustic signal. A light-emitting diode (LED) is blinked next, to let the human observer know that a broadcast is taking place. The acoustic signal is generated, and immediately the XSRF board and the RF module are set back to receive, which is their default state. The reason the receive mode was chosen as the default state is that most of the time the localization module is trilaterating (receiving).

If a robot does not match its own ID with the broadcast ID, it will start trilaterating. First, it will listen for the synchronization byte, which is followed by the heading and the ID of the robot that is broadcasting. This data is stored in a special data structure, to be used at a later time.

For both broadcasting and trilaterating robots, the next step is to decide who will broadcast next. This is done by incrementing the ID of the last robot that broadcast by one, and *mod* that by the total number of robots. The loop then starts over and the robot that matches the new broadcast ID is the one that has the token and is next to transmit the RF–acoustic pulse combination.

One important aspect of this protocol is how fast the token is passed. Theoretically, we should have at least a 10 ms pause between broadcasts, to allow the sound from the previous broadcast to dissipate and not interfere with the next broadcast. Practically, the token is passed every 120 ms. This limit is imposed by the RF module switching time. According to the specifica-

## Listing 10.2: Synchronization/Communication Protocol—C code

```
...
broadcast_ID = 0;
own_ID       = read_ID();
...
while (TRUE) {
  if (own_ID == broadcast_ID) {    // broadcast
    change_mode(SENSOR_MODE_TX);
    sc1_transmit_byte(SYNC_BYTE);
    sc1_transmit(robots[own_ID].heading);
    sc1_transmit(own_ID);            // equivalent to the RF pulse

// normally we transmit the acoustic pulse next, but we need to give
// more time to the receiving robot to process the data sent over the
// RF; transmitting the next two bytes, provides the needed delay

    sc1_transmit_byte(FILL_BYTE);
    sc1_transmit_byte(FILL_BYTE);
    blink_LED();
    sonar_activate();               // generate the acoustic pulse
    change_mode(SENSOR_MODE_RX);
  }
  else{                             // trilaterate
    unsigned char comm_OK = TRUE; // protocol status flag
    roll = 0;
    while (sc1_receive_byte() != SYNC_BYTE){// sync. and/or recovery
      if (own_ID == 0 && roll > 50){      // 11.36 is a 1 second delay
        comm_OK    = FALSE;               // protocol has failed
        broadcast_ID = NUMBER_OF_ROBOTS - 1;// reset the broadcast ID
        break;
      }
    }
    if (comm_OK) {
      unsigned char temp_heading = sc1_receive();
      broadcast_ID = sc1_receive();
      sonar_trilaterate();
      robots[broadcast_ID].heading = temp_heading;
    }
  }
  broadcast_ID = (broadcast_ID + 1) % NUMBER_OF_ROBOTS;//compute next ID
}
...
```

tions, the AM-RTD-315 transceiver module can switch between transmitting and receiving in less than 5 ms. Although this may be true, we found out that immediately after switching, the data stream is unreliable. After about 100 ms the data stream becomes reliable. We transmit a "filling" byte for about 120 ms before real data is transmitted—just to have a safety margin.

## 10.7.3 Protocol Recovery

Although wireless communication is reliable and has its own indisputable advantages, one of the main issues with it remains the occasional loss of signal. In our case, this translates to data packets never being received, received incompletely or incorrectly. An occasional failure in the protocol is not critical and may go undetected to the human eye, especially since a trilateration happens every 120 ms. It is more important to detect the failure and recover.

The first thing we did was to synchronize the serial data stream between the transmitter and the receiver by introducing a synchronization byte. This byte is carefully chosen to be "33" or "00110011" in binary representation. The importance of this choice lies in the way the RF module can transmit and receive data (Sect. 10.5.4.2). This byte is not encoded, and it is always sent before sending the actual data. When the receiver recognizes it, the synchronization is made. Now the receiver knows exactly what the order of the following bytes will be, and it will interpret them correctly.

To make the protocol truly self-recoverable, we determined what happens with the protocol if certain bytes are never received. If the error happens within one data packet, we already have a receive error detection mechanism from using the Manchester encoding/decoding algorithm. At the time the received data is decoded, we check if the byte is balanced (Sect. 10.5.4.2), and if that is not the case, we raise a communication error flag. In the more unfortunate cases, a whole data packet is lost, and this can happen at any time in the serial stream. When a byte is sent and fails to reach its destination, the receiving board will never know whether that byte was transmitted or not, and it will indefinitely wait for it. This will cause the protocol to "crash." To prevent this from happening, we introduced a software version of a "watchdog timer." This is a timing device that triggers a system reset if the main program, due to some fault condition such as a hang, neglects to regularly service the watchdog, like writing a pulse to it (also referred to as "feeding the watchdog"). In our case, we have an interrupt service routine that increments a counter approximately every 88 ms. This counter is always reset right before the first byte is received. If this counter reaches a certain threshold value, the robot with the lowest own ID assumes the token was lost and it will automatically reinitiate the protocol. The threshold value is chosen mainly based on the total number of robots.

To prevent the robots from performing an erroneous movement, when a protocol failure is detected, they automatically stop moving. We did numerous tests with this protocol, including shutting off robots at random, to simulate a lost token, and it always recovered automatically, without human interaction.

## 10.8 Maxelbot

The power of our trilateration-based localization module is unveiled when it is put to work on teams of real robots, in real outdoor environments. To truly test the capabilities of our localization module, we decided to design and build a suitable robot platform. Our University of Wyoming robot platform MAXELBOT is named after the two graduate students who designed and built it. In this section, we first introduce the two design approaches that this platform went through. We next present the I$^2$C multi-master serial computer data bus that enabled us to dramatically improve the whole robot design. At the end, we introduce the previous and current mobile robot platforms used.

### 10.8.1 Centralized Versus Modularized

The approach for designing our first robot was to make use of the versatile HCS12 Freescale microcontroller that comes with the Minidragon board. The localization module is only using a fraction of the functions this microcontroller has. More exactly, out of its 112 pins, we are using only 14 pins: three timer channel pins used to capture the output from the XSRF boards, two pins to control the transmit/receive mode on one XSRF, one pin to trigger all three XSRFs when in receive mode, two pins for transmitting/receiving over the serial communication, four pins to select the hardware ID, and two pins for power. With so many pins available to us, it is not a problem to control other subsystems of the robot.

After some preliminary tests, we realized that a better architecture was needed. One reason is that we want our robots to be easily modified to perform a variety of tasks. A centralized architecture would limit this capability. A better design would be to modularize most of the subsystems of the robots (e.g., trilateration, communication, motor control and chemical sensors). Another issue that showed us the inefficiency of the centralized approach relates to the high time sensitivity of the trilateration process. The only time when the robot can perform any additional computations (e.g., compute where to move next) is right before broadcasting. With this design, the trilaterating robots are blocked, waiting for the broadcast. The disadvantage of this approach is that only one robot can move at a time, while the other robots just idle.

In response to these limitations, we designed a new modularized architecture. To achieve this modular approach we needed a data bus that would interconnect all the modules and allow them to efficiently exchange information. In Sect. 10.8.2 we present our data bus choice. As seen in Fig. 10.15, the localization module is now separated from the rest of the subsystems. Each subsystem has its own controller and the only interaction between subsystems is done through the data bus. This also solves the time sensitivity of our localization module, which is not interrupted to service other subsystems of the robot. In this type of architecture you can think of any of these subsystems as black boxes. All we need to know is what information we need to provide it with and what information it returns, without worrying how the information is processed inside the module. For example, the localization module can be easily replaced with a GPS module, and as long as the GPS can provide the same data format as our trilateration module, the other subsystems would be unaware of this exchange.

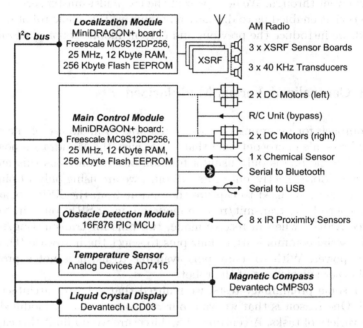

Fig. 10.15: Improved modularized design

## 10.8.2 Inter-Integrated Circuit I²C Bus

After researching the available data buses, our choice was the Inter-Integrated Circuit (IIC or I²C ) Bus. I²C is a multi-master serial computer bus that was invented by Philips Semiconductors in the early 1980s. In 1992 the first standardized version was released [170]. The I²C bus was originally designed to link a small number of devices on the same printed circuit board (e.g., to manage the tuning of a car radio or a TV set). It is a two-wire, bidirectional serial bus that provides a simple, efficient method of data exchange between devices. It minimizes the need for large numbers of connections between devices, and eliminates the need for an address decoder. An important advantage that we have with I²C is the ability to add more I²C-enabled devices for further expansion. The limiting factor is its bus capacitance, which should not exceed 400 picofarads (pF) [70].

The HCS12 microprocessor that the Minidragon comes equipped with has the hardware support for using the I$^2$C bus. The module is able to operate at speeds up to 400 kbps but for our application we will only use 100 kbps. The two pins that have to be connected on the Minidragon are SCL and SDA. The two pins are open-drain, meaning they have to be held in a logic 1 state by obligatory pull-up resistors. For 100 kbps communication on a lightly loaded bus (100 pF), pull-up resistors of no greater than 8 kOhm are recommended [166].

In Sect. 10.8.3 we present the I$^2$C communication sequence that is used to exchange data with our localization module as well as a one-way serial communication alternative.

## 10.8.3 Trilateration Module Connectivity

A localization module that does accurate localization, but is not able to communicate this information, is not very useful. In this section we present two ways to interface with our localization module, I$^2$C, and serial communication.

### 10.8.3.1 I$^2$C Connectivity

To communicate with our localization module using the I$^2$C bus, three wires need to be connected to the following pins on the Minidragon board: pin 85—ground, pin 98—I$^2$C clock line (SCL), and pin 99—I$^2$C data line (SDA). The access to these pins is facilitated through a top connector that encloses these three pins plus an extra pin that can be used to power the module. The voltage provided through the top connector can vary between +5 VDC and +12 VDC. The module can also be powered with +5 VDC only, through pin 84 of the Minidragon board [134].

From the I$^2$C bus point of view, the trilateration module is considered to be a I$^2$C master, and the device that is communicating with it is an I$^2$C slave. The master is the only one that can initiate a communication session. The main reason we programmed our localization module to be a master is related to the time sensitivity of the trilateration process. By being a master, our module can control exactly when a communication session is performed.

The version of the I$^2$C protocol that we are using requires that each device connected to the I$^2$C bus have a unique 7-bit address. Since the "shortest" hardware register is 8-bit, these seven bits occupy only the seven most significant bits, leaving the least significant bit to be used for other purposes. More specifically, when this bit is set to 0, the addressed I$^2$C device is instructed to provide data (transmit). When the bit is set to 1, the addressed $I^2C$ device is instructed to wait for data (receive).

To communicate with the localization module, a second microcontroller that supports $I^2C$ is necessary. This second microcontroller has to be programmed to have an $I^2C$ address of "1010_101" and be prepared to receive a data structure of the form {*four bytes, four bytes, one byte*} for each robot in the team. The first four bytes represent the $x$-coordinate, and the second four bytes are the $y$-coordinate (both in floating point precision) of the sensed robot. The last byte represents the robot's heading, where 0 maps to $0°$ and 255 maps to $360°$. The order in which these sets of nine bytes are received determine the ID of the robot to which this data belongs. For example, for three robots there will be a total of 27 bytes transmitted, and the first set of nine bytes refers to the robot with ID = 0.

First we present the steps that are taking place when data is being transferred from the master to the slave. This is a low-level step-by-step sequence of the communication session. In this sequence we also refer to the fragment of C language code that is used to program our localization module to perform the $I^2C$ communication which is presented in Listing 10.3. Although the program is written in the C language, many variables refer to actual registers (e.g., IBSR, IBCR, IICDR) or bits inside a register (e.g., IBB, TXRX, TXAK, IBIF). This program can also be categorized as an $I^2C$ driver for HCS12 Freescale microcontrollers.

The steps are the following:

1. The master generates a START signal, trying to get exclusive control of the $I^2C$ bus (line 11 in the listing).
2. If the $I^2C$ bus is available, the master switches to transmit mode and transmits the 7-bit address of the slave, "1010_101," with the eighth bit set to one, which instructs the slave to wait for data (line 21).
3. The slave will acknowledge that it is ready to receive data.
4. The master will transmit the first byte (lines 28–34).
5. The slave will acknowledge that it has received the byte.
6. The previous two steps, 4 and 5, will repeat until all the bytes are sent.
7. The master will generate a STOP signal, releasing the $I^2C$ bus (line 36).

### 10.8.3.2 Serial Connectivity

The $I^2C$ bus connectivity is fast and accurate but it may not always be possible to interface with it, one reason being the lack of $I^2C$ interface on the target device. This is why, as an alternative to the $I^2C$ interface, we also offer a standard serial interface. The connection to the Minidragon is done through a RJ11 modular connector. The serial port has to be configured with 9,600 Bd, 8 data bits, no parity (N), and one stop bit (1). Each line of data received over the serial connection provides four numbers: the robot ID, the $x$-coordinate, the $y$-coordinate, and the heading. This data can then

## Listing 10.3: Communicating using I²C—C code

```
...
#define AP_MODULE_ADDR    0xAA
#define AP_INFO_LENGTH    (NUMBER_OF_ROBOTS * sizeof(robot_data))
char* i2c_data;
unsigned char i2c_data_offset = 0;
...
void say_to_other_board(void){
  // wait for IBB to clear - the bus is free
  while (IBSR & IBB);
  // master transmit - also generates the START condition
  IBCR |= (MSSL | TXRX);
  // transmit acknowledge enabled
  IBCR &= ~TXAK;
  // wait for IBB to set - the bus is busy
  while (!(IBSR & IBB))
  // pointer to data packet being sent
  i2c_data = (char*)robots;
  i2c_data_offset = 0;
  // bits 1:7 contain the ID of the secondary
  // bit 0 is  0 - master writes to slave
  IICDR = AP_MODULE_ADDR & 0xFE;
}
...
void i2c_isr(void){
  IBSR |= IBIF;                                   // clear IBIF
  if (IBCR & MSSL){                               // master mode
    if (IBCR & TXRX){                             // master transmit
      if (i2c_data_offset < AP_INFO_LENGTH){      // send robots[]
        IICDR = *(i2c_data + i2c_data_offset++);
      }
      else if (i2c_data_offset == AP_INFO_LENGTH){ // send own_ID
        IICDR = own_ID;
        i2c_data_offset = 0;
      }
      else{
        IBCR &= ~MSSL;                            // generate the STOP signal
      }
    }
  }
}
...
```

be easily parsed, stored, and/or processed by the target device. Figure 10.16 shows a possible setup with a Minidragon board connected to a Serial-to-USB adapter. This setup is useful when there is no legacy serial port on the PC.

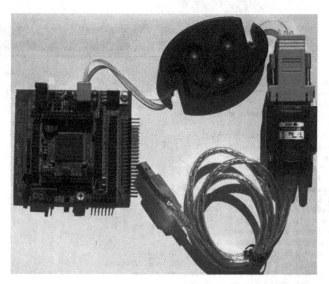

Fig. 10.16: Connecting a Minidragon board to a PC using a Serial-to-USB adapter

## 10.8.4 Mobile Robot Platform

To further test the performance of the localization module, the next step was to build a mobile robot platform that would use it. The MMP-5 robot platform [249] manufactured by The Machine Lab was chosen for this purpose. This platform is very sturdy, weighs 4.5 pounds, can carry an additional 4 pounds, and comes with high-capacity nickel-metal hydride (NiMH), 12 VDC, 4.5 ampere-hour (Ah), rechargeable batteries. These platforms have four wheels, a "tank" style drive, and a zero turning radius.

The MMP-5 uses four motors, one for each wheel. The four motors are connected to the wheels through a high torque gear head. To control them, a pulse of a specific length is required. These four wheels work in pairs (left-side pair and right-side pair), and only one control line is necessary per side. The pulse width sent to the MMP-5 platform must be between 1.0–2.0 ms. A pulse width of 1.0 ms will cause the wheels on the left or right side of the platform, depending on which motors the pulse is sent to, to rotate backwards at full speed. A pulse width of 1.5 ms will cause the wheels to stop, and a 2.0 ms pulse width will cause the wheels to move forward at full speed. As for pulse widths between 2.0 ms and 1.5 ms, the wheels will rotate faster as you move from 1.5 to 2.0 ms, which is the maximum speed. The case is similar for pulse widths between 1.5 and 1.0 ms. A pulse width closer to 1.0 ms will rotate the wheels in reverse faster than one close to 1.5 ms. Linear or exponential response modes to these signals can be selected. The exponential mode softens the control around the zero speed point [54].

We programmed a motor driver, which uses the output comparison system on the Minidragon microprocessor to generate these pulses. The driver is set up so that the direction and the power level, as a percentage, are passed in as arguments. Different function calls are used to control the left and right sides.

The MAXELBOT MMP-5, at a minimum, comes equipped with a localization module and a second Minidragon control board, which can be used to compute the coordinates of the next location to move, based on the control algorithm used (e.g., physicomimetics). This second control board also generates the pulses to control the wheels of the platform. The localization module and the second control board are connected together over the Inter-Integrated Circuit data bus.

The MAXELBOT in Fig. 10.17 is also equipped with an infrared (IR) obstacle detection system as well as a compass module. Section 10.9 will present these and other modules that we use with our robot.

Fig. 10.17: MAXELBOT with MMP-5 platform: rear angle view (left), and top-down view (right)

## 10.8.5 PIC-Based Controller Board

For every digital hardware application or project, there is the need for a microcontroller board. The microcontroller does the interfacing between different electronic components and also executes the logic with which it is programmed. For example, our localization module uses the Minidragon board. Although it has numerous other capabilities that our localization module is not making use of, the Minidragon board was chosen because we are familiar with the Freescale architecture. Our modular robot design requires independent hardware modules that are interconnected with an I$^2$C data bus. Each of these hardware modules has to feature a specialized microcontroller that has to, among other things, communicate over the I$^2$C data bus. Microchip Technology Inc. offers a wide array of CMOS FLASH-based 8-bit microcontrollers that are specialized for different tasks. All we needed was a printed circuit board to accommodate this type of microcontroller and its supporting hardware [134].

We designed the Programmable Intelligent Computer (PIC) board to accommodate the Microchip PIC16F series microcontrollers that come in a 28-pin Plastic Dual-In-line Package (PDIP), Small-Outline Integrated Circuit (SOIC), Shrink Small Outline Package (SSOP) form factor. As seen in Fig. 10.18, the board features a total of four connectors. First, a 30-pin dual row male connector allows access to most of the pins on the PIC microprocessor and provides future expansion through the use of a daughter-board. Second, the power connector is a three-pin single-row male connector that accepts from +5 VDC up to +12 VDC. The middle pin is ground and both outside pins are power pins. Third, a four-pin single-row male connector provides I$^2$C connectivity as well as +5 VDC, which allows for other I$^2$C modules to be connected. Jumpered pull-up resistors are provided for the I$^2$C data bus. Finally, the programming of the PIC is done through a five-pin connector that fits the female connector of the specialized MPLAB ICD 2 programmer.

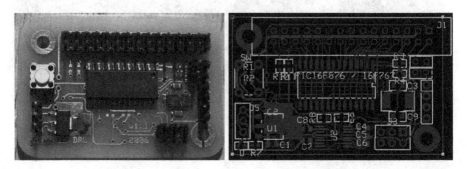

Fig. 10.18: Our PIC board top view (left) and its board layout (right)

To write applications for the PICs, we are using the Microchip MPLAB IDE version 7.20, together with the Hi-Tech PICC Toolsuite version 9.50 C language compiler. The upload of the program to the PIC is done using an MPLAB ICD 2 USB programmer.

We have built PIC-based controller boards using either PIC16F767 or PIC16F876 microcontrollers. Our first hardware module that uses this PIC

board is an obstacle detection module (Sect. 10.9.1), which requires eight
analog to digital inputs. Depending on the application, other PIC microcontrollers are available.

## 10.9 Add-Ons

During the experiments with our seven MAXELBOTS (see "4robots.avi" and
"7robots.avi" for outdoor experiments) we realized that additional sensing
capabilities were necessary. Besides knowing the position in a team of robots,
detecting obstacles and knowing the ambient temperature or the heading of
the robots represent just a part of the information about the surrounding
environment that we can use when testing our control algorithms. Due to
the modularized design of our robots, the addition of new sensing modules
is straightforward. In this section we present the modules with which we can
equip our robots, and how the interfacing is done.

### 10.9.1 Obstacle Detection Module

The ability to detect obstacles in the environment is an important aspect
that needs to be handled when building robotic platforms. Different technologies like radio frequency (used in radars), ultrasound (used in sonars),
infrared (IR), and digital cameras are used in the design of an obstacle detection system. The choice of a certain technology over another may depend on
factors like price, accuracy, power requirements, or the medium in which it
will function. For our obstacle detection we decided to use IR sensors, more
precisely, the Sharp GP2D12 (see "odm.avi"). They are relatively inexpensive
(US $12.50 at the time of writing) and can provide obstacle detection from
10 cm up to 80 cm. One of the drawbacks of this sensor is the relatively high
power consumption, having an average current consumption of 33 mA [216],
but from our own tests it can go as high as 100 mA. This limits us to using
at most eight sensors simultaneously, but this is more than we require for our
applications.

As shown in Fig. 10.19 (left), our obstacle detection module (ODM) hardware consists of a PIC-based board (Sect. 10.8.5), a daughter-board and up
to eight IR sensors. We selected a PIC board equipped with the PIC16F767
microprocessor, since it can handle at least eight 10-bit analog to digital
inputs [160]. This PIC features an $I^2C$ interface, which allows the intercon-
nection of this ODM to our modularized architecture. The daughter-board
is necessary for connection purposes. It does the interfacing between the IR
sensors and the specific pins on the PIC board. It also allows for 12 VDC and

Fig. 10.19: ODM close-up (left) with three IR sensors mounted on the robot (right)

$I^2C$ connectivity in a twin connector setup, which allows for chaining other devices.

The output voltage from the IR sensors varies from 0 VDC up to around 2.6 VDC, depending on the distance to the object [216]. A 2.6 VDC corresponds to about 10 cm, and 0.4 VDC corresponds to approximately 80 cm. This sensor should never be closer than 10 cm to an object, since it outputs voltages lower than 2.6 VDC for distances shorter than 10 cm. In our application we placed the IR sensors in the center of the robot, so they are at least 10 cm to the edge of the platform (Fig. 10.19, right). For each IR sensor connected to it, the PIC will continuously read its voltage. This is translated to a 10-bit number, such that 0 VDC is mapped to 0 and 5 VDC to 1023. For example, a voltage of 2.6 VDC, corresponding to a distance of about 10 cm, will have a reading of 532. The ODM does not have any software filters, and the readings are raw, as they are detected by the analog to digital system.

### 10.9.1.1 $I^2C$ Communication Sequence

Currently, the ODM is designed so that it will only report the IR sensor readings when interrogated. From the $I^2C$ data bus point of view, the ODM is considered to be a slave and the module that requests the information from the ODM is the master. For example, on our MAXELBOT the localization module is the only master on the $I^2C$ bus; all the other modules are slaves.

The software to support the $I^2C$ communication between the PIC-based ODM and a Minidragon board was fully developed in-house. There are two bytes transmitted over the $I^2C$ data bus for each sensor, for a total of 16 bytes. The $I^2C$ communication sequence is described below:

- The master generates a START signal trying to get exclusive control of the $I^2C$ bus.

- The master switches to transmit mode and sends the 7-bit address of the slave, 0010_000, and the eighth bit is set to one, instructing the slave to expect data from it.
- The slave will acknowledge that is ready to send the data.
- The master will switch itself to receive and then signal the slave that it is ready to receive.
- The slave will transmit the first byte as the least significant byte of the sensor 0 reading.
- The master will receive the first byte and acknowledge to the slave that it successfully received the byte.
- The slave will transmit the second byte as the most significant byte of the sensor 0 reading.
- The master will receive the second byte and acknowledge to the slave that it successfully received the byte.
- The above four steps are repeated seven more times, and after the last byte was received, the master will not acknowledge the slave; this is a signal for the slave that the transmission is over.
- The slave will turn itself to receive, which is the default state.
- The master will generate a STOP signal, releasing the $I^2C$ bus.

One of the issues that we had getting this to work was related to the way the PIC processor does the clock stretching. Both the master and the slave share the same clock line (SCL), but only the master drives this line. The slave can only keep it low, preventing the master from driving it. This way the slave is signaling to the master that it needs more time to prepare the data to be transmitted or to process the data just received. Specifically, the CKP bit (SSPCON<4>) should be set to one when the slave is ready to communicate. Another aspect related to the $I^2C$ communication is the way the end of the $I^2C$ address cycle is signaled between the Freescale and the Microchip architectures. The PIC provides a bit signaling the end of the address cycle, but this bit does not exist in the HCS12 microprocessor (Minidragon) and has to be implemented in software.

## 10.9.2 LCD

To enhance debugging of the system in the field, we added LCD screens to the robots. These screens are produced by Robot Electronics [51] and are $I^2C$ enabled, meaning the data to be displayed can be transmitted to the LCD via the $I^2C$ data bus. The $I^2C$ address of this LCD is 0xC6, and it can display ASCII characters from 32 to 255.

These screens allow us to see in real time the performance of the localization module and the control algorithm. Back-lighting is available if external light is low. In Fig. 10.20 (left), a MAXELBOT is supposed to be 26 in directly behind another MAXELBOT, and it is maintaining that position. The power to the left and right motors is very low, reflecting that the robot is at the desired position.

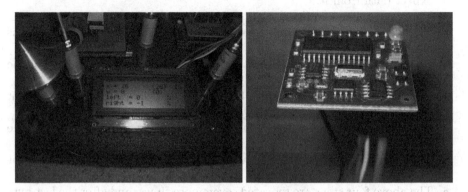

Fig. 10.20: Robot Electronics LCD03 (left) and CMPS03 Compass (right), both installed on the MAXELBOT

### 10.9.3 Digital Compass

It is important to be able to determine the heading of the robot relative to the earth's magnetic poles when a global goal is desired or when it is necessary for robots to align their egocentric coordinate systems. We are using the CMPS03 digital compass (Fig. 10.20, right) produced by Robot Electronics, which incorporates a Philips KMZ51 magnetic field sensor [50] sensitive enough to detect the earth's magnetic field. This compass is also $I^2C$ enabled, and its $I^2C$ address is "1100_000." From the $I^2C$ point of view, the compass is a slave, and it can transmit the bearing as a byte, 0–255 for a full circle, or as a word (two bytes), 0–3599 for a full circle, representing 359.9°. In our application, one byte bearing provides sufficient accuracy. The localization module on the MAXELBOT is the $I^2C$ master that interrogates this module.

### 10.9.4 Digital Temperature Sensor

Another useful add-on for the MAXELBOT is a digital temperature sensor. The information provided by this sensor can be used to simply detect if the platform is overheating or, in more complex applications, for a team of robots to follow rising temperatures towards a heat source (the goal). In our application we are using the 10-bit AD7415 digital temperature sensor from

Analog Devices. As seen in Fig. 10.21, this is a miniature sensor and comes in a Small-Outline Transistor (SOT-23) form factor. The accuracy of AD7415 is $\pm 0.5°C$ and the temperature range varies from $-40°C$ to $+125°C$ [4]. The AD7415 is $I^2C$ enabled and its slave address is "1001_000." An example of $I^2C$ communication between our PIC-based board equipped with a PIC16F876 microprocessor (see Sect. 10.8.5) and this sensor is shown in Listing 10.4.

Fig. 10.21: Digital temperature sensor setup (the actual microprocessor is located on the bottom)

## 10.9.5 Serial Over Bluetooth

One of the last additions to our robots is the capability to send data in real time, wirelessly, to a PC. We are using the serial over Bluetooth adapter made by Wireless Cables Inc. (Fig. 10.22). This adapter connects directly to the serial connector on the Minidragon, and on the PC side, standard Bluetooth software allows for installation of a standard serial port. After the adapter and the PC are paired, the adapter is a transparent pass-through device from the Bluetooth link to the RS232 port on the PC [273]. Up to seven adapters can be paired with a PC at one time. Figure 10.22 shows a possible setup using a Minidragon board and the Bluetooth adapter (top right).

## 10.10 Summary

The focus of this chapter is on the design and development of our trilateration-based localization module. In the process of creating this localization module, we developed the XSRF sensor board, a novel ultrasonic range finder, by building on an existing design. Finding a way to efficiently channel the acoustic energy into a plane, adding the wireless communication capabilities, and selecting a suitable microprocessor to perform the trilateration calculations, are some of the problems we had to address. We have developed a synchro-

Listing 10.4: Communicating with the temperature sensor using I²C—C code

```c
...
#define AD7415_ADDR 0x90 // AD7415-0 temperature sensor I2C address
...
// read the data from the temperature sensor
float read_ad7415(void){
    float temperature = 0.0;
    while (STAT_RW||ACKEN||RCEN||PEN||RSEN||SEN);// while I2C bus busy
    SSPIF = 0;                // clear the SSP interrupt flag
    SEN = 1;                  // take over I2C bus by generating START signal
    while (!SSPIF);           // wait for the START signal to complete
    SSPIF = 0;
    SSPBUF = AD7415_ADDR; // AD7415 I2C address with the last bit zero
    // next master writes to slave (AD7415)
    while (!SSPIF);           // wait for the data to be transmitted
    SSPIF = 0;
    while (ACKSTAT);//check for acknowledgement from the slave-SSPCON2<6>
    /* transmit the address of register to read from the slave:
        0 - temperature value 10-bit register (read-only)
        1 - configuration register (read/write) */
    SSPBUF = 0;
    while (!SSPIF);    // wait for the data to be transmitted
    SSPIF = 0;
    while (ACKSTAT);
    RSEN = 1;          // transmit a repeated START signal
    while (!SSPIF);    // wait for the repeated START signal to complete
    SSPIF = 0;
    SSPBUF = AD7415_ADDR + 1;//transmit the I2C address with last bit one
    // next master reads from slave (AD7415)
    while (!SSPIF);            // wait for the data to be transmitted
    SSPIF = 0;
    RCEN = 1;                  // enable receive mode for master
    while (!SSPIF);            // wait for the first byte to be received
    SSPIF = 0;
    temperature = SSPBUF << 2;// store the eight MSB received
    ACKDT = 0;                 // receive acknowledge enabled
    ACKEN = 1;                 // start acknowledge sequence
    while (!SSPIF);            // wait for acknowledge sequence to complete
    SSPIF = 0;
    RCEN = 1;
    while (!SSPIF);            // wait for the second byte to be received
    SSPIF = 0;
    temperature += (SSPBUF >> 6);// store the two LSB received
    ACKDT = 1;                 // acknowledge disabled for the last byte
    ACKEN = 1;                 // start acknowledge sequence
    while (!SSPIF);            // wait for acknowledge sequence to complete
    SSPIF = 0;
    PEN = 1;                   // send a STOP signal
    while (!SSPIF);            // wait for the STOP signal to complete
    return (temperature / 4) * 9 / 5 + 32; // return temperature in F
}
...
```

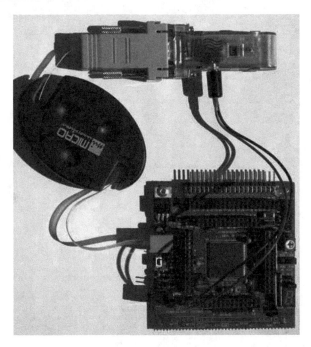

Fig. 10.22: Bluetooth connectivity setup—similar to the one used on the MAXELBOT (the Bluetooth adapter is located on the top right)

nization protocol that allows multiple localization modules to localize each other. The communication with the localization module can be performed using two different communication protocols.

To test our localization module, we have designed and built the MAXELBOT, a modularized robotic platform. By using the I$^2$C bus, it is easy to extend the functionality of the MAXELBOT with the addition of new modules. We designed and built a PIC-based microcontroller and we presented an obstacle detection module equipped with it. We introduced several add-ons, but the list is far from being exhaustive. The only requirement for adding new capabilities to the MAXELBOT is that the new module be I$^2$C-compatible.

# Chapter 11
# Uniform Coverage

Paul M. Maxim

## 11.1 Introduction

Robotic coverage can be defined as the problem of moving a sensor or actuator over all points in a given region [101]. *Uniform* coverage means that all
points in the region are visited equally often. Robotic demining operations,
snow removal, lawn mowing, car-body painting, reconnaissance, surveillance,
search-and-rescue operations, and ship hull cleaning are some of the multitude of tasks to which coverage algorithms can be applied. Coverage is a task
that is well-suited for a swarm of robots to tackle, and it is also addressed in
Chaps. 4 and 7 of this book. In fact, Chap. 4 provides a brief introduction
to the material in this chapter on uniform coverage. A coverage algorithm
must generate what is called a coverage *path*, which is a sequence of motion
commands for a robot. These algorithms can be classified as either *complete*
or *heuristic* and randomized. By definition, complete algorithms guarantee
a path that completely covers a free space [33]. Heuristic algorithms, on the
other hand, use simple rules that may work very well but they have no provable guarantees to ensure the success of coverage.

As presented in Chap. 4, Douglas Gage introduced a randomized search
strategy that provides uniform coverage [73] and defines certain properties
that a path generator algorithm must take in consideration, like (1) the path
should not be predictable (so that an observer cannot predict the searcher's
future path), (2) the search path must not be determined by a fixed reaction
to the environment, and (3) the periphery of the search space should not be
neglected. Gage's algorithm is based on diffuse reflection—how light reflects
off a matte surface—which is then generalized to any convex search area.
However, in order to implement this algorithm, "the searcher *must* be able

Paul M. Maxim
Wyoming Department of Transportation, Cheyenne, WY 82009, USA, e-mail: paul.
maxim@dot.state.wy.us

W.M. Spears, D.F. Spears (eds.), *Physicomimetics*,
DOI 10.1007/978-3-642-22804-9_11,

to determine its position within the search area, and the distance to the opposite boundary in all directions." Gage also states that it is impossible to extend this strategy to generalized non-convex areas, unless more restrictions are added.

We have invented a novel, yet simple, algorithm that meets Gage's conditions, but works in a broad class of environments that can be tiled with square cells. The environment can be convex or concave. The algorithm is randomized and complete.

Section 11.2 introduces the simulation tool we developed to test our algorithm, the performance metric and the test environments. Two versions of our uniform coverage algorithm are also presented in this section. In Sect. 11.3 we present the theory that supports our algorithm. Validation of our algorithm with actual robot experiments is presented in Sect. 11.6. Section 11.7 summarizes this chapter.

## 11.2 Simulation

Our simulation is programmed in the C language, and we use the Free OpenGL Utility Toolkit (Freeglut) version 2.4.0 library [173] to add graphic display capabilities. Freeglut is an open-source alternative to the OpenGL Utility Toolkit (GLUT) library. We run our simulation on platforms with Linux-based operating systems, but it can easily be ported to different platforms and operating systems.

### 11.2.1 Environment Setup

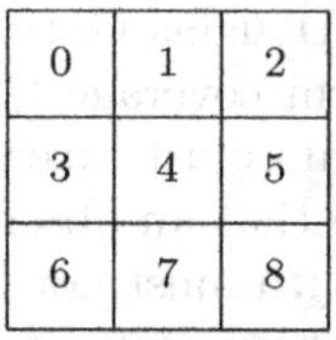

| 0 | 1 | 2 |
|---|---|---|
| 3 | 4 | 5 |
| 6 | 7 | 8 |

Fig. 11.1: A 60 × 60 environment tiled with nine cells of size 20

One important aspect of our simulation is the shape and the size of the environment. To be able to monitor how uniformly the robot is covering the environment, we split the environment into equal-sized square cells (Fig. 11.1). For example, a square environment that is 60 × 60 can be split into nine equal cells, each cell being $20 \times 20$ (and we will henceforth adopt the convention that a cell that is $c \times c$ will be referred to as a cell of size $c$). This way we can monitor how often each cell was visited and measure how well the algorithm is performing. To ensure that our algorithm is scale invariant, we also test on scaled-down versions of the same environment. Using the same $60 \times 60$ square environment as an example, the scaled-down versions would be $45 \times 45$, $30 \times 30$ and $15 \times 15$ with the corresponding cell sizes of 15, 10, and 5 units. Note that this does not change the number of cells, since the whole environment is scaled the same way.

## 11.2.2 Configuration and Control

The configuration and control of the simulation tool can be performed at two levels: before and during the simulation. Before the simulation is started, a configuration file allows the user to set certain parameters to their desired values. Some of these parameters control the simulation environment, like the size of the window, the simulation speed and the simulation length, while other parameters control the actual algorithm and the robot, or agent, being simulated. Examples are the obstacle detection sensor range and the distance at which the robot will react to the obstacles.

The actual simulation window has three distinct areas. The top area displays important real-time data about the current simulation. The middle area is the largest and it shows the outline of the environment together with the robot that navigates through the environment. The robot is shown as a small disk, with a black dot marking its current heading and a short line that shows the position and the current range of the obstacle detection sensor. A legend at the bottom of the screen informs the user of the commands that can be performed while the simulation is running, like increasing or decreasing the simulation speed or the obstacle detection sensor response range. Figure 11.2 shows a snapshot of the simulation window that features a complex 84-cell environment. The robot is shown close to the lower left corner of this environment.

A file with the environment configuration must be provided to the simulation tool. Then a robot is run for 10 million time steps inside this environment, recording to a file (every 10,000 time steps) the number of visits the robot has made to each of the cells in the environment. We are particularly interested in the final distribution of visits per cell. Initially, for each environment we varied the starting position of the robot from being in a corner, a side, or an interior cell. Also the obstacle detection sensor response range was set to 1.0 to meet Gage's condition that the periphery should not be neglected.

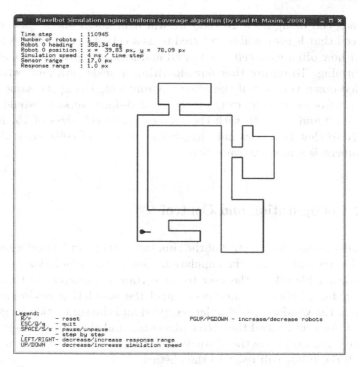

Fig. 11.2: Uniform coverage simulation engine snapshot featuring a complex 84-cell environment (cell size = 20)

### 11.2.3 Kullback–Leibler Divergence

After we run our simulation tool on an environment, we get a distribution that shows us how many visits each cell in the environment has received. Rather than just relying on visual inspection, we use a metric to formally measure the difference between this observed distribution and the optimum *uniform distribution*. With a uniform distribution, all cells are visited equally often.

The frequency of cell visitation can be considered as a *probability distribution*, which is defined (assuming discrete values) to be the probabilities of a random variable assuming different values. In our particular case, the random variable is the position of the robot, the values are the specific cells in the environment, and the probabilities are the frequencies of cell visitation.

The *Kullback–Leibler (KL) divergence* is a non-commutative measure of the difference between two probability distributions $P$ and $Q$. Typically, $P$ represents the "true" distribution of data, observations or a precisely calculated theoretical distribution. The measure $Q$ typically represents a theory, model, description or approximation of $P$ [133]. In our case, $P$ represents the observed distribution, and $Q$ represents the optimum uniform distribution. The minimum value of the divergence is 0.0, meaning the observed distribution is precisely the same as the uniform distribution. The KL divergence of $Q$ from $P$ is defined to be:

$$D_{\mathrm{KL}}(P\|Q) = \sum_i P(i) \log\left(\frac{P(i)}{Q(i)}\right) .$$

(11.1)

A KL divergence of 0.01 is considered to be a very good match between two distributions. For our results with real robots our goal is approximately 0.01. For the simulation results our goal is 0.001 since we can run the simulation for a greater number of steps.

## 11.2.4 Markov Chain Analysis

*Markov chains* are used in mathematics to describe the future states of a stochastic process based on the present state [81]. Given the present state of the system, the future states are assumed to be independent of the past state and are reached through a probabilistic process. A change of state is called a *transition* and the probability that governs this change is called a *transition probability*. A probability matrix $Q$ defines the probability of transitioning to a state $j$ in the next step, given that the system is currently in state $i$ (i.e., $Q(i,j)$).

A Markov chain is called *ergodic* if all states are recurrent, aperiodic, and communicate with each other. A state $i$ is *transient* if there exists a state $j$ that is reachable from $i$, but state $i$ is not reachable from $j$. If a state is not transient, it is *recurrent*. Also, a state $i$ is *periodic* if all paths leading from $i$ back to $i$ have a length that is a multiple of some integer $k > 1$. If this is not the situation, the state is *aperiodic*. All states communicate with each other if there is a path from any state $i$ to any state $j$ (and vice versa).

Every ergodic Markov chain has a *steady-state distribution* that describes the long-term behavior of the system. The existence of a steady-state (equilibrium) distribution indicates that the initial state of the system is of no importance in the long term. Furthermore, if $Q$ is *symmetric*, i.e., ($Q(i,j) = Q(j,i)$), then the steady-state distribution is the uniform distribution. This latter property is crucial to achieving a good uniform coverage algorithm.

Here we assume that a *state* is simply the cell that the robot is in. The robot transitions from cell to cell as it moves. Our final algorithm for moving the robot can be described as an ergodic Markov chain with a symmetric probability transition matrix. The steady-state distribution must be uniform and in this case the algorithm is correct. The algorithm is also complete if the environment is topologically connected.

Consider the nine-cell environment shown in Fig. 11.1. Let us also consider an idealized situation where a robot can reside at the center of cells. For example, suppose the robot is at the center of cell 4. The robot could then transition to adjacent cells 1, 3, 5, or 7. The probability of transitioning directly to cells 0, 2, 6, or 8 is essentially zero (i.e., the robot would have to pass through cells 1, 3, 5, or 7 first to get to the other cells). Hence we will ignore the "diagonal" neighbors. An ideal algorithm is as follows. The robot turns uniformly randomly and then attempts to move. It might stay in the same cell (if the cell is along the border of the environment), or it might enter an adjacent cell. If it enters an adjacent cell we assume it then moves to the center of that cell, so that the algorithm can continue.

The Markov chain for this idealized algorithm is trivial to compute. If a cell has $b$ periphery edges, then there is a $b/4$ probability of staying in that cell. The remaining "probability mass" (which is analogous to physical mass) is uniformly distributed among the adjacent cells.

The probability transition matrix $Q$ is

$$
\begin{bmatrix}
0.50 & 0.25 & 0.00 & 0.25 & 0.00 & 0.00 & 0.00 & 0.00 & 0.00 \\
0.25 & 0.25 & 0.25 & 0.00 & 0.25 & 0.00 & 0.00 & 0.00 & 0.00 \\
0.00 & 0.25 & 0.50 & 0.00 & 0.00 & 0.25 & 0.00 & 0.00 & 0.00 \\
0.25 & 0.00 & 0.00 & 0.25 & 0.25 & 0.00 & 0.25 & 0.00 & 0.00 \\
0.00 & 0.25 & 0.00 & 0.25 & 0.00 & 0.25 & 0.00 & 0.25 & 0.00 \\
0.00 & 0.00 & 0.25 & 0.00 & 0.25 & 0.25 & 0.00 & 0.00 & 0.25 \\
0.00 & 0.00 & 0.00 & 0.25 & 0.00 & 0.00 & 0.50 & 0.25 & 0.00 \\
0.00 & 0.00 & 0.00 & 0.00 & 0.25 & 0.00 & 0.25 & 0.25 & 0.25 \\
0.00 & 0.00 & 0.00 & 0.00 & 0.00 & 0.25 & 0.00 & 0.25 & 0.50
\end{bmatrix}
$$

for the nine-cell environment. The rows represent the current state (i.e., the cell in which the robot presently resides) and the columns represent the next state. This matrix is ergodic, and the steady-state distribution (obtained by calculating $\lim_{t \to \infty} Q^t$) indicates that each cell is visited precisely $1/9^{\text{th}}$ of the time, which is what we desire.

It is important to point out that this is an existence proof that an algorithm may exist that provides uniform coverage for all topologically connected environments that can be tiled with square cells. The algorithm given above is not realistic. However, the argument above does provide some guidance on how to create a viable algorithm: the resulting Markov chain must be ergodic and have a symmetric probability transition matrix.

## 11.2.5  First Algorithm

Our first algorithm requires only that the robot be able to sense an object (wall) in front of it. The robot starts by moving forward. When it senses an object directly in front of it, the robot stops. Then the robot makes a random turn. If there is still an object in front, the robot turns again. The robot continues to turn until it detects no obstacle in front of it. Then the robot moves forward again. This algorithm does not avoid the periphery, it is not predictable, and it works in all connected environments. Proof of this claim lies in the assumption of tiling the environment with square cells. The resulting Markov chain is ergodic and will have a steady-state distribution. However, it is not clear whether the steady-state distribution is uniform.

## 11.2.6  Environments

The first test of our algorithm was on a simple $40 \times 40$ square environment, with four cells of size 20. As seen in Listing 11.1, the distribution is very close to uniform. This is reasonable, since all four cells are corner cells and can be considered equivalent. There is no reason for the algorithm to favor one corner cell over any other.

Listing 11.1: Visits in the four-cell square environment

```
cell 0: 2503625
cell 1: 2500184
cell 2: 2494401
cell 3: 2501790
```

The next tested environment is the $60 \times 60$ square environment shown in Fig. 11.1, with nine cells. This provides us with different categories of cells, such as corners, sides and an inner cell. The results for this environment are presented in Listing 11.2. We can see that the distribution is *not* uniform. In fact, the center cell (4) has the fewest number of visits. The main reason is the inner position of this cell. The robot does not detect any obstacles in this cell and spends the least amount of time there because it moves straight through it. Clearly, this first algorithm is not adequate, since the probability of transitioning from the center cell to any other cell is higher than transitioning from any other cell to the center cell. A similar difference exists between the corner cells and side cells.

Listing 11.2: Visits in the nine-cell square environment (1st algorithm)

```
cell 0: 1242516
cell 1: 1048036
cell 2: 1229541
cell 3: 1042531
cell 4:  888209
cell 5: 1052790
cell 6: 1225855
cell 7: 1041632
cell 8: 1228890
```

Taking random turns only when detecting a wall is not sufficient. This tells us that in certain situations symmetry does not hold, i.e., $Q(i,j) \neq Q(j,i)$. An improvement has to be brought to this algorithm. This improvement is inspired by physics, as we describe next.

## 11.2.7 Second Algorithm

The prior results on the nine-cell environment indicate that the transition probabilities are not symmetric. In fact, the results indicate that the robot needs to spend more time in cells with fewer peripheral edges. This can be achieved using the concept of *mean free path*. In physics, a molecule in a gas moves with constant speed along a straight line between successive collisions. The mean free path is defined as the average distance between such successive collisions [85].[1] Depending on the number of molecules in the space available to them, the mean free path can range from zero to infinity.

In the first algorithm, the only time the robot changes direction is when it senses a wall. Hence it spends less time in cells with fewer peripheral edges. In order to counteract this, we want the robot to act as if it collides with another virtual object occasionally. Hence we want the robot to randomly change direction after it has moved a certain distance, namely, its mean free path. Because the mean free path needs to be normalized by the cell size, we

---

[1] The mean free path is also a central concept in the coverage algorithm described in Chap. 7.

**Listing 11.3:** Visits in the nine-cell square environment (2nd algorithm)

```
cell 0: 1098245
cell 1: 1110451
cell 2: 1112045
cell 3: 1124196
cell 4: 1127813
cell 5: 1124090
cell 6: 1089300
cell 7: 1111144
cell 8: 1102716
```

assume that the mean free path will be $f \times c$, where $f$ is some real number and $c$ is the cell size.

We empirically determined that $f \approx 0.6$ works best. If we rerun the same experiment with the nine-cell environment, the results are shown in Listing 11.3. We can easily see the improvement of this version of the algorithm over the previous version. By using the Kullback–Leibler (KL) divergence metric we can measure how close this distribution is to the uniform distribution. For this experiment, the KL divergence is 0.000060. Remember that the closer this metric value is to 0.0, the better. To provide a better understanding of how the KL metric value changes over time, Fig. 11.3 shows the graph of the KL divergence metric every 10,000 time steps for 10 million time steps. This graph is not smooth because it represents only one simulation run.

If a Markov chain is ergodic then it is known that the time average of a process (algorithm) is the same as the ensemble average that would be obtained by running the process multiple times. Hence, it isn't necessary to perform multiple runs. The long-term behavior of the algorithm represents the ensemble average and more realistically describes the behavior of a robot, which we assume has a very long-term presence in the environment, performing the desired uniform coverage task.

## 11.3 Derivation of the Mean Free Path

The mean free path can be thought of as the average distance traveled by a robot to move from one cell to another adjacent cell. Consider Fig. 11.4, with two adjacent cells. Each cell is $1 \times 1$ in size. Without loss of generality, assume robot A is located inside the left cell at coordinates $(x_1, y_1)$. We would

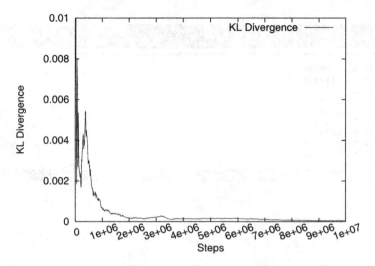

Fig. 11.3: KL divergence metric for the nine-cell square environment

like to compute the average distance traveled by the robot to barely enter
the right cell at location $(1 + \varepsilon, y_2)$.

Fig. 11.4: What is the average distance traveled by A to enter the right cell?

The simplest way to compute this distance is via simulation, as shown
in Listing 11.4. Function U(0,1) returns floating point values uniformly randomly from 0 to 1. This simulation yields a mean free path of 0.6517.

A more formal derivation can be accomplished by computing the average
distance using a definite integral:

$$\int_0^1 \int_0^1 \int_0^1 \sqrt{((1 - x_1)^2 + (y_1 - y_2)^2)} \, dx_1 dy_1 dy_2 . \qquad (11.2)$$

The solution to the indefinite integral is (thanks to Mathematica and Jim
Partan of the University of Massachusetts):

**Listing 11.4: Mean free path simulation—C code**

```c
int main( int argc, char** argv ) {
    int i, N;
    double sum, x1, x2, y1, y2;
    N = 10000000;
    sum = 0.0;
    for (i = 1; i <= N; i++) {
        x1 = U(0,1);
        x2 = 1.001;
        y1 = U(0,1);
        y2 = U(0,1);
        sum = sum + sqrt((x2 - x1)*(x2 - x1) + (y2 - y1)*(y2 - y1));
    }
    printf("Average length = %f\n", sum/N);

    return 0;
}
```

$$F(x_1, y_1, y_2) = \tag{11.3}$$

$$\left[ 3y_2{}^4 - 28y_1 y_2{}^3 + 42y_1{}^2 y_2{}^2 \right.$$

$$- 36(1 - x_1)y_2{}^2 \sqrt{y_2{}^2 - 2y_1 y_2 + (1 - x_1)^2 + y_1{}^2} - 28y_1{}^3 y_2$$

$$+ 48(1 - x_1)^3 y_2 \ln\left(-y_2 + y_1 + \sqrt{y_2{}^2 - 2y_1 y_2 + (1 - x_1)^2 + y_1{}^2}\right)$$

$$+ 72(1 - x_1)y_1 y_2 \sqrt{y_2{}^2 - 2y_1 y_2 + (1 - x_1)^2 + y_1{}^2}$$

$$- 12(y_2 - y_1)^4 \ln\left(1 - x_1 + \sqrt{y_2{}^2 - 2y_1 y_2 + (1 - x_1)^2 + y_1{}^2}\right)$$

$$+ 48(1 - x_1)^3 y_1 \ln\left(y_2 - y_1 + \sqrt{y_2{}^2 - 2y_1 y_2 + (1 - x_1)^2 + y_1{}^2}\right)$$

$$+ 24(1 - x_1)^3 \sqrt{y_2{}^2 - 2y_1 y_2 + (1 - x_1)^2 + y_1{}^2}$$

$$\left. - 36(1 - x_1)y_1{}^2 \sqrt{y_2{}^2 - 2y_1 y_2 + (1 - x_1)^2 + y_1{}^2} \right]\left[\frac{-1}{288}\right].$$

Given this, we can evaluate the definite integral to be:

$$F(1,1,1) - F(1,0,1) - F(0,1,1) + F(0,0,1) -$$
$$F(1,1,0) + F(1,0,0) + F(0,1,0) - F(0,0,0) = 0.651757. \tag{11.4}$$

This is in excellent agreement with the value computed using Listing 11.4. In experiments, we noted that $f = 0.60$ works very well, which is slightly lower than theory. The difference is caused by the fact that the robot must stay one unit away from the walls, which in turn makes the length of the mean free path slightly smaller. For example, we generally use cells of size 20. Thus, having a wall response range of one provides a 5% buffer zone. Assuming a 5% buffer, the theoretical mean free path is lowered to 0.6079, which is extremely close to our empirically derived value.

It is important to note that the concept of mean free path has allowed us to make the probability transitions symmetric, i.e., $(Q(i,j) = Q(j,i))$. This can be observed by noting that robot A could be in any cell of any square-tiled environment, and the mean free path is the average distance traveled to enter any adjacent cell, regardless of whether these cells are on the periphery or not.

## 11.4 Further Confirmation

We ran our simulation over a large variety of environments, and we generally obtained a KL divergence number of 0.001 or lower, which indicates that the behavior is extremely close to the optimum (averaged over all environments the KL divergence was 0.00066). Each simulation was run 40 million time steps. Table 11.1 provides results over the different environments and cell sizes. The environments are described in [156, 239].

Table 11.1: KL divergence metric for different environments and cell sizes

| Environment | Cell Size | | | | |
| --- | --- | --- | --- | --- | --- |
| | 10 | 15 | 20 | 25 | 30 |
| 3x1 | 0.00011 | 0.00005 | 0.00003 | 0.00003 | 0.00002 |
| 4cellL | 0.00026 | 0.00022 | 0.00019 | 0.00016 | 0.00022 |
| 4cellRoll | 0.00090 | 0.00082 | 0.00068 | 0.00066 | 0.00071 |
| 7cell | 0.00078 | 0.00075 | 0.00064 | 0.00056 | 0.00061 |
| 3x3 | 0.00018 | 0.00010 | 0.00007 | 0.00007 | 0.00006 |
| 15cell | 0.00019 | 0.00010 | 0.00015 | 0.00016 | 0.00019 |
| 22cell | 0.00088 | 0.00124 | 0.00188 | 0.00137 | 0.00145 |
| 35cell | 0.00138 | 0.00079 | 0.00067 | 0.00067 | 0.00106 |
| 6x6 | 0.00021 | 0.00011 | 0.00009 | 0.00008 | 0.00009 |
| 82cell | 0.00223 | 0.00114 | 0.00269 | 0.00097 | 0.00059 |
| 84cell | 0.00210 | 0.00309 | 0.00067 | 0.00156 | 0.00064 |
| 10x10 | 0.00051 | 0.00057 | 0.00009 | 0.00057 | 0.00065 |

Figure 11.5 shows snapshots in time with one simulated robot uniformly covering the rather complex 84-cell environment. The grayscale denotes the frequency that a particular cell is occupied (where darker means less often). The uniformity of the gray cells in the last snapshot (lower right) indicates excellent uniform coverage. The KL divergence graph for this environment is shown in Fig. 11.6 and at the last step in the simulation its value is 0.00067.

## 11.5 Theory for Multiple Robots

One robot does very well at uniformly covering an environment using our algorithm. We have shown that if a connected environment is composed of $C$ square cells, then in the long term there is a $\frac{1}{C}$ probability of the robot being in each cell.

Let us assume we have a team of $N$ independent robots performing the same algorithm in a large environment, such that their interactions are minimal (e.g., they merely have to avoid each other on rare occasions). For example, $N = 10$ to $50$ is very reasonable for systems that are usually deployed in the field. The probability of one robot being in one of the cells is $\frac{1}{C}$. The probability of one robot *not* being in that cell is $1 - \left(\frac{1}{C}\right)$. The probability of all $N$ robots *not* being in that cell is then $\left(1 - \left(\frac{1}{C}\right)\right)^N$. Hence, the probability that at least one robot is in that cell is

$$1 - \left(1 - \left(\frac{1}{C}\right)\right)^N . \tag{11.5}$$

Now suppose we require some minimum probability such that some robots will be in any cell at any time step. Let us call this probability $P_m$. Then we need to set the number of robots $N$ high enough to ensure that

$$\left(1 - \left(1 - \left(\frac{1}{C}\right)\right)^N\right) > P_m . \tag{11.6}$$

This allows us to easily compute the number of robots $N$ that we need in order to satisfy the constraints imposed by the task.

## 11.6 Hardware Implementation

Moving from the simulation world to the real-world application with hardware robots is not necessarily a small step. Before we can make this move, we need to carefully choose the real-world equivalents of the two most important simulation measuring units. First, the unit to measure distance in the simu-

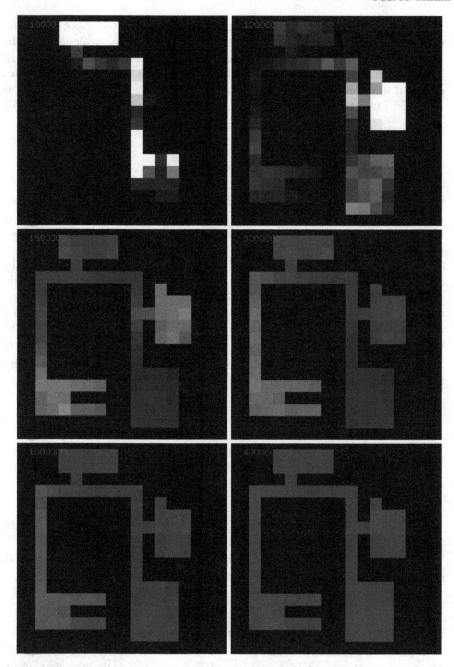

Fig. 11.5: One simulated MAXELBOT obtaining uniform coverage in a complex 84-cell environment—snapshots taken at 10,000, 100,000, 1,500,000, 5,000,000, 10,000,000, and 40,000,000 time steps

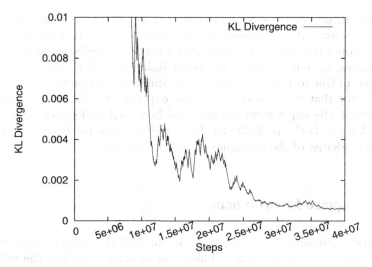

Fig. 11.6: KL divergence metric for a simulated 84-cell environment

lation world is one pixel, and we translate that to one inch in the real world. Second, time is measured in "steps" in the simulation, and we translate that to seconds in the real world. Having defined these basic units of measure, we can easily translate the important variables, like the maximum speed of the robot and the obstacle detection response range.

The hardware demonstration utilizes our trilateration-based localization module in a novel way. The robots used are MAXELBOT platforms, which are described in detail in Chap. 10. Our trilateration-based approach to localization is also presented in depth in Chap. 10. We assume that one MAXELBOT is moving inside the environment, while stationary MAXELBOTS, embedded in the "walls" of the environment, record the position of the moving robot at all times.

The trilateration module on the moving MAXELBOT emits one RF–acoustic pulse every 120 ms. This translates to 8.33 pulses per second or a frequency of 8.33 Hz. This is important in order to compute the time passed from the number of trilateration readings. For example, 100 trilateration readings are equivalent to roughly 12 seconds of run time. To detect obstacles, the moving MAXELBOT is equipped with our Obstacle Detection Module, which, in this setup, uses only two infrared (IR) sensors. The two IR sensors are placed at the front corners of the MAXELBOT and are aimed straight ahead.

The embedded stationary MAXELBOTS monitor the position of the moving robot and send the data in real time to a PC over a serial wired connection. This data allows us to verify that the moving robot does indeed visit all the "cells" equally often. All the environments presented next are broken into virtual cells, all cells having the same size, which is 2.5 ft × 2.5 ft. Each of the corners of these virtual cells have their coordinates preset. The number of visits to one of these virtual cells translates to how many trilateration readings were performed inside the perimeter of the cell. This allows us not only to count how many times the perimeter of any cell was crossed, but also the amount of time the robot has spent inside the cell perimeter.

Next, in this section we present three different environments of increasing complexity that were chosen for these experiments. For each environment we present the experimental setup and how well uniformity is achieved— by looking at both the Kullback–Leibler divergence metric and a graphical offline rendering of the frequency of cell occupancy.

## 11.6.1 Square Environment

The first environment in which we test our uniform coverage algorithm is a small, 5 ft × 5 ft, square environment. As mentioned earlier, the cell size is 2.5 ft, which yields a 2 × 2 cell environment.

Fig. 11.7: Square environment (5 ft × 5 ft)

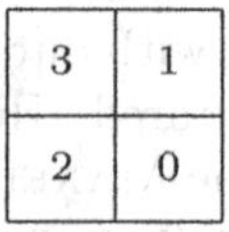

|   |   |
|---|---|
| 3 | 1 |
| 2 | 0 |

Fig. 11.8: Cell division of the 5 ft × 5 ft environment

Figure 11.7 shows a picture of the experimental setup that we built in our Distributed Robotics Laboratory at the University of Wyoming. Each tile on our laboratory floor is exactly 12 in × 12 in, and this makes it easy for the placement of the "walls," which are pine wood beams, 15 ft long, 6 in wide and 10 in tall. For this particular experiment we also used the top of a wood table, which is exactly 5 ft long, for one of the walls. In the lower right corner of the picture there is one stationary MAXELBOT (inside the outline) that is performing the localization, based on the pulses emitted by the other moving MAXELBOT (shown in the same picture). The stationary MAXELBOT is connected to a serial port on a PC in order to transmit real-time data on the position it senses from the moving MAXELBOT. The video "serial.avi" shows an example of a stationary MAXELBOT connected via a serial port to a PC, while monitoring the position of a moving MAXELBOT.

One MAXELBOT was run inside the environment for about 20 minutes, which generated a little over 9,700 trilateration readings. Figure 11.8 shows the virtual cell division for this environment, which was used during the data analysis. A trace of the path taken by the moving MAXELBOT is shown in Fig. 11.9. The snapshots were taken at 0.2, 1, 3, and 20 minutes and show that after only 3 minutes, most of the environment was already covered.

Fig. 11.9: MAXELBOT path trace in the 5 ft × 5 ft environment—snapshots taken at 12, 60, 180, and 1,200 seconds

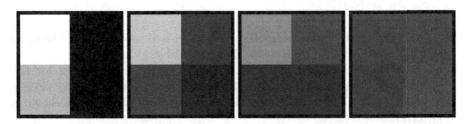

Fig. 11.10: Cell occupancy in the 5 ft × 5 ft environment—snapshots taken at 12, 60, 180, and 1,200 seconds

Another useful graphical representation of this data is shown in Fig. 11.10. Each of the four snapshots were taken at the same time the path trace snapshots in Fig. 11.9 were taken, and show the frequency of occupancy of the

Table 11.2: Specific cell visits for the 5 ft × 5 ft environment

| Cell number | Time—minutes (readings × 100) | | | |
|:---:|:---:|:---:|:---:|:---:|
| | 0.2 (1) | 1 (5) | 3 (15) | ~20 (97) |
| 0 | 0 | 58 | 270 | 2576 |
| 1 | 0 | 104 | 391 | 2464 |
| 2 | 43 | 116 | 271 | 2253 |
| 3 | 57 | 222 | 568 | 2407 |
| KL divergence | – | 0.110353 | 0.050299 | 0.001158 |

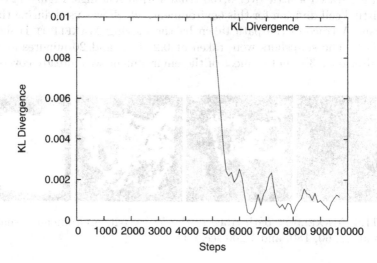

Fig. 11.11: KL divergence for the 5 ft × 5 ft environment

four cells (where brighter means the cell was visited more often). The exact number of cell visits for each cell is shown in Table 11.2.

The value of the Kullback–Leibler divergence metric is 0.0012 for this experiment, which shows that our cell occupancy distribution is very close to the uniform distribution. The evolution over time of this metric is shown in Fig. 11.11.

## 11.6.2 Rectangular Environment

A larger 5 ft × 15 ft environment was the second environment in which we
tested our uniform coverage algorithm. This configuration yields a 2 × 6
cell environment (with 2.5 ft cell size). Figure 11.12 shows how we numbered
each of the 12 cells. Although it is possible for one MAXELBOT to "hear" the
RF–acoustic pulse generated by the moving MAXELBOT, the accuracy of the
localization reduces drastically above 10 ft. The solution to this problem is to
use two MAXELBOTS for monitoring, one at each end of the rectangular environment. The picture of the experimental setup is shown in Fig. 11.13. The
two MAXELBOTS used for monitoring are highlighted with the circular outlines and the moving MAXELBOT is featured in the center of the environment.
Both stationary MAXELBOTS monitor the whole environment.

| 6 | 0 |
|----|---|
| 7 | 1 |
| 8 | 2 |
| 9 | 3 |
| 10 | 4 |
| 11 | 5 |

Fig. 11.12: Cell division of the 15 ft × 5 ft environment (the upper MAXELBOT
monitors cells 0, 1, 2, 6, 7, and 8 and the lower MAXELBOT monitors cells 3,
4, 5, 9, 10, and 11)

During this experiment there are two sets of trilateration readings that
are generated. Since data can be lost occasionally, when sent over the RF,
it is possible for the two data sets to lose synchronization. This would make
it very difficult to merge the data from the two stationary MAXELBOTS.
In order to help with the synchronization between the two data sets, we
introduce an identification number to the RF signal. Each time the moving
MAXELBOT is broadcasting the RF–acoustic pulse combination, a number
is also transmitted over the RF. This number is incremented with every
pulse. The receiving MAXELBOTS attach this number to the trilateration
reading performed. This tuple is then transmitted in real time to PCs that
are connected to the receiving MAXELBOTS with serial cables.

The duration of this experiment is approximately 2 hours. During this
time, 60,000 trilateration readings were performed by each of the two moni-

Fig. 11.13: Rectangular environment (15 ft × 5 ft)

toring MAXELBOTS. Figure 11.14 shows a graphical representation of the two data sets before merging. We found this to be a good test of the range and accuracy of our localization system. The localization system on one of the MAXELBOTS performs better than the other one, especially at distances over 10 ft. (Note, especially, the impossible locations outside of the environment.)

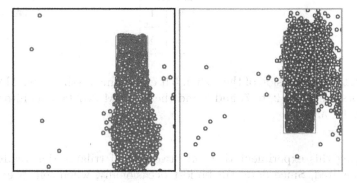

Fig. 11.14: Individual localization readings performed by the two fixed MAX-ELBOTS

To merge the data and reduce the noise, we divided the environment into halves. Then we selected coordinates from the upper MAXELBOT that are in the upper half of the environment. Similarly, we selected coordinates from the lower MAXELBOT that are in the lower half. We used the time stamp to resolve ambiguities. The results are shown in Fig. 11.15. Although we do not filter the points shown in this figure, for data analysis we discard all the points that are outside the outline of the environment.

Fig. 11.15: Merged localization readings performed by the two fixed MAXEL-BOTS

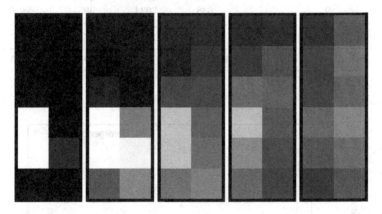

Fig. 11.16: Cell occupancy in the 15 ft × 5 ft environment—snapshots taken at 0.2, 5, 30, 60, and 120 minutes

The frequency of the occupancy of the cells for this environment is shown in Fig. 11.16. The brighter the cell the higher the number of visits to that cell. The snapshots are taken at 0.2, 5, 30, 60, and 120 minutes, and the exact numbers of visits they represent are shown in Table 11.3. After approximately 5 minutes into the experiment, all the 12 cells in the environment were reached, with cells in the center having a larger number of visits compared with the cells close to the two ends. As the data shows, this discrepancy is drastically reduced 120 minutes into the experiment.

The final Kullback–Leibler divergence metric value was approximately 0.015, which is quite reasonable. The graph of this metric for the entire length of this experiment is shown in Fig. 11.17. As an aside, the KL divergence metric depends on the cell granularity. For example, if this environment is treated as being composed of three cells of size 5 ft × 5 ft, the KL divergence drops

Table 11.3: Specific cell visits and KL divergence for the 15 ft × 5 ft environment

| Cell number | Time—minutes (trilateration readings × 100) | | | | |
|---|---|---|---|---|---|
| | 0.2 (1) | 5 (25) | 30 (150) | 60 (299) | ~120 (598) |
| 0 | 0 | 35 | 779 | 1616 | 5242 |
| 1 | 0 | 49 | 899 | 2472 | 6102 |
| 2 | 0 | 24 | 1131 | 2701 | 5665 |
| 3 | 0 | 256 | 1355 | 2580 | 6612 |
| 4 | 5 | 402 | 1437 | 2327 | 5908 |
| 5 | 0 | 265 | 1561 | 2103 | 5168 |
| 6 | 0 | 41 | 660 | 1397 | 3512 |
| 7 | 0 | 6 | 572 | 1944 | 4487 |
| 8 | 0 | 82 | 1105 | 2800 | 4489 |
| 9 | 41 | 638 | 2034 | 3951 | 5807 |
| 10 | 54 | 498 | 2005 | 3248 | 5206 |
| 11 | 0 | 204 | 1462 | 2761 | 3877 |
| KL divergence | – | 0.469953 | 0.069357 | 0.035933 | 0.015255 |

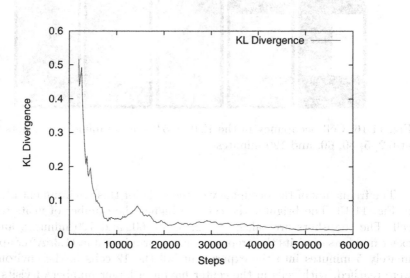

Fig. 11.17: KL divergence metric for the 15 ft × 5 ft environment

considerably to 0.002. Decreasing the size of cells means it takes longer to achieve uniformity.

## 11.6.3 L-Shaped Environment

We continue by testing our uniform coverage algorithm in a more complex environment, specifically, an L-shaped concave environment. Recall that achieving uniform coverage on concave environments is considered to be difficult, if not impossible [73]. The two sides of the L-shape are each 15 ft long, and the width is constant at 5 ft. With a cell size of 2.5 ft, this environment yields a total of 20 equal cells. The cell division and how the cells are numbered are shown in Fig. 11.18.

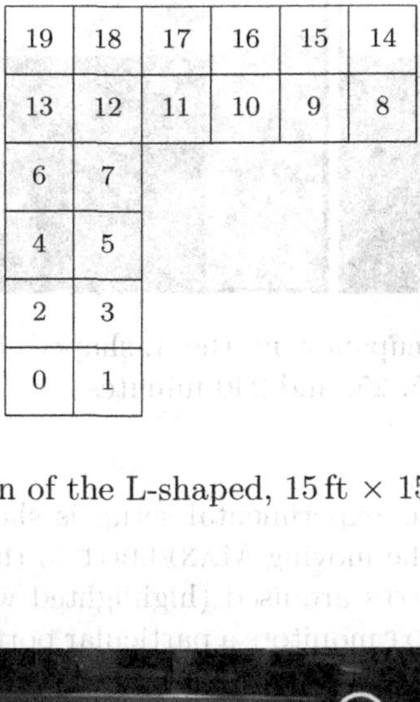

| 19 | 18 | 17 | 16 | 15 | 14 |
|----|----|----|----|----|----|
| 13 | 12 | 11 | 10 | 9  | 8  |
| 6  | 7  |    |    |    |    |
| 4  | 5  |    |    |    |    |
| 2  | 3  |    |    |    |    |
| 0  | 1  |    |    |    |    |

Fig. 11.18: Cell division of the L-shaped, 15 ft × 15 ft, environment

Fig. 11.19: L-shaped, 15 ft × 15 ft, environment

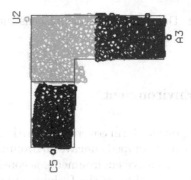

Fig. 11.20: Merged trilateration readings from three MAXELBOTS for the L-shaped, 15 ft × 15 ft, environment

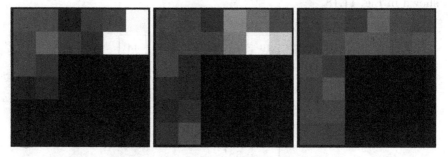

Fig. 11.21: Cell occupancy in the L-shaped 15 ft × 15 ft environment—snapshots taken at 5, 25, and 240 minutes

The picture of the experimental setup is shown in Fig. 11.19. To monitor the location of the moving MAXELBOT in this larger environment, three stationary MAXELBOTS are used (highlighted with circles). As before, each stationary MAXELBOT monitors a particular portion of the environment. This helps to ensure that we are using accurate trilateration readings. The environment outline in Fig. 11.20 shows the three regions, as well as a graphical representation of the approximately 118,000 merged readings, that were collected during the four hour long experiment. The positions of the three MAXELBOTS are shown by the three circles located outside the perimeter and are labeled "A3," "U2" and "C5." We use these labels to distinguish between the seven MAXELBOTS we have. C5 is responsible for cells 0, 1, 2, 3, 4, and 5. MAXELBOT U2 monitors cells 6, 7, 11, 12, 13, 17, 18, and 19. Finally, we used the data collected by MAXELBOT A3 for cells 8, 9, 10, 14, 15, and 16.

Table 11.4: Specific cell visits and KL divergence for the L-shaped, 15 ft ×
15 ft, environment

| Cell number | Time—minutes (readings × 100) | | | |
|---|---|---|---|---|
| | 0.2 (1) | 5 (25) | 25 (125) | ~240 (1180) |
| 0 | 0 | 0 | 430 | 5547 |
| 1 | 0 | 0 | 701 | 5313 |
| 2 | 0 | 0 | 286 | 6156 |
| 3 | 0 | 0 | 430 | 6228 |
| 4 | 0 | 34 | 396 | 5721 |
| 5 | 0 | 56 | 342 | 7164 |
| 6 | 0 | 121 | 628 | 6291 |
| 7 | 0 | 90 | 332 | 5140 |
| 8 | 40 | 674 | 1211 | 7145 |
| 9 | 0 | 392 | 1366 | 6737 |
| 10 | 0 | 77 | 964 | 7314 |
| 11 | 0 | 108 | 619 | 7361 |
| 12 | 0 | 163 | 643 | 6646 |
| 13 | 0 | 124 | 607 | 5200 |
| 14 | 59 | 259 | 653 | 5898 |
| 15 | 0 | 117 | 862 | 6372 |
| 16 | 0 | 106 | 995 | 7931 |
| 17 | 0 | 28 | 396 | 5090 |
| 18 | 0 | 132 | 597 | 5978 |
| 19 | 0 | 135 | 655 | 5292 |
| KL divergence | – | – | 0.090440 | 0.008751 |

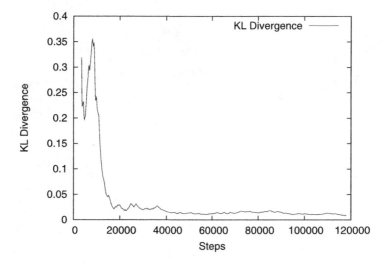

Fig. 11.22: KL divergence metric for the L-shaped, 15 ft × 15 ft, environment

The number of visits to each of the cells can be seen both as specific numbers in Table 11.4, and as a graphical representation in Fig. 11.21. For this environment it takes the moving MAXELBOT about 7 minutes to visit all the cells.

The final KL divergence from the uniform distribution is 0.0088. As shown in Fig. 11.22, the value of the KL metric drops suddenly after approximately 30 minutes, meaning that a good uniform coverage is already achieved. After that, the KL value continues to drop, but at a much slower rate.

## 11.7 Summary

In this chapter we have presented our uniform coverage algorithm. It is a randomized, theory-based algorithm that is complete, unpredictable, and covers the periphery. The algorithm requires no knowledge about the environment (e.g., estimating the cell size can be based on the robotic platform's sensing capabilities, rather than on environmental considerations). Our algorithm also extends naturally to multiple robots (see the NetLogo simulation "uniform_mfpl.nlogo"). The performance of our algorithm was shown both in simulation and real-world experiments. The Kullback–Leibler divergence metric we used confirms that our algorithm is very close to achieving optimal uniform coverage.

# Chapter 12
# Chain Formations

Paul M. Maxim

## 12.1 Introduction

One important problem in cooperative robotics is the self-organization of robotic chain formations in unknown environments. Surveillance and reconnaissance in sewers, ducts, tunnels, and caves or narrow passageways, are some of the applications for this type of formation. These tasks are greatly augmented via the use of multi-robot teams. First, multiple robots can cover more territory than one. Second, if the multiple robot system is designed properly, there is no single point of failure. The failure of one or more robots will degrade performance, but the task can still be accomplished. Chain formations are also addressed in Chap. 4, where a brief introduction to the material in this chapter is presented.

The scenario we envision is as follows. A robot is placed at the entrance of an environment that is to be explored autonomously by robots. This first robot will remain stationary and will wirelessly communicate information to a laptop. Other robots are released into the environment, by being placed in front of the first robot. The task of the robots is to self-organize into a chain structure that reflects the shape of some portion of the environment. Because our localization technique (see Chap. 10) combines localization and data communication, the robots can use this self-organized communication link to provide their positions and orientations in the environment back to the user. The user will see a diagram of the positions and orientations of the robots, relative to the position and orientation of the first robot, on the

Paul M. Maxim
Wyoming Department of Transportation, Cheyenne, WY 82009, USA, e-mail: paul.
maxim@dot.state.wy.us

©2008 IEEE. Portions reprinted, with permission, from S. D. Hettiarachchi, Paul M. Maxim, W. M. Spears, and D. F. Spears, Connectivity of Collaborative Robots in Partially Observable Domains, ICCAS 2008.

W.M. Spears, D.F. Spears (eds.), *Physicomimetics*,
DOI 10.1007/978-3-642-22804-9_12,
© Springer-Verlag Berlin Heidelberg 2011

laptop screen. The real-time generated diagram informs the user of the shape of this portion of the environment.

The main challenge in generating fully connected chain formations is to maintain connectivity across corners (Fig. 12.1). The robots will pause in corners, due to line-of-sight considerations. However, the robots should pop out of the corners when enough robots are queued up behind them—to promote exploration. Similarly, to promote robustness, we want the robots to reconnect across corners when breaks occur. We want our algorithm to balance these two main requirements *without treating corners as special objects*.

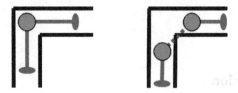

Fig. 12.1: Connectivity across corners

Modifications we made to the existing physicomimetics framework are presented in Sects. 12.2, 12.3, and 12.4. An elegant modification to our algorithm that significantly improves its performance is emphasized here. Explanations regarding the use of *evolutionary algorithms (EAs)* to optimize the physicomimetics parameters used in our algorithm are then presented.[1] Our approach for creating a robot-accurate simulation engine is shown in Sect. 12.6, with our experimental results. This simulation engine includes a reasonable model of our MAXELBOT (see Chap. 10), the robot platform developed by us for exactly these sorts of tasks. Sect. 12.7 describes how we ported our algorithm to the MAXELBOTS. The summary and conclusions are presented in Sect. 12.8.

## 12.2 Modification to the Force Law

While the original behavior of the robots in the physicomimetics framework creates formations that uniformly spread 360° in an environment, this is not suitable for narrow passageways, where chain formations are desired. To promote self-organized chain formations, we build on the physicomimetics framework by changing the inner circular disk (which represents the region of repulsion, as shown in Fig. 4.6) of the split Newtonian force law to be an *elliptical* disk [230]. Each robot has a virtual ellipse that surrounds it as shown in Figs. 4.7 and 12.2. The major axis of the ellipse is aligned with the

---

[1] This task was sufficiently complex that we were unable to deduce all parameter settings from theory—a topic for future research.

heading of the robot. Robot $p$ feels repulsion from any other robot $k$ that enters the ellipse of robot $p$. A robot is attracted to any robot that is outside of the ellipse and is closer than $1.5D$. In other words, on the boundary of the ellipse there is a "switch" from repulsion to attraction.

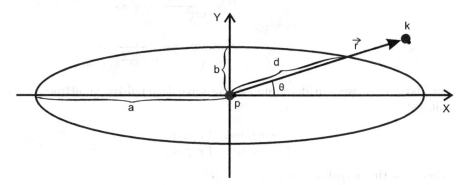

Fig. 12.2: Elliptical disk example

To mathematically compute this repulsive/attractive switch, let us assume that we have the situation presented in Fig. 12.2, shown from the point of view of robot $p$'s local coordinate system. The $x$-axis is the forward direction of robot $p$, and robot $k$ is entering the sensing range of robot $p$. A vector $r$, shown as a directed edge from robot $p$ to robot $k$, corresponds to the actual sensed distance between robot $p$ and robot $k$, and the angle $\theta$ at which robot $p$ senses robot $k$ (which is precisely the information that trilateration provides). We need to decide if the force exerted by robot $k$ on robot $p$, which is computed using physicomimetics, is attractive or repulsive.

By knowing the angle $\theta$, we can compute the distance $d$ from $p$ (along the vector $r$) to the point where $r$ intersects the ellipse. If we consider $a$ to be the semi-major axis and $b$ to be the semi-minor axis of the ellipse mentioned above, then the equation of this ellipse in the Cartesian coordinate system is

$$\frac{x^2}{a^2} + \frac{y^2}{b^2} = 1 \, . \tag{12.1}$$

By rewriting Eq. 12.1 in polar coordinates in terms of $d$ and $\theta$, we obtain

$$\frac{d^2 cos^2\theta}{a^2} + \frac{d^2 sin^2\theta}{b^2} = 1 \, . \tag{12.2}$$

Next, from algebra, we get

$$\frac{d^2 cos^2\theta}{a^2} + \frac{d^2 sin^2\theta}{b^2} = 1 \Rightarrow$$

$$d^2 \left( \frac{cos^2\theta}{a^2} + \frac{sin^2\theta}{b^2} \right) = 1 \Rightarrow$$

$$d^2 = \frac{1}{\frac{cos^2\theta}{a^2} + \frac{sin^2\theta}{b^2}} \Rightarrow$$

$$d = \frac{1}{\sqrt{\frac{cos^2\theta}{a^2} + \frac{sin^2\theta}{b^2}}} \,.$$

By knowing $d$, we can determine if the force is repulsive or attractive as follows:

$$\begin{cases} repulsive, \text{ if } r < d\,, \\ attractive, \text{ if } r \geq d\,, \end{cases} \tag{12.3}$$

where $r$ is the magnitude of $\boldsymbol{r}$, i.e., $r = |\boldsymbol{r}|$.

We consider this modification to the original framework to be an elegant and minimal solution, since the change in the actual simulation code is limited to the addition of only a few lines of mathematical code. In essence, all that needs to be done is the calculation of the distance $d$. This is then compared with $r$ (i.e., the sensed distance between the robots), which in turn decides the sign of the force.

Let us consider the behavior of two robots that are within sensing range. If both of them are heading directly towards each other, then they will be repelled and move directly away from each other. In the case where they are not heading directly towards or away from each other, they will be attracted. As soon as that happens, they will head towards each other. Then they will be repelled and move away. The interesting fact is that this also works with large numbers of robots—as an emergent behavior. You can try this by running the NetLogo simulation "formation_chain.nlogo".

## 12.3 Backflow Force

One issue that needs to be addressed when creating chainlike formations is that once the robots start repelling each other, they will continue to repel until they are no longer within sensor range. Once out of sensor range, they no longer interact.

To solve this problem we make use of a "backflow" force. This force drives a robot towards the last known location of its closest "upstream" neighbor. Rather than being applied continuously, this force is only applied when the connectivity is lost with the upstream neighbor. Once the connectivity is re-established, the force is disabled. This feature of our algorithm also automatically compensates for robots that cease to operate. The robot "downstream"

Listing 12.1: Front tangential force—C code

```
...
if ( s == 0 ) { // if the front sensor has detected a wall
   if ( U(0,1) > 0.5 ) { // 50% of the time go left
      wall_force.x += params.front_tangential_force * \
                      sin ( agent->heading + params.PI / 2.0 );
      wall_force.y += params.front_tangential_force * \
                      cos ( agent->heading + params.PI / 2.0 );
   }
   else {                  // 50% of the time go right
      wall_force.x += params.front_tangential_force * \
                      sin ( agent->heading - params.PI / 2.0 );
      wall_force.y += params.front_tangential_force * \
                      cos ( agent->heading - params.PI / 2.0 );
   }
}
...
```

of the disabled robot will eventually connect with the next "upstream" neighbor. This way the chain will become a bit shorter, but connectivity along the entire chain will be restored.

## 12.4 Front Tangential Force

To help with the expansion of the chain formation across corners, we introduce a front tangential force. When a robot senses a wall with its front sensor, it feels a tangential force to the left or to the right. The choice between the two directions is randomly chosen. The magnitude of this force is one parameter that we optimize using evolutionary algorithms, as presented in the following section. A code fragment (in the C programming language) showing how the front tangential force is applied is shown in Listing 12.1. Function U(0,1) returns a random real value between 0.0 and 1.0.

## 12.5 Parameter Optimization with Evolutionary Algorithms

One way of improving the performance of our algorithm is to optimize the physicomimetics parameters using an evolutionary algorithm (EA). EAs are inspired by natural evolution and are often used for optimization purposes. We use *mutation* and *recombination* operations on a population of candidate solutions (individuals). Each individual is a vector of real-valued physicomimetics parameters. Mutation randomly perturbs the parameters in an individual, and recombination combines parts of two existing individuals to generate new individuals. We are using a maximizing EA that considers higher *fitness* solutions to be better, where fitness is a measure of solution quality. Every generation, all individuals are evaluated with respect to how well the robots performed in an environment, such as the one shown in Fig. 12.3. Here, the fitness is the average of an individual's reachability, sparseness and coverage values (as defined in Sect. 12.6.2). Only the individuals with the best fitness are used to create the next generation of individuals. One important advantage of using this population-based stochastic algorithm (i.e., an EA) is that it quickly generates individuals that have high performance. To show that the best solutions are not only high in quality, but are also general, these solutions have been tested on additional environments described in Sect. 12.6.3.

Based on our experiences with the uniform coverage algorithm (Chap. 11), we consider one pixel to be equivalent to one inch and one simulation time step to be equivalent to one second. A MAXELBOT can move one foot/second, which implies that the maximum speed of the robot in simulation should be ≤ 12.

We split the parameters into two categories: those that we consider fixed and those that we want to evolve using the EA. We kept the following seven parameters fixed: the total number of robots (10), the desired distance between robots (40), the maximum speed a robot can have (three), the number of infrared (IR) sensors used for wall detection (three, at sensing angles $0°$, $90°$, and $270°$), the amount of motor noise (10%), the rate at which a new robot is added to the environment (i.e., one robot every 2,000 time steps), and the total simulation time (100,000 time steps). The evolved parameter set contained 11 parameters, as shown in Listing 12.2.

Next, we present our robot-faithful simulation engine, followed by different experiments that were performed using the 11 parameters optimized by the evolutionary algorithm.

## 12.6 Simulation

First, we describe our robot simulation engine, which includes a reasonable model of our MAXELBOT robot. Then we introduce the metrics that we use to evaluate the performance of our chain formation algorithm. We continue with a presentation of the multiple environments we use to test our algorithm. In the last part of this section we elaborate on our experiments that demonstrate the excellent performance of our algorithm despite sensor and motor noise.

**Listing 12.2: The final set of evolved parameters**

```
11 parameters optimized by EA

Parameter description: final evolved value

 1. Friction of the system:                        0.579306
 2. Wall detection range:                   7.381481
 3. Ellipse major axis:               119,382.723558
 4. Ellipse minor axis:                     95.740130
 5. Back-flow force magnitude:               0.371287
 6. Back-flow friction:                 0.569767
 7. Newtonian law robot-robot gravitational constant G:   1,895.527010
 8. Newtonian law robot-robot power of the denominator p:     3.022846
 9. Newtonian law robot-wall gravitational constant G:    1,218.359939
10. Newtonian law robot-wall power of the denominator p:     5.832936
11. Front tangential force magnitude:            2.082259
```

## 12.6.1 A Maxelbot-Faithful Simulation Engine

Our simulation engine is programmed in the C programming language. The Graphical User Interface (GUI) is generated with the open-source alternative to OpenGL Utility Toolkit—GLUT, namely the Free OpenGL Utility Toolkit—Freeglut version 2.4.0 library [173]. This simulation is intended to be platform independent, but it was developed on a Linux-based platform.

Features like robot orientation, the ability to turn on a dime, and a maximum velocity constraint are introduced in this simulation to model the physical aspects of our MAXELBOT robots [155]. We also have reasonable models of the trilateration accuracy and obstacle detection sensors by introducing noise coefficients that match the real hardware performance. The simulated IR sensors also feature adjustable limits on both the maximum range at which they are able to detect an obstacle and the threshold after which they report when an obstacle is detected. We also add in noise to the movement of the robot, both in terms of the angle turned and the distance moved.

### 12.6.1.1 Simulation Description

The GUI of our simulation is represented by a window that is $750 \times 750$ pixels in size. The top-left part of the window, as shown in the snapshot of our simulation tool in Fig. 12.3, displays important real-time information about the status of the simulation. Time steps in the simulation and the current number of robots are two examples. In the top-right corner, the four metrics that we use to evaluate the performance of our algorithm are displayed. Exact numbers, as well as graphical sliding bars, show the value of each metric.

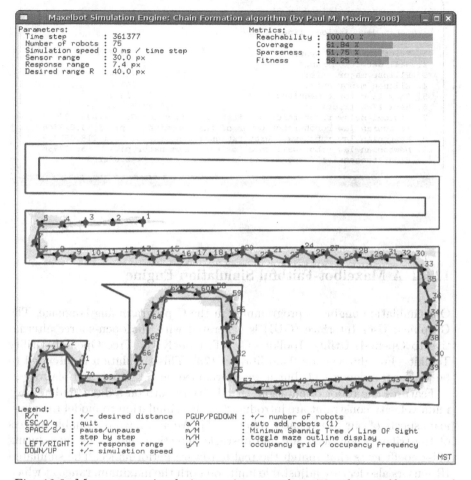

Fig. 12.3: MAXELBOT simulation engine snapshot: 78 robots self-organized in a chain formation that extends over 11 corners (where the occupancy grid is shown in yellow)

The largest part of the simulation window is in the middle, where the outline of the simulated environment as well as the robots are displayed. The robots that navigate inside the environment are drawn as blue disks, nine pixels in diameter. A black dot indicates the current heading of the robot. The position and the response range of the IR sensors are shown as red lines coming out of the robot. Each robot is assigned a unique number and, for easy identification on the screen, this number is shown above and to the right of the disk that represents the robot.

It is easy to see when an IR sensor from a robot detects a wall since a red dot will be drawn on the screen, which represents the exact intersection point between the wall and the beam of the sensor. (Six such occurrences can be identified in Fig. 12.3—see robots 67 and 74 for two examples.) When more robots are added, an edge is drawn between robots that sense each other. To help with visualizing where the robots have already been in the environment, all the "visited" pixels are colored in yellow. Our simulation has several other useful graphical features that are presented later in this section.

A *legend* provides information on what keys are used to control the simulation. It is displayed on the bottom of the window.

We will continue with a description of how our simulation is configured, followed by a presentation of the key(board) combinations that are used to control it.

### 12.6.1.2 Simulation Configuration

We built this simulation engine mainly for research purposes; hence, it is highly configurable with both control flags inside the C code and configuration files that are provided to the simulation at run time. While having a graphical user interface (GUI) is important for visual feedback regarding whether the algorithm is performing correctly, a GUI also significantly increases the execution time of the simulation. We run our simulation on a system that features an Intel Centrino Core 2 Duo T5600 CPU running at 1.83 gigahertz (3661.40 bogomips). When we simulate 50 robots for 100,000 time steps, it takes our GUI version more than an hour to finish. This is not practical if we want to run our simulation on multiple environments, each with multiple different parameter configurations. Because of this, we also have a command-line interface (CLI) version that is significantly faster than the GUI version and provides exactly the same final output. For comparison, it takes the CLI version simulation approximately 3 minutes to run for 100,000 time steps with 50 robots. Being able to run the simulation at least 20 times faster makes a significant difference. To switch between the two versions is as simple as setting a GRAPHICS flag to true or false, which is provided in our code.

The simulation requires two configuration files to be provided at run time. The first configuration file provides the simulation engine with a set of parameters that are considered *fixed*, meaning their values do not change often. An example of such a file is presented in Listing 12.3. The first four lines configure the initial number of infrared (IR) sensors each robot has and their positions; in this case, there are three IR sensors. Other interesting parameters that can be configured here are the coefficients for noise introduced by the localization module, IR sensors, and motors. An interesting parameter presented here is the maximum speed a robot can have in the simulation. It is difficult to give a measuring unit for it because, for example, miles per

Listing 12.3: Fixed parameters configuration file

```
sensor_number       3        # number of infrared (IR) sensors
0                            # 1st IR sensor angle position (degrees)
90                              # 2nd IR sensor angle position (degrees)
270                             # 3rd IR sensor angle position (degrees)

world_width     750    # window size (pixels)
world_height    750

agent_number       1      # initial number of robots
agent_number_max   10     # total number of robots
sensor_range       30     # the fixed IR sensor range (pixels)

trilat_noise      0.1    # coefficient of localization noise
IR_noise       0.1    # coefficient of IR sensor noise
motor_noise    0.1    # coefficient of motor noise

release_rate       500    # 1 agent per X time steps
total_sim_time     100000 # total simulation time steps

sim_speed_ms       0      # control simulation speed (ms): graphics
agent_radius       4.5    # size of the agent (pixels): graphics
wall_width      3       # width of wall (pixels): graphics

R       40.0   # desired distance between robots(pixels)
max_V       3.0    # maximum agent velocity - physics
step        1.0    # time step - physics

PI        3.141592653589793238462643
```

hour would not make sense in our simulation world. We will address this issue later in this chapter, when we present how we port our algorithm to the MAXELBOTS. For now, robot speed is in terms of pixels per simulation step.

The second configuration file stores the information about the environment that is to be used in the simulation. Listing 12.4 shows the configuration file for the environment that is presented in Fig. 12.3 (earlier in this chapter). The total number of vertices is declared on the first line, followed by the coordinates for each of them. Although our MAXELBOT-faithful simulation engine allows an environment to have at most one loop inside it, this particular example does not have one (i.e., see the last line in Listing 12.4). Still, as presented in Sect. 12.6.3, we also experiment with loop environments.

### 12.6.1.3 Simulation Control

Control of the GUI portion of our simulation is done exclusively with the keyboard. All the keys that can be used are presented in the *legend* section

### Listing 12.4: Environment configuration file

```
vertex_number    35
-190.0     -60.0
-180.0      20.0
-100.0      20.0
-110.0     -30.0
 -40.0     -30.0
 -40.0      80.0
 -80.0     100.0
 120.0     100.0
 114.0     -40.0
 370.0     -30.0
 370.0     125.0
-190.0     130.0
-190.0     205.0
 380.0     200.0
 380.0     225.0
-190.0     230.0
-190.0     300.0
 400.0     305.0
 400.0     275.0
-170.0     280.0
-170.0     250.0
 400.0     255.0
 400.0     180.0
-170.0     175.0
-170.0     150.0
 400.0     155.0
 400.0     -70.0
  96.0     -60.0
  90.0      80.0
  10.0      80.0
 -10.0     -50.0
-130.0     -60.0
-130.0       0.0
-160.0       0.0
-150.0     -60.0

loops             0
```

at the bottom of the simulation window (see Fig. 12.3). Since most of the
actions are intuitive, we only explain those that are not so obvious. The *auto
add robots* feature simulates how the robots are inserted into the environment,
by placing a new robot in front of the stationary robot. This is similar to how
we envision deploying robots in the field. The frequency with which the robots
are added can be set in the configuration file presented earlier. The value (0)
for *off* or (1) for *on* is displayed on the screen, next to the description of this
feature, to show which mode is selected.

The *m/M* key is used to toggle between a *line-of-sight* or a *minimum
spanning tree* display. In the line-of-sight mode, an edge is drawn between all

pairs of robots that detect each other. The robots cannot detect each other through the walls of the environment or if they are separated by a distance greater than $1.5D$. The minimum spanning tree mode only draws the edges of the minimum spanning tree with the first stationary robot as root and all the other robots as nodes. We will present more details on this in Sect. 12.6.2. The acronym *LOS*, for line-of-sight, or *MST*, for minimum spanning tree, is displayed in the lower-right corner of the simulation window to let the user know which mode is currently displayed. An example of the differences between the two display modes is shown in Fig. 12.4: the top-left snapshot is using the MST mode (Fig. 12.4 (a)) and the top-right is using the LOS mode (Fig. 12.4 (b)). In LOS mode, more edges are shown when the robots are in close proximity to each other. Because we cannot have more than one path to reach the same node in a minimum spanning tree, in this example there is no edge drawn between robots 12 and 13 when the MST mode is selected. In this simulation there are 20 robots that have self-emerged into a chain formation.

The outline of the environment can be hidden by using the $h/H$ key. This feature creates an experience similar to what would happen in the field. When the robots are deployed inside an unknown environment, the user only sees the robots that are deployed, slowly advancing inside the environment. Generally, it is easy to estimate the shape of the environment merely by looking at the shape of the robotic chain formation within it. To help the system user to more clearly discern the true environment outline, each time a sensor on a robot detects a wall, that detection point remains permanently displayed on the screen. In this way, a sketch of the walls is slowly created as more sensors detect the surrounding walls. The outline of the environment is only shown in the bottom-right snapshot (Fig. 12.4 (d)). As seen from this example, revealing the outline of the environment does not reveal much more shape information than we already know.

Given the different shapes an environment can have and how the chain formation self-organizes, the robots may not always be in close proximity to the environment walls. This in turn will not accurately reflect the shape of the environment. To help with this issue, each time a robot hits a point in the environment, that point is permanently displayed on the screen, eventually creating an occupancy grid. Furthermore, we decided to not only show this occupancy grid, but also allow the user to visualize which areas are visited more frequently than others. In this mode, darker areas are those that have been more frequently visited. The switch between the two modes is done with the $y/Y$ key. Again, the differences between these modes can be seen by examining the snapshots in Fig. 12.4. The top-left snapshot (Fig. 12.4 (a)) displays the occupancy frequency (shown in grayscale), and the bottom-left snapshot (Fig. 12.4 (c)) displays the occupancy grid (shown in yellow).

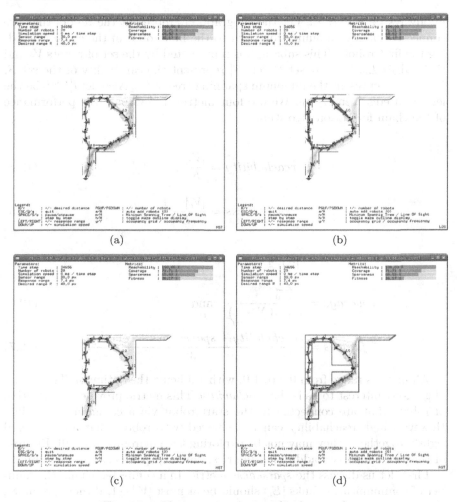

(a)                                              (b)

(c)                                              (d)

Fig. 12.4: Various simulation display modes

## 12.6.2 Performance Metrics

We now define the metrics used for evaluating the performance of the chain formations in the environments. An *optimal chain formation* is one in which all robots are connected via communication links and they are extended into the environment as fully as possible. This means that the total number of edges is minimized and the lengths of the edges are maximized. We define this formally as follows.

Each team of $N$ robots defines a weighted undirected graph $G$ with a set of robots (the vertices) $V$ and set of communication edges $E$. The weights are the lengths of the communication links between robots. $G$ may not necessarily always be *connected*. Let the subgraph $G_c$ represent the robots that are connected via a communication path to the first robot in the chain (including the first robot). This subgraph is represented by the set of robots $V_c$ and their edges $E_c$. For any set $S$, let $|S|$ represent the cardinality of the set $S$. Let $G_{\text{MST}}$ represent the minimum spanning tree of $G_c$. Also, let $l_i^{G_{\text{MST}}}$ be the length of edge $i$ in $G_{\text{MST}}$. We use four metrics to evaluate the performance of the chain formation algorithm:

$$reachability = \frac{|V_c|}{N}\,, \tag{12.4}$$

$$sparseness = \frac{|V_c| - 1}{|E_c|}\,, \tag{12.5}$$

$$coverage = \frac{\sum_{i=0}^{|V_c|-2} l_i^{G_{\text{MST}}}}{R_{\max}(N-1)} \quad \text{and} \tag{12.6}$$

$$fitness = \frac{reachability + sparseness + coverage}{3}\,. \tag{12.7}$$

All metrics range from 0.0 to 1.0, with 1.0 being the optimum. The metric of greatest interest to us is the *reachability*. This metric provides the fraction of robots that are connected to the start robot via a communication link. However, high reachability can be achieved with robots that are clustered tightly together—they may not be exploring the environment at all. Because of this, we added two additional metrics, namely, *sparseness* and *coverage*.

First, let us discuss the *sparseness* metric. In a connected chain, the number of communication links $|E_c|$ should be at most $|V_c| - 1$. If there are more, the chain has redundant links, and the sparseness metric will be less than 1.0. In other words, we want the chain formation to have as few edges as possible, while remaining connected.

Second, we want the robots to extend as far as possible into the environment, while remaining connected. If the maximum separation between robots is $R_{\max}$, then the maximum distance that can be covered is $R_{\max}(N-1)$. If we sum the lengths of all the edges in the minimum spanning tree of $G_c$, this provides a reasonable metric of the distance covered by the robots in $G_c$.

Finally, it is clear that these metrics can be difficult to simultaneously optimize. We provide a *fitness* metric to indicate the average reachability, sparseness, and coverage over time. This was the metric used by the EA.

We consider these metrics to be user-oriented and, hence, they are always computed from the point of view of the stationary robot placed at the front of the environment. For example, if we have 10 robots self-organized in a perfect chain formation, but they manage to get out of range of the first robot, then those robots are worthless. We also make use of the total number of robots placed in the environment in these metrics. We consider anything after the first *unbroken* chain to be lost, and this represents a performance degradation.

### 12.6.3 Experimental Test Environments

In order to test our algorithm more thoroughly we created a suite of nine environments (Fig. 12.5). We built these environments to have a gradual increase in difficulty. All environments feature different corridor widths (e.g., Figs. 12.5 (b) and (e)), both going from narrow to large (Fig. 12.5 (a)) and vice versa (Fig. 12.5 (c)). Various degree angles for the bends are used and two environments have difficult concave corners with angles less than 90° (Figs. 12.5 (g) and (i)). We designed some of these environments to also have branches (Figs. 12.5 (d), (e), (h), and (i)) and loops (Figs. 12.5 (d), (h), and (i)). We are interested in environments with loops, since they may lead to a particular situation in which the robot at the farther end of the chain formation loops around and reconnects with another robot from the chain. Our chain formation algorithm is not specifically designed to branch, but we want the formation to handle this particular case. In the experiments, the chain formation did branch—especially when a large number of robots occupied the same region that had multiple exit points. This was a desirable emergent property of the system.

A spiral environment (Fig. 12.5 (f)) as well as a 15-corner environment (Fig. 12.5 (g)) are part of this suite. We want to evaluate the chain expansion both across corners and along straight corridors. Ideally, in the same environment, the distance between robots should get smaller at corners, creating redundant communication links and extending to the allowed maximum along the straight segments.

The entrance in each of these environments is positioned next to the lowest and leftmost corner and it is marked by the shape of a robot. The wall at the entrance is virtual and the initial stationary robot is placed there.

### 12.6.4 Experimental Design

In this section we present details on how the simulation tool is configured to test our chain formation algorithm. To speed up the simulation execution time, the graphics are turned off. We use the optimized EA parameters presented earlier in Listing 12.2.

Each robot has three Sharp IR sensors for range detection, looking forward, left and right. We model noise introduced into the system by the localization,

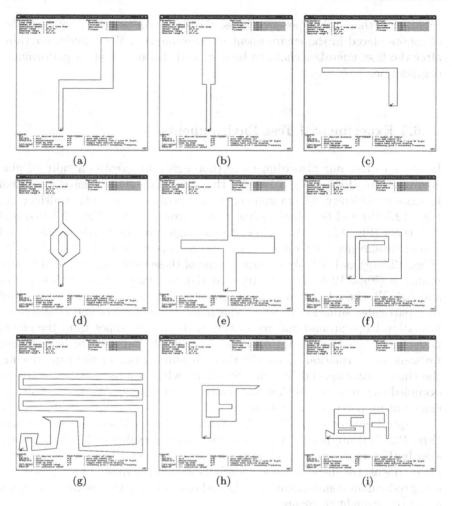

Fig. 12.5: Chain formations—environment suite (each environment is labeled separately)

IR sensors, and motors as a multiplicative uniform random variable $[1-\delta, 1+\delta]$. For example, 10% noise means that the random variable has range $[0.90, 1.10]$. Motor noise of 10% means that 10% noise is applied to both the robot turn and the distance the robot travels. Similarly, sensor noise of 10% means that 10% noise is applied to the Sharp IR distance sensor reading, as well as the range and bearing reading from our trilateration module.

Each pair of robots within detection range of each other respond to the elliptical split Newtonian force law (see Chap. 3) with parameters $p$ and $G$. As a reminder, "split" refers to the fact that the force law can be attractive and repulsive. Similarly, robots will respond to walls with a repulsive Newtonian force law with a separate set of $p$ and $G$ parameters. There is no guarantee and our algorithm does not require that a robot be able to sense both sides of a corridor. If a robot senses a wall in front, it will "feel" a tangential force to the left or right. As mentioned above, in order to promote robustness in the formation, a backflow force is used.

We created one rather complex environment, presented earlier in this chapter (Fig. 12.3), for performing the experiments that we present next. Initially, there is only one robot at the beginning of the environment. This robot is stationary, and new robots are added in front of this robot at periodic intervals. As the robots are added, they start exploring the environment while maintaining line-of-sight with their upstream neighbor. The desired distance between the robots is set at $D = 40$ and the maximum velocity is $V_{\max} = 3.0$. If not otherwise specified, localization, motor and IR sensor noise are set to 10% and the total number of robots is $N = 10$. The robots can sense each other within a distance of $1.5D$.

We next present three experiments that were performed. In the first, we examine the performance as the number of robots $N$ is increased (scalability). In the second, we change the amount of sensor and motor noise. Finally, we examine the performance as the number of robots $N$ increases and then decreases (robustness).

## 12.6.5 Experiment I: Variable Number of Robots

Recall that only 10 robots are used during training with the EA. To test the scalability of our algorithm, we perform a series of experiments where the number of robots varies from five to 50, with an increment of five. Experiments have been run for 100,000 time steps, in order to examine the stability of the chain formations (i.e, to make sure they don't collapse or break apart). Each data point is an average over 10 independent runs. The results of all four metrics are shown in Figs. 12.6–12.9.

Figure 12.6 shows good reachability results as $N$ increases, indicating that the evolved parameter set is working well with larger numbers of robots. Reachability is generally better earlier in the simulation, but remains stable after 50,000 time steps.

Figures 12.7 and 12.8 indicate that good reachability comes at the expense of somewhat reduced coverage and sparseness, as $N$ increases. This is quite reasonable. As the number of robots increases, the probability of a connectivity link failure in the chain also increases (see the analysis in Sect. 12.6.9). To help offset this problem, the resulting chain formations are not stretched as much as possible and possess redundant links.

Figure 12.9 indicates that the average of all three metrics is quite acceptable. Average performance falls slowly with increased $N$, and the chain formations are very stable (i.e., time has almost no effect).

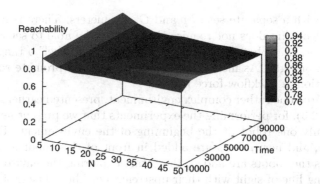

Fig. 12.6: Reachability as a function of $N$ and time

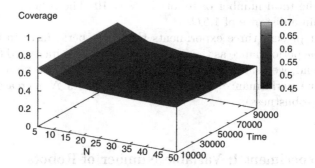

Fig. 12.7: Coverage as a function of $N$ and time

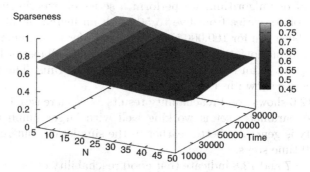

Fig. 12.8: Sparseness as a function of $N$ and time

## 12.6.6 Experiment II: Effect of Motor and Sensor Noise

Our simulation takes into consideration the noise introduced by certain robot
hardware components. Noise ranges from 0% to 50%. Sources of noise include

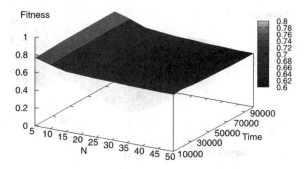

Fig. 12.9: Fitness as a function of $N$ and time

the Sharp IR obstacle detection sensors, our trilateration hardware, and the
MAXELBOT motors. Figures 12.10–12.13 illustrate how the performance is
affected when noise is introduced in the system. Interestingly, the results
paint a similar picture to that shown in the prior experiment.

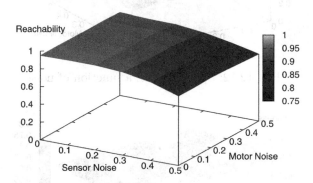

Fig. 12.10: Reachability as a function of noise

Once again, reachability (Fig. 12.10) is maintained very well, despite
the noise. This is achieved because the chain formation does not stretch
to its maximum limit (i.e., the robots are compressed to some extent—see
Figs. 12.11 and 12.12). The overall fitness (Fig. 12.13) degrades slowly as
noise increases, and is least affected by motor noise.

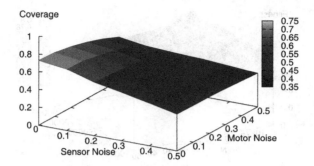

Fig. 12.11: Coverage as a function of noise

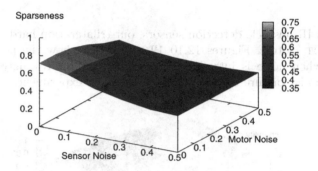

Fig. 12.12: Sparseness as a function of noise

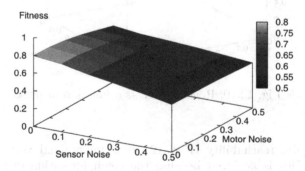

Fig. 12.13: Fitness as a function of noise

## 12.6.7 Experiment III: Robustness

In this experiment, sensor and motor noise are set to 10%. The simulation starts with one robot. Then one robot is added every 500 time steps for a total of 24,500 time steps (yielding 50 robots). The simulation then runs for 500 time steps with the 50 robots. Finally, a robot is randomly removed (i.e., a random robot fails) every 500 time steps until only two robots remain, to test robustness. All metrics are averaged over 10 independent runs. The results are shown in Fig. 12.14.

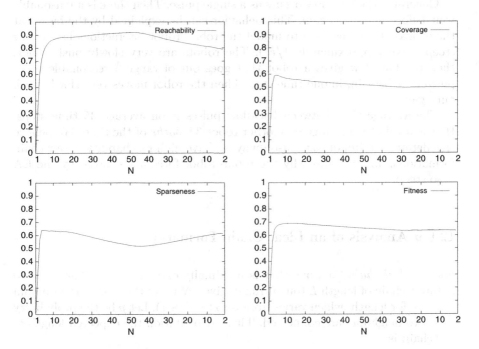

Fig. 12.14: Robustness experimental results

As would be expected, there is a reduction in reachability when robots fail. Recall that reachability represents the length of the chain formation, starting with the first robot. Random robot failures break the chain, reducing reachability. However, the reduction is not severe. The reason for this can be seen in the sparseness graph. A sparseness of 0.5 indicates that there are redundant communication links (edges) in the chain. These help compensate for breaks in the chain. Note that sparseness increases as the number of robots is decreased. This is because the number of redundant links has been reduced due to the robot failures.

What is surprising is the relative flatness of the fitness graph. Again, the fitness graph represents the average of the other three metrics. Despite a rather rapid loss of robots (where they are lost as fast as they were introduced), the fitness is remarkably stationary, indicating excellent robustness in the chain formation.

## 12.6.8 Backflow Force Analysis

Next, we run our simulation engine with only two robots in a straight corridor. The purpose is to compute how often the backflow force is actually applied. For this experiment, we run the simulation for 4,000 time steps.

Generally, the backflow occurs as a single pulse. Then there is a reasonable wait before it pulses again. This behavior can be explained by the fact that the evolved force law used to model the robot–robot interactions has a very steep curve (approximately $1/r^3$). The robots are very slowly pushed out, then the backflow gives a robot that goes out of range a reasonable pulse pushing it back in, in one time step. Then the robot moves very slowly back out again.

The average time between backflow pulses is on average 35 time steps. Hence, a robot is in range of another robot $34/35ths$ of the time. We believe this detail is an important reason why we have high reachability in our chain formations, and explains why the robot–robot force law evolved by the EA is so steep.

## 12.6.9 Analysis of an Ideal Chain Formation

For an *ideal chain* (i.e., one that is maximally extended), how hard is it to obtain a chain of length $L$ (on average), given $N$ robots? Let $L$ be the random variable for length, which varies from zero to $(N-1)$. Let $p$ be the probability that two neighbors are connected. Then by definition the expected length of the chain is:

$$E\,[L] = \sum_{l=0}^{(N-1)} l \times Prob(l)\,. \tag{12.8}$$

The probability that $l = 0$ is $(1-p)$.
The probability that $l = 1$ is $p(1-p)$.
The probability that $l = 2$ is $p^2(1-p)$.

$\cdots$

The probability that $l = (N-2)$ is $p^{(N-2)}(1-p)$.
The probability that $l = (N-1)$ is $p^{(N-1)}$.

So,

$$E\,[L] = 0(1-p) + 1p(1-p) + \cdots +$$
$$(N-2)p^{(N-2)}(1-p) + (N-1)p^{(N-1)}\,.$$

Hence,

$$E\,[L] = \left[\sum_{l=1}^{N-2} lp^l(1-p)\right] + (N-1)p^{(N-1)}\,. \tag{12.9}$$

Now let us find a closed form solution:

$$S = \sum_{l=1}^{N-2} lp^l(1-p) = (1-p)\sum_{l=1}^{N-2} lp^l$$

$$= (1-p)\left[1p + 2p^2 + 3p^3 + 4p^4 + \cdots + (N-3)p^{(N-3)} + (N-2)p^{(N-2)}\right]$$

$$= 1p + 2p^2 + 3p^3 + 4p^4 + \cdots + (N-3)p^{(N-3)} + (N-2)p^{(N-2)} - p^2$$
$$-2p^3 - 3p^4 - \cdots - (N-4)p^{(N-3)} - (N-3)p^{(N-2)} - (N-2)p^{(N-1)}$$

$$= p + p^2 + p^3 + p^4 + \cdots + p^{(N-2)} - (N-2)p^{(N-1)}\,.$$

Then,

$$E\,[L] = S + (N-1)p^{(N-1)}$$
$$= p + p^2 + p^3 + p^4 + \cdots + p^{(N-1)}$$
$$= \sum_{l=1}^{N-1} p^l$$
$$= \frac{p - p^N}{1-p}\,.$$

Or,

$$E\,[L] = \frac{p - p^N}{1-p}\,. \tag{12.10}$$

As an example with 50 robots, achieving an average chain length of 37 ($\approx$ 75%) requires that each link be "up" 99% of the time. For another example, consider the prior backflow analysis, which showed that $p \approx 34/35 \approx 0.97$. For $N = 50$ robots the expected chain length (i.e., reachability) is roughly 25.785. This is 51.6% of the 50 robots. However, if one examines the reachability graph in Fig. 12.14, the reachability is well above 51.6%. The reason for the improved reachability lies once again in redundant links (which can be seen in the sparseness graph). If sparseness is 50%, then there are twice as many communication links as required. One link failure will not break the chain. However, two link failures can break the chain. The probability of a simultaneous failure of two communication links that break the chain is roughly $(1 - p)^2 = (1/35)^2 \approx 0.001$, so the effective $p$ has been increased to approximately 0.999. In this situation the expected chain length is 47.8, which is 95.6% of the 50 robots. This is reasonably close to the maximum reachability achieved with 50 robots.

## 12.7 Hardware Implementation

Because the simulation results are so promising, the next step is to port the chain formation algorithm to the MAXELBOTS. Our goal is not only to get the MAXELBOTS to self-organize in a chain formation while exploring an unknown environment, but also to build a graphical tool that generates an image of the shape of the environment that they are exploring.

Let us come back to the scenario we presented in Sect. 12.1. A stationary MAXELBOT is placed at the entrance of an unknown environment that needs to be mapped as well as explored. The stationary MAXELBOT is connected to a laptop. As more MAXELBOTS are released into the environment, they start self-organizing into a chainlike structure that takes the shape of a portion of the unknown environment. The communication link created by the MAXELBOTS is then used to transmit their positions back to the first MAXELBOT. A graphical application running on the laptop will use these positions to generate a diagram that informs the user in real-time of the shape of the portion of the environment that the MAXELBOTS are currently exploring. *These sorts of tasks are precisely why we have invented our localization technique.*

We start by presenting how we modified our current synchronization protocol to accommodate the data transmission necessary for this application. We continue with a presentation of the graphical tool we developed to translate the transmitted data back by the robots in the field, in an easy to understand graphical representation. An analysis of how the performance of our algorithm is affected by the protocol performance is shown in Sect. 12.7.4. The results from the experiments that we conducted on the MAXELBOTS are introduced next. In the experiments shown in Sect. 12.7.5 all the MAXELBOTS are stationary and we are testing both the new *chain formation list (CFL)* real-time generation algorithm and the new data communication protocol. The MAXELBOTS are moving in the experiments presented in Sect. 12.7.6. This set of dynamic experiments is used to evaluate the chain formation algorithm performance.

All of the MAXELBOT experiments presented in this section were performed indoors. As part of the future work we plan to test our algorithms on MAXELBOTS running outdoors.

## 12.7.1 Modified Communication Protocol

To prevent any confusion, let us mention that we do not plan to modify the *synchronization* protocol as presented in the earlier Chap. 10; instead, we only want to adapt the *communication* part of the synchronization protocol to our new needs.

As part of the localization process, RF communication is used to broadcast a signal that synchronizes the transmitting and receiving localization modules. From the beginning, we implemented this signal to also contain minimum information regarding the broadcasting robot. As a reminder, three bytes of information were transmitted: the current heading, the chemical density, and the ID of the robot. The heading is provided by an onboard digital compass. A volatile organic compound (VOC) sensor was used to get chemical density readings from the surrounding environment. This information was used in a different research project involving chemical source localization (see Chap. 9). For our application we need a different set of data to be transmitted. We present both the version we implemented and an alternative simplified version.

### 12.7.1.1 Implemented Version

Before developing a new communication protocol, we have to determine what information we need and how to use it in order to generate a diagram of the robots in the chain formation. From the experience we gained with our simulation work, we decided to generate this diagram as a "chain formation list" (CFL), where the stationary robot is the beginning of the list.

Our approach is to "build" the CFL in a distributed fashion, which can be done in $\Theta(N)$ time, where $N$ is the number of deployed robots. Let us explain this approach by using the undirected weighted graph presented in Fig. 12.15. The nodes are robots and the edge weights are the distances between the robots. Each robot is represented by a gray disk, and its ID is written inside this disk. The heading of the robot is represented by both a small black disk and the actual degree. The distances between all the robots are also shown. The information shown in Fig. 12.15 becomes available to all robots *after* the synchronization protocol token "visits" every robot. The X–Y coordinate system of the robot relative to its heading is also shown in Fig. 12.15 (top right). A heading of $0°$ means the forward side of the robot is aligned with *north*. Also, to change its heading, for example from $0°$ to $90°$, the robot makes a clockwise rotation.

The total number of robots in the formation, $N$, is known by each robot. Each robot locally stores the information it learns from other robots in a two-dimensional vector called *robots*[]. *Robots*[$i$] is of the form [*x-coordinate, y-coordinate, heading*] where $i$ is the ID of the robot this information belongs

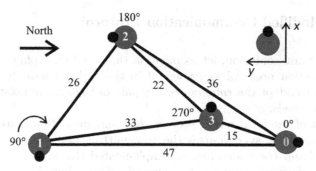

Fig. 12.15: Communication protocol example

to. The coordinates are generated by the localization module and are relative to the coordinate system of the receiving robot.

Besides the *robots*[] vector mentioned above, each robot also has two vectors that help generate the CFL. $CFL[]$ is a one-dimensional vector that only stores robot IDs. The order of the IDs represents the computed CFL up to the current robot. The second vector, $CFL_{info}[]$, is two-dimensional, and each of its elements is also a vector of the form [*x-coordinate, y-coordinate, heading*] for each robot in the CFL. The information stored at each $CFL_{info}[i]$ position corresponds to the robot whose ID is stored at the analogous $CFL[i]$ position. The heading is essential to align the various coordinates to a global coordinate system when generating the CFL graphical image. (However, alignment of the coordinate systems can be done without the global compass, but requires twice the communication.) All of these vectors are initialized to zero. Our CFL algorithm requires one initialization pass through all $N$ robots—in order for each robot to know its neighbors. The CFL is built in the second and subsequent passes.

When the protocol is first initialized, the only information each robot knows is its own heading and ID. Since robot 0 is always the CFL root, it starts by writing its ID at $CFL[0]$ (and hence $CFL[] = [0, -, -, -]$, because $N = 4$). Robot 0 also updates its heading at $CFL_{info}[0]$. The corresponding $x$- and $y$-coordinates of robot 0 at $CFL_{info}[0]$ are both always zero, since robot 0 represents the CFL origin. Next robot 0 tries to find its closest neighbor, but since it has no information on any robot yet, it defaults to adding the ID = own_ID + 1 = 0 + 1 = 1 at $CFL[1]$; hence, $CFL[] = [0, 1, -, -]$. Robot 0 has no localization information on robot 1, so $CFL_{info}[1]$ is not updated. Next, robot 0 broadcasts both the $CFL_{info}[]$ and $CFL[]$ vectors, as well as its ID, over RF. Every robot that is in the range of the RF will receive this information. For ease of exposition, we assume that all four robots are in range of each other in our example.

The next step for all the listening robots is to decide which robot will add the next node to the list, and that is the robot whose ID was last added to

**Listing 12.5: Information each robot knows after the first protocol round**

```
Robot 0: robots[0] = [    0.0,    0.0,    0]
         robots[1] = [ -47.0,    0.0,   90]
         robots[2] = [ -30.0,   20.0,  180]
         robots[3] = [ -14.0,    4.0,  270]

Robot 1: robots[0] = [    0.0,   47.0,    0]
         robots[1] = [    0.0,    0.0,   90]
         robots[2] = [ -20.0,   16.0,  180]
         robots[3] = [  -5.0,   32.0,  270]

Robot 2: robots[0] = [ -30.0,   20.0,    0]
         robots[1] = [  16.0,   20.0,   90]
         robots[2] = [   0.0,    0.0,  180]
         robots[3] = [ -16.0,   16.0,  270]

Robot 3: robots[0] = [  -4.0,  -14.0,    0]
         robots[1] = [  -5.0,   32.0,   90]
         robots[2] = [  15.0,   16.0,  180]
         robots[3] = [   0.0,    0.0,  270]
```

the $CFL[]$. To determine which ID was last added, the listening robots use
the ID of the robot that just transmitted and do a search in the received
$CFL[]$. The robot whose ID is positioned after the ID of the robot that just
broadcast is the next robot to perform the computations. In our example,
robot 1 is now looking for the nearest neighbor from the robots whose IDs
are not yet added to the $CFL[]$. Robot 1 only has no localization information
on the remaining robots 2 and 3, so by default it writes the ID = own_ID +
$1 = 1 + 1 = 2$ at $CFL[2]$, and hence $CFL[] = [0, 1, 2, -]$. Again, $CFL_{info}[2]$
is not updated—since no information is known on robot 2. Finally, robot 1
broadcasts its own ID together with the $CFL[]$ and $CFL_{info}[]$ vectors over
RF.

The above step repeats two more times for robots 2 and 3. After robot 2,
$CFL[] = [0, 1, 2, 3]$. Then robot 3 determines it is positioned at $CFL[3]$;
hence, there is no new node that it can add to the CFL. To make the programming simple, robot 3 still broadcasts the two unchanged vectors, together with
its ID, over RF.

The second time around, every robot knows about its neighbors and Listing 12.5 shows what information each robot has in its *robots*[] vector. This
information is in terms of each robot's local coordinate system. The data is
pertinent to our example setup shown in Fig. 12.15. As we continue with our
step-by-step example to see how the CFL is distributively computed, please
refer to the weights of the undirected graph shown in Fig. 12.15 as well as
the information each robot knows, as presented in Listing 12.5.

Again, it is robot 0's turn to start constructing the list. First, all values in both $CFL[]$ and $CFL_{info}[]$ are set to zero, with the exception of $CFL[0]$ and $CFL_{info}[0]$, which are unchanged since last time. Second, robot 0 finds robot 3 to be the closest and updates the CFL information to $CFL[1] = 3$ and $CFL_{info}[1] = [-14.0, 4.0, 270]$. Finally, robot 0 broadcasts the entire current $CFL[]$ and $CFL_{info}[]$ over RF. At this point $CFL[] = [0, 3 , - , -]$.

Robot 3 determines that it is the last one added to $CFL[]$ and it finds robot 2 to be closest. Again, robot 3 can only select the closest neighbor from the robots that are not yet added to $CFL[]$. Next, robot 3 updates the CFL information to $CFL[2] = 2$ and $CFL_{info}[2] = [15.0, 16.0, 180]$, and transmits the entire two vectors over RF. Remember that the localization process still takes place as before. The only change we are doing to the protocol is in the data communication part. Now, $CFL[] = [0, 3 , 2 , -]$.

The next robot to determine that it is the last one added to $CFL[]$ is robot 2. The closest neighbor of robot 2 is robot 1; hence the following updates are done by robot 2: $CFL[3] = 1$, $CFL_{info}[3] = [16.0, 20.0, 90]$. The entire content of the two vectors is again transmitted over RF. At this stage, $CFL[] = [0, 3 , 2 , 1]$.

Since the ID of robot 1 is placed in the last position in $CFL[]$, there are no more robots to be added to the list. Robot 1 simply retransmits the two vectors over RF. Again, $CFL[] = [0, 3 , 2 , 1]$.

The token arrives back at robot 0 and the entire process starts again. The current hardware installed on the MAXELBOTS allows an RF transmission to occur approximately every 120 ms, which is equivalent to a frequency of approximately 8 Hz or 8 times per second.

Although every robot in the formation receives the RF broadcast information at every time step, we are particularly interested in making sure that robot 0 receives it. Remember from the scenario we presented at the beginning of this section that robot 0 is the only robot connected to a laptop. In Sect. 12.7.2, we present the format in which this information is sent to the user, and how we further use it to generate a real-time graph on the laptop screen.

### 12.7.1.2 Alternative Communication Protocol

Since we assume that every robot can receive the information broadcast over RF, we also designed a slightly different communication protocol. As with the implemented version, this algorithm does not require that every robot can receive the acoustic signal. This version of the protocol significantly reduces the transmitted information.

Each robot still stores $CFL[]$ and $CFL_{info}[]$, but the only information transmitted over RF at every time step is $[ID_1, ID_2, x_2, y_2, h_1]$. First, $ID_1$ represents the ID of the robot that is transmitting. Second, $ID_2$, $x_2$, and $y_2$ represent the ID of the last robot that is added to $CFL[]$ and its coordinates from the transmitting robot point of view. Last, $h_1$ is the heading of the transmitting robot.

This information is enough to allow each robot, including robot 0, to construct its own $CFL[]$ and $CFL_{info}[]$, as presented earlier in the implemented communication protocol. As part of our future work, we plan to implement this more efficient communication protocol.

## 12.7.2 Real-Time Graphical Representation Tool

Similarly to what was done with our simulation engine presented above, we developed our graphical visualization tool using the C programming language and the Free OpenGL Utility Toolkit version 2.4.0 [173]. The GUI window is 750 × 750 pixels in size (Fig. 12.16). The top part of the window presents information about the current status of the diagram displayed, such as the number of data samples received so far from the MAXELBOT and the current scaling factor.

A *legend* is displayed on the bottom that lets the user know what the controls are for this tool. To improve the visualization, the user can zoom the image, both in and out. The scaling factor and a graphical scale (in pixels) are shown in the top part of the window. Because the update of the screen is relatively fast, the screen can be frozen if desired. The update speed can also be adjusted, and a step-by-step visualization is available.

The real-time generated diagram of the robots is displayed in the center of the window. The snapshot in Fig. 12.16 shows a four-robot diagram generated from the data set obtained at the last step in the communication protocol example above. Let's take a closer look at the data set passed to our graphical tool by the stationary MAXELBOT: [0 3 2 1 0.0 0.0 0 −14.0 4.0 270 15.0 16.0 180 16.0 20.0 90]. The first four numbers, [0 3 2 1], represent the IDs of the robots (the CFL). The order of the IDs is important since it shows how to correctly parse the following values in the data set. The remainder of the data set represents four groups of three values each. Information about robot 0 is enclosed in first group: [0.0 0.0 0]. The first two numbers represent the position of robot 0 from its own point of view, followed by the current heading of robot 0. The first two values in the second group, [−14.0 4.0 270], represent the coordinates of robot 3 from robot 0's point of view. The last value in this group shows the heading of robot 3 as was transmitted over RF. We know this information is about robot 3 because ID 3 follows after ID 0 in the first four values in the data set (the CFL). In a similar fashion, from the third group we find out the coordinates of robot 2 from robot 3's point of view, as well as the current heading of robot 2 (180). The coordinates are (15.0, 16.0). From the last group we extract the current heading of robot 1, which is 90, and its coordinates from robot 2's point of view, (16.0, 20.0).

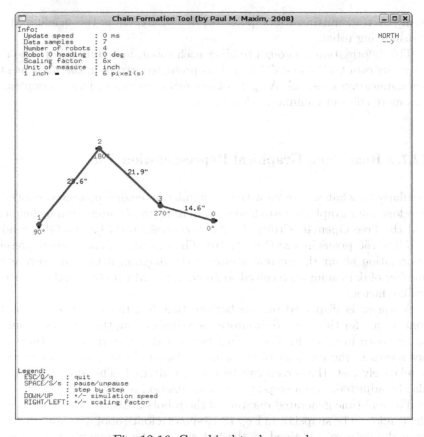

Fig. 12.16: Graphical tool example

All of the information parsed above is used to generate the diagram in the snapshot. Each robot is drawn as a blue disk with a smaller black disk to show its heading. The value of the heading, as well as the ID, are displayed next to each robot. The communication link created by robots is shown by the edges that connect the robots. The number next to each edge represent the length of the edge in inches. In this snapshot we used 6× magnification.

## 12.7.3 Connectivity Options

We present three different ways (which we tested) for communicating between our application running on a laptop and the stationary MAXELBOT. All three use the serial port on the Minidragon (see Chap. 10). The first two are wired, and the last one is wireless.

The first connection approach is straightforward. A serial cable is used to connect the RJ11 connector on the Minidragon to a standard nine-pin serial connector on the laptop. A widely available nine-pin male RJ11/RJ45 modular adapter is also needed. The serial port on the laptop has to be configured with 9,600 Bd, 8 data bits, no parity (N), and one stop bit (1).

Laptops that come equipped with the standard nine-pin serial port are more difficult to find on the market at the present time. For this reason, we present a second wired connection choice that uses a widely available serial-to-USB adapter. It still may be necessary to have a serial cable and a nine-pin male RJ11/RJ45 adapter to connect the Minidragon to the serial-to-USB adapter. The setup of the virtual serial port on the laptop has to be configured with the same parameters as presented above.

Finally, we tested a wireless connection that uses Bluetooth technology. Specifically, we used an AIRcable Serial3 serial-to-Bluetooth adapter to essentially replace the serial cable. The laptop has to be equipped with a Bluetooth adapter, which upon configuration will make a virtual serial port available. This port has to be configured with the same parameters as above. We had successful connections up to 30 ft away from the robot with line-of-sight.

Another possible connection between the MAXELBOT and the laptop is through the use of the $I^2C$ data bus. We never tried this approach, but it can be implemented if $I^2C$ compatible hardware is available.

### 12.7.4 Communication Protocol Performance Analysis

As presented in Chap. 10, our communication protocol is performed on inexpensive radio frequency modules. One drawback of these modules is that data packets which are transmitted wirelessly occasionally get lost. This is similar to considering that the token used by our protocol gets "lost." When this happens, the protocol fails to perform along the entire robotic chain, which in turn does not provide complete information on all the deployed robots. Because our protocol has an automatic recovery mechanism built in, when the token gets lost the first robot in the chain generates a new token.

It is important to formally understand how these protocol failures affect the performance of our algorithm in general. Let us assume that the token successfully passes from one robot to another with probability $p$. Then protocol failures occur with probability $(1 - p)$. Assuming independence, the probability of getting no errors with $N$ robots in a row is $p^N$.

As $N$ increases, $p^N$ drops quickly, unless $p$ is very high. For example, if $p = 0.9$, then with even five MAXELBOTS the probability of getting the token all the way through the chain is only 59%.

## 12.7.5 Static Maxelbot Experiments

The purpose of this first set of experiments is to test the performance of the
CFL algorithm and the new data communication protocol. A total of five
MAXELBOTS were used during all the static experiments. Three different formation shapes were tested: *line*, *L-shaped*, and *U-shaped*. For each formation
shape the robots were placed on the floor in predetermined positions. Three
separate experiments were performed for each formation shape, and we varied the inter-robot distances between each of these three experiments. The
inter-robot distances tested were: two, four, and six feet. For each of the static
experiments the distances between any two adjacent robots were equal. Since
each MAXELBOT has a unique hardware ID, we made use of it to maintain
the order of the MAXELBOTS constant for all the static experiments: 0, 4, 3,
1, 2. The MAXELBOTS in the formation have different headings, and these
headings are constant for all the static experiments. The following list shows
the heading of each robot, together with the corresponding ideal compass
reading shown between the parentheses:

- MAXELBOT 0 – north (0° or 360°)
- MAXELBOT 1 – northwest (315°)
- MAXELBOT 2 – southwest (225°)
- MAXELBOT 3 – northeast (45°)
- MAXELBOT 4 – southeast (135°)

The first robot in the chain transmits in real-time, via Bluetooth, to a
Bluetooth enabled laptop, the positions and the orientations of all five robots.
The graphical tool presented in Sect. 12.7.2 runs on this laptop and parses
the incoming data stream, generating a diagram representing the shape of
the MAXELBOT chain formation. This diagram can be viewed on the laptop
screen. There are a least 320 data sets in the data stream received from the
first MAXELBOT for each of the static experiments. For more details on the
data set structure see Sect. 12.7.2.

### 12.7.5.1 Summary of the Line Formation Experiments

The robot placement for the two-foot inter-robot distance line formation
experiment is shown in Fig. 12.17. For the four- and six-foot inter-robot
distance experiments we maintained the rightmost MAXELBOT fixed and we
expanded the formation to the left.

A snapshot of the graphical tool for the four-foot inter-robot distance experiment is shown in Fig. 12.18. For improved visualization, a "3X" scaling
factor was used to draw the diagram shown in this figure (i.e., one inch is
three pixels). Due to the fact that there is a 10.5° average error in the digital compass readings, the orientation of the line formation is not exact. The
alignment of the digital compass with the body of the robot may also be a

Fig. 12.17: Experimental setup for the two-foot inter-robot distance line formation (the rightmost MAXELBOT is the first robot in the chain; north is to the left)

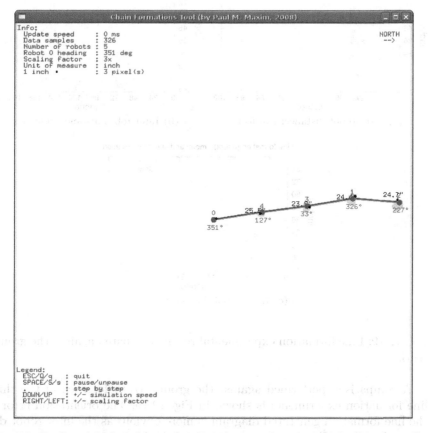

Fig. 12.18: Graphical tool snapshot for the two-foot inter-robot distance line formation experiment (the first robot in the chain is shown at the center of the window; north is to the right)

source of heading error since we did not perform a rigorous alignment. The quality of the reported distances between robots reflects the good performance of the localization modules. The yellow areas surrounding the four rightmost robots represent the oscillations in the trilateration readings. The farther along the chain, the bigger the error. Despite the fact that the error is compounded from robot to robot, the oscillations in the readings remain small.

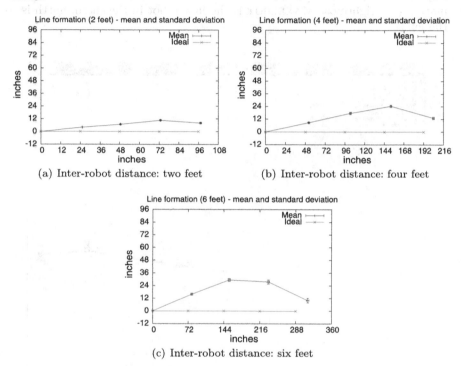

(a) Inter-robot distance: two feet        (b) Inter-robot distance: four feet

(c) Inter-robot distance: six feet

Fig. 12.19: Line formation experimental results compared against the ground truth

A comparison performed against the ground truth for each of the three line formation experiments is shown in Fig. 12.19. The orientation error of the line formation generated diagram is more obvious as the inter-robot distances increase. The distances reported by the localization modules on the MAXELBOTS tend to be larger than the actual distances, especially when the inter-robot distances are large. The standard deviation, both on the $x$- and $y$-axes, remains very small.[2]

For each experiment, we recorded the number of communication protocol failures, meaning the token failed to "visit" all five robots in one pass (i.e., the

---

[2] Each experimental data point is displayed with a small horizontal and vertical error bar to show the standard deviation at each point.

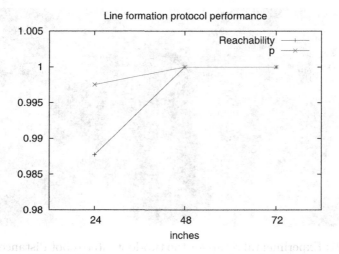

Fig. 12.20: Line formation protocol performance

token got "lost"). We then divided the total number of tokens that made it all the way to the end of the robotic chain by the total number of tokens that were broadcast by the first robot in the chain. This gives us the probability that the token can make it all the way to the end of the robotic chain (see Sect. 12.7.4). Let $p_e$ denote this probability. If we consider $p$ as being the probability that the token can make it from one robot to the next, then $p_e = p^5$, since we have five robots in the formation. We can then compute the exact value of $p$. For these static experiments we will consider $p_e$ as being the *reachability* metric.

The graph in Fig. 12.20 shows both the reachability and the probability $p$ for all three line formation experiments. For the two-foot inter-robot distance experiment, there were only a few protocol failures, but for the other two experiments all the broadcast tokens made it all the way to the end of the chain.

### 12.7.5.2 Summary of the L-Shaped Experiments

We continue our testing with a set of L-shaped formation experiments. The placement of the robots for the two-foot inter-robot distance L-shaped formation experiment is shown in Fig. 12.21. The top left MAXELBOT is the first robot in the chain. For the four- and six-foot inter-robot distance experiments we kept robot 3 (the center robot) in place, and we moved the others to expand the formation.

Fig. 12.21: Experimental setup for the two-foot inter-robot distance L-shaped formation (the leftmost MAXELBOT is the first robot in the chain)

In order to simulate a real-world situation we placed obstacles to obstruct the acoustic line-of-sight between some of the robots. Although the ultrasonic waves are deflected, the robots can still communicate over RF.

The comparisons between the generated diagrams of the L-shaped formations and the ideal robot positions are presented in Fig. 12.22. Similarly to the line formation experimental results, the graphs in Fig. 12.22 show that the orientation of the generated diagrams are off to a modest degree due to the error in the digital compass readings. The distances reported are reasonably accurate, showing a tendency to be slightly larger than the actual distances. The standard deviation in the trilateration readings, both on the $x$- and $y$-axes, remains very small.

The protocol performance for the L-shaped formation experiments is shown in Fig. 12.23. We recorded protocol failures in all three L-shaped formation experiments, but the lowest probability $p$ of a token to make it from one robot to the next is still excellent at approximately 0.99.

### 12.7.5.3 Summary of the U-Shaped Experiments

The U-shaped formation experiments are the last set of static experiments to test the performance of the CFL algorithm and the new communication protocol. Fig. 12.24 shows the setup for the four-foot inter-robot distance experiment. The first robot in the chain is shown at the top-left in this figure. Obstacles were placed on the floor to obstruct the acoustic line-of-sight between some of the robots. This setup is very similar to real-world situations where the robotic chain may extend over multiple corners that may partially

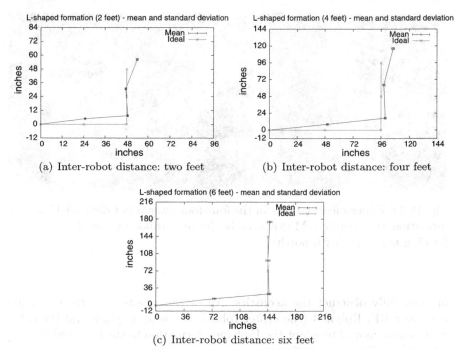

Fig. 12.22: L-shaped formation experimental results compared against the ground truth

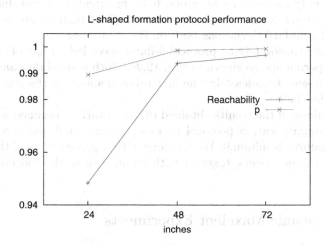

Fig. 12.23: L-shaped formation protocol performance

Fig. 12.24: Experimental setup for the four-foot inter-robot distance U-shaped formation (the top-left MAXELBOT is the first robot in the chain, and its heading is aligned with north)

or even fully obstruct the acoustics. All of the robots can still communicate over RF. Robot 3 (the center robot) was kept in place and the other robots were moved to adapt the U-shaped formation to the two- and six-foot inter-robot distance experiments.

As shown is Fig. 12.25, the quality of the U-shaped formation generated diagrams is generally maintained. The diagrams almost overlap the ideal diagrams, only showing more errors, both in orientation and distances, for the six-foot inter-robot distance experiment. The trilateration readings have very small standard deviations both on the $x$- and $y$-axes.

There were absolutely no protocol failures recorded for any of the three U-shaped experiments, as shown in Fig. 12.26. Both reachability and $p$ overlap as all the tokens broadcast by the first robot made it all the way to the last robot in the chain.

The analysis of the results obtained in these static experiments shows that the new communication protocol performs as expected and the number of protocol failures is minimal. We continue with a presentation of the dynamic MAXELBOT experiments, together with an analysis of the collected data.

## 12.7.6 Dynamic Maxelbot Experiments

We next present three dynamic experiments that we performed in order to test the performance of the chain formation algorithm. Five MAXELBOTS were tested in three different environments: *open, straight corridor*, and *L-shaped corridor*. The open environment was not constrained to a certain shape, but as the chain expanded a cardboard wall was used on one side to prevent the

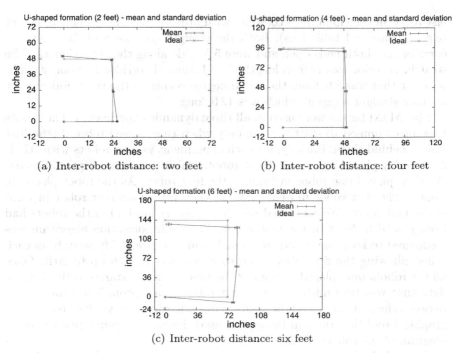

(a) Inter-robot distance: two feet          (b) Inter-robot distance: four feet

(c) Inter-robot distance: six feet

Fig. 12.25: U-shaped formation experimental results compared against the ground truth

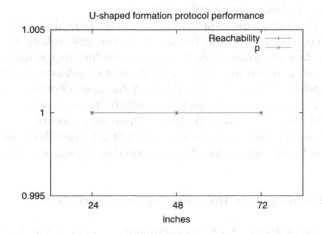

Fig. 12.26: U-shaped formation protocol performance

last robot in the chain from colliding with objects that are difficult to detect (e.g., a chair and table legs). Both the straight corridor and the L-shaped corridor had both ends open and were 5 ft wide along the entire length. The straight corridor was 14 ft in length. The L-shaped corridor had one straight segment that was 8 ft long, then a 90 degree corner to the right, followed by another straight segment which was 12 ft long.

Five MAXELBOTS were used in all three dynamic experiments. The robots were programmed to start moving only when they sensed robot 0 (the first robot) behind them. Initially, for each experiment we had robots with IDs 1, 2, 3, and 4 placed behind the first robot, which we always kept stationary. We then placed one robot in front of the first robot. As the robot placed in front of the first robot moved away, we then placed another robot in front of the first robot. We continued the same process until all of the robots had been placed in front of the first robot. The chain formation algorithm was configured to keep the robots at a maximum distance of 5 ft away from each other, allowing the five-robot chain to potentially stretch up to 20 ft. Once all the robots were placed in front of the first robot, we started collecting the data that was transmitted by the first robot to a laptop. The connectivity between the first robot and the laptop was done wirelessly via Bluetooth. The graphical tool that runs on the laptop used this data to generate a real-time diagram of the robotic chain.

Each of the four moving MAXELBOTS was equipped with three infrared (IR) sensors facing forward, left, and right. The sensing range of the obstacle detection sensors was set at 9 in from the edge of the robot in all three directions (just slightly greater than the value found by the EA, as shown in Listing 12.2, to better avoid wall collisions).

For all three dynamic experiments, we programmed the robots to move in lock step. Every time the token got to the first robot, all the robots, except the first stationary robot, moved according to the output from the chain formation algorithm. Then the first robot broadcasts the token and the process repeats. A token was broadcast by the first robot every 4.25 seconds to give enough time to each robot to complete its movement.

We used the data collected during these experiments to compute the *sparseness* and *coverage* metrics (see Sect. 12.6.2 for details on these metrics). The protocol performance for all three experiments was also analyzed.

### 12.7.6.1 Summary of the Open Environment Experiments

The experimental setup for this experiment is shown in Fig. 12.27. The first robot is shown at the bottom, and the cardboard wall mentioned earlier is visible in the top-right corner of this figure. The cardboard wall prevents the last robot from colliding with objects difficult to detect. During this entire experiment the chain expanded to at most 14 ft out of the 20 ft possible.

Fig. 12.27: Open environment chain formation experimental setup (the bottom MAXELBOT is the stationary robot, and its heading is aligned with north; the picture was taken after the introduction of all five MAXELBOTS)

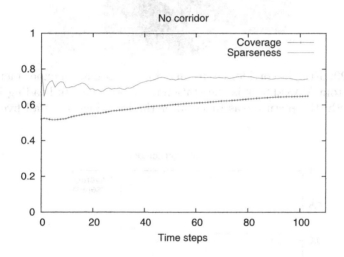

Fig. 12.28: Open environment experimental results

The evolution of the sparseness and coverage metrics is shown in Fig. 12.28. In the beginning the sparseness shows the inter-robot distances vary more and eventually as the formation stabilizes the sparseness remains relatively constant. The coverage keeps increasing, showing the robots' tendencies to reach for the maximum allowed inter-robot distances. The actual values of the two metrics are similar to the values obtained in simulation when large amounts of motor noise were introduced (see Sect. 12.6.6).

Fig. 12.29: Straight corridor environment chain formation experimental setup (the bottom MAXELBOT is the stationary robot, and its heading is aligned with north; the picture was taken after the introduction of all five MAXEL-BOTS)

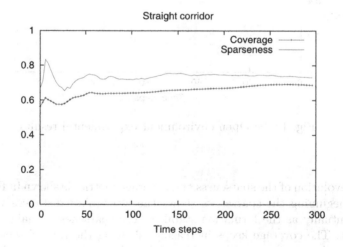

Fig. 12.30: Straight corridor environment experimental results

### 12.7.6.2 Summary of the Straight Corridor Environment Experiments

We constrained the environment the robots could move in to a straight corridor for this experiment. Figure 12.29 shows the actual experimental setup. The first robot is shown at the bottom of this figure. The picture shown in this figure was taken towards the end of the experiment after the formation stabilized. The robots tend to stay in the center of the corridor, and this is due to the fact that the first robot was placed in the center (and it is important to emphasize that the robots could not simultaneously detect both sides of the corridor). The chain expanded to at most 16.5 ft out of the 20 ft possible during this experiment.

The two-metric graph in Fig. 12.30 shows that the formation stabilized early in the experiment and kept excellent stability for the remainder of the experiment. This experiment was run over a period of time almost three times longer than the open environment experiment presented earlier, and we can see that the metrics improve even more over time.

### 12.7.6.3 Summary of the L-Shaped Corridor Environment Experiments

Figure 12.31 shows the setup we built for this experiment, as well as the self-emerged five-robot chain formation. The bottom robot is the first robot in the chain. The chain expanded to 17.5 ft out of the 20 ft possible, and this length was measured along the chain formed by the robots. The maximum corridor length covered by the robots was 19.5 ft, measured at the center of the corridor. Figure 12.32 shows a picture of the environment, as reported back to the first robot via the communication chain.

Fig. 12.31: L-shaped environment chain formation experimental setup (the bottom MAXELBOT is the stationary robot, and its heading is aligned with north; the picture was taken after the introduction of all five MAXELBOTS)

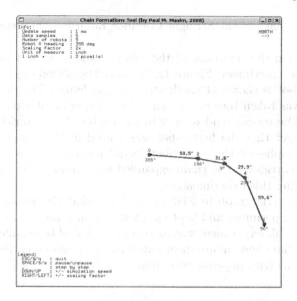

Fig. 12.32: A picture of the environment, as communicated via the chain formation back to the first MAXELBOT

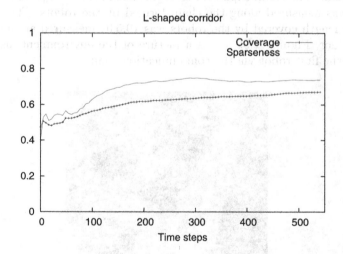

Fig. 12.33: L-shaped environment experimental results

The robotic chain tends to "cut" the corners, rather than have a MAXEL-BOT placed right in the corner (see Fig. 12.31). The MAXELBOTS tend to stay closer to the right corridor wall (inner corner) than to the left corridor wall

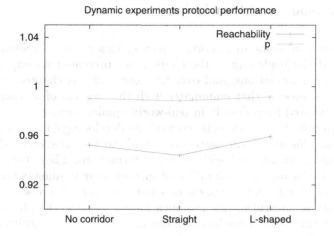

Fig. 12.34: Dynamic experiments protocol performance

(outer corner). The inter-robot distances are smaller at the corner and get larger on the second straight corridor. A four-robot formation "collapse" followed by a recovery was observed. During the collapse the robots exchanged positions in the chain. Changing positions does not affect the chain formation algorithm. All the observations mentioned above were also present in the simulations.

Due to hardware problems, at least two MAXELBOTS did not come to a complete stop when programmed to do so, but continued to move slowly forward. (We suspect the Sabertooth motor controller inside the MMP-5 platform to be the source of the problem.) This problem (which is a form of constant motor noise) did not seem to affect the chain formation unless the robots were in close proximity to each other.

The chain formation inside the L-shaped corridor maintains excellent stability over time, and this is reflected in both metrics shown in Fig. 12.33. The presence of the corner prevents the robots from maintaining maximum separation and this explains why the coverage metric value is smaller than it is for the straight corridor experiment presented earlier.

### 12.7.6.4 Protocol Performance

Similarly to the static experiments, we recorded the number of protocol failures during all three dynamic experiments, and the results are shown in Fig. 12.34. Despite the fact that the robots are moving and additional electromagnetic noise is introduced by the robots' motors, the reachability is still high and the probability $p$ that a token successfully passes from one robot to the next is excellent at approximately 0.99 for all three experiments.

## 12.8 Summary and Conclusions

This chapter presents an important part of this multi-disciplinary research project. At the beginning of the chapter, we presented an important task, namely, chain formations, and over the course of this chapter we showed a step-by-step process that culminates with the creation of a complex "tool" that can be used immediately in real-world applications.

Most of this chapter presents our work on developing a multi-robot control algorithm. This algorithm successfully leads to the creation of distributed, self-emerging, robust, and scalable chain formations. These types of formations are useful for the exploration of unknown environments that are long and narrow in nature. Sewers, ducts, caves, and tunnels are some examples.

We start by presenting the modification we made to the physicomimetics framework that is the foundation of our chain formation algorithm. We continue by building a simulation engine to test our algorithm. This simulation faithfully models most of the real capabilities of our MAXELBOT.

A graphical tool is presented next, together with both a modified data communication protocol and an algorithm that distributively constructs a chain formation list whose nodes are the exploring robots. Combining all these with our MAXELBOTS leads to map generation of the explored environment.

The experiments with the MAXELBOTS described at the end of this chapter confirm the simulation results and make it easy to think about many real-world applications for which this tool can be used.

# Chapter 13
# Physicomimetic Motion Control of Physically Constrained Agents

Thomas B. Apker and Mitchell A. Potter

## 13.1 Introduction

Artificial physics provides an intuitive, simple, and robust scheme for controlling a group of robots without having to explicitly specify a trajectory for each agent. This is accomplished by specifying behavior rules in the form of artificial force fields that define the interactions among agents. In most instances, the direction of the force is defined along a vector between each pair of agents and its magnitude by the distance between them. A typical formulation is based on an inverse square law with an attract–repel (AR) boundary, across which the direction of the force changes.

The trajectory of each agent is determined by the sum of the artificial forces acting on its virtual mass. Forces are determined based only on the position of the agent relative to other points in the environment. These may be other robots, observed obstacles, user-specified objectives, or other points computed to improve the swarm's ability to accomplish its mission. This inherent flexibility and scalability allows heterogeneous swarms to accomplish complex missions without computing complete robot trajectories in advance.

Thomas B. Apker

NRC/NRL Postdoctoral Fellow, Washington DC, USA, e-mail: apker@aic.nrl.navy.mil

Mitchell A. Potter

US Naval Research Laboratory, Washington DC, USA, e-mail: mpotter@aic.nrl.navy.mil

W.M. Spears, D.F. Spears (eds.), *Physicomimetics*,
DOI 10.1007/978-3-642-22804-9_13,
© Springer-Verlag Berlin Heidelberg 2011

### 13.1.1 Challenges for Physical Implementation

Physicomimetics is an example of a *guidance* algorithm. Given an estimate
of a vehicle's current state and the surrounding environment, it computes
the next desired state, generally in terms of a velocity. This is also known as
"outer-loop" control because it wraps around the "inner loop" when drawn as
a block diagram such as Fig. 13.1. Physicomimetics allows the user to specify
environment information such as the location of other particles and their
interactions, while allowing each agent to compute its own desired states.

Poorly designed guidance systems can excite unstable modes in the inner-
loop/plant system, or request states that the controller cannot achieve.
Chen [31] developed a robust guidance law for swarming systems with dynamic constraints, but with the caveat that physical agents often cannot meet
the desired kinematic performance specified by control laws such as physicomimetics. In this light, a robust control strategy must incorporate physical
vehicle dynamics as well as artificial physics. Several authors have addressed
this for wheeled vehicles [232, 263]. This chapter will examine the more general problem of defining the interaction of vehicle dynamics and artificial
physics (physicomimetics) frameworks.

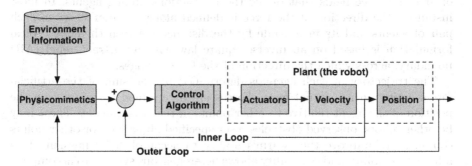

Fig. 13.1: Block diagram showing physicomimetics' place in a physical robot
control system

### 13.1.2 Defining Physically Useful Artificial Physics

Particle-based physicomimetics assumes that each agent has no restriction
on its direction of movement. For example, our MAXELBOTS can "turn on
a dime," allowing us to model them as particles. However, many physical
vehicles are extended bodies with finite actuators and mass, and so cannot
achieve the rapid, small motions of point-mass particles in simulation. Thus,

physical implementations of physicomimetics involve a correction factor of some kind, although these are generally *ad hoc*. A systematic approach to this problem would provide a means to achieve highly intuitive and robust multi-vehicle control while ensuring that the guidance laws do not make unacceptable demands on the vehicle or its inner-loop controller.

The rest of this chapter will describe how to define outer-loop control algorithms for vehicles based on physicomimetics in a way that accounts for the dynamic constraints of the vehicle under control. In doing so, it separates the concepts of "agent" and "particle," and introduces a new type of physicomimetic force law.

## 13.2 The Extended Body Control Problem

This section defines dynamic constraints on two common robotic platforms and describes how particle-based physicomimetics can result in undesirable behavior. It will introduce the concept of the "reachable space" of the vehicle in terms of its physical properties and inner-loop controller, which can then be compared to the commands generated by a physicomimetic controller. While the analysis used in the following sections works in the velocity space of a vehicle moving in a plane, $\mathbb{R}^2$, the analysis techniques presented can be extended to three dimensions.

Given the nature of physical actuators such as wheels and rudders, *tangential speed* $u_t$, and *turn rate* $\Omega$, are common outer-loop control variables for a variety of vehicles. We will work in a space defined by $u_t$ and possible heading changes, $\Delta\Psi$, of the vehicle. The tangential speed is a measure of how quickly the vehicle is moving forward or backward. We use the *tangential* to emphasize that we are performing force calculations in a tangential–normal frame defined by the vehicle's current trajectory, as opposed to a the velocity vector in global coordinates, which we denote by $\boldsymbol{v}$. The heading change $\Delta\Psi$ is the product of the instantaneous turn rate, $\Omega$, and the time between outer-loop control updates, $\Delta t$, and denotes how far the vehicle has turned in a single time step. Analysis in three dimensions requires a second angle denoting the change in flight-path angle relative to level flight.

When we discuss point masses, we assume that the direction of the instantaneous velocity vector defines a heading, and that the maximum heading change is $\pi$. In other words, the agent can reverse its direction instantly, and its maximum speed is limited only by the simulation designer. Figure 13.2 shows a plot of the reachable space of a point mass particle in simulation with $\Delta t = 0.1\,\mathrm{s}$ and maximum speed, $u_{t,\mathrm{max}} = 1.0\,\mathrm{m/s}$. With no restriction on turn rate or acceleration, a point mass subject to artificial physics constraints can instantly change from any velocity with magnitude below $u_{\mathrm{max}}$ to any other in the presence of an arbitrarily large force.

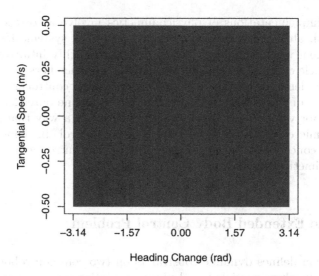

Fig. 13.2: Diagram of the reachable space (shaded area) of point mass agent with a maximum speed of $u_{t,\max} = 1.0\,\mathrm{m/s}$ and time step $\Delta t = 0.1\,\mathrm{s}$

Equation 13.1 shows the general algorithmic definition of the split Newtonian gravitational force rule defined by Spears et al. [235]. It depends on five parameters, three of which determine the magnitude of the force, and two that define the direction and range of the interaction. Given particle masses of $m_i$ and $m_j$, the force $F_{ij}$ along a line between them is scaled by the gravitational constant $G$, mass power $p_m$, and distance power $p_d$. This force is positive (repulsive) if the distance $r$ between them is less than the attract–repel boundary $D$, and negative (attractive) if it is between the attract–repel boundary and maximum range of the force, $D_{\max}$. The quality of point-mass formations depends on the fact that they can instantly change direction as they cross $D$ and rapidly jump back and forth across the boundary, allowing their net displacement to approach zero in an ideal case. The force is

$$F_{ij} = \begin{cases} -\,G\dfrac{(m_i m_j)^{p_m}}{r_{ij}^{p_d}} & \text{if } r_{ij} < D \,, \\[2mm] G\dfrac{(m_i m_j)^{p_m}}{r_{ij}^{p_d}} & \text{if } D \le r_{ij} \le D_{\max} \,, \\[2mm] 0 & \text{otherwise} \,. \end{cases} \qquad (13.1)$$

The virtual force produced by this guidance law is shown as a function of position in Fig. 13.3. In this plot, the force exhibits a sharp discontinuity at

$D = 1\,\text{m}$ and $D_{\text{max}} = 4\,\text{m}$. The sharp and immediate change from repulsive to attractive force at $D$ represents a significant problem for physical systems.

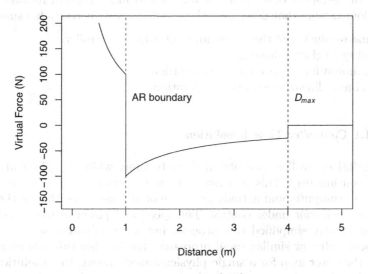

Fig. 13.3: Plot of virtual force versus distance for the standard physicomimetic inverse square rule

Point-mass particles in simulation are only subject to actuation constraints imposed by the designers. In most cases, this is limited to a maximum allowable speed. In these lightly constrained cases, formations quickly emerge for most combinations of force laws that are easily re-formed after passing obstacles or encountering simulated disturbances [232].

Traditional physicomimetics uses *first-order forward Euler integration (FOFE)* of the virtual forces to determine the next velocity command $v_{n+1}$ based on the current velocity and a $\Delta v$ computed based on the artificial forces $\Sigma F$, particle mass $m_p$, and time step $\Delta t$. If $\Sigma F$ is very large, such that $\|\Delta v\| \gg v_{\text{max}}$, then Eq. 13.2 reduces to Eq. 13.3.

$$v_{n+1} = v_n + \Delta v \ . \tag{13.2}$$

$$\Delta v = \frac{\sum F}{m_p}\Delta t \ .$$

$$v_{n+1} = \frac{\sum F}{\|\sum F\|}v_{\text{max}} \ . \tag{13.3}$$

## 13.2.1 Reachable Space for Real Vehicles

From the perspective of their guidance algorithms, physical robotic systems have four factors that determine the topology of their reachable space:

1. Time resolution of the inner- and outer-loop controller
2. Ability to change heading
3. Maximum linear speed and acceleration
4. Minimum linear speed and acceleration

### 13.2.1.1 Controller Time Resolution

All digital controllers operate in discrete time, while the physical vehicles move continuously. This is a major issue for inner-loop control as it must measure, compute, and actuate at a rate of at least twice that of the fastest dynamic behavior under control. The speed and power of modern electronics has greatly simplified this process, but most robotic systems employ a microcontroller or similar small processor that handles this role exclusively.

At the outer loop for a single physicomimetic agent, the resolution of the controller determines how likely the agent will be to detect changes such as when it has crossed an AR boundary. In simulations, the product of the time step $\Delta t$ and maximum speed $v_{max}$ determines the amount of "jumpiness" one will observe for agents near an AR boundary, and a similar effect can be seen with physical vehicles. However, unlike the point-mass case, $\Delta t$ is determined by the needs of the inner-loop controller and computation available to the agent—making this a factor that must be accounted for in the implementation of a physicomimetics system.

With multiple interacting agents, finite time resolution means commands generated for a particular arrangement of agents will continue to be applied after the agents have moved from their positions. As a result, a swarm of vehicles may pass through stable formations and force the swarm into oscillations about "low energy" configurations. Often, the agents will be reacting to stimuli that change at, but are not synchronized with, their update rates. As a result, distributed outer-loop controllers cannot achieve the *Nyquist criteria* for stabilizing formations on their own, and each agent must have internal logic to damp out undesired motion.

### 13.2.1.2 Changing Heading

The defining feature of point-mass particles is that they do not have an orientation, and thus can change from any velocity vector to another in one time step given an arbitrarily large force. An object that can instantaneously change its direction of motion is called *holonomic*. Real vehicles are typically *not* holonomic; they have both mass and rotational inertia, which make sudden changes in heading impossible. In addition, they are often designed to move in a particular direction, such as in the direction of the wheels of a ground robot or along the longest axis of an aircraft or boat. The ability to change direction of motion is generally defined by the ability to reorient the vehicle, which is a process much slower than applying a force to achieve some $\Delta \boldsymbol{v}$.

Consider differential drive robots such as the Pioneer P3-DX (left) and Talon (right) shown in Fig. 13.4 and schematically in Fig. 13.5. Their tangential speed $u_t$ is determined by the average speed of their wheels or treads (Eq. 13.4), and they achieve a particular turn rate, $\Omega$, by increasing the speed on one side while retarding the other. The resulting turn rate can be expressed as a function of the difference between the right and left drive speeds, $v_r$ and $v_l$, respectively, and the distance between the wheels and the rotation center, $r_{\text{diff}}$, in Eq. 13.5. If the maximum speed of each side, $v_{\text{max}}$, is limited, then the maximum forward speed given a desired change in direction can be found using Eq. 13.6. Figure 13.6 shows a plot of the reachable space of a vehicle with $\Delta t = 0.1\,\text{s}$, $v_{\text{max}} = 0.5\,\text{m/s}$, and $r_{\text{diff}} = 0.3\,\text{m}$.

Fig. 13.4: Photographs of common differential drive robots, the Pioneer P3-DX (left, used by permission of Adept MobileRobots) and Talon (right)

$$u_t = \frac{(v_r + v_l)}{2}. \tag{13.4}$$

$$\Omega = \frac{(v_r - v_l)}{2\,r_{\text{diff}}}. \tag{13.5}$$

$$u_{t,\text{max}} = v_{\text{max}} - \Omega\,r_{\text{diff}}. \tag{13.6}$$

In Fig. 13.5, note that the rotation center does not automatically coincide with the particle location. This can be due to the design of the robot's navigation system, inherent in the properties of skid steered systems or both. For example, laser-map navigation allows reasonably precise localization of the laser sensor itself, and thus this location is used by the physicomimetics

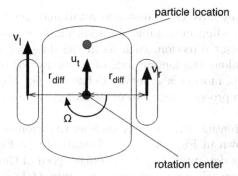

Fig. 13.5: Schematic of a differential drive robot showing the relationship between wheel speeds, robot movement, and particle location

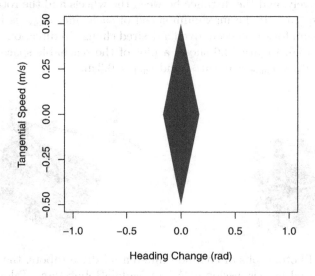

Fig. 13.6: Plot of the reachable space of a ground vehicle with differential drive

system to determine the next heading and velocity. However, as was the case in [263], the laser was not mounted at the rotation center of the vehicle, resulting in a mismatch between the expected and actual motion of the robot. In addition, particularly for tracked and multi-wheeled vehicles, the position of the rotation center depends on the condition of the treads or wheels and the ground over which they roll, any or all of which may not be constant [149].

### 13.2.1.3 Maximum Speed and Acceleration

Safety, thermodynamics, and material properties determine the maximum change and magnitude of vehicle kinetic energy. This effect is typically included as a "maximum speed" in simulations, and has a powerful damping effect on virtual particle motion, especially near the AR boundary. The most profound effect of this constraint is that it creates an inherent coupling between the motion states of the vehicle. A differential drive is a robotic constraint applied to each wheel, each of which has a maximum speed $v_{max}$, and so the maximum forward speed is determined in part by the desired rotational speed according to Eq. 13.6. A previous study to systematically address this problem was conducted by Ellis and Wiegand [58] who modified a "friction" term to limit forward speed during large heading changes. For a simple lattice formation, this actually improved settling time and limited the loss of formation quality as vehicle constraints grew more restrictive.

For aircraft and boats, the stress on components grows with the square of the vehicle speed. As a result, the maximum allowable speed is a function of both the power of its propulsion system and the strength of its structure. Maneuvering increases these stresses, and so all vessels, particularly aircraft, must have rules in their guidance laws to ensure that the turn rate does not exceed the safety threshold for the current speed.

### 13.2.1.4 Minimum Speed and Acceleration

Another interesting problem for robot guidance is caused by the minimum speed and acceleration of a vehicle. For example, mechanical friction and electrical inductance determine the minimum current, and thus speed, of a wheeled robot's drivetrain. The minimum turn rate of a fixed-wing aircraft is determined by the smallest nonzero bank angle the control system can detect and hold, while the airplane's stall speed sets the minimum rate of forward motion. On all platforms, finite resolution of digital to analog (D/A) converters guarantees that continuous actuation is impossible.

From a guidance perspective, this problem creates two challenges. First, when combined with the controller's $\Delta t$, it defines the smallest movements allowable for robots settling into formation, and thus the range of lattice quality, assuming the constraints allow lattice formation at all. Second, it means agents are likely to get stuck in "low energy" formations that prevent them from achieving other goals because forces along a particular axis do not result in enough motion to escape the force fields generated by other particles.

Its kinematics are described mathematically in Eqs. 13.4–13.6. The reachable space of the Pioneer 3-AT used in [263] is shown in Fig. 13.7, with $r_{diff} = 0.3\,\text{m}$, $\Delta t = 0.1\,\text{s}$ and $v_{wheel} \in [0.1, 0.5]\,\text{m/s}$. Note that while the wheels may come to a complete stop, there is a sizable gap in achievable speeds around zero forward or backward rotational motion, which makes small changes in position impossible to achieve.

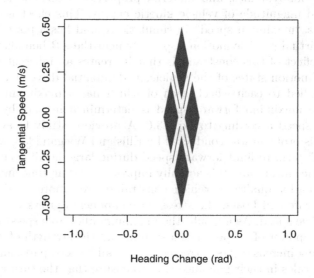

Fig. 13.7: Plot of the reachable space of a Pioneer 3-AT robot with minimum achievable wheel speed

## 13.2.2 Case Study: Wheeled Robots

Perhaps the most common research platform for mobile robotics is the differential-drive ground robot. Because these robots are not holonomic like point-mass particles, the equations of motion of the artificial physics must be translated into vehicle commands involving speed and turn rate. There are several ways to accomplish this, but all effectively treat the portion of the force along the tangential vector as a speed command $u_t$ and in the normal direction as a turn rate $\Omega$. A friction factor $\mu$ was added to help damp out oscillations. A simple expression of this is shown in Eqs. 13.7–13.8, where $m_p$ is the mass of the particle, $\hat{t}$ denotes the tangential direction, and $\hat{n}$ the normal direction along the axles of the robot.

$$u_{t,\text{new}} = \left( u_{t,\text{old}} + \frac{\sum \boldsymbol{F} \cdot \hat{\boldsymbol{t}}}{m_p} \right) \mu \,. \tag{13.7}$$

$$\Omega_{\text{new}} = \left( \Omega_{\text{old}} + \frac{\sum \boldsymbol{F} \cdot \hat{\boldsymbol{n}}}{m_p} \right) \mu \,. \tag{13.8}$$

This arrangement produced a command space that was similar to the reachable space of the vehicle, as shown in Fig. 13.6. It also meant that, unlike the particle case, a force along the robot's normal direction resulted in zero instantaneous change in position. This had significant consequences for wheeled vehicles attempting to move in formation by introducing lag into their reaction times that was not accounted for in the design of their guidance laws.

In addition, the controller's update rate of 10 Hz and the limited speed of the wheels required that the sharp change from attraction to repulsion be suppressed to limit its impact on the vehicle's drivetrain. A "damping" term, $\eta$, was computed to create a smoother boundary between the two. An example of this from [263] is shown in Eqs. 13.9–13.10.

$$\eta = \left( \frac{\|(D - r_{ij})\|}{D} \right)^{p_d} \quad \text{if } r_{ij} < 2D \,. \tag{13.9}$$

$$\boldsymbol{F}_{\text{used}} = \boldsymbol{F}\eta \,. \tag{13.10}$$

Figure 13.8 shows an example of the modified force versus distance plotted as in Fig. 13.3. This rule spreads out the "low-energy" portion of the robots' force fields, which caused them to slow down as they approached the AR boundary and gave the outer-loop controller time to detect that the net force on the robot had fallen below a preset threshold, at which point the robot would stop moving. Together, these heuristics allowed the robots to settle into fixed lattice formations that resembled simulation results.

However, there are two substantial drawbacks to the point-mass approach. The first is that it ignores vehicle orientation, which is a significant problem in cases where there is a need to point a sensor, such as a camera or laser rangefinder, at a target. As a result, one of the primary benefits of the physicomimetics framework, specifically, the ability to make *local* navigation decisions, is compromised by the need for an external system to provide information about the environment.

In addition, this inability to control the orientation can lead to situations that make formation control impossible. For example, three robots in an equilateral triangle can prevent each other from responding to forces that affect only one of them. This formation is known as the "triangle of death" because when encountered, it prevents the robots from achieving any other objective without manual intervention. In this situation, all three robots are positioned around an equilateral triangle with their axles facing the center of the formation; see Fig. 13.9.

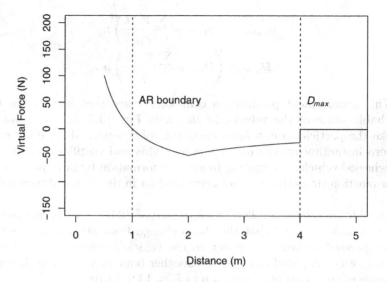

Fig. 13.8: Plot of the force versus distance for the attraction laws used in [263]

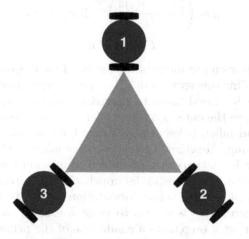

Fig. 13.9: Physical robot arrangement in the triangle of death

Consider the force field seen by robot number 3 in Figs. 13.10 and 13.11. There is almost no force in the direction the robot can move, but it is surrounded by forces that will turn the robot back toward facing its zero-movement orientation. If the applied force has a large component normal to its wheel (which is typically true when trying to attract the whole formation to a point), then robot 3 will rotate slightly. If its estimate of its position in space happens to coincide with the center of robot rotation, then the robot will simply rotate until it faces in the direction of the attractor and then roll in that direction, dragging its neighbors along. However, if the rotation causes the point location to leave the local minimum and instead encounter an opposing turning force, then the robot may stop in the new orientation or, worse, enter a stable oscillation of minimum speed rotations and never generate enough virtual force in a tangential direction to move.

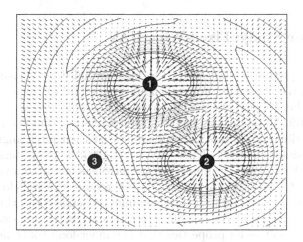

Fig. 13.10: Triangle of death force field from the perspective of robot 3

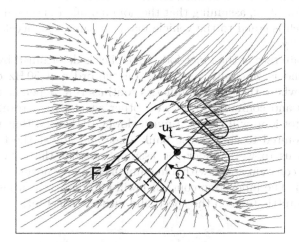

Fig. 13.11: Close-up of the triangle of death force field at robot 3's position with an additional force applied parallel to the axle

In summary, the point mass agent is not an appropriate model for a differential-drive ground robot due to the ambiguity of the turning logic, inability to define orientation of the robot and its mounted sensors, and pathological cases in which formation management becomes impossible. The nonlinear constraints on the vehicle motion need to be accounted for in the design of the guidance system.

## 13.2.3 Case Study: Fixed-Wing UAV

One of the platforms that offers the most promise for sensor-network deployment is the fixed-wing unmanned air vehicle (UAV). These vehicles use a small propeller to generate enough thrust to maintain airflow over their wings to provide lift. They turn by banking their wings to point their lift vector in the desired direction of motion, and so increase their pitch angle and/or airspeed to maintain altitude. There is a substantial body of literature devoted to proving that relatively simple linear controllers can keep a vehicle upright and able to follow trajectories that do not require the vehicle to exceed its linear behavior [244]. From a physicomimetics perspective, this means that the reachable space of the vehicle is not defined by its physical properties, but by the subset of its properties that the inner-loop controller allows. For a typical linear controller, this means that the forward speed $u$ will be held constant or within a small range, and the maximum allowable bank angle $\Phi_{max}$ will be limited to approximately $20°$. These factors allow us to define the set of achievable commands, starting with the normal acceleration component in Eq. 13.12, assuming that the nominal lift, $L$, is equal to the vehicle weight, which is equal to the vehicle mass, $m$, times the acceleration due to earth's gravity, $g$.

The update rate of the inner-loop controller is determined by the vehicle's dynamics and is generally very fast—on the order of $50\,\mathrm{Hz}$ or more. The outer loop, where the physicomimetics routines run, depends on more data-rich sensors and communication with other agents. As a result, we assume that the outer loop will run at a much slower $10\,\mathrm{Hz}$, and that we can ignore the time required to change the bank angle and adjust the forward speed. This allows us to work under the assumption that the UAV is a particle traveling at constant speed with a maximum change in normal velocity shown in Eq. 13.13. The inner-loop controller will automatically adjust the aircraft pitch and throttle to keep the speed and altitude constant.

$$F_n = L \sin(\Phi) \,. \tag{13.11}$$

$$a_n = \frac{L}{m} \sin(\Phi) = g \sin(\Phi) \,. \tag{13.12}$$

$$\Delta v = \begin{bmatrix} 0 \\ \Delta t \, g \sin(\Phi) \end{bmatrix} \,. \tag{13.13}$$

$$\Omega_{\max} = \frac{g \sin(\Phi_{\max})}{2v} \,. \tag{13.14}$$

Consider, for example, a small fixed-wing UAV with an inner-loop controller that holds its speed at 13 m/s and has a maximum allowable bank angle of 20°. The reachable space of this system consists of a short horizontal line segment as shown in Fig. 13.12. Unlike a wheeled robot, it cannot stop, and unlike a virtual particle it cannot instantly change direction. This is important because the two ways that agents can reduce their average speed, or "lose energy," and establish lattice-type formations are not available to fixed-wing UAVs. As a result, point-mass physicomimetics *cannot be used for fixed-wing UAV formation control.*

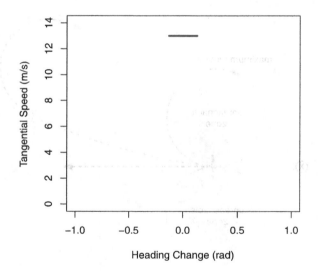

Fig. 13.12: Plot of the reachable space of a small, fixed-wing UAV with a linear inner-loop controller

This constant speed behavior combined with an attractive force law can make it impossible to do waypoint navigation as well. The operator's objective in most physicomimetics designs is to drive the agents towards the AR boundary of a nearby particle of interest and, for robots that can stop as they reorient themselves, this behavior is guaranteed even in the presence of significant motion constraints [58]. Depending on the position of the attractor, the aircraft's maximum speed and turn rate, the aircraft will either exit the attractive radius or enter an orbit around the attractor at a radius determined by its forward speed, not the AR boundary. The reason for this lies in the fact that the only influence an attractor has on this type of agent is to force it to turn in the direction of the attractor. In other words, only Eq. 13.8 applies.

This will cause a UAV to do one of three things based on the relative position and orientation of the aircraft and attractor. If the attractor is directly behind the UAV, for example, located at the black ⊗ in Fig. 13.13, the UAV will fly away from the attractor, because the normal component of the force is zero. If the attractor is outside of the two unreachable zones, shaded red in Fig. 13.13, but not directly behind the UAV, the UAV will eventually overfly it in a continuous loop, as indicated by the blue ⊗ and corresponding blue trajectory. Otherwise, if the particle is located within an unreachable zone, as is the red ⊗, the UAV will fly in a continuous orbit that gets as close as possible to the attractor, but never reach it.

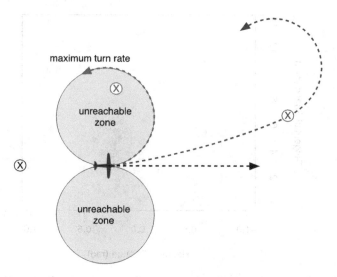

Fig. 13.13: Possible flight paths of a single UAV attracted to a single attractor particle in various positions indicated by ⊗

## 13.2.4 Summary

The advantages of physicomimetics as a guidance algorithm are its intuitive human interface, simple and distributed trajectory generation, and physics-based analyzability. Particle-based rules, however, do not provide adequate control of physical robots. The challenge for guidance system designers, therefore, is to find a way to go beyond simple particle models without losing the benefits of physicomimetics.

This approach can take several forms, but each form must account for all four of the factors discussed in Sect. 13.2.1. In order to deal with the outer-loop controllers' update rate, the formulation must be smooth enough numerically that it avoids jitter. Constraints on vehicle heading rate require the guidance law to directly account for vehicle heading, instead of just its position. Maximum speed and acceleration place restrictions on how quickly the agent can respond to changes in the forces applied to it. Minimum speed constraints mean the designer must be wary of regions the agents cannot reach directly and "potential wells" that can wreck formations.

## 13.3 Beyond Point-Mass Particles

The challenge of extending the physicomimetics framework to agents with constrained motion requires prioritizing which system features to preserve. From the operator's perspective, the most important is that it provides an intuitive approach to waypoint navigation and formation generation. For the mission designer, it is important to be able to design, instead of just tune, formation shapes to achieve objectives such as the placement of sensors. Less important is the computational simplicity offered by the point-mass agent representation, as long as the reactive local-knowledge based approach that enables distributed control and scalability is preserved.

This section defines how to extend physicomimetics to real robots while preserving its most beneficial properties and respecting the motion constraints of the agents. At the fundamental level, two changes are required to make this work. First, vehicle heading must be incorporated into a physics-based guidance rule using torques instead of forces, allowing direct control of the orientation and direction of the vehicle. Second, it is important to separate the concepts of "particle" and "agent."

## 13.3.1 Incorporating Vehicle Heading

In the UAV case study in the previous section, we examined the impact of using gravity-based force laws to generate trajectories for a class of vehicles that fly at a constant speed and have a maximum turn rate. We found that this approach does not allow a constant-speed vehicle to overfly a single attractor point if the attractor is in the agent's unreachable zone, and will simply fly away if the attractor is directly behind the vehicle as shown in Fig. 13.13. Since point-mass physicomimetics does not have direct access to the only controllable state of this agent, it is effectively impossible to use it for anything more than approaching attractor points. Also, regardless of factors such as particle mass, applied force and friction, the point-mass approach effectively reduces physicomimetics to being a *bang–bang directional controller*, i.e., one that switches between only two states. This is not always undesirable, as discussed in the UAV case study below, but its inflexibility negates user and mission benefits described above.

Recovering these benefits involves incorporating the heading angle as a third vehicle state. This "heading particle" approach replaces the point mass with an extended body, having both mass and rotational inertia, that lets the user specify the relationship between changes in speed and turn rate. For constant speed agents, this approach gives direct access to the controllable state. In addition, it allows a new kind of force law based on virtual torsion springs between agents as shown in Fig. 13.14.

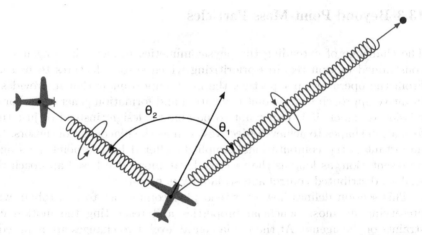

Fig. 13.14: Schematic view of the physicomimetics heading agent with applied torques

The torque law can take two forms, an attract/repel pair used for waypoint navigation and site monitoring as shown in Fig. 13.14, and a co-heading law that is useful for formation management. An attractor point exerts a virtual torque of $-\kappa\theta$ as shown in Eq. 13.15, drawing the agent towards itself, while a repulser point applies a torque of $\kappa\theta$, where $\kappa$ is the virtual spring constant. The co-heading law applies a torque based on the neighbor's or a specified compass heading, if the agents have access to a global reference such as a magnetic compass, and allows formation management similar to the "boids" swarming rules within physicomimetics.

$$\tau_{ij} = \begin{cases} -\kappa\theta_{ij} & \text{if } r_{ij} \leq D_{\max} , \\ 0 & \text{otherwise} . \end{cases} \tag{13.15}$$

Just as the sum of the forces determined the change in commanded velocity in Eqs. 13.2–13.3, the sum of the applied torques can be used to determine the turn rate of the vehicle. Specifically, each agent $i$ applies a torque to agent $j$, and the results are summed according to Eq. 13.16. These are used to compute a change in turn rate, given an artificial mass moment of inertia, $I$, using Eq. 13.17. The new turn rate is simply the sum of the current turn rate and the torque-induced change, as shown in Eq. 13.18.

$$\tau = \sum_i \tau_{ij} . \tag{13.16}$$

$$\Delta\Omega = \frac{\tau}{I}\Delta t . \tag{13.17}$$

$$\Omega_{\text{new}} = \Omega_{\text{old}} + \Delta\Omega . \tag{13.18}$$

## 13.4 Multi-particle Agents

The ability to control both orientation and position requires a combination of means to apply forces and torques to an agent. A simple means to accomplish this is to use the dumbbell agent shown in Fig. 13.15. It consists of two particles, $p_{\text{front}}$ and $p_{\text{back}}$, at fixed distances from the vehicle's center of rotation. The vehicle's orientation defines the front particle's heading angle, allowing it to be controlled by torque laws in addition to conventional force laws.

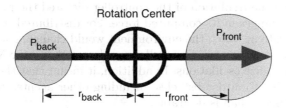

Fig. 13.15: Diagram of the dumbbell agent model with front and back particles

The trajectory of the agent as a whole is determined by the sum of the forces, $F$, and moments, $M$ applied to it as shown in Eqs. 13.19 and 13.20.

The next speed and velocity commands are determined using Eqs. 13.21 and 13.22. This formulation explicitly separates the friction related to turning, $\mu_\Omega$, from the forward speed, $\mu_{u_t}$, providing the user with a pair of "knobs" to stabilize the agents' behavior internally.

$$F_{\text{agent}} = \sum Fp_f + \sum F_{p_b} \,. \tag{13.19}$$

$$M_{\text{agent}} = T_{\text{applied}} + r_{p_f} \times \sum Fp_f + r_{p_b} \times \sum Fp_b \,. \tag{13.20}$$

$$u_{\text{new}} = \left( u_{\text{old}} + \frac{\sum F \cdot \hat{t}}{m} \right) (1 - \mu_{u_t}) \,. \tag{13.21}$$

$$\Omega_{\text{new}} = \left( \Omega_{\text{old}} + M I^{-1} \right) (1 - \mu_\Omega) \,. \tag{13.22}$$

However, this "more realistic" agent alone is not enough to guarantee that physical swarms controlled via physicomimetics will be able to accomplish the designer's goals. The four criteria mentioned above still must be taken into account in the design of specific vehicle controllers. Case studies for wheeled robots and small, fixed-wing UAVs are presented below.

### 13.4.1 Case Study: Wheeled Robot

This section provides a description of the physicomimetics-based controller for a Pioneer 3-AT robot whose reachable space is shown in Fig. 13.7. The following paragraphs describe how the authors systematically approached each of the four factors defining the difference between simulation and real robots.

#### 13.4.1.1 Outer-Loop Time Resolution

At 10 Hz, the time resolution of the controller dictated the responsiveness of the robot to changes in its computed force. For distributed control, this also dictated the rate at which the environment would change, guaranteeing that the interactive portion of the controller could not achieve the Nyquist criteria required to stabilize oscillations. In addition, it meant that the vehicle would continue moving towards obstacles, including other agents, up to one tenth of a second longer than is desirable.

Removing oscillations requires some kind of damping; however, too much damping impairs reactivity. The easiest way to get around this is to have the system "learn" when it is oscillating and increase the appropriate friction coefficient. We implemented this by using a pair of integrators to compute $\mu_{u_t}$ and $\mu_\Omega$. When the sign of the command changed, the effect of the friction grew. When the sign was constant or the command was zero, the effect of friction on the appropriate command was reduced according to Eq. 13.23. For the Pioneer 3-AT, the friction growth rate to reduction rate ratio ($k_{\text{more}}/k_{\text{less}}$) must be greater than 10.0 to stop oscillations between mutually attracted robots.

$$\mu_{\text{new}} = \mu_{\text{old}} + \begin{cases} k_{\text{more}}(1 - \mu_{\text{old}}) & \text{if } u_{\text{new}} * u_{\text{old}} < 0 \text{ or} \\ -k_{\text{less}}\mu_{\text{old}} & \text{if } u_{\text{new}} * u_{\text{old}} \geq 0 . \end{cases} \tag{13.23}$$

### 13.4.1.2 Changing Heading

The dumbbell agent allows the designer to explicitly specify the relationship between the effect of normal and tangential forces on the vehicle's orientation and direction of motion. For three to five Pioneer 3-ATs forming an oriented ring around a single attractor point with their front-mounted cameras pointed inward, the swarm's performance is best if they arrive at their correct points in the proper orientation. This is because fine heading changes are actually easier while the robot is moving (in the shaded part of Fig. 13.7) than when it is stopped and must deal with the minimum speed gap.

As a result, the ratio between virtual mass and moment of inertia in Eqs. 13.21 and 13.22 should favor allowing the vehicle to move in its tangential direction relatively quickly. For the Pioneer 3-AT, a ratio of 1:1 gave good performance. With these parameters, not only do the agents generally arrive quickly and in the proper orientation, but if they are attracted to a point directly behind them, they will automatically execute a three-point turn, moving the bulk of the robot to keep the specified point as close as possible to the shortest path to the objective.

### 13.4.1.3 Maximum Speed and Acceleration

The maximum speed of a physicomimetic agent is the most easily described and handled parameter in any implementation, because in the most extreme cases the agent velocity commands simply reduce to the direction of the local net force. As Ellis and Wiegand describe, this does not inhibit physicomimetics guidance for lattice formations [58]. Additionally, the maximum speed of the Pioneer 3-AT is faster than most operators are willing to drive a ground robot in a cluttered or indoor environment. However, as with the previous example, $u_{\text{new}}$ and $\Omega_{\text{new}}$ were always capped by $u_{\text{max}} = 0.5\,\text{m/s}$ and $\Omega_{\text{max}} = 0.4\,\text{rad/s}$.

Acceleration, however, was a serious problem for obstacle avoidance. No matter how great the force exerted by a particle's repulsion field, a finite number of time steps was required to slow the vehicle from $u_{\text{max}}$. To account for this, we used an adaptive approach similar to the approach we used to dampen oscillations—adjusting the maximum speed we used to clamp the commanded speed by having it grow up to 0.5 m/s when there were no objects nearby, and shrink to 0.1 m/s as more objects were detected within 1 m. This was implemented using Eq. 13.24 for each of the laser or sonar range measurements, $r_m$. A ratio of $k_{less}/k_{more}$ greater than 10.0 prevented collisions between the Pioneers and obstacles, while still allowing them to move quickly in open areas.

$$u_{max,new} = u_{max,old} + \begin{cases} k_{more}(.5 - u_{max,old}) & \text{if } r_m > 1.0\,\text{m} \text{ or} \\ -k_{less}(u_{max,old} - .1) & \text{if } r_m \leq 1.0\,\text{m} . \end{cases} \qquad (13.24)$$

### 13.4.1.4 Minimum Speed and Acceleration

The adaptive approach to friction removed most of the low-speed jitter between interacting agents, but certain cases could arise, especially when their position estimation was poor, when there could be long-term oscillatory behaviors at minimum speed. A "static friction" check on the applied forces and moments, agent mass, and acceleration due to gravity, $g$, as shown in Eq. 13.25 for the tangential speed prevented this, as it guaranteed that when the vehicle was oscillating strongly, a significantly greater force would be required to cause a change in position. By using a damping factor on the force laws as described in Eq. 13.23, we were able to guarantee that a large applied force meant that the environment had changed enough that moving the robots was the correct action. This rule generally allowed the Pioneer 3-ATs to settle into fixed positions in two or three passes across their mutual AR boundaries, and remain in place until a new force was applied.

$$u_{new} = \begin{cases} \left(u_{old} + \frac{\sum F \cdot \hat{t}}{m}\right)(1 - \mu_{u_t}) & \text{if } \mu_{u_t} mg < \sum F \cdot \hat{t} m , \\ 0.0 & \text{otherwise} . \end{cases} \qquad (13.25)$$

## 13.4.2 Case Study: Fixed-Wing UAV

For this study, we wanted the agents to pass directly over randomly placed targets in a specified area, the central rectangles in Fig. 13.16, simulating a surveillance mission. Achieving this goal using physicomimetics requires two types of particles: a fixed target particle and a mobile agent particle. The target is assumed to be an attractor with no AR boundary, so it always attracts agents using either an inverse square or torque law. The agents repel each other using the same type of law as the targets. This allowed us to directly compare the behavior of point-mass and extended-body agents.

The classic point-mass approach results in disorganized motion about the attractors, as shown in the top plot in Fig. 13.16. The agents in this swarm are almost always turning at their maximum turn rate to try and align themselves with a constantly changing force vector. As a result, they spread out and cover a broader area than required by the mission.

If the agent uses the torque law described above with spring constants set so that no single particle pair interaction will cause a command that exceeds the maximum turn rate, we see in the bottom plot of Fig. 13.16 that the swarm quickly settles into a series of overlapping loops that pass over the target area once per orbit. This approach substantially increases the time-over-target for the agents and reduces the amount of turning, and thus energy consumption, required by the swarm.

## 13.5 Conclusions

Physicomimetics is an effective means of distributed, agent-based swarm guidance if the guidance rule and mission designers properly account for the physical constraints of their vehicles. These can be defined in terms of four major criteria that separate real robots from simulated point-mass particles. First, physical robots are controlled by digital hardware with a finite up-date rate that is generally slower than the dynamics of the vehicle and not necessarily synchronized with the other agents. Second, an extended body cannot instantly change its orientation or direction of motion the way a point-mass particle can. Third, the acceleration and velocity of physical agents is bounded. Finally, the minimum speed and acceleration of the vehicle, not just the force laws, determine what formation types can be achieved and with what precision.

Early efforts to use physicomimetics for wheeled robot navigation showed that large, discontinuous changes in the sign of an applied force resulted in poor swarm shape formation. A partial solution to this problem involved the use of a damping factor to drive the force at the AR boundary to zero continuously, rather than abruptly. This change commands the agents to slow down as they approach desirable positions in a formation, and allows them to use a minimum force threshold to determine when to stop.

However, realizing the full benefits of physicomimetics for swarm guidance on physically constrained robots is more complicated. Guidance laws must directly account for the nonholonomic nature of the agents they control or they will not give commands that the inner-loop controller can follow. In certain cases, this can result in the agents getting "stuck" in unexpected places or even activating unstable modes such as the Dutch Role mode of a fixed-wing aircraft. As a result, using physicomimetics on motion-constrained vehicles requires that the agent model itself and that the agent's controller incorporate more knowledge of the real behaviors of the physical agent under

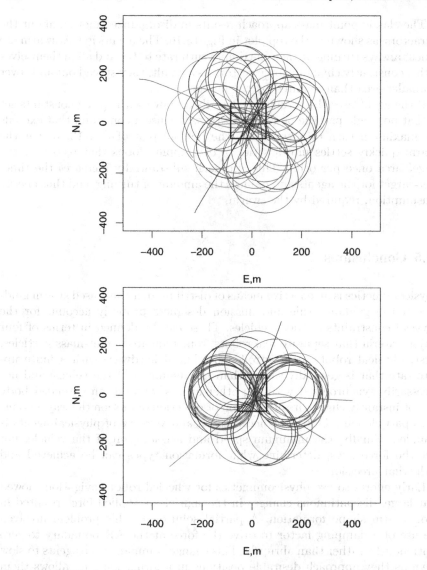

Fig. 13.16: Flight-path trace of five planes seeking 10 targets from a single simulation using a gravity law (upper) and torque law (lower). Target extents are shown as small central rectangles

control. To do this, the guidance law designer should follow the procedure outlined below:

1. Develop a dynamic model of the vehicle inner-loop controller system.
2. Determine the "reachable space" of the vehicle in terms of motion commands available to the vehicle.

   • These will generally be tangential speed and heading angle change.

3. Develop an agent model that includes "heading particles" placed and oriented appropriately for the vehicle and missions.

   • A vehicle that can only turn requires only a single "heading particle."
   • A vehicle that must go forward and backward requires a particle in front of and behind its center of rotation.

4. Use a friction model to stabilize each individual agent's motion and prevent undesired behavior.

   • Friction has little impact on formation shape, but strongly affects oscillatory and obstacle avoidance behaviors.
   • It can be programmed to tune itself if there is a rigorous description of the undesirable behavior.

Unfortunately, discarding the point-mass agent with its holonomic behavior makes it difficult to use some of the tools designed for physicomimetic swarm analysis. Real agents will exhibit substantially more complicated behaviors, and will therefore require more sophisticated agent models such as those used in molecular dynamics simulations, which can be used to study complicated state transitions of materials like water. Keeping this complexity growth in check requires the guidance system designer to have a clear idea of what motions are required for the intended missions, and to design the agent models and force laws to maximize the likelihood that the agents will follow them.

**Acknowledgements** This work was performed under Office of Naval Research Work Order N0001409WX30013.

# Part IV
# Prediction, Adaptation, and Swarm Engineering

# Chapter 14
# Adaptive Learning by Robot Swarms in Unfamiliar Environments

Suranga Hettiarachchi

## 14.1 Introduction

Although the goal of this book is to design swarms methodically, it is not always possible to theoretically determine the values of all parameter settings a priori. In this situation, the behavior of the swarm can be improved via optimization of the parameters, as we did in Chap. 12. This chapter assumes that optimization is done with an evolutionary algorithm (EA). Traditionally, the evolution of swarm behaviors occurs *offline* (i.e., in simulation prior to actual task execution), and the swarm must be retrained whenever environmental conditions change. However *online* adaptation (i.e., during the process of task execution) is much more practical. Offline approaches are unrealistic because, in general, the environment cannot be anticipated a priori due to numerous factors such as sensor noise, environmental dynamics, obstructed perception (i.e., partial observability), and a lack of global knowledge. Furthermore, time-intensive retraining is unreasonable for many time-constrained tasks. Retraining to adapt to changes is especially problematic when the environment changes often. In this chapter, we demonstrate the utility of an online swarm adaptation technique, called DAEDALUS, that is combined with physicomimetics. The task addressed here is that of a swarm of robots that needs to learn to avoid obstacles and get to a goal location. Online learning facilitates flexible and efficient task accomplishment in dynamic environments. Here, "learning" is performed both for adaptation to the unexpected and for effective performance on the task. By employing our DAEDALUS online learning approach in conjunction with physicomimetics, swarms exhibit fault-tolerance, adaptability and survivability in unfamiliar

Suranga Hettiarachchi
Indiana University Southeast, 4201 Grant Line Road, New Albany, IN 47150, USA, e-mail:
suhettia@ius.edu

W.M. Spears, D.F. Spears (eds.), *Physicomimetics*,
DOI 10.1007/978-3-642-22804-9_14,

environments. The success of our approach is greater than that of previous approaches that have been applied to this task.

## 14.2 Adaptive Swarms

The accomplishment of complex robotic tasks has traditionally involved expensive, preprogrammed robots. These robots are typically trained and tested in highly structured environments, under predefined and static conditions. This approach is unsatisfactory for many tasks, for at least two important reasons. First, as the requirements of the task increase in complexity, the robot's ability to adapt accordingly diminishes. Second, when the robots are exposed to unfamiliar environments with conditions for which they were not trained, the robots may get stuck (e.g., in cul-de-sacs), thereby resulting in task failure.

In traditional approaches such as remotely controlled robots, the robot is entirely dependent upon the human operator.[1] The performance feedback from human to robot is vital, and any delay or perturbation of this feedback loop could jeopardize the mission. Without constant involvement of the human observer, the robots may be incapable of performing the task; consequently, removing the human (e.g., a global observer or operator) could be fatal to the success of the mission. The lack of autonomy in these robots is problematic for many tasks, especially those for which constant human supervision is either unavailable or too expensive. The autonomous swarm robotics approach is an improvement over these traditional robotic approaches because it combines effectiveness with robustness. Robot swarms are highly effective because they can perform tasks (such as regional coverage) that one expensive robot cannot. They are robust in the sense that if some robots fail, the swarm can still achieve the task. Furthermore, *adaptive* swarms are even more desirable for accomplishing tasks, since they can provide robustness and fault tolerance even when the swarm operates in unfamiliar environments.

The specific objective of this chapter is to address the task of swarm navigation through a field of obstacles to get to a goal location while maintaining a cohesive formation under difficult conditions, *even if the environment changes*. We provide an analysis of how well we have achieved this objective within the context of adaptive learning of *heterogeneous* swarms in offline and online situations. The swarms are called "heterogeneous" because not all individuals use the same control laws. Our swarms are controlled by physicomimetics using the Lennard-Jones force law (see Chap. 3). We introduce our DAEDALUS paradigm for online learning by heterogeneous swarms, and provide an analysis of swarm behavior in unfamiliar environments. Our long-term

---

[1] In fact, it is debatable whether a remotely controlled device or vehicle should even be called a "robot." The focus of this chapter, as well as this entire book, is about autonomous vehicles only—not remotely controlled ones—because the former are clearly robots.

goal is to present results that will contribute extensively to the understanding of general swarm robotics issues for dynamic, real-world environments.

## 14.3 Lennard-Jones Force Law

We derive a generalized Lennard-Jones force law from the Lennard-Jones potential function invented in [136]. The Lennard-Jones potential function was first proposed by John Lennard-Jones in 1924. This potential function models two distinct forces between neutral molecules and atoms. These forces are based on the distances between molecules. At long ranges the attractive force makes the molecules move closer, and at short ranges the repulsive force makes the molecules move apart; this results in the molecules maintaining a natural balance. The Lennard-Jones potential function can be given by the expression:

$$LJP_r = 4\varepsilon \left[ \left( \frac{\sigma}{r} \right)^{12} - \left( \frac{\sigma}{r} \right)^{6} \right], \tag{14.1}$$

where $r$ is the distance between particles, $\varepsilon$ is the depth of the *potential well*, and $\sigma$ is the distance at which the potential is zero. The "potential well" is where the potential energy is minimized. The bottom of the potential well occurs at $r = 2^{1/6}\sigma$.

Figure 14.1 shows the potential function when $\varepsilon = 1$ and $\sigma = 1$. When the separation distance $r > 2^{1/6}\sigma$, the two molecules are attracted to each other. When $r < 2^{1/6}\sigma$, they are repelled. The Lennard-Jones potential function is ideal for modeling inter-robot interactions, as well as interactions between robots and their environment.

The Lennard-Jones force function can be derived from the potential function by taking the negative derivative of the potential, i.e.,

$$F = - \left( \frac{\mathrm{d}\,(LJP_r)}{\mathrm{d}r} \right), \tag{14.2}$$

and the force between robots $i$ and $j$ can be expressed as

$$F_{ij} = -4\varepsilon \left[ \frac{-12\sigma^{12}}{r^{13}} + \frac{6\sigma^{6}}{r^{7}} \right], \tag{14.3}$$

where $D = 2^{1/6}\sigma$ is the desired distance between two robots. We derive (and generalize) the force function for the interaction between two robots as

$$F_{ij} = 24\varepsilon \left[ \frac{d\sigma^{12}}{r^{13}} - \frac{c\sigma^{6}}{r^{7}} \right]. \tag{14.4}$$

Following the precedents established in this book, $F \leq F_{\max}$ is the magnitude of the force between two robots, and $r$ is the actual distance between

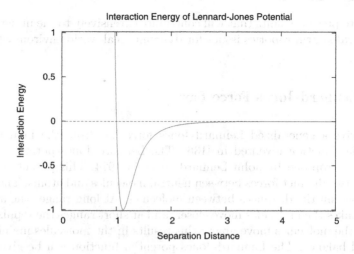

Fig. 14.1: Interaction potential of Lennard-Jones with $\varepsilon = 1$ and $\sigma = 1$

the two robots. The variable $\varepsilon$ affects the strength of the force, while $c$ and $d$ control the relative balance between the attractive and repulsive components. In order to achieve optimal behavior, the values of $\varepsilon$, $c$, $d$ and $F_{\max}$ must be determined, as well as the amount of friction. Our motivation for selecting the Lennard-Jones force law is that (depending on the parameter settings) it can easily model crystalline solid formations, liquids, or gases.

## 14.4 Swarm Learning: Offline

Given generalized force laws, such as the Lennard-Jones force law that we discussed in Sect. 14.3, it is necessary to optimize the force law parameters to achieve the best performance. We accomplish this task using an evolutionary algorithm (EA). In particular, we develop an evolutionary learning (EL) system for a swarm of robots learning the obstacle avoidance task [234]. Our intention is to have the robots learn the optimal parameter settings for the force laws (rules); thus they can successfully maintain a formation while both avoiding obstacles and getting to a goal location in a complex obstacle-laden environment.

An EA is a population-based stochastic optimization algorithm inspired by natural evolution. In this chapter, we assume that the members (individuals) of the population are candidate parameter values for three Lennard-Jones force laws. One force law describes the robot–robot interaction. Another describes the robot–obstacle interaction. Finally, a third describes the robot–goal interaction. In particular, every individual *chromosome* in the population is a vector of values for the real-valued parameters, and it represents one instantiation of the three force laws. The EA initializes the starting population with randomly chosen force law parameter values. Then it applies *genetic operators*, in particular, *mutation* and *recombination* (described below) to modify these candidate solutions (individuals) based on their performance in an environment. Performance evaluation results in giving a *fitness* value to each individual that indicates its performance score (i.e., its level of success at the task). Finally, the EA creates a population of offspring from the parent population of candidate solutions. This process repeats until a desired level of fitness is reached. An overview of EAs can be found in [234].

Choosing an efficient representation for an individual is one of the important EA design decisions. There are many ways to represent an individual, and the choice of representation is strongly dependent on the problem that we need to solve. When choosing a representation we should bear in mind that this decision determines how the fitness of the individual will be evaluated, and what type of genetic operators should be used to create a new population for the next generation from the current population. The creation of the next generation is done with a process called *selection*. There are three commonly used representations for individuals: bit strings, integers, and real values.

The Lennard-Jones force law parameters are represented as a list of real values, restricted by upper and lower bounds (see Table 14.1). Each force law contains the following parameters: the interaction strength, a non-negative attraction constant, a non-negative repulsion constant and a force maximum. In addition to these four parameters, we evolved the friction of the system. The parameters we optimized are

- $\varepsilon_r$ – strength of the robot–robot interactions,
- $c_r$ – non-negative attractive robot–robot parameter,
- $d_r$ – non-negative repulsive robot–robot parameter,
- $F_{\max_r}$ – maximum force of robot–robot interactions,
- $\varepsilon_o$ – strength of the robot–obstacle interactions,
- $c_o$ – non-negative attractive robot–obstacle parameter,
- $d_o$ – non-negative repulsive robot–obstacle parameter,
- $F_{\max_o}$ – maximum force of robot–obstacle interactions,
- $\varepsilon_g$ – strength of the robot–goal interactions,
- $c_g$ – non-negative attractive robot–goal parameter,
- $d_g$ – non-negative repulsive robot–goal parameter,
- $F_{\max_g}$ – maximum force of robot–goal interactions and
- $Fr$ – friction in the system.

EAs use genetic operators to create offspring from one or more individuals in the current population. The two main genetic operators are mutation and recombination. Based on the type of EL approach, the use of these operators may vary. Genetic operators (and especially mutation) help to maintain the genetic variation of a population, thus avoiding *premature convergence* (i.e., a lack of diversity leading to stagnation) of that population.

Since we represent force law parameters with real values, we use *Gaussian mutation* as one of the genetic operators in our EA. Gaussian mutation is standard for real-valued representations. This method replaces the value of a *gene* (i.e., one of the values within an individual) with the sum of the current gene value and a perturbation drawn randomly from a Gaussian distribution with mean 0.0 and a predetermined standard deviation (see Table 14.1).

| Parent | 1.3 | 4.1 | **2.4** | 4.7 | 0.2 | 2.2 | **1.5** | 4.1 | 2.6 |
|---|---|---|---|---|---|---|---|---|---|
| Gaussian Step | 1.3 | 4.1 | **2.4+0.7** | 4.7 | 0.2 | 2.2 | **1.5-0.3** | 4.1 | 2.6 |
| Offspring | 1.3 | 4.1 | **3.1** | 4.7 | 0.2 | 2.2 | **1.2** | 4.1 | 2.6 |

Table 14.1: Before and after Gaussian mutation of a real-valued representation, where two genes are modified

The other widely used genetic operator is recombination, also referred to as "crossover." Recombination creates one or more offspring by merging genes from one or more parents. The most popular recombination methods are *one-point, multi-point* and *uniform* recombination. Our EA uses one-point crossover to produce offspring [233].

Selection determines the survival of individuals from generation to generation. It is a process of selecting which individuals will get to be parents for the next generation. Our EA uses *fitness proportional selection*. In fitness proportional selection, individuals are given a probability of being selected that is directly proportional to their fitness. Our prior observations have shown that fitness proportional selection produces very desirable behavior for the obstacle avoidance task. This may be due to the fact that selection pressure depends on our evaluation (fitness) function and the population dynamics, as shown by Sarma [206].

We carefully designed a fitness function to evaluate the force law individuals. This fitness function captures our three objectives, which correspond to the three subtasks that must be accomplished in the context of the overall navigation task. The objectives are as follows:

- **Avoid obstacles:** The robots are capable of sensing the virtual repulsive forces of obstacles from a distance of $R_o+20$ from the center of the obstacle. $R_o$ is the radius of an obstacle, which is 10. The radius of the robot is one. If the distance between the center of the robot and the center of the obstacle is less than $R_o+1$, a collision will occur. Such collisions should be avoided.
- **Maintain formation:** The robots are expected to form a hexagonal lattice formation and maintain this formation while navigating. A robot is attracted to its neighbors if the neighbors are farther than $D$ away and

is repelled by its neighbors if the neighbors are closer than the desired distance $D$.

- **Reach a goal:** The robots should get to a desired goal, while maintaining their formation and avoiding collisions with obstacles. Robots can sense the global attractive force of the goal at any distance. They are given a limited amount of time to reach the goal. This time limit reduces the practicality of time-intensive offline retraining.

Our fitness function captures all three of the above objectives, and therefore it is called a *multi-objective fitness function*. Because we do not change our objectives during the optimization of Lennard-Jones force laws, we use the same evaluation (fitness) function throughout this project. The biggest challenge is how to develop a compromise between our three objectives, so that the evaluation function is capable of providing us with an optimal solution that is not completely biased toward any one of the objectives to the exclusion of the others.

We introduce a weighted multi-objective fitness function with penalties to evaluate the force law individuals in our population of force laws. The weights reflect the relative importance of the three objectives:[2]

$$fitness = w_1 P_{\text{Collision}} + w_2 P_{\text{NoCohesion}} + w_3 P_{\text{NotReachGoalInPermittedTime}} \cdot$$

The weighted fitness function consists of three components:

- **A penalty for collisions:** For many swarm tasks, avoiding collisions with obstacles in the task environment is important. In our simulation world, there are no safety zones around the obstacles as in [8]. The maximum sensing distance of the repulsive force on a robot from the center of an obstacle is set at $R_o + 20$. A collision occurs if the center of the robot is within the perimeter of an obstacle. We add a penalty to the fitness score if the robots collide with obstacles. All robots in the simulation world are evaluated for collisions, and a penalty is added at each discrete time step within the permitted time interval.
- **A penalty for lack of cohesion:** Maintaining a cohesive formation is another important aspect, especially during tasks such as chemical plume tracing (see Chaps. 8 and 9), terrain mapping (see Chap. 6), pre-assembly alignment (see Chap. 5), snaking through tunnels (see Chap. 12), or target tracking in nonstationary environments (see Chap. 19). With cohesive formations, robots maintain uninterrupted communication paths between one another, thus allowing efficient distribution of resources. The formation to be accomplished here is a hexagonal lattice/grid.

We penalize the robots for a lack of cohesion. The cohesion penalty is derived from the fact that in a good hexagonal lattice formation, the interior robots should have six local neighbors, each at distance $D$. A penalty

---

[2] $w_1 = w_2 = w_3 = 1/3$ works quite well.

occurs if a robot has fewer or more neighbors than six at the desired distance, and the value of the penalty is proportional to the error in the number of neighbors. This fitness pressure prevents the robots from forming tight clusters with overlap; it also prevents fragmentation of the entire formation, which in turn may cause the swarm to form sub-swarms when navigating through obstacles. All of the robots in the simulation world are evaluated for lack of cohesion, and a penalty is added at each discrete time step within the permitted time interval.

- **A penalty for robots not reaching the goal:** We also introduce a penalty to our fitness function for robots not reaching the goal within the permitted time limit. In time critical search-and-rescue missions, or defense-related missions, having the swarm achieve the goal in a limited permissible time interval is extremely important. We added a penalty if less than 80% of the robots from the initial swarm did not reach the goal within the permitted time interval. At the end of the permitted time interval, the EA evaluates the number of robots that reach the goal, and if this number is less than 80%, a penalty is added. A robot has reached the goal if the robot is within a $4D$ radius of the center of the goal.

The fitness function uses positive penalties, and the optimization problem consists of trying to minimize its value. Fitness evaluation occurs at every time step, for every individual, within the permitted time limit.

Our overall EL architecture consists of an environment generator, an EA (i.e., a training module), and a performance evaluator (i.e., a testing module) for a swarm of robots to learn the obstacle avoidance task. The environment generator creates task environments to test the force laws. The environment consists of robots, randomly positioned obstacles, and a goal. Each force law is tested on $n$ different environment instances created by the environment generator.

Our two-dimensional simulation world is $900 \times 700$ units in size, and it contains a goal, obstacles, and robots. Although we can use up to a maximum of 100 robots and 100 static obstacles with one static goal, our current simulation has 40 robots and 90 obstacles in the environment when using the training module. The goal is always placed at a random location on the right side of the world (environment), and the robots are initialized in the bottom left area. The obstacles are randomly distributed throughout the environment, but are kept 50 units away from the initial location of the robots and the goal—to avoid initial proximity collisions. Each circular obstacle has a radius $R_o$ of 10, and the square-shaped goal is $20 \times 20$. When 90 obstacles are placed in the environment, roughly 4.5% of the environment is covered by the obstacles (similarly to [8]). The desired separation $D$ between robots is 50, and the maximum velocity $V_{\max}$ is 20 units/s. Figure 14.2 shows 40 robots navigating through randomly positioned obstacles. The larger circles are obstacles, and the square on the right is the goal. Robots can sense other robots within a distance of $1.5D$, and can sense obstacles within a distance of $R_o + 20$. The goal can be sensed at any distance.

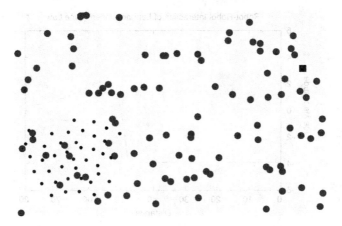

Fig. 14.2: Example performance *before* EA learning

Each robot has a copy of the force law, which it uses to navigate towards the goal while avoiding obstacles. Recall that robots are given a limited amount of time to accomplish the obstacle avoidance task and reach the goal while maintaining the formation. We refer to one complete task execution, where the time limit has been reached, as an *evaluation run*.

The global observer, which exists solely for the purpose of evaluating system performance, uses the fitness function to evaluate the performance of the force law in each particular environment. It assigns a fitness value, $\mathcal{R}_i$, to each individual based on its performance at the end of each evaluation run $i$. The final fitness, $fitness_{\text{ind}}$, of an individual is computed after $n$ evaluation runs have been completed:

$$fitness_{\text{ind}} = \mathcal{C} - \frac{\mathcal{R}_1 + \mathcal{R}_2 + \cdots + \mathcal{R}_n}{n}. \tag{14.5}$$

As stated earlier, this EA uses fitness proportional selection, which maximizes fitness. Since the goal is to minimize the penalties, the penalties are subtracted from a large constant value $\mathcal{C}$, creating a fitness value that is higher when the penalties are smaller. Once the termination criterion of the EA has been met, the EA outputs the optimal parameter setting for the force law that is being optimized. The termination criterion of our EA is the completion of $G$ generations.[3]

Figure 14.3 shows the evolved Lennard-Jones robot–robot force law. The force is repulsive when the distance between robots is less than 50 (i.e., $D$ is set at 50), and it is attractive when the distance is greater than 50. The evolved $F_{\text{max}_r}$ takes effect when the distance between robots is less than 45.

---

[3] Not to be confused with the gravitational constant.

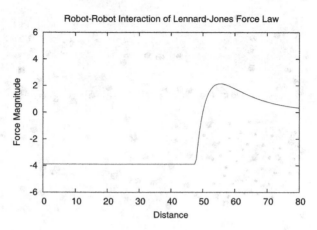

Fig. 14.3: Evolved Lennard-Jones force law for robot–robot interactions

Figure 14.4 shows the evolved Lennard-Jones robot–obstacle force law. The maximum sensing distance of the repulsive force on a robot from the center of an obstacle is set to $R_o + 20$. At a distance of zero, the robot is closest to the obstacle and $F_{\max_o}$ is in effect, which makes the force between a robot and an obstacle highly repulsive. Beyond distance 14, the effect of the repulsive force diminishes as the robot moves away from the obstacle.

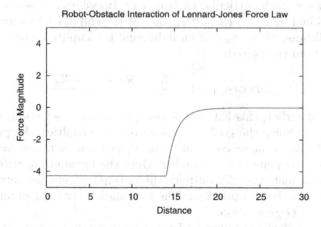

Fig. 14.4: Evolved Lennard-Jones force law for robot–obstacle interactions

Figure 14.5 shows the evolved Lennard-Jones robot–goal force law. A robot can sense the goal force globally. The evolved force law for the robot–goal interaction is constant, regardless of the distance from the robot to the goal.

The robots sense the maximum attractive force of $F_{\max_g}$ from the goal. This evolved robot–goal interaction is capable of avoiding any clustering of the robots once they reach the goal, thereby preserving the swarm lattice formation.

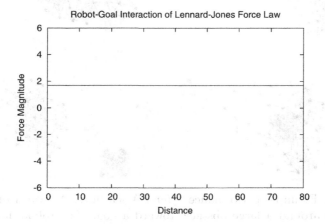

Fig. 14.5: Evolved Lennard-Jones force law for robot–goal interactions

Observation of the system behavior shows that the formation acts like a viscous fluid, rather than a solid. Although the formation is not rigid, it does tend to retain much of the hexagonal structure as it maneuvers around obstacles. Deformations and rotations of portions of this multi-robot fluid are temporary perturbations to the lattice, imposed by the obstacles. It is important to note that the added flexibility of this formation has a significant impact on the swarm behavior. Specifically, the optimized Lennard-Jones force law provides low collision rates, very high goal reachability rates within a reasonable period of time, and high swarm connectivity [92]. Interestingly, Chap. 15 shows that this optimized Lennard-Jones force law also leads to superior swarm predictability.

Figure 14.6 shows a sequence of snapshots of 50 robots navigating around a large obstacle using the optimized Lennard-Jones force law. Robots act as a viscous fluid while avoiding the obstacle. In the first snapshot, robots are in a fully connected sensor network and are navigating towards the goal, but the robots have not yet encountered the obstacle. The second snapshot shows the swarm starting to flow around the obstacle on two fronts while maintaining 100% connectivity. The third snapshot shows the robots on the two fronts merging back together. In the final snapshot, the robots are back in a cohesive formation when they have reached the goal.

We observe that when the swarm reaches the obstacle, it navigates around the obstacle as a viscous fluid while maintaining excellent connectivity and

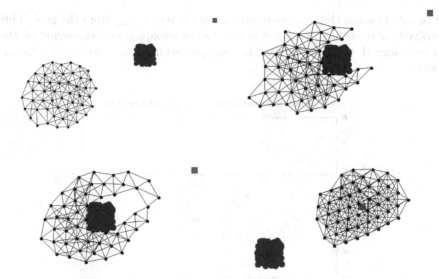

Fig. 14.6: Example performance *after* EA learning. A swarm of 50 robots navigates around a large obstacle toward a goal. The robots maintain full connectivity while avoiding the obstacle by acting as a viscous fluid, using the evolved Lennard-Jones force law

reachability. This fluid type property of the Lennard-Jones force law is an emergent behavior of the swarm.

A summary of our experimental results, averaged over 50 runs, produced by the testing module of the EL system and shown in Table 14.2, indicate that the Lennard-Jones controlled robots learn to perform well in their environment. This is because the emergent behavior of the Lennard-Jones controlled swarm is to act as a viscous fluid, generally retaining good connectivity while allowing for the deformations necessary to smoothly flow through the obstacle field. Despite being trained with only 40 robots, the emergent behavior scales well to larger numbers of robots in offline environments.

These are excellent results for offline learning. Nevertheless, when the swarm is introduced to unfamiliar environments, its success rate is drastically reduced. Performance degrades on all three metrics, especially when we introduce multiple changes between the learned environment and a new unfamiliar environment (e.g., with more obstacles) for testing the swarm. A solution to this problem requires a new adaptive online learning paradigm, due to various reasons that we will discuss in the next few sections.

| | Robots | | | | |
|---|---|---|---|---|---|
| | 20 | 40 | 60 | 80 | 100 |
| Collisions | 0 | 0 | 0 | 2 | 4 |
| Connectivity | 12 | 22 | 36 | 54 | 67 |
| Reachability | 99 | 98 | 98 | 98 | 98 |
| Time to goal | 530 | 570 | 610 | 650 | 690 |

Table 14.2: Summary of results for 20 to 100 robots, with 100 obstacles. Collisions: the number of robots that collided with obstacles. Connectivity: the minimum number of robots that remain connected. Reachability: the percentage of robots reaching the goal. Time to goal: the time for 80% of the robots to reach the goal

## 14.5 The DAEDALUS Paradigm

Unforeseen and changing environmental conditions present challenges for swarms. There are several advantages to utilizing online learning in numerous applications where the swarm's performance could be affected by changing environmental conditions. A swarm that can learn to adapt online does not have to depend on a global controller during its operations, it can compensate for delayed responses by swarm members due to environmental noise, and can often eliminate the problem of having to process delayed rewards. The traditional "go back and learn in simulation" approach to adaptation is not applicable to swarms in frequently varying and unpredictable environments. In response to this liability of offline learners, we have invented a novel paradigm for online swarm adaptation. This paradigm, called "Distributed Agent Evolution with Dynamic Adaptation to Local Unexpected Scenarios" or DAEDALUS, is coupled with physicomimetic robot control in our research, although it is general enough to be used for swarm adaptation with other control methods as well.

With the DAEDALUS paradigm, we assume that swarm members (agents) interact with each other while interacting with their environment. The swarm agents may adopt different roles when accomplishing the task. The objective is to evolve agents' task behaviors and interactions between agents in a distributed fashion, such that the desired global behavior is achieved. Assume that each agent has some procedure to control its own actions in response to environmental conditions and interactions with other agents. The precise implementation of these procedures is not relevant; they may be implemented with programs, rule sets, finite state machines, real-valued vectors, force laws or any other procedural representation. Agents have a sense of self-worth, or "fitness." Agents are rewarded based on their performance, either directly or indirectly. In particular, agents can get direct rewards based on their experiences, and these rewards translate to higher fitness. Other agents may not experience any direct reward, but may in fact have contributed to the success of the agents that did receive the direct reward. Those robots with higher fitness give their procedures to agents with lower fitness. Evolutionary recombination and mutation provide necessary perturbations to these procedures, thereby improving the overall swarm performance as well as its ability to respond to environmental changes. Different species may evolve different procedures, thus filling a variety of environmental niches.

We have successfully applied our DAEDALUS paradigm, combined with physicomimetics, to accomplish the task of avoiding obstacles while seeking a goal. The paradigm was successful despite noise and occluded sensing leading to partial observability of the environment. The success can be attributed to DAEDALUS providing an online learning capability to the robot swarm that enabled it to achieve our performance objectives of reduced collisions while maintaining swarm cohesion.

## 14.6 Swarm Learning: Online

Online swarm evolution is challenging, both conceptually and in terms of how it is applied. Evolving the aggregate behavior of a collective via deliberations with neighbors and with limited sensory capabilities is one of the challenges. These deliberations among robots may occur at different stages of the evolution under different constraints. Furthermore, the learning difficulties increase when the swarm has to adapt to unfamiliar or changing environments (see Fig. 14.7). Our aim is to provide the swarm with intelligent adaptive capabilities and to enable the robots to execute these capabilities in a distributed fashion, while evolving desirable aggregate swarm behaviors for task accomplishment.

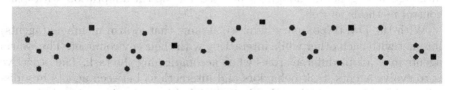

Fig. 14.7: A long corridor with randomly placed obstacles and five goals

In our DAEDALUS paradigm, each robot in the swarm is considered to be an individual in an EA population. Evolution occurs when individuals cooperate through interactions. When the population is initialized, each robot is given a *slightly mutated* copy of the optimized force law rule set found with offline learning. These mutations ensure that the robot population is not completely homogeneous. Furthermore, this slight heterogeneity has another advantage, namely, when the environment changes, some mutations will perform better than others. The robots that perform well in the environment will have higher fitness than the robots that perform poorly. Those robots with high fitness give their procedures to agents with lower fitness. Hence, better performing robots share their knowledge with their poorer performing neighbors. Additionally, evolutionary recombination and mutation operators provide necessary perturbations to these procedures, thus improving the swarm's performance and its ability to respond to environmental changes. Different species may evolve different procedures, reflecting the different niches they fill in the environment.

When we apply DAEDALUS to obstacle avoidance, we focus on two specific versions of our performance metrics: reducing robot–obstacle collisions and maintaining the cohesion of the swarm. Robots are penalized if they collide with obstacles and/or if they leave their neighbors behind. The second metric is relevant when the robots are left behind in cul-de-sacs due to a larger obstacle density. This causes the cohesion of the formation to be reduced. The work presented in this chapter focuses primarily on swarm survival.

In our experimental setup, as mentioned above, each robot in the swarm is initialized with a slightly mutated (1% mutation rate) copy of the optimized Lennard-Jones force law rule set found with offline learning. Again, the force law rules are mutated with Gaussian mutation. All of the robots have the same initial fitness (or "worthiness"), which is 1000 at the start. This fitness value does not correlate with any other system parameters. Instead of one goal, we have increased the challenge and presented five goals that must be reached while traversing a long corridor. Between each randomly positioned goal there is a different obstacle course, with a total of 90 randomly positioned obstacles in the environment. Robots that are left behind (due to obstacle cul-de-sacs) do not proceed to the next goal, but robots that collide with obstacles and make it to the goal are allowed to proceed to the next goal. We assume that damaged robots can be repaired once they reach a goal.

We previously described the simulated environment used for offline learning, which will be called the "offline environment." The size of the two-dimensional world for online learning, which is called the "online environment," is $1600 \times 950$ units; this is larger than the offline environment. Figure 14.8 shows an example of this more difficult environment. In this environment, each obstacle has a radius of 30, whereas the obstacle radius was 10 for the offline learning environment. Therefore more than 16% of the online learning environment is covered with obstacles. Compared to the offline environment, the online environment triples the obstacle coverage. We also increase the maximum velocity of the robots to 30 units/s, allowing the robots to move 1.5 times faster than in the offline environment. The Lennard-Jones force law learned in offline mode is not sufficient for this unfamiliar environment; it results in collisions with obstacles due to the higher velocity and, furthermore, the robots never reach the goal (i.e., swarm survival is

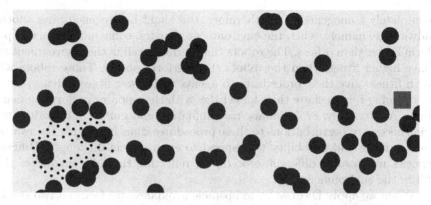

Fig. 14.8: This figure shows 60 robots moving to the goal. The larger circles represent obstacles, while the square at the right represents the goal. The large obstacles make this environment very difficult for the robots to traverse

not achieved) due to the high percentage of obstacles. The remaining system settings are kept the same as with the offline methodology.

Although noise may be present in dynamic environments, it is not specifically modeled in our simulation. However, it is worthwhile noting that it has been shown that when the physicomimetics framework is applied to actual robots, the performance is robust despite noise [240]. In fact, noise can actually improve performance by overcoming local optima in the behavior space [151, 240].

Figure 14.9 shows our first attempt at applying DAEDALUS. The curve labeled "Survival in offline" shows the number of robots that survive to reach a goal when the swarm is tested in an environment similar to the training environment. After five goals, the survival rate is about 98%. This is expected since the swarm is in a familiar environment. The curve labeled "Survival in online" shows the performance results in the new environment specifically designed for online learning (described above). The survival rate of the swarm after five goals is 51.66%. Although not shown in the graph, it is important to point out that the collision rates were low; related work can be found in [92]. Apparently the DAEDALUS paradigm allows us to improve the survival rate via mutation of the robot–robot and robot–goal interactions.

The results show that surviving in one environment with the force laws learned in another environment is difficult. In the offline environment the robots did not learn to avoid cul-de-sacs; the obstacle density did not produce cul-de-sacs, leaving sufficient space for the robots to navigate through. Therefore in the online environment the robots are not capable of completely avoiding cul-de-sacs, and they get stuck behind these cul-de-sacs. Although we were able to improve the robot survival rate via mutation, we were unable to achieve the survival of all of the robots that were initially fielded. Our

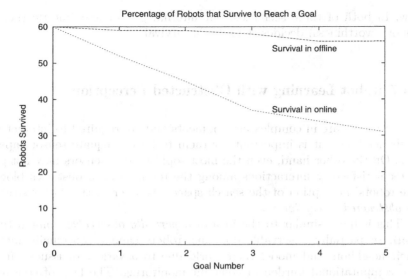

Fig. 14.9: A comparison of (a) the number of robots that survive when the rules are learned and tested in an offline environment, and (b) the number of robots that survive when learned in an offline environment and tested in an online environment

intention is to maintain their survival at a healthy level so that our swarm maintains its diversity.

To improve the swarm diversity, the robots in a swarm can be categorized into groups based on their differences. We could relate these differences to the robot's physical structure, algorithmic differences, mechanical properties, skills, behavioral aspects, and possibly many more. These types of classifications do not provide a guarantee that a robot belongs to an exact category due to the complex nature of a swarm that can react and behave distinctly. Though these classifications of swarms are popular, traditionally swarms have been classified as either heterogeneous or homogeneous, depending on the differences previously mentioned. One of the most important units of classification of robots into either heterogeneous or homogeneous groups is the metric used for assessing the swarm's performance.

We presented a classification of robots in a swarm based on their genetic makeup (the force law) in [93]. This can be coupled with a classification of our robots using their mutation rate. Each robot is assigned a predefined mutation rate and the robot mutates its copy of the force law based on the circumstances it faces in the environment. As an example, a robot may mutate its force law with an assigned mutation rate if the robot gets stuck behind a cul-de-sac, but at the same time another robot may completely avoid the same cul-de-sac without any mutation due to the higher quality of its force law. In both of these scenarios, our internal performance metric (i.e., the robot's worthiness) decides the robot's survival.

## 14.7 Robot Learning with Obstructed Perception

When robots are in complex environments and are required to interact with their neighbors, it is important for them to have adequate sensor capabilities. On the other hand, even the most sophisticated sensors may not guarantee satisfactory interactions among the robots due to obstacles blocking the robots' perception of the search space. We refer to this latter situation as *obstructed perception*.

This is quite similar to the idea of a *partially observable* domain (problem). In partially observable domains, robots cannot continually monitor their neighbors and model their world due to a variety of issues, such as the computational burden of constant monitoring. The lack of continuous monitoring and modeling leads to increased uncertainty about the state of other agents [254]. This partial observability can introduce uncertainty in the agent's information about the current state of the world, thus causing degradation in the robot's performance. If the cause of the robot's inability to continually monitor its environment is due to a sensory impediment, then we refer to it as *obstructed perception*. Obstructed perception follows a strict interpretation of sensor failure or limitations.

Robots face increasing amounts of sensor obstruction when they act in environments with large obstacle densities. This causes the robots to either observe their environment partially or not at all. Here, we assume that obstructed perception due to obstacles prevents robots from sensing other robots. Another reason for partial observation is the limited sensing distance of our robots. Sensors with limited range can reduce the robot's interactions with other robots. A decrease in interactions among swarm members causes a reduction in the population diversity, thus making it more difficult for the swarm to improve its task efficiency.

We use our DAEDALUS paradigm to improve swarm efficiency in accomplishing a task with obstructed perception. DAEDALUS is designed to improve the swarm's performance when the robots are in complex, challenging environments. We apply the DAEDALUS paradigm to improve the swarm's survivability in high obstacle density environments where the robots have their perception obstructed.

Recall that large obstacles can prevent robots from sensing each other. This occurs when a robot's line-of-sight to another robot passes "through" the obstacle. On the other hand, when a robot's line-of-sight lies along an edge of an obstacle, the robots are capable of sensing each other. Surprisingly, this is not generally modeled in prior work in this area [8]. Figure 14.10 shows an example scenario of obstructed perception. The larger circle represents an obstacle, and $A$ and $B$ represent robots. We define $minD$ to be the minimum distance from the center of the obstacle to the line-of-sight between robots $A$ and $B$, and $r$ is the radius of the obstacle. If $r > minD$, robot $A$ and robot $B$ have their perception obstructed.

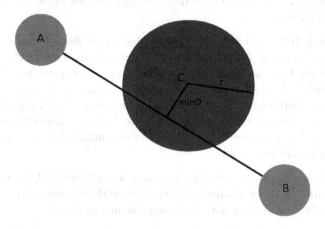

Fig. 14.10: The sensing capability of two robots ($A$ and $B$) is obstructed by a large obstacle $C$

Let's formalize this. We utilize a parameterized description of a line segment [84] to find the value of $minD$:

$$term_1 = (((1-q)X_A + qX_B) - X_C)^2 ,$$

$$term_2 = (((1-q)Y_A + qY_B) - Y_C)^2 ,$$

$$minD = \sqrt{term_1 + term_2} , \tag{14.6}$$

where $X_A$ and $X_B$ are the $x$ positions of robots $A$ and $B$, $Y_A$ and $Y_B$ are the $y$ positions of robots $A$ and $B$, $X_C$ and $Y_C$ are the $x$ and $y$ positions of the center of the obstacle, and $0 \leq q \leq 1$. The value of $q$ that minimizes Eq. 14.6 is defined by

$$\frac{(X_C - X_A)(X_B - X_A) + (Y_C - Y_A)(Y_B - Y_A)}{(X_B - X_A)^2 + (Y_B - Y_A)^2} . \tag{14.7}$$

## 14.8 Survival Through Cooperation

We now present two different approaches based on swarm diversity for improving the swarm survival rate in the presence of partial observability in unfamiliar environments. These heuristic approaches make use of the diversity in the swarm.

- **Approach 1:** If agent $i$ is not moving (due to an obstacle in the way) and a neighboring agent $j$ is moving, then agent $i$ receives agent $j$'s agent–agent interaction force law, including agent $j$'s mutation rate (see [93]).
- **Approach 2:** If agent $i$ is at a goal, agent $i$ computes the mean (average) of its better performing neighbors' mutation rates. When doing this, agent $i$ only considers its neighbors who have higher self-worth (or "fitness") than agent $i$. Then agent $i$ uses this mean mutation rate as its own when moving to the next goal(s).

Both of these approaches are implemented as optional in DAEDALUS. In the experiments below we evaluate DAEDALUS without these approaches, with each one alone, and then with both approaches together.

### 14.8.1 Initial Experiments

First, we compare our results with DAEDALUS (without the above two approaches) to those without DAEDALUS, in three control studies. In the first control study, we train the agents with an offline EA on small obstacles, and then test them again on small obstacles to verify their performance. In the second control study, we train the agents with an offline EA on large obstacles and test them on large obstacles. The purpose of these two control studies is to assess the difficulty of the task. Finally, in the third control study, we train the agents with an offline EA on small obstacles and test them on large obstacles. The purpose of this study is to see how well the knowledge learned while avoiding small obstacles transfers when navigating in environments with large obstacles. The results of all control studies presented in this section are averaged over 25 independent runs.

Figure 14.11 shows the results of our first experiment where the two approaches explained in this section are not utilized. In all the figures presented in this section, the $y$-axis shows the number of agents that survive to reach a goal (and recall that there are five goals) at each stage of the long corridor. The top performance curve is for the first control study. Note that learning with small obstacles in the offline mode is not hard, and the agents perform very well in the online environment. This is due to the fact that the small obstacles make the environment less dense, thereby providing the agents with sufficient space to navigate. Out of the 60 initial agents released into the online environment, 93.3% survive to reach the last goal. This is similar to the 98% survival rate in Fig. 14.9. With such small obstacles (which are the norm in the majority of the related literature), obstructed perception is not an important issue.

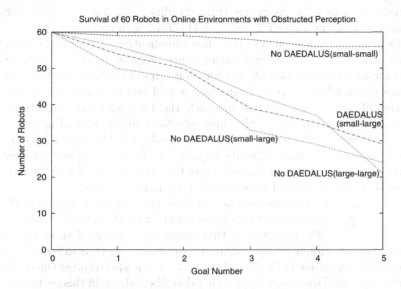

Fig. 14.11: The number of agents that survive to reach a goal. Agents are not allowed to receive mutation rates from other agents during navigation. Agents do not receive the mean mutation rate at a goal

| Agents | Mutation Rate | | | | |
|--------|------|-----|-----|-----|-----|
| Survive | 1% | 3% | 5% | 7% | 9% |
| 60–start | 12 | 12 | 12 | 12 | 12 |
| 54–goal 1 | 11 | 11 | 11 | 11 | 10 |
| 50–goal 2 | 10 | 9 | 11 | 10 | 10 |
| 39–goal 3 | 8 | 6 | 10 | 9 | 6 |
| 35–goal 4 | 8 | 5 | 9 | 8 | 5 |
| 29–goal 5 | 6 | 4 | 9 | 7 | 3 |

Table 14.3: The number of agents that survive to reach a goal and their mutation rates. Agents are not allowed to receive mutation rates from other agents during navigation. Agents do not receive the mean mutation rate at a goal

For the DAEDALUS performance curve labeled "DAEDALUS (small–large)," we divide the agents into five groups of equal size. Each group of 12 agents is assigned a mutation rate of 1%, 3%, 5%, 7%, or 9%. In this control study, agents are not allowed to receive mutation rates from other agents, and the agents do not receive the mean mutation rate at goals. Out of the initial 60 agents, 29 or 48.3% survive to reach the final goal. Surprisingly, this is better than when the agents are trained offline to handle the large obstacles, as shown with the curve labeled "No DAEDALUS (large–large)."

Because the mutation rate (i.e., the classification of swarm diversity) has a major effect on the swarm's performance [93], we decided to compare the performance as a function of the mutation rates of the five groups. Table 14.3 shows the number of agents that reach a goal and their mutation rates. Out of the 29 agents that survive to reach the final goal, nine agents or 31% had a 5% mutation rate. Only three agents or about 10% of agents with a mutation rate of 9% survive to reach the final goal. The agents with 3% and 9% mutation rates have difficulty coping with the changed environment and partial observability; the performance of the agents with 1% and 7% mutation rates is similar to that of agents with a 5% mutation rate.

In an attempt to improve the survivability, we again initialize the five agent groups, as in the previous experiment, with the mutation rates of 1%, 3%, 5%, 7%, or 9%. However, in this second experiment, if an agent receives rules from a neighbor (which, again, occurs if that agent is in trouble, for example if it gets stuck behind a cul-de-sac), then it also receives the neighbor's mutation rate. This is our Approach 1 described above in this section. In this experiment, the agents do not receive the mean mutation rate of their neighbors at a goal. In other words, they do not employ Approach 2. The curve with the label "DAEDALUS (small–large)" in Fig. 14.12 shows the results of this study.

Out of the initial 60 agents, 37 or 61.6% survived to reach the final goal. The number of agents reaching the final goal is a 13.3% improvement over the results seen in Fig. 14.11, where we do not allow the agents to receive other agents' mutation rates. This improvement is statistically significant, based on a two-tailed $t$-test with a significance level of $\alpha = 0.05$. It is also of *practical* significance for various swarm applications.

Table 14.4 shows the number of agents that reach a goal and their mutation rates for the second experiment. In this implementation, agents in trouble not only change their rules, but also their mutation rate. The agents do not receive the mean mutation rate of their neighbors at a goal.

Once again, it is clear that the agents favor 5% and 7% mutation rates over 1%, 3%, and 9%. Interestingly, this adaptation shown in Table 14.4 is not rapid enough to create a significant impact on the mutation rate of the swarm. Due to partial observability, the agent–agent cooperation is limited during navigation. This lack of cooperation limits the agents' abilities to make informed decisions about better performing mutation rates.

Our results suggest that for the agents to make informed decisions on good mutation rates to be received, an agent in trouble needs to cooperate with a large number of other agents. This requirement has limitations dur-

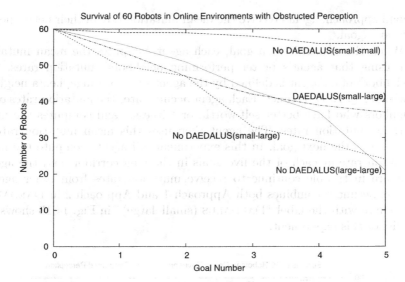

Fig. 14.12: The number of agents that survive to reach a goal. Agents are allowed to receive mutation rates from other agents during navigation. Agents do not receive the mean mutation rate at a goal

| Agents | Mutation Rate | | | | |
| Survive | 1% | 3% | 5% | 7% | 9% |
|---|---|---|---|---|---|
| 60–start | 12 | 12 | 12 | 12 | 12 |
| 54–goal 1 | 10 | 10 | 11 | 12 | 11 |
| 47–goal 2 | 9 | 8 | 11 | 10 | 9 |
| 42–goal 3 | 6 | 7 | 11 | 10 | 8 |
| 39–goal 4 | 6 | 6 | 10 | 9 | 8 |
| 37–goal 5 | 5 | 6 | 10 | 9 | 7 |

Table 14.4: The number of agents that survive to reach a goal and their mutation rates. Agents are allowed to receive mutation rates from other agents during navigation. Agents do not receive the mean mutation rate at a goal

ing navigation due to "obstructed perception" and dynamic variations in the formation cohesiveness. Due to the behavior of the Lennard-Jones force law, the fluid-like motion learned by the swarm causes the swarm formation to stretch, thereby reducing the formation cohesiveness when navigating around obstacles. On the other hand, when the agents are at a goal, "obstructed perception" is minimal due to a lack of obstacles around goals, and the swarm cohesion is high due to agent–goal interactions. We take these observations into consideration in our Approach 2 explained above in this section. In this second approach, agents receive the mean mutation rate of their better neighbors at a goal.

When the swarm is at a goal, each agent computes the mean mutation rate using that agent's better performing neighbors' mutation rates. The "neighbor" of an agent is defined by the agent's sensor range, i.e., a neighbor is within $r \leq 1.5D$ distance. Each agent accumulates the mutation rates of its neighbors who have better self-worth, or "fitness," and computes the mean of these mutation rates. The agent then uses this mean mutation rate to navigate to the next goal. In this experiment, all agents compute the mean mutation rate at each of the five goals in the long corridor, and the agents that are in trouble continue to receive mutation rates from other agents. This experiment combines both Approach 1 and Approach 2 in DAEDALUS. The curve with the label "DAEDALUS (small–large)" in Fig. 14.13 shows the results of this experiment.

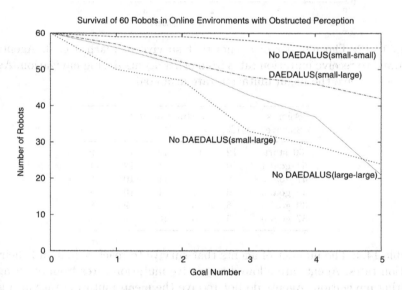

Fig. 14.13: The number of agents that survive to reach a goal. Agents are allowed to receive mutation rates from other agents. Agents receive the mean mutation rate at a goal

Out of the initial 60 agents, 41 or 68.3% survive to reach the final goal. This is a statistically as well as practically significant gain over the results in Fig. 14.11. The two-tailed $t$-test was used with $\alpha = 0.05$.

The results in Table 14.5 clearly show that the swarm performs better with the mutation rate ($\mu$) in two ranges, $3\% < \mu \leq 5\%$ and $5\% < \mu \leq 7\%$, than with the other three mutation rate ranges. The agents that use the mean of the better performing neighbors' mutation rates at a goal rapidly adapt to

their online environment, thus causing the swarm to survive over the corridor
with five goals.

| Agents | Mutation Rate Ranges | | | | |
|--------|------|------|------|------|------|
| Survive | $\leq 1\%$ | $\leq 3\%$ | $\leq 5\%$ | $\leq 7\%$ | $\leq 9\%$ |
| 60–start | 12 | 12 | 12 | 12 | 12 |
| 54–goal 1 | 11 | 11 | 11 | 11 | 10 |
| 51–goal 2 | 1 | 1 | 27 | 22 | 0 |
| 48–goal 3 | 0 | 1 | 30 | 17 | 0 |
| 45–goal 4 | 0 | 1 | 27 | 17 | 0 |
| 41–goal 5 | 0 | 2 | 24 | 15 | 0 |

Table 14.5: The number of agents that survive to reach a goal and their mutation rates. Agents are allowed to receive mutation rates from other agents
during navigation. Agents receive the mean mutation rate at a goal

This approach does not introduce new global knowledge into our swarm
system; instead, it allows the swarm behavior to emerge through local interactions. The results suggest that "obstructed perception" has a significant
impact on swarm survival. The "obstructed perception" limits the agents'
ability to receive mutation rates from other agents during navigation, but
once the swarm is at a goal, the agents have improved access to their neighbors due to the swarm cohesion. This cohesive formation at a goal increases
the number of neighbors each agent encounters and allows higher agent–agent
interactions.

To better understand the impact of the combined approaches presented in
Fig. 14.13, we conduct an experiment with only Approach 2. In this experiment, the agents compute and use the mean mutation rate at each goal, but
do not receive the mutation rates from other agents during navigation. The
results of this experiment are shown in Fig. 14.14.

Out of the initial 60 agents, 34 or 56.6% survived to reach the final goal.
The outcome is quite similar to the second control study in Fig. 14.12, where
37 agents survived to reach the final goal. In summary, although each approach improves performance, the combination of the two works best.

The results in Table 14.6 clearly show that the swarm again performs
better with the mutation rate ($\mu$) in two ranges, $3\% < \mu \leq 5\%$ and
$5\% < \mu \leq 7\%$, than it performs with the other three mutation rate ranges.
Utilizing a single approach improves the survival, but utilizing the two approaches together provides the most improvement in swarm survival in partially observable environments.

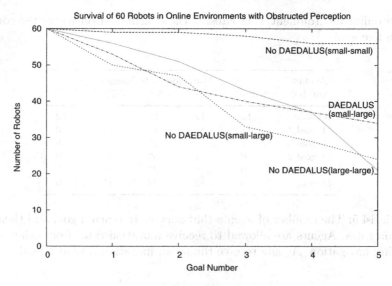

Fig. 14.14: The number of agents that survive to reach a goal. Agents are not allowed to receive mutation rates from other agents. Agents receive the mean mutation rate at a goal

| Agents | Mutation Rate Ranges | | | | |
|---|---|---|---|---|---|
| Survive | $\leq 1\%$ | $\leq 3\%$ | $\leq 5\%$ | $\leq 7\%$ | $\leq 9\%$ |
| 60–start | 12 | 12 | 12 | 12 | 12 |
| 53–goal 1 | 11 | 10 | 11 | 10 | 11 |
| 44–goal 2 | 0 | 1 | 17 | 26 | 0 |
| 40–goal 3 | 0 | 0 | 16 | 24 | 0 |
| 37–goal 4 | 0 | 0 | 13 | 24 | 0 |
| 34–goal 5 | 0 | 0 | 17 | 17 | 0 |

Table 14.6: The number of agents that survive to reach a goal and their mutation rates. Agents are not allowed to receive mutation rates from other agents. Agents receive the mean mutation rate at a goal

## 14.8.2 Survival in High Obstacle Density Environments

We further explore the performance of DAEDALUS in high obstacle density environments. In this situation, the robots are only capable of observing the environment partially—due to sensor obstruction. The swarm once again maintains its diversity with different mutation rates, and the robots cooperate using Approach 1 presented earlier in this section.

The results from four separate experiments are presented here. The first experiment (Table 14.7) shows the number of agents that survive to reach a goal without DAEDALUS and without obstructed perception. The second experiment (Table 14.8) shows the number of agents that survive to reach a goal without DAEDALUS and with obstructed perception. The third experiment (Table 14.9) shows the number of agents that survive to reach a goal with DAEDALUS and with obstructed perception. The fourth experiment (Table 14.10) shows the number of agents that survive to reach a goal with DAEDALUS and without obstructed perception. Each experiment consists of several control studies, where each control study is conducted with a different obstacle density. The results from all control studies show the number of agents surviving to reach a goal, along with the obstacle density of the environment. All results are averaged over 25 independent runs. The initial swarm size is 40 robots.

Table 14.7: The number of agents that survive to reach a goal without DAEDALUS and without obstructed perception

| | Goal Number | | | | |
|---|---|---|---|---|---|
| Density | Goal 1 | Goal 2 | Goal 3 | Goal 4 | Goal 5 |
| 5% | 39 | 38 | 37 | 37 | 36 |
| 10% | 38 | 37 | 34 | 31 | 30 |
| 15% | 35 | 33 | 28 | 22 | 21 |
| 20% | 32 | 27 | 20 | 15 | 13 |
| 25% | 31 | 20 | 13 | 10 | 7 |
| 30% | 22 | 10 | 6 | 3 | 0 |

The results of experiment 1 in Table 14.7 show the number of agents that survive to reach a goal without DAEDALUS and without obstructed perception in the online environment. Out of 40 agents, 36 agents or 90% survive to reach the last goal when the obstacle density is 5%. The agents do not have significant difficulty performing in this environment since this is similar to the offline behavior learned using the EA. This result is statistically significant (with a two-tailed $t$-test and $\alpha = 0.05$). The 10% (four agents) decrease in survival can be mostly attributed to the difference in the online (in a corridor with five segments) versus offline obstacle course. When the obstacle density is 15%, 52% or 21 of the agents reach the final goal. None of the agents survive to reach the final goal when the obstacle density is 30%, in which case the density is six times higher than it is in the offline environment.

In our second experiment, we introduce obstructed perception to agent–agent interactions. The results are shown in Table 14.8. The results show the number of agents that survive to reach a goal without DAEDALUS and with

Table 14.8: The number of agents that survive to reach a goal without
DAEDALUS and with obstructed perception

| | | | Goal Number | | |
|---|---|---|---|---|---|
| Density | Goal 1 | Goal 2 | Goal 3 | Goal 4 | Goal 5 |
| 5% | 38 | 38 | 38 | 36 | 36 |
| 10% | 37 | 36 | 35 | 31 | 29 |
| 15% | 32 | 32 | 27 | 13 | 12 |
| 20% | 27 | 26 | 22 | 5 | 5 |
| 25% | 13 | 11 | 8 | 2 | 0 |

obstructed perception in the online environment. Out of the 40 agents, 36
agents or 90% survive to reach the last goal when the obstacle density is 5%.
The number of agents reaching each goal in this control study is very similar
to the number in the first control study of the first experiment. When the
obstacle density is 15%, only 30% or 12 of the agents reach the final goal,
and this is a 22% (nine agents) decrease in agent survival compared to the
results of the same control study in the first experiment. These differences in
agent survival show that obstructed perception has a major effect on swarm
adaptation in online environments, and requires a DAEDALUS-like paradigm
to overcome the navigational difficulties. No agents survive to reach the last
goal when the obstacle density is 25%.

We next apply our DAEDALUS paradigm to further improve the swarm
survival rate. The results of this third experiment are presented in Table 14.9;
they show the number of agents that survive to reach a goal with DAEDALUS
and with obstructed perception.

Table 14.9: The number of agents that survive to reach a goal with DAEDALUS
and with obstructed perception

| | | | Goal Number | | |
|---|---|---|---|---|---|
| Density | Goal 1 | Goal 2 | Goal 3 | Goal 4 | Goal 5 |
| 5% | 39 | 38 | 37 | 36 | 36 |
| 10% | 38 | 37 | 35 | 33 | 31 |
| 15% | 37 | 36 | 34 | 28 | 25 |
| 20% | 35 | 33 | 30 | 27 | 22 |
| 25% | 32 | 26 | 24 | 16 | 13 |
| 30% | 28 | 23 | 21 | 13 | 9 |
| 35% | 22 | 14 | 11 | 5 | 0 |

In the first control study, out of 40 agents, 36 agents or 90% survive to reach the last goal when the obstacle density is 5%. Again, the agents do not have significant difficulty when performing in this environment. For obstacle densities of 15% through 25% the swarm's performance is significantly better than it is in experiment 2 (see Table 14.8). When the obstacle density is 15%, more than 62.5% or 25 of the agents reach the final goal. In the same control study in the second experiment where DAEDALUS is not utilized, only 30% or 12 reach the final goal, and this is a 32.5% (13 agents) improvement. The control studies conducted with 20% and 25% also show significant improvements in agent survival over the same control studies in the experiment without DAEDALUS. No agents survive to reach the last goal when the obstacle density is 35%.

To better understand the impact of obstructed perception on agents, we conducted a fourth experiment. The results are presented in Table 14.10; they show the number of agents surviving to reach a goal with DAEDALUS and without obstructed perception.

Table 14.10: The number of agents that survive to reach a goal with DAEDALUS and without obstructed perception

| Density | Goal Number | | | | |
|---|---|---|---|---|---|
| | Goal 1 | Goal 2 | Goal 3 | Goal 4 | Goal 5 |
| 5% | 38 | 37 | 37 | 36 | 36 |
| 10% | 38 | 35 | 35 | 31 | 27 |
| 15% | 37 | 34 | 34 | 29 | 26 |
| 20% | 35 | 31 | 30 | 25 | 23 |
| 25% | 34 | 29 | 27 | 24 | 21 |
| 30% | 29 | 23 | 19 | 14 | 12 |
| 35% | 26 | 21 | 18 | 10 | 8 |
| 40% | 21 | 15 | 12 | 5 | 0 |

Out of 40 agents, 36 agents or 90% survive to reach the last goal when the obstacle density is 5%. Once again, the number of agents reaching each goal in this control study is very similar to the number in the first control study of the first experiment. When the obstacle density is 15%, 65% or 26 of the agents reach the last goal, which is similar to 62% or 25 of the agents reaching the last goal in the same control study of the third experiment. At 35% obstacle density, 20% or eight of the agents survive to reach the last goal. No agents survive to reach the last goal when the obstacle density is 40%. These results show that DAEDALUS effectively improves the agent survival in obstacle-dense environments, even though adapting to new environments is much more difficult when the agent interactions are obstructed by obstacles.

## 14.9 Summary of Research

In this book chapter, we have addressed issues and concerns that are related
to our task, i.e., the task of getting a swarm of robots to reach a goal while
avoiding obstacles and maintaining a cohesive formation, despite environment
changes. We provided an analysis of this task within the context of adaptive
learning by heterogeneous swarms in offline and online environments. Our
swarms were controlled by physicomimetics using the Lennard-Jones force
law. We introduced our DAEDALUS paradigm for heterogeneous swarm learning in online environments, and provided extensive experimental analyses
of swarm behavior in partially observable as well as highly obstacle-dense
environments.

Traditionally, accomplishing complex robotics tasks involves an expensive
and possibly remotely controlled robot. This traditional approach overwhelms
our robotic resources with an ever increasing complexity of task requirements.
The traditional approaches do not support complete autonomy or the distributed computing capabilities of modern robots. The robots depend continuously on human operations. Therefore, performance feedback becomes
vital and any delay or perturbation of the feedback loop due to environmental constraints may jeopardize the task. In other words, because of the
need for constant human supervision, any absence of a global observer could
be fatal to the mission's success. Furthermore, traditional swarm robotics
tasks are learned and tested offline, often using evolutionary learning approaches. These offline evolutionary learning approaches are time-consuming.
For swarms adapting to perform tasks in unfamiliar environments, especially
for time critical applications, our online DAEDALUS approach can be considerably more efficient and effective than earlier traditional approaches.

## 14.10 Related Work

DAEDALUS is conceptually similar to an approach developed independently
by [259], called "embodied evolution." In embodied evolution, robots virtually
mate and exchange genetic information using a "probabilistic gene transfer
algorithm." As with our concept of worthiness, robots can reproduce according to their ability to perform a task. Interestingly, the authors examine the
task of locating a light source (phototaxis). Unlike our work, each robot uses
a neural network control algorithm. Also, there is no concept of a swarm;
each robot can perform the task by itself. Finally, there are no obstacles,
which makes the task much simpler.

In the specific context of obstacle avoidance, the most relevant papers
are [7, 8, 69]. Balch [7] examines the situation of four agents moving in formation through an obstacle field with 2% coverage. In [8] he extends this
to an obstacle field of 5% coverage, and also investigates the behavior of 32

agents moving around one medium sized obstacle. Fredslund and Matarić [69] examine a maximum of eight agents moving around two wall obstacles. To the best of our knowledge, we are the first to systematically examine larger numbers of robots and obstacles using a swarm robotics approach.

Schultz presents a projective planning algorithm for real-time autonomous underwater navigation through obstacles [212]. He uses "SAMUEL," an off-line learning system based on genetic algorithms, to learn high-performance reactive strategies for navigation and collision avoidance. He presents results in simulation with a single autonomous underwater vehicle and shows that SAMUEL can achieve a success rate of 96% on randomly generated mine fields, which is far better than the success rate of a human-designed strategy; the latter has an average success rate of only 8%.

Simmons presents a new method for local obstacle avoidance by indoor mobile robots that formulates the problem as one of constrained optimization in velocity space [222]. The robot chooses velocity commands that satisfy all of the desired constraints (such as the physical limitations of the robot and the environmental limitations), and that maximize an objective function which trades off speed, safety, and goal-directedness. He demonstrates the utility of this algorithm using a single physical robot in a controlled laboratory environment.

Another related method for obstacle avoidance is the *dynamic window approach (DWA)*. This approach is mostly suited for robots moving at a high speed and is derived directly from the robot's motion dynamics [68]. With the DWA, the search for commands that control the robot is carried out in the space of velocities. The robot only considers those velocities that are "safe" with respect to obstacles. The authors present their results with tests done using an autonomous robot called "RHINO" that uses proximity sensors to compute the dynamic window.

Borenstein et al. present an approach that permits the detection of unknown obstacles simultaneously with the steering of a mobile robot to avoid collisions and advance toward a target [20]. They use a *virtual force field* method that uses potential fields for navigation and certainty grids for representing obstacles. They show that this method is especially suitable for noisy and inaccurate sensor inputs. They also addressed the *local minimum trap* problem, where robots can get stuck in a U-shaped obstacle.

O'Hara uses an embedded network distributed throughout the environment to approximate the path-planning space and uses the network to compute a navigational path with a framework called "GNATs" when the environment changes [172]. The dynamism of the environment is modeled with an opening and closing door in the experimental setup. However, the embedded network is immobile, whereas our network is completely mobile.

Passino introduces a set-based stochastic optimization method [176] and provides a discussion of the algorithm using a proportional-derivative controller for the tanker ship design as an application. He shows how his online

472 Suranga Hettiarachchi

optimization algorithm can achieve a type of online evolution of controllers to achieve real-time evolutionary adaptive control.

The approaches presented in [118, 184, 246] utilize different multi-agent learning paradigms. Tan [246] uses reinforcement learning to address agent learning by sharing instantaneous information, episodes, and learned policies. The task environment is a $10 \times 10$ grid world with a maximum of four agents, two hunters, and two prey. Our work is conceptually similar and was developed independently. The cooperative learning discussed in [246] is fundamentally offline, whereas in our approach learning is both offline and online, and agents continue to adapt to their changing environment through cooperation. The idea of agents depending only on offline learning can be problematic due to the complex nature and unexpected changes in the environment. The work in [118] utilizes fuzzy automata to address autonomous mobile robot learning on a reactive obstacle avoidance task, using two robots. The robots share their experiences while learning the task simultaneously. Their results clearly show that sharing experience makes learning faster and more repeatable than individual learning. This confirms the observations that we presented in this chapter. Pugh and Martinoli [184] explore how varying sensor offsets and scaling factors affect parallel swarm robotic learning of obstacle avoidance behaviors using both a genetic algorithm and particle swarm optimization. In our work, all agents have the same sensors, and we assume that the sensor readings have no variations. Their results show that both algorithms are able to withstand small variations in sensor offsets and large variations in sensor scaling factors, while showing poor performance with high offset variations. We intend to utilize sensor offsets and scaling factors in our future work.

## 14.11 Future Work

A significant portion of this work is dedicated to exploring issues related to obstructed perception. However, there are significant issues that arise with respect to "wall following methods" and "local minimum trap" problems that we have not yet adequately addressed. We have observed the "local minimum trap" problem in our work, but we did not address this issue in detail. We intend to introduce a hybrid liquid and gas model, combined with our DAEDALUS approach, as a solution. The robots will switch to a gas model as presented in Chap. 7 to avoid the trap at a local minimum. Once the robots have escaped, they can continue using the previous force law.

The performance metrics we defined in this book chapter can provide future researchers with meaningful benchmarks of swarm behavior. A possible extension to these metrics could specify a distribution of sub-swarms and the number of robots for each sub-swarm.

Also, another possible avenue for improving the performance of our DAEDALUS approach lies with reward sharing (i.e., *credit assignment*) techniques. Current work on classifier systems uses mechanisms such as the *bucket brigade* or *profit sharing* algorithms to allocate rewards appropriately to individual agents [80]. However, these techniques rely on global blackboards and assume that all agents can potentially act with all others, through a bidding process. We intend to modify these approaches so that they are fully distributed and appropriate for online learning by heterogeneous swarms.

# Chapter 15
# A Statistical Framework for Estimating the Success Rate of Liquid-Mimetic Swarms

Diana F. Spears, Richard Anderson-Sprecher, Aleksey Kletsov, and Antons Rebguns

## 15.1 Introduction

This chapter presents a novel statistical framework for risk assessment of robotic swarm operations. This research has arisen from the crucial problem of predicting (before actual deployment) what portion of the swarm will succeed in accomplishing its mission. Our scouts framework uses a small group of expendable "scout" agents to predict the success probability for the entire swarm, thereby preventing a loss of a large portion of the swarm. The framework is based on two probability formulas—the standard Bernoulli trials formula and a novel Bayesian formula that was derived from fundamental probabilistic principles. For experimental testing, the framework is applied to a navigation task, where simulated agents move through a field of obstacles to a goal location. For more about this task, see Chap. 14. The performance metric used to analyze the results is the mean squared error, which compares the experimental prediction results of both formulas to ground truth. Our statistical scouts framework shows both accuracy and robustness. The control algorithm for the motion of the swarm is implemented with physicomimetics. The use of a Lennard-Jones force law (which produces liquid-mimetic

Diana F. Spears
Swarmotics LLC, Laramie, Wyoming, USA, e-mail: dspears@swarmotics.com

Richard Anderson-Sprecher
Department of Statistics, University of Wyoming, Laramie, Wyoming, USA, e-mail: sprecher@uwyo.edu

Aleksey Kletsov
Department of Physics, East Carolina University, Greenville, North Carolina, USA, e-mail: kletsov@gmail.com

Antons Rebguns
Department of Computer Science, University of Arizona, Arizona, USA, e-mail: anton@email.arizona.edu

W.M. Spears, D.F. Spears (eds.), *Physicomimetics*,
DOI 10.1007/978-3-642-22804-9_15,
© Springer-Verlag Berlin Heidelberg 2011

swarm behavior) to model inter-agent forces plays a favorable role in the effectiveness of our framework. This theoretical framework is potentially useful for military operations such as search-and-destroy, search-and-rescue, and surveillance tasks.

## 15.1.1 The Main Idea and Research Goals

The main thrust of the research described in this chapter is simple. We address the question of how to predict the risk of failure based on a small-sample estimation of this risk. The need for predicting the risk of failure (or, equivalently, the probability of success) arises frequently in both military and civilian swarm robotic applications, such as search-and-rescue, *demining* (which is defined as the process of removing land mines from a hazardous region), surveillance (see Chaps. 6 and 7), and foraging [153]. In all of these applications, the loss of a substantial fraction of the swarm of task-oriented agents would be costly or might prevent completion of the operation.

This chapter has two main goals:

1. To show how mathematics and physics (physicomimetics), working hand-in-hand, have led to the development of an important and practical theoretical framework for predicting the odds of success for swarm operations.
2. To demonstrate that our framework is effective by applying it to a generic task of navigating through obstacles to a goal.

## 15.1.2 Success Probability and Risk Assessment

Estimating the probability of task success/failure belongs under the umbrella of *risk assessment*. Risk assessment is a preliminary phase of any operation that involves potential threat to the agents performing the operation. Based on the information gathered during risk assessment, the operation supervisor (e.g., Commander-in-Chief) decides what deployment strategy to use, how many agents to deploy in order to ensure a desired operation success probability, and whether it is worthwhile to start the operation at all. Risk of the operations inherently depends on two constituents: the agents themselves and the environment where they operate. For example, for a swarm of military robots performing a search-and-destroy mission, risk could be caused by hardware failures (in electrical circuits and mechanical systems of robots) as well as by the environment (such as rough terrain, water, unexpected obstacles, or hazards such as explosives).

## 15.1.3 Scouts for Assessing Risk

We propose a novel method to assess the risk of failing an operation—by deploying a group of expendable agent *scouts* ahead of the entire swarm, and finding the fraction of scouts that successfully reach the goal. This fraction is then used to predict the probability that a desired percentage of the swarm will reach the same goal. In other words, the essential element of our framework is a rigorously derived formula for predicting the probability of swarm success based on the fraction of scouts succeeding at the task. The idea comes from statistical sampling [71]. Given the fraction of successes in a sample, the success rate of the population can be estimated.

Our framework is practical for making important swarm deployment decisions. For example, if the predicted success rate is within acceptable bounds, then the full swarm should be deployed. Alternatively, our framework can be used to determine the number of agents that need to be deployed in order to ensure that a desired number will reach the goal. To the best of our knowledge, scouts have never previously been used for these purposes.

In fact, the prior research on using agent scouts at all—even for other purposes—is scarce. We now describe what little we have found in the literature. Most studies have appeared in the *natural* science (rather than "agents") literature—because scouts are a popular mechanism for assisting biological swarms. For instance, Greene and Gordon published a study of red harvest ants that use a group of "patrollers" as scouts before sending out the full swarm of ants [79]. These patrollers establish a path for foraging. When they do appear in the "agents" literature, scouts adopt the role of "trail blazers," which is particularly relevant for search-and-rescue [200], remote reconnaissance, distributed sensor networks [250], and swarm "nest assessment" [201]. Scouts are also used in another chapter of this book (Chap. 6); their purpose there is to find the safest route for a covert Naval operation. Similarly to these other approaches, our approach in this chapter also uses scouts for assessing the safety of the environment for the swarm. On the other hand, unlike these other approaches, our framework uses scout agents specifically for the purpose of predicting the success probability of the mission performed by the entire swarm.

## 15.1.4 Physicomimetics As the Control Algorithm

Although our framework does not depend on any agent control algorithm in particular, here we assume that physicomimetics is used. Recall from earlier chapters of this book that the use of physicomimetics makes the analysis of simulated systems much easier than it is with other (e.g., biomimetic) control algorithms. In particular, if anything goes wrong in a physicomimetic simulation, then it is straightforward to analyze and fix the problem. This simplicity of analysis played an important role in facilitating code creation for our scouts framework.

However there is another advantage of physicomimetics. With physicomimetic swarm algorithms, we have the opportunity to choose solid-like or liquid-like behavior—because physicomimetics can imitate any "state of matter," as shown in Chap. 3 of this book. This is also advantageous for our research, because we can analyze the tradeoffs between the two states of matter and then determine which state is better for our problem. Results described in this chapter clearly demonstrate that our framework makes better predictions for the obstacle avoidance task when the swarm adopts a liquid-mimetic, rather than a solid-mimetic, control algorithm. This is a truly intriguing result and, for most readers, we suspect it will not be intuitively obvious *why* this is the case. Note that we are stating that the *predictions* are better—not the performance—when liquid-mimetic forces are used.[1] (Go ahead and try to guess why before reading the answer!) Toward the end of this chapter, we will provide the answer.

## 15.1.5 Case Study: Navigation Through Obstacles

Recall that the developed framework has been applied to a particular navigation task for this chapter. For this task, a swarm of robots needs to travel from an initial location to a goal location, while avoiding obstacles. This is a popular task in the robotics literature, and it is also addressed in Chap. 14 of this book. Before deploying (and possibly losing) the entire swarm for this task, we need a certain level of confidence that a desired, predetermined portion of the swarm will successfully reach the goal. To find this level of confidence, a few (less than 20) scouts are sent from the swarm, which may consist of hundreds of agents. A human or artificial agent, called the *sender*, deploys the scouts. Then, an agent (e.g., a person, an artificial scout, or a sensing device), called the *receiver*, counts the fraction of scouts that arrive at the goal successfully. The sender and receiver must be able to communicate with each other to report the scout success rate, but no other agents require the capability of communicating messages. Using the fraction of scouts that successfully reach the goal, a formula is then applied to predict the probability that a desired percentage of the entire swarm will reach the goal. Based on this probability, the sender decides how many (if any) agents to deploy in order to yield a desired probability that a certain number of agents will reach the goal.

---

[1] Although, incidentally, the performance is also better with liquid-mimetic swarms, but that is not our focus here.

## 15.1.6 Chapter Outline

The remainder of this chapter proceeds as follows. We begin in Sect. 15.2 with a short review of the essential material on probability and conditional probability needed to understand the remainder of the chapter. This includes a presentation of Bayes' formula, which is used in our framework. Next, Sect. 15.3 presents the framework, along with two statistical formulas within it, either of which can be used for prediction. Section 15.4 shows a picture of our swarm navigation simulator, along with a description of it, including the physicomimetics force law used for agent interactions. Then Sects. 15.5–15.8 discuss the design of our experiments, followed by the algorithms and a presentation of the experimental results and conclusions. Section 15.7 states our experimental hypotheses and shows graphs presenting the results of testing these hypotheses. Section 15.8 summarizes our experimental conclusions. Then, we revisit the role that physicomimetics plays in this chapter. Earlier in this introduction we mentioned that the capability of adopting varying states of matter with physicomimetics yields an advantage, namely, that of determining which state is most beneficial for predicting the success rate of the swarm. In Sect. 15.9 we explain *why* it is the liquid-mimetic state that leads to the best predictions. Finally, the chapter concludes with a summary and a discussion of the issues to be addressed by those who wish to apply our framework to real robots.

## 15.2 A Brief Review of Probability

The statistical scouts framework described here extensively uses probability theory, and therefore this section is to remind you of a few fundamental probabilistic concepts. Probability is the mathematical science of quantifying uncertainty. We provide a more precise definition below. Statistics, on the other hand, uses the framework provided by probability to guide the collection and interpretation of data. Statistical inference is typically used to draw general conclusions about a population based on samples. This section presents the fundamentals of probability necessary for describing our results. Statistics then enters the picture in the next section, in our two formulas where we use the scouts as samples to predict the performance of the swarm population.

## 15.2.1 Probability and Uncertainty

Probabilistic reasoning arises from uncertainty and allows one to quantify the prevision of future events. The uncertainty is a consequence of three major causes:

1. Unnoticed or doubtfully recorded past
2. Partially known present
3. Unknown future

The uncertainty associated with risk assessment of swarm operations is uncertainty of the second and third types, i.e., we have to make an inference about a situation from incomplete information about a partially known present and a completely unknown future. Incorporating incomplete information of the present is best done using a Bayesian approach. The Bayesian approach is based on a somewhat different view of probability than is classically put forth, and we next clarify the essential issues regarding the definition of probability.

Two views of probability dominate the literature, one which may be called "classical" and the other which is called "subjective." Classically, *probability* is defined as the attribute of an uncertain event that quantifies the likelihood that this event will occur. In this sense, "probability" refers to the likelihood of possible outcomes of an experiment, which can be calculated as the frequency of occurrence of a particular outcome relative to the total number of all possible outcomes. Formally, $p = k/n$, where $p$ is the probability, $k$ is the number of particular outcomes, and $n$ is the total number of all outcomes that occurred. The interval of possible values for $p$ is thus $[0, 1]$.

The subjectivist view, which includes the Bayesian approach, allows one to apply probabilities not just to events, as in the classical view, but to any uncertainty about which one has a belief (which is possibly completely neutral). To assure that allowable beliefs are at least self-consistent, the well-developed rules of probability theory are imposed upon beliefs. "Subjective" is unfortunately a contentious descriptor, because subjectivity is not generally encouraged in science, but this broadened view of probability can be both rational and practical when properly used. In particular, experts' opinions can be formally incorporated into statistical analyses, and this can be done in ways that are both justified and conservative.

## 15.2.2 Bernoulli Trials and the Binomial Probability Distribution

A *probability distribution* is most easily seen as an assignment of probabilities to all possible outcomes of a random experiment (assuming, for now, that the distribution consists of discrete values). For example, if there exist $k$ possible outcomes, then the discrete *uniform distribution* assigns equal values $1/k$ to all of these $k$ outcomes. For any distribution, we can calculate its *mean*, or average, and its *variance*, which measures the extent to which the numbers in the distribution vary from each other, i.e., their "spread." One of the most important discrete probability distributions is the *binomial distribution*. The binomial distribution results from independent *Bernoulli trials*, which are trials (experiments) with only two possible outcomes (which are generally coded as 1 for "success" and 0 for "failure"). One assumes that success probabilities remain the same during the time of trials, such as would be the case if we do not change the conditions of the experiment. The probability of "success" for any single trial is symbolized by $p$ and the probability of "failure" by $q = 1 - p$.

The aggregate of all possible distinct outcomes constitutes a *sample space*, and the probabilities associated with these outcomes make up a probability distribution. The sample space of $n$ sequential Bernoulli trials contains $2^n$ compound events, each representing one possible sequence of outcomes of the compound experiment. If we are interested only in the total number of successes that occurred in a sequence of $n$ Bernoulli trials, but not in their order, we can use the number of combinations of $k$ successes out of $n$ trials to find the probability of $k$ successes. The number of successful outcomes out of $n$ trials can range from zero to $n$, with corresponding probabilities $P(k = 0), P(k = 1), \ldots, P(k = n)$. To determine each of these $n + 1$ probabilities, we note that any particular ordering of $k$ successes and $n - k$ failures will have (assuming independent trials) a probability of the form $ppqqp \ldots qqp = p^k q^{n-k}$. Thus, if we can determine the number of different ways that $k$ objects (e.g., successes) can be distributed among $n$ places, we will be able to determine the overall probability of $k$ successes. The number of ways that "$n$ trials may result in $k$ successes and $(n - k)$ failures" is well known from the combinatorics formula $\binom{n}{k} = \frac{n!}{k!(n-k)!}$. Using the previously determined probability for any particular sequence with $k$ successes, we find that the total probability for all sequences with $k$ successes must be $\binom{n}{k} p^k q^{n-k}$. Putting the pieces together results in the following standard *Bernoulli trials formula* [71]:

$$P(k; n, p) = \binom{n}{k} p^k q^{n-k} \quad \text{for } k = 0, 1, 2, \ldots, n . \tag{15.1}$$

This formula calculates the probability of getting exactly $k$ successes out of $n$ trials. If we want to generalize this formula to calculate the probability of getting $k$ or more successes out of $n$ trials, then we need to take the sum over all possible values. In this case we get

$$\sum_{j=k}^{n} \binom{n}{j} p^j q^{n-j} . \tag{15.2}$$

## 15.2.3 Bayes' Theorem for Calculating the Posterior Probability

A *conditional probability* is the probability of one random outcome, given that one or more other outcomes are true. For example, the probability of rolling "two sixes" using a pair of fair dice is 1/36, but the probability becomes 1/6 if you see that one of the dice has already come up with a "six." Conditional probabilities are written as $P(a\,|\,b)$, which is read as "the probability that event $a$ will occur, given that event $b$ has already occurred."

If, in the original space, event $E$ occurs with probability $P(E)$, we can understand $P(E)$ as the frequency of "success" (from observing $E$) relative to observing anything at all (which has a probability of 1). Looked at in this way, we trivially write $P(E) = P(E)/1$. Now, conditioning on an event $H$ effectively changes the sample space to a subset of the original sample space, in particular, $H$. Thus the baseline relative frequency is now $P(H)$ instead of 1. Also, success now only occurs if $EH$ occurs, that is, if both $E$ and $H$ occur. Thus $P(E\,|\,H) = P(EH)/P(H)$, in the same way that we formerly wrote $P(E) = P(E)/1$. This is the formula for the conditional probability of $E$:

$$P(E\,|\,H) = \frac{P(EH)}{P(H)}\,. \tag{15.3}$$

Analogously, the formula for the conditional probability of event $H$ given $E$ is

$$P(H\,|\,E) = \frac{P(EH)}{P(E)}\,. \tag{15.4}$$

A bit of algebra results in *Bayes' Theorem*:

$$P(E\,|\,H) = \frac{P(E)P(H\,|\,E)}{P(H)} = \frac{P(E)P(H\,|\,E)}{P(E)P(H\,|\,E) + P(E^c)P(H\,|\,E^c)}\,, \tag{15.5}$$

where $E^c$ is the complement of $E$, i.e., the result "anything other than $E$."

Conditional probabilities, including the result in Bayes' Theorem, are important for any formulation of probability, but they play a special role in Bayesian statistics. Recall that Bayesian statistics follows the subjectivist paradigm of probability. Using this point of view, we can let $E$ correspond to a state of nature, not just to an outcome of an experiment, while $H$ corresponds to observed data. Using a Bayesian approach, we want to update our *belief* about $p$, which is the true probability of success (as in Sect. 15.2.2 above on Bernoulli trials). We can use $P(E)$ in Bayes' Theorem to find $P(E\,|\,H)$, which is the probability of $p$ having a particular value, given the observed data $H$. With the Bayesian approach, we represent our uncertainty about $p$ in an intuitive and appealing way—but not in a manner that is allowed when, as in the classical view, probabilities apply only to outcomes of random experiments.

In this formulation, $P(E)$ is called the *prior probability* or simply the *prior*, and $P(E \mid H)$ is called the *posterior probability* or simply the *posterior*. It is worth noting that instead of using Bayes' Theorem to calculate the posterior probability of an event, one can alternatively use it to calculate a *posterior probability distribution* over a set of possibilities. In this case, the posterior is based on being given a *prior probability distribution*. In fact, this is how Bayes' Theorem is used in our Bayesian formula in Sect. 15.3.3. The formula is used to derive the belief $P(E \mid H)$ as a posterior probability distribution of $p$ (where $p$ is the true probability of success), given a prior probability distribution.

## 15.3 Statistical Scouts Framework for Risk Assessment

With our background in probability, we can now move on to our statistical framework, which consists of two alternative predictive formulas for using scout samples to predict the probability of success in the overall swarm population. We will later compare the performance of these two formulas against each other.

## 15.3.1 Stochastic Predictions of Success: From Scouts to Swarms

The objective of our framework is to predict the probability that $y$ or more out of a total of $n$ agents will successfully navigate through an environment with obstacles and reach the goal location within a time limit.[2] This prediction is based on an estimation of the true probability, $p$, that one agent will successfully reach the goal. The true probability, $p$, is essentially a level of environmental difficulty, which in the simplest case is the number of obstacles per unit area of the territory where agents navigate.

To estimate the true probability, $p$, we send a sample of $k$ scouts (where $k$ is much less than $n$) and use them to calculate an estimate of $p$. Note that from now on this chapter will use the symbol $k$ to denote the number of scouts. The $k$ scouts are not included in the group of $n$ swarm agents (although modifying our formulas for the case of the scouts being part of the swarm would be straightforward). The usual estimate of $p$ is $\hat{p} = k_{\text{success}}/k$, where $k_{\text{success}}$ is the number of scouts successfully reaching the goal. Given $\hat{p}$

---

[2] In this chapter we use $n$ instead of $N$ for the number of agents in the swarm, to follow statistical convention.

as input, our formulas output the probability that $y$ or more out of $n$ agents will successfully reach their goal, i.e., the formulas calculate $P(Y \geq y \,|\, \hat{p})$, where $Y$ is a *random variable* that assumes values from zero to $n$. For the purposes of this chapter, all the reader needs to know about a random variable is that it can take on the values of possible outcomes of a random experiment.

## 15.3.2 The Standard Bernoulli Trials Formula

As mentioned above, scouts play the role of the sample, where each scout reaching the goal represents a success, and each scout failing to do this represents a failure. Then the standard (*maximum likelihood*) estimate of success probability for one swarm agent is $\hat{p} = k_{\text{success}}/k$, and the estimate of failure probability for one swarm agent is $1 - \hat{p}$. In this case, the entire process of a swarm reaching the goal can be considered a Bernoulli trials process, where the probability of $y$ out of $n$ agents reaching the goal corresponding to $y$ successes in $n$ independent trials is (similarly to Eq. 15.1)

$$P(Y = y \,|\, \hat{p}) = \binom{n}{y} \hat{p}^y (1 - \hat{p})^{n-y} . \tag{15.6}$$

Then the probability of $y$ *or more* out of $n$ agents reaching the goal is simply the sum of probabilities (as in Eq. 15.2) for every integer number from $y$ to $n$:

$$P_{\text{Bern}} = P(Y \geq y \,|\, \hat{p}) = \sum_{j=y}^{n} \binom{n}{j} \hat{p}^j (1 - \hat{p})^{n-j} . \tag{15.7}$$

Note that 15.7 defines the Bernoulli trials formula called $P_{\text{Bern}}$ that is one of the two formulas used in our predictive framework.

## 15.3.3 Our Novel Bayesian Formula

The Bernoulli formula, $P_{\text{Bern}}$, presented in Sect. 15.3.2 is a valid formula for predicting the success probability of swarm operations with a large number of scouts sent ahead of the swarm. But the problem with using $P_{\text{Bern}}$ is that the number of scouts, $k$, needs to be large, optimally more than 20 scouts, for reasonable predictions. However, in real-world situations it is often the case that practitioners do not want to send out more than a few scouts. For instance, although swarm robotic platforms are typically meant to be inexpensive and expendable, that may not always be the case. Furthermore, there are many other situations in which practitioners may want to minimize the number of scouts and save as many of their platforms as possible for the swarm. For example, it might be the case that swarm coverage or exploration of the region is important in addition to getting to the goal. Alternatively, it may be the case that the scouts must be stealthy—because if one of them gets discovered then it ruins the chances of deploying the swarm. Fewer scouts implies greater stealth.

To address any of the above (or other) reasons which motivate minimizing the number of scouts, in our research we assume a very small number (less than 20) of scout agents. This facilitates the application of our theory to a wide variety of practical situations. Unfortunately, with such a small $k$, the variance of $P_{\text{Bern}}$ tends to be high; there is a substantial chance that either all or no scouts will reach the goal, leading to poor predictions of future swarm behavior. This is both the result of high estimator variance for small sample sizes and a well-known problem of skewness of tiny samples [71]. To overcome this limitation of $P_{\text{Bern}}$, we derived a novel Bayesian formula, $P_{\text{Bayes}}$, for predicting the success of the swarm, starting with Bayes' Theorem for posterior probability, Eq. 15.5. To be clear, there is really no "free lunch" in all cases, but the Bayes estimator allows for great improvement when one has good a priori knowledge of what to expect, and it allows for modest improvements when prior knowledge is limited or lacking.

Our formula $P_{\text{Bayes}}$ assumes a prior probability distribution (representing the prior belief) over the probability $p$ of a single agent reaching the goal. So far in this chapter we have talked about discrete probability distributions, i.e., those assigning probabilities to each element in a discrete set of possible outcomes. But the prior distribution used by $P_{\text{Bayes}}$ must be a *continuous* distribution, because $p$ can assume any of an infinite range of real values on the interval [0,1]. Within this infinite set of possibilities is the true probability $p$, representing the true environmental difficulty, but we do not know beforehand what it is. This prior probability distribution can be initialized with prior information, if such information is available. For example, we may believe that the prior distribution should be roughly bell-shaped (Gaussian) with a mean of 0.5. Any prior knowledge that can be provided in the form of this prior distribution will reduce the variance of the predictions and "even out" the random perturbations of small samples, thereby enabling greater precision with a small $k$. We use a *Beta distribution* $\mathcal{B}(\alpha, \beta)$ over [0,1] for the prior [297]. The Beta distribution is a good choice because it can represent a variety of distributional shapes when the possible values are on the finite interval [0,1]. No other standard family of distributions has these properties. If the Beta distribution parameters $\alpha$ and $\beta$ are both 1, then it is the uniform distribution (which has constant probability over the interval [0,1]). In Sect. 15.5.1 we will describe our approach for methodically varying the parameters $\alpha$ and $\beta$ to achieve the characteristics we desire in the Beta distribution for our experiments.

We now have the necessary tools for presenting our Bayesian formula. Recall that our objective is to find the probability of $y$ or more successes out of $n$ agents. We are allowed to observe $k$ scouts, and their performance yields a fraction $\hat{p}$ of successes that is used to predict the true probability $p$ of one agent reaching the goal. Our Bayesian formula is (where the derivation may be found in [189])

$$P_{\text{Bayes}} = P(Y \geq y \,|\, \hat{p})$$

$$= \int_0^1 \int_0^p M \, z^{y-1}(1-z)^{n-y} p^{r-1}(1-p)^{s-1} \mathrm{d}z \, \mathrm{d}p \,,$$

$$\text{where} \quad M = \{B(r,s)\, B(y, n - y + 1)\}^{-1} \,,$$

$$r = k\hat{p} + \alpha \,,$$

$$s = k(1 - \hat{p}) + \beta \,,$$

$$\text{and } B(r,s) \text{ is the } Beta \; function, \text{ defined as} \int_0^1 x^{r-1}(1-x)^{s-1}\mathrm{d}x \,.$$

In this formula, $p$ takes on all possible values in the outer integral, $[0,1]$, and $z$ is a variable that ranges from zero to $p$.[3] This formula gives the posterior mean of the probability that $y$ or more agents will reach the goal, assuming a Beta prior for $p$ with parameters $\alpha$ and $\beta$. The inner integral gives the probability of $y$ or more successes out of $n$ trials for a given $p$. The outer integral averages (weighted by the posterior distribution) over all possible values $p$ of environmental difficulty, which ranges from zero to one. With a uniform prior and a large sample size, $P_{\text{Bayes}}$ will essentially be the same as $P_{\text{Bern}}$ (Eq. 15.7). The advantage of $P_{\text{Bayes}}$ (even with a uniform prior distribution) over $P_{\text{Bern}}$ is that it is more predictive for situations when the practitioner wants to deploy only a few scouts.

$P_{\text{Bayes}}$ is implemented in our simulation using numerical methods [187]. In particular, it was implemented with the *n-point Gaussian quadrature rule* from [32, 297].

## 15.4 Swarm Navigation Simulator

To visualize and evaluate our solution, we have designed and implemented a simulation (which can be seen in Fig. 15.1) consisting of a distributed, decentralized swarm of agents (the dots), a goal location (square), and a set of obstacles (circles). The locations of all of these entities can be varied. Agents can sense the goal at any distance. They can sense obstacles and other agents in any direction, but only within range $1.5D$, where $D$ is 50 pixels. Their control algorithm is based on physicomimetics, although our approach

---

[3] This formula is not the same formula as one would obtain by plugging the posterior mean of $p$ into the formula for $P_{\text{Bern}}$.

is not dependent on this particular algorithm and potentially can be used with other swarm control algorithms.

For our physicomimetic force law, we use a variant of the Lennard-Jones force between $i$ and $j$:

$$F_{ij} = 24\varepsilon \left[ \frac{d\delta^{12}}{r^{13}} - \frac{c\delta^6}{r^7} \right] ,$$

where the parameter $\varepsilon$ affects the strength of the virtual force between entities, $r$ is the actual distance between $i$ and $j$, $D = 2^{1/6}\delta$ is the desired distance between $i$ and $j$, and $c$ and $d$ control the relative balance between attractive and repulsive forces, respectively (see Eq. 9.5 for further discussion). The Lennard-Jones force law was selected because it is extremely effective for agents staying in geometric formations while navigating around obstacles to get to a goal location. We use optimal parameter settings for this force law that were evolved for the task by Suranga Hettiarachchi in his Ph.D. dissertation research [90]. See Chap. 14 for more about Hettiarachchi's research results. The values of all parameters that were used in our simulation can be seen on the left hand side of Fig. 15.1. To use the Lennard-Jones force law, agents sense the range and bearing to nearby obstacles, neighboring agents, and to the goal location.

Initially, all agents begin at uniformly randomly chosen locations within a starting square of side length 100 pixels, approximately 1,000 pixels from the goal. The width of the simulation area is 800 pixels, and its height is 600 pixels. The starting square for the agents is located in the lower left corner and the goal located in the upper right corner. The simulation ends after $t = 1,500$ time steps.

We can run this simulation in two modes: one where the agents are completely independent of each other, and another where there are virtual Lennard-Jones inter-agent forces. In this book chapter we focus only on agents with forces, e.g., Fig. 15.1 shows the simulation with inter-agent forces enabled. The screen view uses lines for the purpose of visualizing these virtual inter-agent forces. When these forces are employed, the agents stay in a geometric lattice formation while avoiding obstacles and navigating to the goal location. Such formations are especially useful for applications like search-and-rescue, demining, surveillance, chemical plume tracing, distributed sensor networks, or other tasks where data sharing is required.

It is worthwhile noting that when agents have their virtual forces enabled, the swarm (and the scouts) violate the independence assumptions (i.e., independent "trials") underlying both of our predictive formulas, $P_{\text{Bern}}$ and $P_{\text{Bayes}}$. Our prior research showed that with independent agents, the predictive accuracy of these formulas is excellent, as expected in the case where the assumptions are satisfied [189]. However, in this chapter we focus on the case of inter-agent forces, which violates the formulas' assumptions but is more realistic for most swarms. (In fact, in Chap. 17, Kazadi *defines* a "swarm"

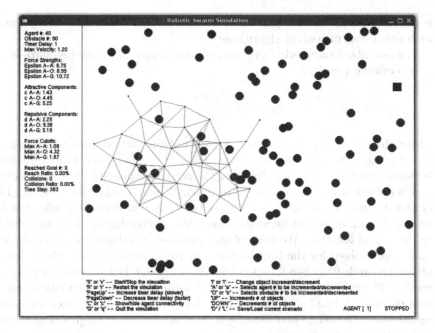

Fig. 15.1: Swarm simulation screen shot, with inter-agent forces. Large circles represent obstacles, the square in the upper right represents the goal, and the small dots represent agents [189]. (©2010, IGI Global. Used with permission)

as necessarily having inter-agent interactions.) The development of formulas with the interactions defined within them is beyond the scope of this chapter. Instead, in the remainder of the chapter we apply our existing formulas and determine how predictive they are—despite the violation of their independence assumptions.

## 15.5 Experimental Design

In this section we formulate and test several hypotheses about the effectiveness of our statistical scouts framework to predict the swarm operations success. The testing of hypotheses is accomplished by analyzing the error between the predictions and ground truth, in simulation experiments on the task of navigating through obstacles to a goal location. The Lennard-Jones force law is used for inter-agent virtual forces. To test each hypothesis, the parameter of interest is varied and other parameter values are held constant. We state that a hypothesis is confirmed if it holds true for overall trends in the graphs, even if it does not necessarily hold true for every point on every graph. After displaying the experimental results, general conclusions are drawn based on these results.

## 15.5.1 The Environments and the Prior Distribution for the Bayesian Formula

The experiments described in this chapter are designed to determine how the predictive accuracy of our two formulas varies as a function of changes in the independent variables (which are the parameters of our framework). One of these parameters that can be varied is the difficulty of the environment, e.g., the number of obstacles in it. We designed three different levels of environmental difficulty, which is similar to what was done in the bioluminescence problem in Chap. 6. Recall from that chapter that we chose low, medium, and high density environments to vary the challenge for the agents. We do likewise here, but for this application we vary the density of the number of obstacles instead of the density of the ambient chemical material. The three different levels of environmental difficulty used in this chapter are $p = 0.1$, 0.5, and 0.9. These levels of environmental difficulty were hand-crafted—by choosing the goal location, and the number, sizes, shapes, and locations of the obstacles, and the location of the starting square for the agents.

Another parameter that can be varied is the prior distribution in the formula $P_{\text{Bayes}}$. This distribution may be either more or less confident, and more or less accurate. Confidence is measured in terms of the prior *strength*, and accuracy is measured in terms of the prior *correctness*. Basically, a stronger prior distribution has a smaller variance, and a more correct prior distribution has a mean that is closer to the true probability of success in the environment, namely, $p$. Although in real-world situations the value of $p$ is not known, when experimentally evaluating our formulas we do know what it is. In addition to correct and incorrect priors, we also include a uniform prior, which is called "non-informative" because it is neither correct nor is it incorrect. Various forms of non-informative priors exist, even for the binomial probability model, but the uniform prior is intuitive and fairly represents a lack of prior knowledge about the success rate.

The prior distributions are given a level of correctness based on how well their means match the true value of environmental difficulty, which is measured as $p$. Therefore, now that we have selected our three values of $p$, we can define our prior distributions and how they vary along the strength and correctness dimensions. We agreed on seven final priors that provide nice variation along the strength and correctness dimensions for the different levels of environmental difficulty. Note that instead of designing correct priors for each environment, we used "almost correct" priors. The reason for this is that it is usually too hard to manually design environments that have a precise level of difficulty. Furthermore, priors used in practice are hopefully close to "correct" but are never expected to be fully correct, and small differences from ground truth have little impact on performance. The final seven prior distributions are:

1. The uniform distribution, which is the weakest prior. In this case, $\alpha = \beta = 1.0$.
2. A weak, almost correct prior distribution for $p = 0.1$. $\alpha = 2.1$ and $\beta = 18.9$.
3. A strong, almost correct prior distribution for $p = 0.1$. $\alpha = 10.9$ and $\beta = 98.1$.
4. A weak, almost correct prior distribution for $p = 0.5$. $\alpha = 2.5$ and $\beta = 2.5$.
5. A strong, almost correct prior distribution for $p = 0.5$. $\alpha = 14.5$ and $\beta = 14.5$.
6. A weak, almost correct prior distribution for $p = 0.9$. $\alpha = 18.9$ and $\beta = 2.1$.
7. A strong, almost correct prior distribution for $p = 0.9$. $\alpha = 98.1$ and $\beta = 10.9$.

Note that a prior that is correct for a particular value of $p$ will be incorrect for any *other* value of $p$. For example, a prior that is correct for $p = 0.1$ will be incorrect if the true $p$ is 0.3, even more incorrect if the true $p$ is 0.5, and even worse if the true $p$ is 0.9. This incorrectness is exacerbated if the prior is strong.

## 15.5.2 Experimental Parameters

Recall that one of our experimental objectives is to vary the parametric conditions and measure how the prediction accuracy varies as a function of the parametric variations. A *factorial experimental design* has been performed. In a factorial experimental design, we obtain results for *all* combinations of values of *all* parameters. Factorial experiments are notoriously computationally lengthy. To make our experiments tractable, we select only a few representative values for each parameter to vary.

The following list enumerates the parameters, i.e., the independent variables, that have their values varied in the experiments:

- **Total number of agents.** One variable is $n$, the total number of agents in the swarm. The values of $n$ are 100, 200, 300, 400, and 500.
- **Number of scouts.** Perhaps the most important independent variable is $k$, the number of scouts. It is important because it is typically under user control. The integer values of $k$ are three through 15 inclusive, but different graphs corresponding to different experiments focus on a subset of these values.
- **Desired number of successes.** Here, $y$ is the desired number out of the $n$ swarm agents that are required to succeed. This varies exhaustively from one to $n$. The first graph of results (below) varies the value of $y$ along the horizontal axis.

- **Environmental difficulty.** An important parameter that we vary is $p$, the "true" probability of one agent succeeding, which is found using 1,000 runs of the simulation of $n$ agents each. In Sect. 15.5.1 we learned that three environments were hand-crafted in order to get the following specific values of $p$: 0.1, 0.5, and 0.9.
- **Strength of the prior distribution.** Recall that we use a Beta prior probability distribution for $P_{\text{Bayes}}$. The "strength" of this prior is quantified as $1/\sigma^2$, where $\sigma^2$ denotes the variance. In other words, a lower variance yields a "stronger" prior. The reason we have selected this parameter to vary is because (a) the choice of prior is under control of a system user and therefore the hope is that the graphical data will provide advice to the user on how best to select a prior and (b) a strong prior can dominate the $k$ samples in influencing the value of $P_{\text{Bayes}}$. Therefore, if the system user is confident that his/her choice of prior is reasonably well-matched to the truth ($p$), then a stronger prior would be a wise choice, and vice versa.
- **Correctness of the prior distribution.** Another independent variable that we vary is the prior "correctness," which is the degree to which the prior is well-matched to the truth ($p$). Correctness is measured as the absolute value of the difference between the mean of the prior distribution and the true value of $p$. The smaller the difference, the more correct the chosen prior distribution is considered to be. A more correct prior should result in better predictive accuracy. In fact, the best accuracy is found by choosing the strongest correct prior. A user's belief may be close to or far from the true environmental difficulty (which is unknown to the user), but it is quite valuable to see how performance varies as a function of different combinations of strength and correctness. The methodology that was used to vary the strength and correctness (by varying $\alpha$ and $\beta$) of the priors was described above. For our experiments, we have selected six priors that vary along the strength and correctness dimensions, plus the weak uniform prior; these priors are listed in Sect. 15.5.1.

It is important to emphasize that a factorial design was used because of potential interactions between the independent variables. In particular, parameters related to prior selection are expected to be related to each other. Although, in theory, the prior will simply represent the actual belief of the user, in practice most users need guidance regarding the risks and benefits relative to the interplay between strength and correctness.

## 15.5.3 The Performance Metric

Our primary experimental objective is to evaluate and compare the quality (predictive accuracy) of the standard Bernoulli trials formula, which we call $P_{\text{Bern}}$, and our Bayesian formula, which we call $P_{\text{Bayes}}$. Evaluating the quality of a formula implies comparing its value against a form of ground truth. The comparison must be fair, so when comparing either $P_{\text{Bern}}$ or $P_{\text{Bayes}}$ with ground truth, we ensure that all three use the same values for all parameters.

We use the *mean squared error (MSE)* for our performance metric here because it includes both the *bias* and the variance of the error, and is therefore highly informative.[4] The $MSE$ is a function of the "absolute error," which is the difference between the estimate and the truth. Assuming $X$ is the true value and $\hat{X}$ is an estimate of $X$, the $MSE$ is defined as:

$$MSE(\hat{X}, X) = E[(\hat{X} - X)^2],$$

where $E$ denotes the *expected value*. The expected value of a discrete random variable is the weighted average of all of its possible values, where the weights are the probabilities of the values. The definition for a continuous random variable is analogous, though a bit more complicated. The expected value is sometimes called the "expectation" or the "mean." It becomes the "average" in the common sense of the word if all probabilities are equal.

To measure the $MSE$, we need a form of ground truth. Our ground truth, which we call $P_{\text{Truth}}$, is the fraction out of 1,000 runs of the simulation in which $y$ or more of the $n$ agents reach the goal [188].

The expectation $E$ is calculated as an average over 1,000 runs. For each run, we randomly vary the agent locations within the starting square. The dependent variable $MSE$ is calculated as follows. Let $P_B$ be $P_{\text{Bern}}$ or $P_{\text{Bayes}}$, where these algorithms are defined in the next section. Then

$$Bias = E[P_B] - P_{\text{Truth}},$$

$$Variance = E[(P_B - E[P_B])^2],$$

$$MSE = E[(P_B - P_{\text{Truth}})^2] = Bias^2 + Variance.$$

## 15.6 The Experimental Algorithms

In this section, we present the algorithms used in our experiments. Each experiment measures the $MSE(P_{\text{Bern}}, P_{\text{Truth}})$ and $MSE(P_{\text{Bayes}}, P_{\text{Truth}})$ with a single choice of values for all parameters for multiple runs.

Before we present the algorithms, it is important to consider how we will calculate the value of $\hat{p}$, which is the estimate of $p$ found using $k$ scouts, to insert into the formulas $P_{\text{Bern}}$ and $P_{\text{Bayes}}$. One way to obtain $\hat{p}$ in our experiments would be to deploy the $k$ scouts in the simulation to get $\hat{p}$ as the fraction out of the $k$ scouts that succeeded, and then plug $\hat{p}$ into the formulas for $P_{\text{Bern}}$ and $P_{\text{Bayes}}$. This is what would be done in real-world situations

---

[4] In statistics, the bias of an estimator is the difference between its expected value (which will be defined shortly) and the true value.

with the actual robots. The problem with doing this for our experimental comparisons, however, is that it does not predict performance *in expectation*. For the experimental evaluation of our framework, our methodology needs to be adapted for fair experimental comparisons. The correct solution for the experiments is to compute $\hat{p}$ over multiple runs. Now we are ready to present the algorithms.

The algorithm for $P_{\text{Truth}}$ was just described in Sect. 15.5.3. The following is the algorithm for $P_{\text{Bayes}}$ and $P_{\text{Bern}}$:

1. Using 100 runs of the simulation with $k$ scouts, obtain an expected value of $\hat{p}$, which is an estimate of $p$.
2. Apply the implemented formula for $P(Y \geq y \,|\, \hat{p})$, using the expected value of $\hat{p}$ obtained from Step 1.

In the following section we formulate and test hypotheses. Because we have performed a factorial experimental design, the results are voluminous, i.e., there are hundreds of graphs. This chapter only has room to show a few of the most interesting and representative results. In the graphs, the horizontal axis is one of the independent variables and the vertical axis is the (average) $MSE$. Each curve shows the $MSE$ estimated by averaging over 1,000 runs.

## 15.7 Experimental Hypotheses and Results

Throughout this chapter, we assume that *both* the swarm and the scouts use the Lennard-Jones force law. This ensures that the scout behavior is representative of the swarm behavior.

Here, we formulate and test four main hypotheses and two subsidiary hypotheses. To test each hypothesis, the parameter of interest is varied and the other parameter values (except $y$) are held constant. When $y$ is not on the horizontal axis, each curve is an average over all values of $y$. Conclusions about both the $MSE(P_{\text{Bern}}, P_{\text{Truth}})$ and $MSE(P_{\text{Bayes}}, P_{\text{Truth}})$ are presented. We begin with a hypothesis that is specific to $P_{\text{Bayes}}$ only:

**Hypothesis 1 (about the prior distribution for $P_{\text{Bayes}}$):** *(1) It is better to have a correct rather than incorrect prior. (2) If the prior is incorrect, it is better to be weak rather than strong. Also, the non-informative uniform (weakest) prior is better than any incorrect prior. (3) If the prior is correct, it is better to be strong than weak.*

All three parts of Hypothesis 1 have been confirmed by our experimental results. For example, the two curves for the strong and weak almost correct priors in Fig. 15.2 confirm part 3 of Hypothesis 1 over most values of $y$, where $y$ is varied along the horizontal axis. In addition to confirming part 3 of Hypothesis 1, Fig. 15.2 shows $P_{\text{Bayes}}$ with a uniform prior, and $P_{\text{Bern}}$, for the sake of comparison. The interesting *bimodal* ("double-humped") shape of the curves is explained in [189]. For the sake of conciseness of comparisons of $P_{\text{Bayes}}$ versus $P_{\text{Bern}}$, after this hypothesis we will show curves that average over all values of $y$ and vary the parameter of interest along the horizontal axis.

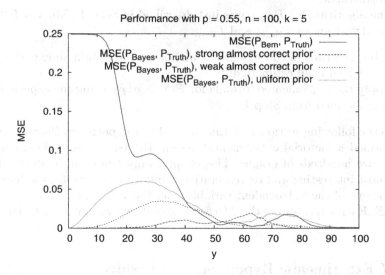

Fig. 15.2: Graph for Hypothesis 1, part 3, with $p = 0.55$, $n = 100$ and $k = 5$ [189]. (©2010, IGI Global. Used with permission)

Overall, we have found that a strong, correct prior is optimal, and a uniform prior is the best choice when prior knowledge is lacking. Although these results are not expected to be interesting to readers familiar with Bayesian reasoning, we include them in this book chapter because we anticipate that many of our readers will not have studied this topic in depth before.

Next, we consider hypotheses that pertain to *both* of the formulas $P_{\text{Bern}}$ and $P_{\text{Bayes}}$ in our framework. The first is about increasing the number of scouts while the swarm size remains constant. Then, we will discuss the situation of increasing the swarm size.

**Hypothesis 2 (about the number of scouts):** *Increasing the number of scouts, $k$, will improve the predictive accuracy of the formulas.*

Although it seems quite intuitive that this hypothesis should hold, the results are mixed for both formulas, $P_{\text{Bern}}$ and $P_{\text{Bayes}}$. Figure 15.3 exemplifies situations where Hypothesis 2 is violated. In this figure the swarm size $n$ is held constant and the number of scouts $k$ is varied. Observing the simulation reveals *why* Hypothesis 2 is violated. Increasing the number of scouts (i.e., the sample size) does not increase the predictive accuracy of our formulas in general because, due to the Lennard-Jones forces, the scouts stay in formation and act as a quasi-unit, just like the swarm does. Because of this quasi-unit behavior, the number of (scout or swarm) agents is irrelevant for easy to moderate environments. On the other hand, for the specific case of especially difficult environments (where $p$ is close to 0.0), adding more scouts does tend to help the predictions. This is because the quasi-unit tends to lose formation connectivity in difficult environments (e.g., with many obstacles), and therefore a larger sample size is needed to make accurate predictions of swarm behavior.

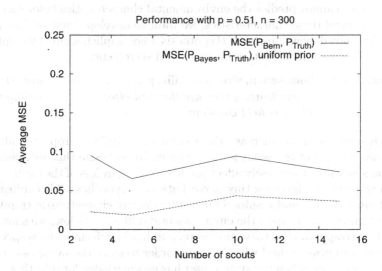

Fig. 15.3: Graph of MSEs with $p = 0.51$, $n = 300$, $k$ ranging from 3 to 15 [189]. (©2010, IGI Global. Used with permission)

To reiterate this point, in easy to moderate environments, very few scouts are needed for a very good assessment of the swarm success rate, and adding more scouts does little to improve the predictions. This is a fortunate result because it implies that just a few scouts can be enough to obtain a good estimate of the swarm success rate. Nevertheless, it does *not* hold in difficult environments with many obstacles. This can be summarized in the following two hypotheses:

**Hypothesis 2.1 (about the ratio of scouts to swarm size):** *In easy to moderate environments, a small number of scouts is predictive, regardless of the swarm size.*

**Hypothesis 2.2 (about the ratio of scouts to swarm size):** *In difficult environments, the number of scouts needs to be proportional to the
number of agents in the swarm, in order to maintain predictive accuracy.*

In other words, for harder environments, the sample size (i.e., number of
scouts) needs to increase as the swarm gets bigger—because obstacles in the
environment fracture the formation. Hypothesis 2.2 is based on a statistical argument; see [189] for details. Both Hypotheses 2.1 and 2.2 have been
confirmed by the data.

Since we cannot predict the environmental characteristics beforehand, nor
can we control these characteristics, we want to develop a general hypothesis
that does not depend on them. Hypothesis 3 accomplishes this by combining
Hypotheses 2.1 and 2.2 into a more general statement:

**Hypothesis 3 (about swarm size/scalability):** *Increasing the swarm size does
not significantly increase the prediction error, assuming the ratio $k/n$ is held constant.*

In other words, if we increase the swarm size, *and* we proportionally also
increase the size of the scout group, *then* we hypothesize that increasing the
swarm size will not adversely affect predictions, regardless of the environment.

In general (i.e., ignoring tiny perturbations), Hypothesis 3 is confirmed by
all of our experimental results. Our results display characteristic trends, and
these trends depend upon the environmental difficulty. Below, we show three
trends corresponding to easy, moderate, and hard environments, respectively.
The uniform prior is used with $P_{\mathrm{Bayes}}$ in order to show the worst case results.
In the worst case, when the system user has no knowledge for selecting a prior,
he/she can always use the uniform distribution. Showing this case gives the
reader an idea of a lower bound on the accuracy of predictions possible using
our $P_{\mathrm{Bayes}}$ formula.

Figure 15.4 exemplifies the trend in easy environments, where $p = 0.9$.
This figure shows that for easy environments the $MSE$ actually *decreases*
when going from 100 to 300 agents, and then it increases somewhat when
going from 300 to 500 agents. However, with 500 agents the $MSE$ is still less
than with 100 agents. This assumes a constant $k/n$ ratio of 0.03.

Figure 15.5 shows the performance trend in moderate environments, where
$p = 0.5$. There is a slight increase in $MSE$ for $P_{\mathrm{Bayes}}$ as the swarm size scales
up, while the ratio of scouts to swarm agents is held constant. However, the
maximum increase in $MSE$ in either curve is very small and not particularly
noteworthy.

Finally, Fig. 15.6 exemplifies the trend in the most difficult environments,
where $p = 0.1$. As the number of agents increases while the ratio is main-
tained, the $MSE$ is reduced for $P_{\mathrm{Bern}}$ and reduced or maintained for $P_{\mathrm{Bayes}}$
(if we focus on the overall trends and ignore the small spikes for both al-
gorithms at 12/400). This result is both surprising and significant. It shows

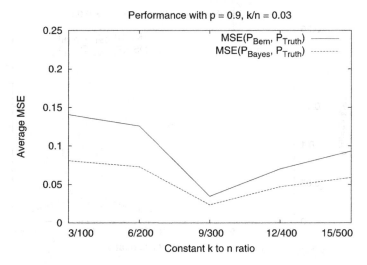

Fig. 15.4: Graph of $MSE$ with varying $n$ and $k$, $k/n = 0.03$, $p = 0.9$, and a uniform prior for $P_{\text{Bayes}}$ [189]. (©2010, IGI Global. Used with permission)

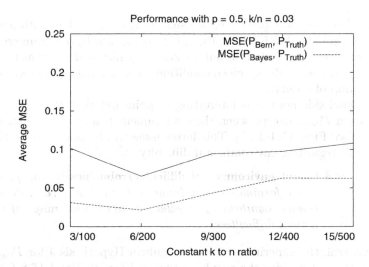

Fig. 15.5: Graph of $MSE$ with varying $n$ and $k$, $k/n = 0.03$, $p = 0.5$, and a uniform prior for $P_{\text{Bayes}}$ [189]. (©2010, IGI Global. Used with permission)

that in the most difficult environments, predictability can actually improve with our framework.

Based on the graphs, Hypothesis 3 is almost always true—for both the $MSE(P_{\text{Bayes}}, P_{\text{Truth}})$ (with the uniform prior, though the results with a

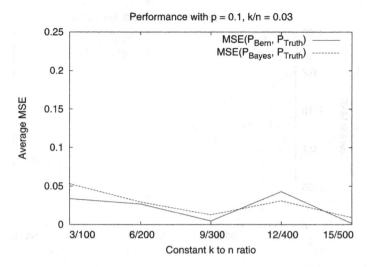

Fig. 15.6: Graph of $MSE$ with varying $n$ and $k$, $k/n = 0.03$, $p = 0.1$, and a uniform prior for $P_{\text{Bayes}}$ [189]. (©2010, IGI Global. Used with permission)

strong, almost correct prior are, as usual, much better than with a uniform prior) and for the $MSE(P_{\text{Bern}}, P_{\text{Truth}})$. In other words, our framework consisting of swarm prediction formulas shows very nice scalability as the swarm size increases, under the specified conditions (i.e., when using a proportionally larger group of scouts).

As a final side note, it is interesting to point out that $P_{\text{Bayes}}$ has a lower $MSE$ than $P_{\text{Bern}}$, except when the environment is maximally difficult with $p = 0.1$ (see Figs. 15.4–15.6). This latter issue will be addressed in the next hypothesis regarding environmental difficulty.

**Hypothesis 4 (about environmental difficulty/robustness):** *The predictions of the formulas in our framework are robust (i.e., they do not change significantly in value) across a wide range of environmental difficulties.*

In general, the experimental results confirm Hypothesis 4 for $P_{\text{Bayes}}$, but not for $P_{\text{Bern}}$. Consider the graphs shown in Figs. 15.7 and 15.8 (and note that environmental difficulty *decreases* when going from left to right along the horizontal axis). When varying the environmental difficulty along the horizontal axis of the graphs, two different trends/patterns can be seen. Figure 15.7 is typical of the first pattern. $P_{\text{Bayes}}$, even with a uniform prior, is considerably more robust than $P_{\text{Bern}}$ over the range of environments. In this figure, $MSE(P_{\text{Bayes}}, P_{\text{Truth}})$ never goes significantly above 0.05. On the other hand, the predictive ability of $P_{\text{Bern}}$ degrades considerably in easy environments, e.g., $MSE(P_{\text{Bern}}, P_{\text{Truth}}) \rightarrow 0.09$ as $p \rightarrow 0.9$ in Fig. 15.7. $P_{\text{Bayes}}$ with a uniform (but not strong correct) prior performs worse than $P_{\text{Bern}}$ in the most difficult environment ($p = 0.1$), *but* both perform surprisingly well at that extreme.

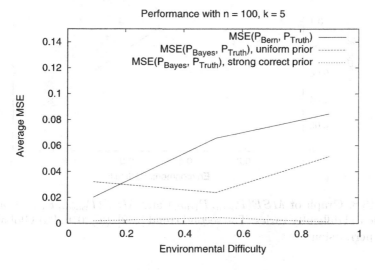

Fig. 15.7: Graph of $MSE(P_{\text{Bern}}, P_{\text{Truth}})$ and $MSE(P_{\text{Bayes}}, P_{\text{Truth}})$, as environmental difficulty varies. The first trend [189]. (©2010, IGI Global. Used with permission)

Figure 15.8 is typical of the second trend. In the set of graphs typifying the second trend, both $P_{\text{Bern}}$ and $P_{\text{Bayes}}$ with uniform prior curves have a similar shape, but the $MSE(P_{\text{Bern}}, P_{\text{Truth}})$ is much higher for moderate environments than the $MSE(P_{\text{Bayes}}, P_{\text{Truth}})$ with a uniform prior. Again, $P_{\text{Bayes}}$ with a uniform prior performs worse than $P_{\text{Bern}}$ when the environment is toughest, but otherwise it performs better than $P_{\text{Bern}}$.

We conclude that when averaged over all environments (which is an uncontrollable parameter), $P_{\text{Bayes}}$ is superior to $P_{\text{Bern}}$ and it is robust. $P_{\text{Bern}}$, on the other hand, shows a lack of robustness that can make it barely acceptable in moderate to easy environments.

Finally, note that both Fig. 15.7 and Fig. 15.8 reinforce the value of a strong, almost correct prior—if feasible to obtain. The $MSE$ is always lower, and often substantially lower, when the prior is well-matched to the environment.

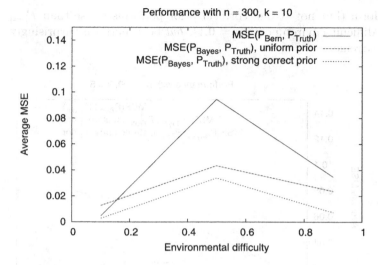

Fig. 15.8: Graph of $MSE(P_{\text{Bern}}, P_{\text{Truth}})$ and $MSE(P_{\text{Bayes}}, P_{\text{Truth}})$, as environmental difficulty varies. The second trend [189]. (©2010, IGI Global. Used with permission)

## 15.8 Experimental Conclusions

The following conclusions hold, assuming (as we did throughout this chapter) that the scouts and swarm both use the Lennard-Jones force law for physicomimetics:

1. **Accuracy:** *Both formulas usually have good predictive accuracy.*
2. **Scalability:** *For both formulas, predictive accuracy is largely maintained when increasing the swarm size, if there is a proportional increase in the number of scouts.*
3. **Robustness of the Bayesian formula:** $P_{\text{Bayes}}$ *is "robust" across wide variations in environmental difficulty. By "robust" we mean that the* $MSE$ *does not substantially increase as the environmental difficulty is varied.* $P_{\text{Bern}}$ *lacks robustness.*
4. **Superiority of the Bayesian formula:** $P_{\text{Bayes}}$ *with a uniform prior is usually a better predictor than* $P_{\text{Bern}}$, *and is often much better.*

The above conclusions are based on factorial experiments run in our simulation on the task of navigating through obstacles to get to a goal location.

Let us further elaborate on the superiority of $P_{\text{Bayes}}$ over $P_{\text{Bern}}$. In most cases, the predictions obtained using $P_{\text{Bayes}}$ with a uniform prior have a much lower mean squared error (i.e., they are much more accurate) than those of $P_{\text{Bern}}$. All graphs comparing $P_{\text{Bayes}}$ with $P_{\text{Bern}}$ support this conclusion. Furthermore, $P_{\text{Bayes}}$ is much more robust than $P_{\text{Bern}}$. These conclusions have significant practical implications. They imply that even in the case when prior knowledge is absent (which is a worst-case situation for the Bayesian formula), $P_{\text{Bayes}}$ typically outperforms $P_{\text{Bern}}$ and is more robust. The primary advantage of the Bayesian formula is its superior predictive accuracy *when there are few scouts* (and, theoretically, we know that as the number of scouts gets large, the difference between $P_{\text{Bayes}}$ and $P_{\text{Bern}}$ will vanish). Because of the pragmatic value of deploying only a few scouts (in terms of cost, stealth, and other considerations), the superiority of $P_{\text{Bayes}}$ for this situation is potentially very significant for swarm practitioners.

In summary, our novel Bayesian formula is generally superior to the standard Bernoulli trials formula for the situation of sending out only a few scouts to determine the probability of swarm success. This, along with our mathematical framework for risk assessment, are our contributions to the field of swarm robotics.

## 15.9 The Role of Physicomimetics

Throughout our research, we used the Lennard-Jones force law for both the scout agents and the swarm agents. Toward the end of this project, we decided to explore physicomimetics with a different force law. Recall that one of the advantages of physicomimetics is that it is possible to achieve behavior analogous to either a fluid or a solid. As discussed in the earlier chapters of this book, a split Newtonian force law results in solid-mimetic behavior, whereas a Lennard-Jones force law results in liquid-mimetic behavior. To gauge the performance of our statistical framework for solid-mimetic scouts and swarms, we applied it with the split Newtonian force law.

When we compared the *predictive accuracy* (and remember that we are focusing on the accuracy of the predictions, not the success rate of task performance) of our framework using split Newtonian versus Lennard-Jones, the accuracy was considerably better with the latter. Our results, which agree with those in Chap. 14, demonstrate that with the split Newtonian force law small groups of agents can squeeze through narrow passageways between obstacles but larger solid-mimetic swarms get stuck [92]. The implication is that scouts and swarms often behave differently when using the split Newtonian force law, whereas they behave quite similarly when using the Lennard-Jones force law. This latter observation can be seen with any ordinary liquid, e.g., a small quantity of water behaves very similarly to a large body of water in terms of flowing around obstacles.

In summary, although our scouts framework does not require physicomimetics for agent control, using it gave us an advantage. With physicomimetics, we were able to draw conclusions about the compatibility of our framework with a particular mimicked state of matter. This should be helpful to users of our framework in terms of better understanding and applying it.

## 15.10 Issues for Porting to Real Robots

The theoretical scouts framework presented in this book chapter is a general approach. It models agents at an abstract level, i.e., without any detailed modeling of the vehicle characteristics, including sensing and actuation specifics as well as noise in sensing or actuation. But in terms of the probability of success, which is the meta-parameter of concern here, noise that causes performance degradation is, at an abstract level, equivalent to a "hard" environment, which we do model in our simulation. Furthermore, as shown in Chap. 19 of this book, physicomimetics is especially adept at handling noise.

We therefore believe that porting our scouts approach from simulation to real robots will be straightforward, provided a sampling assumption holds. This assumption states that the scouts are "typical" of the swarm. In other words, the scout vehicle characteristics, including their imperfections such as noisy sensors, must closely match those of the swarm vehicles. If all of the vehicles are designed to be homogeneous, and the scouts are randomly selected from the pool of swarm vehicles, then this assumption will likely hold.

## 15.11 Summary

This book chapter has presented a statistical framework for predicting swarm success rate based on sampling. It is currently the only framework in the field of swarm intelligence that utilizes scouts to predict the success probability of the operations to be performed by a swarm of agents, prior to the actual deployment of the swarm. The real advantage of this framework is its capability of saving the swarm in difficult or dangerous situations, i.e., when sending the entire swarm without preliminary reconnaissance may result in a loss of a substantial part of the swarm. Our framework has practical value for military and civilian tasks such as search-and-rescue, search-and-destroy, demining, foraging, and surveillance. In a broader sense, the statistical scouts framework may be implemented for risk assessment for many different tasks where hazardous conditions may be encountered. For example, a very different sort of application might be to "test the waters" prior to deploying a swarm of software advertisements.

A key element of our framework is a novel Bayesian formula for estimation of the swarm success probability, derived *ab initio* from probability theory principles. This formula is a Bayesian version of the standard Bernoulli trials formula, but it derives superior performance in the case of few samples due to its use of a prior probability distribution that reduces the variance and, as a consequence, increases the robustness of the formula's predictions over its Bernoulli formula predecessor.

We have also designed and implemented a swarm simulator that uses both the Bernoulli and Bayesian formulas within our framework. This navigation simulator has parameterized capabilities for running rigorous experiments to test the framework. A full factorial experimental design was implemented and executed in the context of the navigation task. The results were analyzed with the mean squared error (MSE). Our main conclusions are:

1. *Both formulas are reasonably accurate; they are also scalable if the size of the scout group is proportional to the size of the swarm.*
2. *The Bayesian formula is more accurate and robust in the case of few (less than 20) scouts, and is therefore recommended in this situation.*

In addition to these conclusions, it is important to point out the practical value of the Bayesian formula. Consider the following illustrative scenario. Suppose that a swarm practitioner requires at least half of the swarm to succeed at the task. He sends out a group of scouts and finds out that not enough of them succeed to be confident that half of the swarm will also succeed. With the Bernoulli formula, his prediction will be based only on the scouts' success rate. In this case, he will not deploy the swarm. But with the Bayesian formula he can potentially do better. Perhaps he started with no prior knowledge about the environmental difficulty. Again, based on the prediction, he decides not to deploy the swarm. But suppose that while he is waiting, he receives updated information that the environment is expected to be quite easy. Then he can insert that information into his prior distribution and rerun the calculations with the Bayesian formula, and there is a reasonable chance that the prediction will suggest that he *should* deploy the swarm. Furthermore, this can be done without using any more scouts. Such added flexibility can make a significant difference to the practitioner.

In conclusion, we believe that our novel statistical scouts framework, especially when used with our Bayesian formula, has the potential to be quite useful for various risk-averse applications and tasks in the field of swarm intelligence. It can serve as a risk assessment tool that precedes other strategies for accomplishing the task.

Finally, note that our statistical framework is not specific to one type of agent control strategy. In the future we would like to implement it with a variety of control strategies. Here, we implemented it with physicomimetics. Two different force laws were compared and we found a special synergy between our framework and the Lennard-Jones force law. This discovery was explained as being related to the liquid-mimetic behavior of physicomimetic scouts and swarms using Lennard-Jones. The general lesson to be learned is that physicomimetic control laws lead to swarms with behavior that can mimic different states of matter, and this allows the practitioner to select the state of matter most compatible with his/her task and performance objectives.

# Chapter 16
# Physicomimetic Swarm Design Considerations: Modularity, Scalability, Heterogeneity, and the Prediction Versus Control Dilemma

R. Paul Wiegand and Chris Ellis

## 16.1 Introduction

While this book focuses on a particular subset of swarm-based approaches—and primarily on applications of such to coordinated control of large multi-agent teams—it's worth stepping back a bit to examine more generally some of the broader issues surrounding *how* one goes about the process of developing such solutions in the first place. A cursory examination uncovers two quite different perspectives on developing multi-agent control systems: *top-down, knowledge-rich* approaches and *bottom-up, knowledge-poor* approaches, each with some advantages and disadvantages.

For example, many traditional top-down methods from the fields of *artificial intelligence* and *operations research* have quite rigorous approaches to developing coordinated multi-agent team behaviors based on a well-formulated, systems-oriented perspective (e.g., *control engineering* approaches [123]). Unfortunately, such methods are particularly difficult to implement in multi-body control settings, and often require a great deal of effort from human engineers to elicit precise mathematical models. In multi-agent problems, the number of interactions between agents can render top-down approaches prohibitive for realistic problems.

In response, bottom-up approaches, such as swarm-based solutions, have tended to be knowledge-poor, relying instead on the idea that complex solutions can *emerge* from nonlinear interactions among relatively simple agents. While this idea of emergent behavior works quite well in simple settings, it is

R. Paul Wiegand
Institute for Simulation & Training, University of Central Florida, 3100 Technology Pkwy., Orlando, FL, USA, e-mail: wiegand@ist.ucf.edu

Chris Ellis
Department of Electrical Engineering & Computer Science, University of Central Florida, 4000 Central Florida Blvd., Orlando, FL, USA, e-mail: chris@cs.ucf.edu

W.M. Spears, D.F. Spears (eds.), *Physicomimetics*,
DOI 10.1007/978-3-642-22804-9_16,
© Springer-Verlag Berlin Heidelberg 2011

not always clear how to construct bottom-up, complex systems based solutions to complicated problems—particularly as problems scale in complexity. Moreover, a hands-free approach to coordinated control appears to ignore a critical source of information about the problem: the engineer.

But we can take at least one more step back. While coordinated control of large multi-agent teams poses particular challenges, the question of how to go about the process of developing complex software solutions is hardly a new one. A well-accepted way to manage complex software development is to develop according to a well-formed *software lifecycle* [227]. There are many lifecycles and many perspectives on lifecycles; some of them aren't that different from the top-down versus bottom-up perspective differences described above (e.g., traditional waterfall versus agile development). However, in *software engineering* nearly all development cycles include one or more steps that typically correspond with some kind of *software design*. Design often involves a number of methods and mechanisms to help bring to bear knowledge about high-level system aspects, such as *modularity* (i.e., the division and reuse of logic) and *scalability* (i.e., how the system performance is affected by increases in problem complexity). Design is also often the place where likely pitfalls and testing strategies are laid out. In some sense, then, the software engineering process itself, which includes components such as *specification*, *design*, *implementation*, *testing*, and *validation*, provides ways for engineers to inject high-level domain knowledge to shape the eventual solution to the problem.

In recent years researchers have begun to consider principled and practical methods for finding a middle ground between top-down and bottom-up approaches to coordinated multi-agent control by incorporating *some* human knowledge into the system, while leaving as much representational flexibility as possible. Analogously, we might imagine a variety of potential *swarm engineering* processes, any or all of which should include some kind of *swarm design* component. While the analogous processes in swarm engineering are not nearly as well developed as in traditional software engineering, there *are* some known methods for designing and testing swarm solutions.

In our view, the process of swarm design is vital to the successful development of realistic swarm solutions to complex, real-world problems. Moreover, engineers designing swarms should give direct and explicit thought to the key properties they need to obtain from their system, as well as to the potential pitfalls they may run into. In this chapter, we take up examples of both of these. First, we discuss some ways an engineer can incorporate domain knowledge during swarm design in order to make knowledgeable and deliberate decisions about several properties that are often necessary, such as the development of *heterogeneous* team behaviors that are scalable and modular. In addition, we discuss some of the issues complicating both the implementation and validation of behaviors based on differences between nature-inspired models and control input. These discussions will include some practical suggestions for certain design motifs for heterogeneous swarms (e.g., *graph-based interaction models*), some ideas for how to construct test cases, as well as a number of running examples to help clarify our points. The goal is to provide a subset of potential tools for what should be a much larger toolkit for swarm engineers to build effective and efficient swarm solutions.

## 16.2 Existing Swarm Engineering Approaches

The term "swarm engineering" has been used in a number of ways. We define it as the process of specifying, designing, implementing, testing and validating swarm behaviors, which is fairly consistent with most interpretations of this phrase. With such a broad definition, it's not surprising that a lot of work falls into such a category. Rather than provide an exhaustive survey of such, we focus on a few researchers who, as we do, highlight the notion that a middle ground between top-down and bottom-up perspectives can and should be sought when designing realistic swarm solutions.

For example, our observation that swarm engineering can be likened to software engineering is by no means a novel perspective. Winfield et al. [271] provide a survey of case studies applying conventional engineering approaches for dependability to swarm design. They point out the general need for better tools for swarm engineering, as we do.

In response to such needs, Kazadi developed a formalism for systems based on swarm intelligence and described an approach to engineering the behavior of swarms according to that formalism (e.g., see Chap. 17 and [112]). His *middle-meeting method* is divided into two high-level steps: (1) Create a global property based on local properties. (2) Construct local behaviors that will lead to the global property. Additionally, Kazadi advocates a rigorous framework in which terms, system properties, goals and behaviors are stated with mathematical clarity [112]. Kazadi's work is perhaps the most complementary with our discussion in the sense that it might be seen as a kind of "bookending" to the development process. In particular, it is useful for establishing specifications, high-level design and validating final behaviors, but not necessarily helpful in directing specific attention to issues such as modularity and scalability.

Chang [30] also describes an approach that combines both top-down macroscopic with bottom-up microscopic swarm design techniques. Much like our discussion later in this chapter on interaction models, his method gives useful guidance to the swarm designer in terms of decomposing the swarm engineering problem. Though the method still leaves open the question of designing low-level behaviors, it promotes similar notions regarding modularity and scalability.

In work contemporaneous with the initial development of physicomimetics, Reif and Wang [190] developed a method called *social potential fields* as a way to program large teams of robots. Like the generalization of physicomimetics discussed below, their method models the agents as particles, provides for multiple types of agents, and generates behaviors through interactions of forces between agents. Reif and Wang propose a hierarchical methodology for determining the set of potential force laws, laying out a step-by-step procedure for developing system behaviors with different levels of interactions. Again, the engineer's task of designing and building the underlying local behaviors is still left open.

In fact, many of the modern works in swarm engineering and swarm design reflect the need to establish a process by which engineers bring domain information to bear in the development of swarm behaviors, and to find some common ground between traditional top-down and bottom-up approaches. This chapter focuses on two critical aspects of the swarm design process: describing interaction patterns such that we can incorporate ideas of scalability and modularity into heterogeneous swarm design, and considering possible pitfalls with respect to challenges generated by differences between natural models and the systems they are meant to control. These ideas are consistent with, and complementary to, a larger swarm engineering process, which we advocate in any case. We suggest the reader explore further works for a comprehensive swarm engineering perspective.

## 16.3 Heterogeneous Behaviors

Historically, most swarm-based multi-agent solutions have relied on *homogeneous behaviors*—those in which the essential control logic in each agent is identical. This makes sense based on the natural inspiration that undergirds most swarm methods. The idea is to keep the individual agent logic as simple as possible and rely on nonlinear interactions to produce aggregate, complex solutions. Still, some problems require swarm engineers to consider slightly more sophisticated solutions that involve different types of agent behaviors. We define swarms with members that contain two or more differentiated, specialized agent types as having *heterogeneous behaviors*.

The computer science, robotics, and biological literature is replete with investigations into issues surrounding the allocation of roles, tasks, or behaviors within teams [26]. The machine learning community in particular has investigated methods for *learning* team behaviors that contain specialization [174]. Indeed, a number of multi-agent learning studies indicate that certain types of problems, such as various cooperative predator–prey type problems, may *benefit* from heterogeneous team behaviors [23, 44, 45, 180] in the sense that design, implementation, and learning can become more tractable for such systems. Moreover, it is likely that many challenging problems will *require* multi-agent teams that contain individuals with specialized behaviors.

As many have observed [44, 96, 263], the degree of heterogeneity of team behaviors is neither a Boolean property (i.e., true or false), nor is it necessarily unidimensional or discrete. For one, a large team with two distinct specialized behaviors is *less specialized* than one in which every team member has some kind of differentiated behavior. For another, many teams have members whose overall behaviors are very similar, save for a few specialized skills (e.g., players on a sports team). That is, while it may be advantageous for certain individuals within a team to have specialized behaviors, it is *also* typically advantageous if these individuals *share* certain sub-behaviors. In this sense, we often want to design teams capable of a kind of *behavioral modularity*.

Of course, agents with the same basic control system can exhibit specialized behaviors—for example, certain insects' behaviors for activities such as nest building are specialized based on the environmental context of the insects [150]. Moreover, a simple thought exercise tells us that in principle we could merely embed *all* role behaviors in a single control system with some kind of switching logic for individual agent behaviors; however, such a solution is typically unsatisfying to an engineer for a number of reasons. For one, it is not entirely clear how one goes about designing and implementing such systems in a way that makes them robust and flexible in terms of variations in the problem. For another, it is likely to pose particular problems for any machine learning methods employed to optimize such behaviors. Further, developing a single complex, multi-role behavior for agents in a multi-agent team seems to violate the very spirit of swarm intelligence, where we wish to keep individual agents as simple as possible.

Nevertheless, when a problem *requires* a heterogeneous solution, then engineers have little choice but to consider the issue of specialized behaviors within a swarm. If the control representation for individual agents is too limited, then the problem may simply not be solvable by a homogeneous solution. If the control representation is quite flexible, then the ultimate solution developed by hand or by optimization may inevitably *become* a complex, multi-role behavior—whether by intent or not. It behooves us to consider the characteristics of the problem up front, and for us to design swarm solutions that incorporate heterogeneous and modular solutions.

## 16.3.1 Physicomimetics

This book focuses on swarm-based methods for the control of coordinated multi-agent teams based on artificial physics models [240]. The resulting behaviors of such systems can be quite intuitive, and traditional analytical tools from physics can be used to help diagnose and predict team behaviors [240]. Physicomimetics is particularly well-suited for tasks that require stable geometric formations such as lattices or rings, and under the proper circumstances one can show that teams will settle into "low-energy" positions provided by such structures. For context, we briefly summarize these methods before explaining how they can be generalized to allow for heterogeneous swarm behaviors.

Recall that in traditional physicomimetics, agents are treated as point-mass $(m)$ particles. Each particle has a position, $x$, and velocity, $v$, that are updated via a discrete time simulation, $\Delta x = v \Delta t$. The change in velocity of the particles is determined by the artificial forces operating on the particles, $\Delta v = F \Delta t / m$, where $F$ is the aggregate force on the particle as a result of interactions with other particles and the environment. Since each particle may also be subject to friction, velocity in the next step becomes $(v + \Delta v) c_f$, where $c_f \in [0, 1]$. For example, see [98, 240].

For simplicity, this chapter focuses on a variation of the split Newtonian force law, defined in Chap. 3. The range of effect of the force is $D_{\max}$, while $D$ is the desired *range of separation* between agents. The gravitational constant is $G$, and there are two parameterized exponents: one for distance, $p_d$, and one for mass, $p_m$. The distance between two particles $i$ and $j$ is $r_{ij}$. The magnitude of the force between particles $i$ and $j$ is computed as follows:

$$F_{ij} = \begin{cases} -G \dfrac{(m_i m_j)^{p_m}}{r_{ij}^{p_d}} & \text{if } r_{ij} \in [0, D) \,, \\ G \dfrac{(m_i m_j)^{p_m}}{r_{ij}^{p_d}} & \text{if } r_{ij} \in [D, D_{\max}] \,, \\ 0 & \text{otherwise} \,. \end{cases} \tag{16.1}$$

A particle repels other particles closer than $D$, attracts particles whose distance is between $D$ and $D_{\max}$, and it ignores particles farther than $D_{\max}$. The force scaling characteristics can be controlled using $p_d$, and the importance of the relative difference in mass can be controlled using $p_m$. In total, there are two parameters associated with each particle $(m$ and $c_f)$ and five parameters associated with their interactions $(G, D_{\max}, D, p_m$ and $p_d)$. The distance variable $r_{ij}$ is an observed phenomenon, and might have been written as $\|x_i - x_j\|$ instead.

Fortunately, these ideas can be easily generalized to allow for heterogeneous swarm solutions. We refer to extensions of artificial physics systems such as those described below as *generalized physicomimetics*.

## 16.3.2 Differentiating Particle Types

In heterogeneous swarms, agents of one type will react differently to their environment and other agents of the same type than they will to agents of a different type. In terms of these artificial physics models, the first step to represent this is to explicitly consider different *particle types* during the swarm design process. The idea is that each distinct *type* of particle has its own mass and coefficient of friction. The particles used within the control system would be instances of one type or another, and there may be many particles of each type in the *particle system*.

Differentiating particle types is sufficient to provide *some* specialized behavior. For example, ring formations of arbitrary radii can be achieved by creating two different particle types: one with a relatively small mass and one with a relatively large mass. Moreover, engineers can control the *level of heterogeneity* by controlling the number of distinct particle types and the ratio of each particle type in cooperative teams. However, without a means of designating *how* different types might interact with one another in a different manner, differentiation of particle types will be insufficient for most realistic heterogeneous problems.

### 16.3.3 Virtual Particles

As we note, the application of artificial physics type controllers leads naturally to the idea that there may be different types of particles. In particular, we often need the swarm to *respond* to things within the environment that are not necessarily a part of the swarm itself (e.g., obstacles, goals). Indeed, it's often useful to explicitly inject some additional information or state capabilities to the swarm system by, in effect, *pretending* that some particle exists even though there may be no obvious correspondence with something in the environment. For example, when we wish agents to lead something they are tracking, we might estimate some point in front of the object it is tracking and use this leading point as a particle in the system. Another example is provided in Chap. 11, where virtual particles are key to our success in uniform coverage.

We refer to these types of particles as *virtual particles* and suggest that swarm engineers make the consideration of virtual particle types an explicit part of the process of differentiating particle types.

### 16.3.4 Differentiating Interaction Types

The next step a swarm engineer should consider when designing heterogeneous behaviors with artificial physics is specializing the different interactions between different particle types. Employing different types of interactions and different particle types facilitates a range of possible complex heterogeneous behaviors. Force interactions between different particle types may also be asymmetric. That is, particle type $A$ may affect particle type $B$ differently than $B$ affects $A$. Furthermore, the underlying force law itself can vary for different interactions. Although this may violate the "equal and opposite" law of real-world physics, we can and should do this if it leads to improved performance on tasks.

These kinds of differentiations are natural when approaching many problems, and they have been implicitly used within applications of artificial physics to accomplish behaviors such as the generation of square lattices (e.g., [240] and Chap. 3) and obstacle avoidance (e.g., [91, 92] and Chap. 14).

## 16.4 Heterogeneous Problems

When problems benefit from or potentially *require* specialized behaviors within a team, we call them *heterogeneous problems*. This designation is admittedly not at all rigorous, and the line between these and what might be respectively called *homogeneous problems* is not clear. In fact, there may be many problem properties that affect the issue of whether or not it is better to use heterogeneous swarm solutions.

Many realistic multi-agent problems are multi-objective in nature. In our view, coordinated multi-agent problems in which there are two or more objectives that are in some way in *conflict* with one another (i.e., optimal team solutions are found on some kind of *Pareto surface* rather than a specific point) will often require or benefit from heterogeneous solutions. This is particularly true when the underlying objectives of the problem are relatively clear to the design engineer since there are obvious behavior decompositions that suggest themselves (e.g., separate behaviors for each objective) [164]. While such decompositions may not be the best approach, they are a natural first consideration in the design process.

We describe several examples of problems with such multi-objective characters. In each case, we also try to indicate where knowledge of the domain can be considered during the design process. These cases will be revisited in Sect. 16.5.5.

## 16.4.1 Example: Centralized Resource Protection

In this problem, a centrally-located, immobile resource must be defended within some defense zone by some number of *protectors*. Some number of slightly faster *intruders* periodically appear in various locations just outside the perimeter and move toward the resource, attempting to avoid protectors and strike the resource. If an intruder is destroyed, hits the resource, or is chased out of the perimeter, it is removed from the simulation, and a new intruder will begin a new run from just outside the perimeter after a short random waiting period. The protectors must keep the resource safe by clearing the defense zone of intruders and preventing the resource from being

struck. This problem is described in more detail in [263]; a diagram of the
problem is shown in Fig. 16.1.

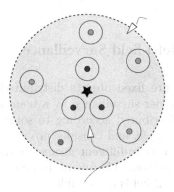

Fig. 16.1: Example resource protection problem. The star marks a resource
to be protected, the outer circle marks the defense perimeter, the concentric
circles indicate protectors (two types), and the triangles indicate intruders
(one type)

Though it may not seem so at first, this problem is multi-objective. The
protectors can reduce the extent of incursions into the defense zone on average by deploying themselves in a large ring just inside the defense perimeter,
but if their numbers are too few to cover the perimeter this may result in
an occasional intruder slipping through the defense and hitting the resource.
Alternatively, they might form a tight ring around the resource itself so that
virtually no intruder could get by, but this comes at the cost of allowing
the intruders to get possibly quite deep into the defense zone. Thus we define two objectives: the average per-step incursion distance of intruders into
the perimeter and the ratio of resource hits taken over the total number of
intruder runs at the resource.

These two objectives can conflict in a variety of ways. By changing aspects of the problem, such as the relative importance of the two objectives,
the number of intruders, or the types of possible intruder behaviors, we can
begin to address questions about how heterogeneous teams of protectors can
help, and what level of heterogeneity is useful in what circumstances. In the
case where intruder behaviors are identical, it is intuitive to design separate
protector roles for each objective. Specifically, an obvious choice is to have
some protectors on the frontier chasing away intruders as soon as they enter
the perimeter, while also having protectors back by the resource to prevent
last-minute strikes.

Moreover, there is domain knowledge that should be considered for likely
performance improvements to simple agents. For example, we can compute
a line from the resource to any detected intruder and use points on this line

to help protectors position themselves *between* the resource and the intruder. While this may or may not be a good idea, it makes little sense to ignore this information during design.

## 16.4.2 Example: Multi-field Surveillance

In this problem there are fixed objects distributed in some bounded area that need to be kept under surveillance by a team of *monitors*. We refer to a pattern of objects distributed according to some relatively geometrically simple distribution as a *field*, and consider problems in which there may be several such fields, each with different geometric distribution patterns that are unknown to the monitors. The monitors must spread out to surveil as many of the objects in all fields as possible.

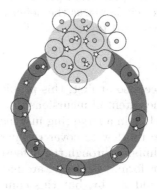

Fig. 16.2: Example multi-field surveillance problem. There are two *fields*, the regions of which are shown using different gray shades. The objects in these fields appear as stars, and the monitors are large points surrounded by a concentric circle indicating the surveillance range

Of course a monitor may simply find the closest object and compete with his fellow teammates for who monitors that object in some way; however, if the individual fields have relatively simple geometric properties, it may be more efficient to exploit their geometry by spreading out over the field rather than myopically focusing on individual objects. Again, the multi-objective aspects of this problem are easily complicated by attenuating problem parameters such as the number and types of different fields, as well as the number of monitors and the range of their surveillance equipment. It may be arbitrarily difficult for a single, simple swarm formation to efficiently cover all fields. If we have some estimate for the number of fields, though, it's quite natural to design one monitor type for each field. This problem is illustrated in Fig. 16.2.

Moreover, we may be able to estimate or learn useful geometric properties of the fields, such as centroid information. If so, it is worth considering how one might make use of that information during design.

### 16.4.3 Example: Covert Tracking

In this problem, *targets* randomly walk about a field and *trackers* attempt to follow the targets as closely as possible without being detected. Each target has some field of vision that leaves a kind of "blind spot" behind it, and optimal trackers will exploit this by tracking the target from within that region. This problem is described in more detail in [58, 181].

If there is more than one type of target, each with its own detection area, then this problem also presents an obvious multi-objective challenge; however, even when there is only one type of target it might be considered multi-objective. For example, minimizing the distance to the object alone will likely result in frequent detections, while staying outside the maximal vision range of the target will fail to obtain the optimal distance.

Again, if we know or can estimate or learn something about the vision range of the targets, it makes little sense to ignore this information. To that end, multi-agent team solutions should explicitly consider this kind of domain knowledge during design. We illustrate a single target and tracker in Fig. 16.3.

Fig. 16.3: Example covert tracking problem. The target is shown at the center of the diagram with its field of vision appearing in gray. The diagram illustrates a tracker successfully tracking the target from its blind spot

## 16.5 Designing Modular and Scalable Heterogeneous Swarms

As we saw in the previous section, it is sometimes necessary to design heterogeneous swarms for certain problems. While generalized physicomimetics is capable of *representing* such solutions, it isn't clear how to *design* them. In particular, it isn't clear how to incorporate concepts of modularity and scalability in the same way we do in other software engineering processes.

This challenge is not trivial. The parameter space of generalized physicomimetics, in which any level of heterogeneity of team members is possible, could be very large. When there are many specialized roles needed, our agent types will require different particle types, each environmental artifact might be represented by a different particle type, and there may be a variety of virtual particles. If we consider all pairwise interactions between these types, the parameter space can grow quadratically with the number of types. Moreover, since there can be strong nonlinear influences between these parameters, designing solutions will become increasingly intractable as the level of heterogeneity increases. Finally, with this system it is unclear how to share successful partial solutions.

With these challenges in mind, we restrict our discussion of design to *interaction design*, i.e., the high-level design of how different types of agents (particles) are permitted to interact with one another. The goal is to describe agent behaviors in terms of *behavior profiles*, which are modular models of interactions for each agent. We provide a principled and practical method of engineering swarm solutions by noticing two key facts: (1) we do not always need every possible interaction, and (2) we can often reuse an interaction's parameters. Reasoning about the types of interactions is necessary for designing successful heterogeneous, swarm-based multi-agent solutions.

*Directed graphs*, also called *digraphs*, are natural and useful tools for this type of reasoning. They are graphs that consist of nodes, along with *directed edges* (arrows) that point from one node to another. Next, we describe our ideas for conducting interaction design using graphs quite generally. Then we apply these ideas to the example problems described above—to be more concrete.

### 16.5.1 Graph-Based Interaction Design

Let each type of particle be a node in a digraph and each interaction be a directed edge in that graph. An edge from one particle type node $u$ to another $v$ indicates an interaction where particles of type $u$ affect the behavior of particles of type $v$. Slightly more formally, edges are associated with a force law as follows. For two particle types, $u$ and $v$, a directed edge $(u, v, \mathcal{F}_{uv})$ denotes an interaction where particles of type $u$ *impart* a force on particles of type $v$ according to the force law defined by $\mathcal{F}_{uv}$. Figure 16.4 illustrates a graph for a two-agent example.

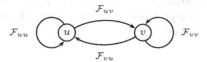

Fig. 16.4: An example force interaction digraph. There are two particle types, $u$ and $v$, and there are separate force laws between every possible pair of particle types

We refer to such digraphs as *interaction models*. Note that interaction models can have isolated nodes and cycles. Omitted edges imply that there is no direct interaction between the particle types represented by those nodes in the graph.

Interaction models in and of themselves are not sufficiently flexible to design any arbitrarily complex team behavior; however, they do represent a simple and obvious way to think explicitly about heterogeneity in terms of modularity and scalability. It is useful to separate two different notions of modularity: *modularity of design* and *behavioral modularity*. In both cases, some simple notational elements can be added to the digraph to aid with these sorts of design issues. Specifically, we use subscripts on particle type labels to index types that will involve repeated design concepts or to indicate when particle types share behaviors. We describe this below in more detail.

## 16.5.2 Design Modularity via Index Variables

When designing something complex, engineers often decompose systems, build and test components separately, and then combine (and test) them. We employ a similar idea for constructing complex multi-agent simulations. Using our graph-based approach, we break the graph into relevant subgraphs, and then consider them in conjunction with one another. When one does this, it often becomes clear that the same basic interaction patterns are *repeated* for a variety of types of particles. It's easy to simplify the design burden and make such repetition quite clear. We use *index variables* in the node labels for this purpose.

We begin by defining a *particle type family* that contains an arbitrary number of particle types for which we'd like to produce a *modular design*. For example, if we were to define $U$ as a particle type family that includes particle types $u_1, u_2, \ldots$ then we could more easily talk about *any* arbitrary particle type of that family using the notation $u_j$.

It is helpful to categorize agents by developing one or more subgraphs that *profile* how agents of some type within the family interact with other agents in the system. So, for example, suppose we wish to indicate that all particle types in the family $U := \{u_1, u_2, \dots \}$ are affected by $v$, $w$, and $x$ (each according to different force laws). We might draw a *single* subgraph to indicate the profile for each $u_j$.

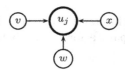

Fig. 16.5: An example of a profile for agents using particle type family $U$. Here $u_j$ is a set of agent types, each of which will respond differently to the $v$-, $w$-, and $x$-types of agents

Note that the example subgraph shown in Fig. 16.5 tells us nothing about the nature of interactions between particles of type $u_j$ with other particles of type $u_j$, nor how particles of *different* types within $U$ interact.

### 16.5.3 Behavior Modularity via Wildcard Indices

In addition to modular design, a swarm design engineer will want to modularize behaviors in a heterogeneous multi-agent team—that is, allow agents to share subsets of behaviors. One way to introduce modularity to generalized physicomimetics is to allow particles to share force laws and parameters. We do this by letting the engineer condense a subgraph by consolidating particle types into a single node.

Here we make use of "wildcard" subscripts (e.g., $u_*$) on particle type node labels indicating particle types that use the *same* force law (including the parameter values) for *all* particle types in the family. In other words, an edge from $v$ to $u_*$ indicates that $v$-type particles will affect particles of types $u_1, u_2, \dots$ in the same way. This is essentially a shorthand way of writing something like $\forall u_j \in U$.

By using index variables and wildcards together, a relatively complex interaction diagram can be managed, and an engineer can consciously and directly design modular behaviors by eliciting a relatively small set of subgraphs. When there is an ambiguity regarding which design concept is conveyed, the more specific concept should be assumed. For example, Fig. 16.6 shows the profile for the particle type family $U$ using two subgraphs. The left subgraph shows that each particle of some type $u_j$ will have distinct force laws in terms of how it interacts with other agents of the same type. The right subgraph shows that $u$-type particles respond to other $u$ types with the same force law. That is, a particle of type $u_1$ will be affected by a particle of type $u_2$ in the same way as $u_3$, but each $\mathcal{F}_{u_j u_j}$ force law is distinct.

Fig. 16.6: An example of a profile for agents of types $u_j$. Here $u_j$ is a set of agent types, each of which will respond the same way to other $u$-type agents, but will respond differently to agents of its *own* type

## 16.5.4 Designing for Scalability with Interaction Models

Thinking in terms of particle type families also gives us a way to clearly discuss and design scaling properties. For example, we can talk about how the parameter space scales *with respect to some family* of particle types. Notice that the example profile from the previous subsection scales linearly with particle type family $U$. Each time a new $u_j$ type is added, a single new force law will be added governing the behavior of particles of that new $u_j$ and other particles of that type. If there are two or more particle type families, one can easily think about scaling in terms of each family or all of them.

## 16.5.5 Example Interaction Model

We now discuss some potential swarm design considerations for the problems described in Sect. 16.4. We construct interaction models that incorporate domain knowledge. Aspects of the problem description, as well as characteristics, are used to inform the modeling. Furthermore, domain relevant features are incorporated into the model by using virtual particles.

### 16.5.5.1 Interaction Model for Centralized Resource Protection

In the centralized resource protection problem, there are essentially three high-level types of objects described: intruders, protectors and a resource. We are not controlling the intruders, and the resource is fixed. However, our protectors must respond to the intruders and the resource, so they will clearly be particle types in our model, i.e., we have types $i$ and $r$, respectively.

We may have many protector types, so we'll define a protector particle type
family $P := \{p_1, p_2, \ldots, p_n\}$.

Recall that we can compute an arbitrary point on a line between some
intruder and the resource. Instead of using the intruder itself, we might (for
example) choose to use a *virtual* point that is some specified distance (line segment) along the line. Indeed, there may be many such schemes that are useful,
and therefore we define another particle type family, $I := \{i'_1, i'_2, \ldots, i'_m\}$. For
the purposes of discussion, we can assume that one of the particle types in
that family is the one that uses a constant factor of 1.0, which is the same
as $i$, so we do not need to duplicate $i$ in our design.

To control parameter growth, we will limit the interaction possibilities of
our control system in the following ways. First, let's assume that protectors of
each protector type are essentially coordinating with protectors of their own
type, with minimal interaction with the other protector types. Thus we will
allow protectors to share the same behaviors in terms of how they interact
with protectors of different types, but each will have a unique behavior with
respect to other particles of the same type. Second, let's assume that each
protector type will respond to only one specific particle type from $I$. These are
natural restrictions. Each protector type will operate from some base stable
ring formation around the central resource and only needs to know how to
avoid other protectors that are not a part of its formation. Furthermore, there
is no good reason for a particle of a given protector type to be attracted to
two or more *different* points along the line between the intruder and the
resource. The interaction model graphs for the protector profile are shown in
Fig. 16.7.

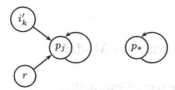

Fig. 16.7: Force interaction model for the resource protection problem. The
two subgraphs describe how forces affect the protector types. Each type of
protector can react to its own type and to specific intruder types in a different
way, but they react identically to all other protectors

This design is simple, natural, and will scale linearly with either $|I|$ or $|P|$.
Moreover, it represents a large category of potential solutions to the problem.
In [263], a solution is developed for this problem in simulation using two protector types and one intruder type ($I = \{i'_1\}$, where $i'_1 = i$, and the intruder
itself is used directly as a particle) for a team of nine agents, three assigned
to an inner ring to protect the resource and six assigned to an outer ring to
reduce average incursion distance. We note that hand-coded solutions to the
problem using this design tended to work better than solutions developed by

optimization methods that are not given the benefit of this kind of modular design.

### 16.5.5.2 Interaction Model for Multi-field Surveillance

For the purposes of this example, we shall assume that the monitor agents have a means of estimating the centroid of each field but cannot discern whether individual objects belong to one field or another. We also assume that we have an upper bound on the number of fields a priori. Therefore our interaction model will require a particle type for objects, a particle type family for monitors, $M := \{m_1, m_2, \ldots, m_n\}$, and a centroid particle type family, $C := \{c_1, c_2, \ldots c_n\}$. The graphs for the monitor profile are shown in Fig. 16.8.

Fig. 16.8: Force interaction model for the multi-field surveillance problem. The two subgraphs describe how forces affect the monitor types. Each type of monitor can react to its own type and the estimated centroid of the field to which it is assigned, and it reacts identically to all other monitors and objects

Notice that there is a one-to-one correspondence between centroid types and monitor types, which is designated in the interaction model graph by using the same index variable, $c_j \mapsto m_j$. Otherwise, this interaction design is similar to the previous example, and has some of the same advantages. For instance, the parameter space scales linearly with $|M|$ (in this case $C$ and $M$ scale together, regardless).

### 16.5.5.3 Interaction Model for Covert Tracking

In the case of covert tracking, it might be reasonable to assume that no "formation of trackers" is necessary as such. The trackers merely need to find the closest target to which they are assigned and then track it covertly. To that end, we identify two potential particle type families: *cats* (trackers) and *mice* (targets). If we assume that we know something about the blind spot of each target type, then we might employ a virtual particle type corresponding to a point offset from the target to its blind spot based on the type of target. We might design this as shown in Fig. 16.9.

Fig. 16.9: Force interaction model for the covert tracking problem. The two subgraphs describe how forces affect the tracker (cat) types. Each type of tracker has an estimated virtual point in the blind spot of its assigned target (mouse) type, and it reacts identically to all targets and trackers

### 16.5.6 Unit Testing the Behavior Modules

One of the many and obvious advantages of designing modular software is the ability to design specific test cases for the modules themselves. A successful outcome from module testing does not guarantee that the software product as a whole functions properly, but it at least verifies that the individual modules behave as they were designed to behave, and it can be a part of wider-scoped tests, such as *systems integration tests*. In software engineering, these module-level tests are sometimes referred to as *unit tests*.

Designing behaviors in a modular fashion suggests some obvious parallels. If one designs a profile for a family of particle types, one can likewise design tests at either the family or particle type level that correspond with the intended behavior. In some cases, this may involve constructing an objective measure of performance in terms of a subset of tasks for the whole problem.

Consider, for example, the centralized resource protection problem that uses two protector types. A measure of incursion distance into the defense zone can be constructed, and the scaling properties of differently sized teams of particles of the first type can be tested using this measure. Separately, one might measure the ratio of hits that the central resource takes during a fixed-time simulation and test the scaling properties of the second type of protector types. Though not reported, this is indeed the process used in [263] for constructing our hand-coded solutions.

Similarly to unit tests in software engineering, these tests in no way ensure that the complete system will function properly. For that, some kind of integration level testing is necessary, where local module-level behaviors are reconciled with global goals and objectives. This is precisely one of the roles that the middle-meeting method for swarm engineering discussed in [117] can play. Nevertheless, like class level modeling in software design, careful interaction modeling allows us to be clearer about how we expect agent roles to relate to one another, and it helps us to design test cases to verify that the agents at least behave as expected.

## 16.6 The Prediction Versus Control Dilemma

Swarm-based control systems are nature-inspired methods and thus involve a computational model of some natural system. In the case of physicomimetics, this is a system of particles. If we consider this idea more closely, we see that control methods based on artificial physics models are essentially performing *two different tasks*: (1) *predicting motion* of particles within a particle-based physics model, and (2) *producing control input* to move agents in some real or simulated world. In simple settings consisting only of agents that are directly treated as simple point-mass particles with no additional constraint on their motion and unforeseen environmental complications, these roles do not conflict. However, for realistic control systems operating in a highly dynamic environment (e.g., multi-robot teams), a conflict between these roles can occur in a number of ways. Moreover, even when there is no obvious conflict, there may be an indirect one due to numerical approximation.

Analyses regarding the stabilization of regular formations, for instance, rely on a temporal element—they predict the dynamics of the particle model itself. When there is a disconnect between the model prediction and control, or between the natural model and its approximation, such analyses may not always be valid since the actualized dynamic will almost certainly differ, potentially quite radically. Additionally, when the parameters of these systems are subject to some optimization procedure, it isn't clear how these model differences might affect the learning gradient.

Yet many swarm-based systems, including and especially physicomimetics, have been successfully applied to a wide range of control problems, some of which are described in this book. Moreover, in many cases control systems have been demonstrated both in simulation and on physical devices with a wide variety of *actuation constraints*—physical limitations on a device's movements. Artificial physics based control systems *can* be affected by actuation constraints on the agents; however, even when this is so, they can be surprisingly robust to such prediction/control disparities. Still, this is not always the case.

Consequently, swarm engineers should be aware of this discrepancy and, if possible, test for it directly. In this section we talk about some of these differences in a bit more detail and use a running example to show a few ways engineers might construct tests to empirically analyze the effects of these differences.

### 16.6.1 Modeling Natural Phenomena

If we are to take the "artificial physics" metaphor to its fullest extent, we must acknowledge a number of obvious ways in which we will have difficulties translating the natural models into computational models. For one, [objects represented by] particles in real systems are *not* mere points but are *systems* of rigid bodies (see also Chap. 13). For another, the difference equations we defined in Sect. 16.3.1 and discussed elsewhere in this book derived from well-known Newtonian laws are more correctly expressed as differential equations. That is, the real dynamics of particles in the natural system are governed by models that operate in continuous time. Additionally, though we may be quite clever in eliciting sophisticated models, there will inevitably be aspects of the environment that are not captured in our natural model.

Indeed, the observation that the nature-based models are continuous time systems is part and parcel with the mathematical formulations that underlay the efforts of Kazadi [112] to formalize terms, behavior, and goal attainment. This work *begins* precisely with the observation that the systems being modeled are governed by differential equations, and indeed sets out a *general swarm engineering condition* expressed in terms of conditions on the differential system.

As has been pointed out in a number of works [30, 112, 138, 240], most research efforts in swarm intelligence are descriptive in nature—very few works offer predictive or prescriptive methodologies for constructing swarms for arbitrary tasks. From a natural systems modeling perspective, this is understandable: though there is obviously a great deal of literature about mathematical modeling of the behavior of natural systems, the goal of such literature typically *isn't* to develop prescriptive techniques.

## 16.6.2 Control Input

On the flip side, most swarm-based systems are designed to control some kind of device, either in software or in hardware, as is the case with distributed multi-robotic control problems. Such platforms typically have constraints imposed by both the design of the systems themselves, as well as physical limitations imposed by reality. For example, most robots have some kind of actuation constraint on their motion, including complications such as turning radius limitations, maximum velocity limitations, and maximum acceleration limitations. In another example, obstacles in the real world obstruct motion of physical devices, whether or not those obstacles are properly modeled by the control systems. So in a very real sense, our swarm models offer merely *suggestions* to the agent on how to move, but do not dictate precisely how they *will* move.

Additionally, we design our systems to respond to many stimuli, often including other mobile agents that are not directly within the control of the system (e.g., human team members). If our goal is to develop stable global behaviors, we have to concede that such global behaviors must be robust to local behaviors that are not necessarily found directly in the feedback path.

### 16.6.3 Summarizing: Three Sources of Challenges

Taking the discussions from the previous two subsections together, we can immediately see a number of potential challenges. Let's summarize the three most obvious.

First, like nearly all particle systems executed in a computer [274], most implementations of physicomimetics amount to explicit approximations of the underlying differential equations—typically using something like *Euler's method*. Euler's method is itself subject to known stability issues. That is, when the step size for updating state variables is too large, the numerical method can be unstable even when the true system is stable. Even when this is not the case, simple explicit numerical integration methods can be quite poor approximations of true system behavior. Of course, more sophisticated explicit techniques can be used (e.g., *Runge–Kutta*), or implicit methods (e.g., *backward Euler*); however, such techniques come at considerable computational cost that may be prohibitive for real-time control systems.

For another, even when the natural models offer some compelling predictive behavior and the numerical approximations are reasonable, differences between what the device is capable of and what the control system suggests can be a source of instability or unpredictability. Indeed, [117] makes this very observation, stating *"Application of this theory in real, physical systems should be direct."* But unfortunately, this is not always possible, and when it isn't, it's not clear how our swarm-based systems will behave. What *is* clear from the general study of nonlinear systems is that these differences may sometimes destabilize what was otherwise a well-designed, stable team behavior.

Finally, in many cases, swarm systems must respond to collaborative dynamic objects within their field that are not directly controlled by the system itself. This can happen when human actors are a participating part of a large team in which the swarm is only a subset, or when the human is a part of the problem description. For example, the centralized resource protection problem described above is very similar to a *VIP protection problem*, save that the "central resource" in the latter problem is a person who moves independently of the swarm. When the dynamics of important components within the system cannot be modeled correctly, designed behaviors may not be as expected.

### 16.6.4 Testing Strategies for Agents with Constrained Motion

Because of the problems just discussed, swarm engineers should consider specific testing strategies to expose any problems that might result from the disconnect between the nature-based model and the control system. By way of example, we summarize our work that examines a swarm system under varying actuation constraints in terms of maximum angular velocity [58].

First, we consider the quality and efficiency with which the system produces its expected low-energy stable lattice positions. To do this, we construct two testing measures: *settling time* and *lattice quality*. The first examines the time it takes a system from random initial conditions to reach a state in which the average scalar acceleration value drops below a certain point. The second measure is determined by computing the *Delaunay triangularization* [82] of the particles after they have settled, then measuring the coefficient of variation in edge lengths within a subset of this graph. The subset includes all nodes within the 80% of the extent of the maximum radius of the swarm as measured from its centroid. In our simple scenarios, lattice quality is not substantially damaged until the constraint is very large—and the settling time is actually *smaller* in the constrained case since the constraint in fact removes energy from the system.

Second, we consider a measure we refer to as *suggested position difference*. This measure simply computes the Euclidean distance between the position *predicted* by the approximated particle system and the *actualized* position obtained by the control system. We examine this distance in the context of an optimization method attempting to learn parameter values for a system that maximizes behavior performance. Note that this is not unlike methods used within numerical integration to estimate errors using trailing order terms in a Taylor expansion. We note that increasing the degree of constraint on actuation *does* result in increased differences between the nature-inspired model and the result of the control system, though the learning system appeared to compensate for this difference.

Another idea not implemented in our study is to implement more sophisticated numerical integration methods, such as Runge–Kutta, for the purposes of testing only. Computing how the traditional Euler-based system departs from a system employing more accurate numerical methods in certain well-chosen test cases might help uncover any substantial approximation errors that could affect the system behaviors.

We are not advocating these specific measures in any general sense, of course. Rather, we are suggesting that the dichotomy between the prediction made by the natural models and the response to the control system should be considered when designing swarm behaviors. More specifically, we advocate the inclusion of some kind of testing strategy that addresses issues that might be introduced by the problems discussed in this section.

## 16.7 Conclusions

In this chapter, we have examined how domain knowledge can be used to guide the development of swarms by addressing some design considerations.

We looked at two basic issues. In the first case, we discussed methods for developing heterogeneous swarms in ways that allow swarm engineers to explicitly incorporate properties such as modularity and scalability into the swarm design process via *interaction models*. In the second case, we discussed some of the issues that can complicate the construction of swarm behaviors that arise out of the dual roles that such systems serve: as models of natural systems and as inputs into actualized control systems. Throughout the chapter, we included a number of example problem domains, as well as testing strategies such as *unit testing* modular behavior profiles and measures for uncovering issues related to numerical approximation errors from system governing equations.

In our design discussion, we described a graph-based method for designing interaction models in physicomimetic systems. This method allows engineers to construct graphs that clearly define what interactions are possible. By using our technique for condensing subgraphs, engineers can think more modularly about the design process and produce reusable behavioral modules, giving the engineer the ability to directly control the scalability of the system. This method has been employed to design heterogeneous, modular solutions to a resource protection problem.

In our nature-based model versus control input discussion, we described three key issues that can crop up because of the essential duality resulting from applying swarm systems to multi-agent control problems: approximation errors in explicit numerical integration, the effects of unmodeled actuation constraints, and the presence of complex collaborating agents that are not governed by the swarm system. We described several measures as examples of empirical "testing" strategies for uncovering critical behavior problems that might result from model discrepancies.

Regardless of whether these or other processes are employed, it is clear that the development of large-scale swarm solutions for sophisticated and realistic problems will require some deliberate process akin to the software engineering process. This suggests that engineers at least consider how they might approach phases like *swarm design* and *swarm testing*. The ideas presented in this chapter can and should be used in conjunction with other specification, design, testing and validation tools.

# Chapter 17
# Using Swarm Engineering to Design Physicomimetic Swarms

Sanza Kazadi

## 17.1 Introduction

Since the mid 1980s, swarms have been a subject of significant computer science and engineering research. Swarms have been defined by [113] as "a set of interacting agents within a system in which one agent's change in state can be perceived by at least one other agent and affects the state of the agent perceiving the change. Moreover, the subset must have the property that every agent has at least one state whose change will initiate a continual set of state changes that affects every other agent in the swarm." This is mathematically indicated by the idea that

$$\frac{\mathrm{d}f_i}{\mathrm{d}x_j} \neq 0$$

whenever $i \neq j$, and where $f_i$ represents the control algorithm for agent $i$ and $x_j$ represents the state of agent $j$. As a result, the state of each agent has an effect on the behavior of each other agent. This communicates the idea that swarms are coherent in some way and that, going forward in time, their interactions remain nonzero and impactful. For this reason, we call it the *swarm interaction equation*.

The idea of *emergence* can also be quite succinctly stated. In particular, given an emergent property $P$, such as the global quality of a swarm lattice, it holds that

$$\frac{\mathrm{d}P}{\mathrm{d}f} \neq 0 \text{ and } \frac{\partial f_i}{\partial P} = 0 \,.$$

In other words, the property is generated inadvertently by the swarm agents; they do not deliberately control or guide $P$. The latter requirement is cap-

Sanza Kazadi
Jisan Research Institute, Alhambra, California, USA, e-mail: skazadi@jisan.org

W.M. Spears, D.F. Spears (eds.), *Physicomimetics*,
DOI 10.1007/978-3-642-22804-9_17,
© Springer-Verlag Berlin Heidelberg 2011

tured by the second equation, which states that the emergent property $P$ does not have any direct effect on the control algorithm of any agent $i$ in the swarm. Because these two equations together define "emergence," the pair is called the *swarm emergence equations*.

The swarm interaction and emergence equations jointly define the character of swarms and have also unraveled the mystique behind swarms. It is quite compelling to think that, in the case of emergent properties, the swarm can do tasks without these things having to be explicitly built into the agents themselves. Moreover, due to the potential for nonlinear and therefore somewhat unpredictable interactions between agents [275], swarms are part of a class of systems that are inherently difficult to design. As a result, the design of swarms has generally been rather difficult.

In the search for methods of bypassing some of these difficulties in designing, in particular, emergent swarms, two general techniques have arisen that have been quite exciting in terms of their applicability. These methods are physicomimetics and *swarm engineering*. In physicomimetics [76, 235], swarms are modeled as "clouds" of particles whose interactions are mediated by a set of physical laws. The laws are adapted from real physical laws, and are used by agents as control algorithms. Agents have the ability to determine their position and the positions of nearby agents. Using these positions, the agents calculate their response from the appropriate physical laws and apply these responses to their own control algorithms. The result is that the physical laws are implemented at the swarm level. What makes this approach so useful is that the vast theory behind well-known physical laws can be applied to make quantitative and qualitative predictions about the swarm's behavior. Moreover, the changing of parameters of the behavior can have predictable effects on the swarm.

With the second method, swarm engineering [112, 113], the swarm design follows two general steps. In the first step a mathematical equation is created that mathematically communicates the goal of the swarm in terms of properties of the system with which the agents can interact and control, such as sensing or changing. This equation is used to derive a swarm condition which indicates what the agents must have the ability to do in order to achieve the goal. Then in the second step, the swarm engineer must create the specific model which accomplishes what the swarm condition indicates.

What makes these two methods (i.e., swarm engineering and physicomimetics) so compatible is that they are both grounded in strong mathematical foundations, and deliver results based on predictable theory. Rather than basing the results on vague arguments about emergence, the two methods deliver specific properties and behaviors. The swarm engineering approach is used to generate specific agent-based goals of the swarm. Physicomimetics then delivers these agent-based goals in a predictable way that can be plugged right into the requirements generated in the first swarm engineering step. Together, they provide not only a mathematically rigorous way of knowing that the global goal will be achieved, but they deliver a relatively simple way of implementing the swarm design.

Other researchers have tackled the problem of generating global structures [109, 110, 272]. However, these papers don't have the predictive quality needed for a true engineering methodology. Moreover, the scope of these approaches seems to be limited; it is not clear that these approaches generalize to generating behaviors once the desired global behavior has been decided on.

The remainder of the chapter is organized as follows. We describe the swarm engineering methodology in Sect. 17.2, and then describe this methodology applied to physicomimetic systems in Sect. 17.3. Next we examine, in detail, two problems from the literature using the combined approach. Sect. 17.4 addresses the hexagonal lattice design problem. Then Sect. 17.5 examines a flocking model, known as the "quark model," that accomplishes obstacle avoidance. Section 17.6 discusses the combined methodology. Finally, Sect. 17.7 concludes the chapter.

## 17.2 Swarm Engineering

When one first encounters swarms, it's easy to think that a swarm is a simple group of agents all collectively colocated either in physical or virtual space. This would seemingly endow many things with the status of being a true swarm—solar systems, galaxies, and people on a beach. However, familiar examples of swarms in nature seem to have another quality to them that is somewhat intangible and hard to miss. This quality, which endows the swarm with life greater than just the dynamic result of a number of things in the same place, comes from the interaction of the agents in the swarm. A crowd on the beach is not a swarm, as it does not enjoy this quality. A flock of pigeons in a square is, though the swarm tends to break up. The quality is very clearly demonstrated when, in response to an oncoming person, the entire flock takes flight as one unit, rather than each pigeon flying away individually as the person gets within that particular pigeon's comfort zone. A school of fish does the same thing, as does a herd of zebras.

The element is not just the ability to react to something and communicate the reaction, but the ability of any individual agent to induce the group behavior through a common communication/reaction scheme. Mathematically, the swarm interaction and emergence equations define a swarm. The nature of the interactions between agents in the swarm can vary widely, can be quite subtle, or can be extremely apparent. However, it is this communication and its ability to endure over the lifespan of the swarm that defines the swarm.

Swarms, as opposed to individual agents, are particularly effective if a coordinated action must be achieved over a spatially extended area. That is, any action that requires coordinated efforts at different locations at the same time, whether these efforts are simple or complex, is a good candidate for the use of a swarm. Swarms utilize bidirectional communication, which allows for constant feedback between agents. When coordinated in a proper way, the effect can be quite striking, and can extend what is capable with conventional agents significantly. In this chapter we are concerned with ways of constructing swarms. Although swarm applications are important, they are not the main focus of this chapter.

What makes the control of a swarm so difficult is the fact that control algorithms for swarms, spread over many agents, generate a complex function of one or more agent-level microcontrollers. These microcontrollers have the potential to generate behaviors that cause unpredictable nonlinear interactions. Such interactions stymie many efforts in swarm engineering, thereby tending to yield swarms that not only don't accomplish the desired global task, but which defy efforts to systematically massage the controllers so as to generate the desired behavior. As a result, any requirement for generating global behaviors can be extremely difficult to satisfy if the swarm practitioner utilizes an intuitive procedure for generating swarm behavior. What is needed is a method for generating these behaviors in such a way that the requisite issues are handled: behaviors, processing requirements, castes, and so on.

Swarm engineering is a method of swarm design that may be used to generate swarms of agents that accomplish a global goal in a provable way. That is, the method allows for the generation of behaviors of agents which may be provably shown to yield the desired global goal. The method creates a set of *swarm properties*. Once these properties have been satisfied by the swarm, the accomplishment of the global goal is mathematically assured. The import of this specific kind of design is that it is constructive as opposed to deductive. That is, rather than deducing the effect of a particular agent design, the design can be developed in such a way that success is guaranteed.

Swarm engineering proceeds via two basic steps. During the first step, abstract global properties, representable as a vector $G$, are equated to a vector function of measurables obtainable by agents in the system, $\{x_i\}_{i=1}^{N_p}$, where $N_p$ is the number of sensor properties available to the agents. These measurables can be anything, from agents' positions in a coordinate system, to temperature, sound levels at different frequencies, light levels, chemical composition levels, radar signals, and so on. Their numerical values become the inputs for the global measurables, such as the average position, variance of position, coordinated formation, time series of sound levels, images, or radar tracking data. Measurables can be obtained at multiple points in time, so as to allow image tracking, velocity profiling, image flow measurements, or other time series data. Once these have been decided on, the global state $G$ can be written as a *global function* $g$ of these sensor inputs:

$$G = g(x_1, x_2, \ldots, x_{N_p}) . \tag{17.1}$$

In fact, we can create a vector of the sensor properties perceivable by the agents and in this way summarize the global function with

$$G = g(x) \, . \qquad (17.2)$$

It is important to note that the vector $x$ represents the sensor properties perceptible to the entire set of agents. In other words, it represents the current state of the swarm system as a whole, rather than of one particular agent from the swarm.

If the vector $G$ has multiple components, it is a convenient way to indicate that there are two or more properties in the global swarm state. $G$ represents a snapshot of the swarm whittled down to the numerical values of the global properties of interest in the swarm. In other words, rather than tracking each individual agent, it may be interesting to reduce the swarm's description to a single vector of numbers that represent abstract swarm properties of interest, such as the swarm lattice quality.

This step is very easily stated, yet remarkably profound. It is difficult to accomplish except in relatively simple cases, particularly if the global state is very abstract and does not represent properties detectable by the agents. For instance, in construction tasks, are there any agent-measurable features that correspond to the correct placement of a beam? How does one determine from the sensor states of the agents that a construction task is complete? These considerations present substantial barriers to the accomplishment of this first step in the system design task. Yet without accomplishing this step, it is generally impossible to create a set of changes to states that can be sensed which accomplish the global goal.

In many of these tasks it is easier to determine sensory states that are affected by progress towards a goal instead of the final goal accomplishment. For instance, in clustering tasks [114], the agents can take a noisy snapshot of the size of the cluster that they are investigating. However, this won't determine whether the final cluster is generated reliably—only that a larger cluster is classified as "large" rather than "small" more often than a smaller cluster, which will yield an average movement of pucks toward the larger cluster. Yet this first step defines the remainder of the swarm engineering approach. It cannot be bypassed.

The second swarm engineering step consists of finding a way for the swarm to go from an initial to a final state. In general, a swarm initially exists in one state, and the goal of the swarm is to have it evolve into a second state. This second state can represent the swarm achieving a goal, carrying out a behavior, or have some other meaning. Because a change in the swarm's global state means moving the swarm from one global state to another, we are interested in the time derivative of Eq. 17.1. This is given by

$$\frac{\mathrm{d}\,G}{\mathrm{d}t} = \sum_{i=1}^{N_p} \frac{\partial g}{\partial x_i} \frac{\mathrm{d}x_i}{\mathrm{d}t} = \nabla g \cdot \left( \frac{\mathrm{d}x_1}{\mathrm{d}t}, \frac{\mathrm{d}x_2}{\mathrm{d}t}, \ldots, \frac{\mathrm{d}x_{N_p}}{\mathrm{d}t} \right) \, . \qquad (17.3)$$

In Eq. 17.3, the left hand side is the overall global properties' changes with respect to time. The right hand side is the vector of functional couplings of the properties to the function dotted into the vector of agent observable states. Differing control algorithms will result in different values for the $\frac{dx_i}{dt}$s. The goal of swarm engineering is to develop appropriate strategies for a sequence of swarm state changes, i.e., $\frac{dx_i}{dt}$s, that will achieve the global goal.

The swarm design problem therefore consists of navigating through the state space in a direction that eventually leads to the achievement of a final global goal state $G_f$ having a vector of values $G_{f.vals}$. A number of different strategies may be employed so as to lead the values $G_{vals}$ of the global states $G$ in the correct direction.

A global function is called "*well-defined and continuous*" if it maps to only one global state with numerical values $G_{vals}$ and it is continuous around the point that generates the numerical values $G_{vals}$. In general, it is not advisable to use a global function unless it is well-defined. The use of a such a function that is not well-defined would result in degeneracies in the final swarm state. This could cause problems in generating the final swarm state. A well-defined and continuous global function satisfies the following:

1. The swarm has one set of variables for which the function takes on the desired value.
2. Given the final agent state $x_f$ and a positive number $\delta > 0$, there is a positive number $\varepsilon$ such that any vector $y$ with the property that $|x_f - y| < \varepsilon$ implies $|g(x_f) - g(y)| < \delta$.

Given a well-defined and continuous global property, we define the set of states $y$ such that $|g(x_f) - g(y)| < \delta$ to be the set of $\delta$-*nearby states*. Clearly, these exist within an $\varepsilon$-ball[1] for an appropriately chosen value of $\varepsilon$.

The need for the global function $g$ stems from the practical value of its inverse, i.e., our desire to map the numerical value of the global state back to a unique point in the agents' state space. This point determines the final point, $x_f$, toward which the system must evolve. As a result, the system design consists of finding a pathway through state space from the set of all initial points $\{x_i\}$ to the final state $x_f$. Because this is a designed system, we are free to determine any pathway through the state space with no expectation that $G$, or the ability to determine the value of $G$, is necessary for the completion of the design task.

Transformations of the state of the swarm that do not affect the global state are defined as *null swarm transformations*. These result when a transformation, such as a swarm rotation, the addition of a vector to the position of an agent, or a renumbering of the agents, causes no change in the numerical values of the global state properties in $G$. Such transformations, as we shall

---

[1] An $\varepsilon$-ball is a concept from real analysis and topology and is defined as the set of all vectors whose distance (for some appropriately defined distance function) is less than $\varepsilon$ from a center point. Typically, an $\varepsilon$-ball is depicted as B$(x; \varepsilon)$, where $x$ is the center of the ball and $\varepsilon$ is the distance.

see, potentially lead to independent global states. While these do result in degenerate states when considering the global properties in $G$ alone, it is not necessarily the case that the consideration of all resulting global properties yields degeneracies; adding another global property that results from the null swarm transformation can generate a nondegeneracy when a degeneracy otherwise exists. In such a case, the swarm is *ill-defined* without the additional global property.

Aside from determining the initial and final states in state space for the system, it is also necessary to determine for what system configurations the system is *infeasible*. These are states that the system cannot adopt, e.g., due to some physical constraint of the system. The pathway through state space must avoid these infeasible regions in order to be valid. Any method of traveling in state space along this pathway is valid and will result in the desired final system configuration. As such, the result is *model independent* and holds for any model that follows the designated feasible pathway.

Note that it is possible for a state space to have no pathway from the set of initial system points and the desired final point. In such a case, the system is impossible to build, and this result is independent of the agent model employed and is therefore a model-independent negative result. In the case that such a system is identified, it is possible to rule it out as a possibility.[2]

Therefore, the task of the swarm engineer is to design a global function, with corresponding global states, that is well-defined and continuous. The inverse of the final (desired) global state can then be used to determine the final (desired) swarm state, while initial swarm states determine the point in state space from which the system evolves. Once both swarm states are known, and the direction $x_f - x_i$ is known, the design problem for the swarm engineer consists of finding a feasible pathway through the state space, if it exists. Otherwise, it is equally important to be able to say that no pathway exists, or to determine for which system configurations a pathway does exist.

## 17.3 Swarm Engineering Artificial Physics Systems

In this section, we use the swarm engineering approach to explore the hexagonal configuration problem. This problem centers around the correct placement of agents in a perfect hexagonal grid. In order to create a hexagonal grid, each agent must take its place in a larger hexagonally shaped structure. The task has been accomplished previously [76, 235] by Spears and colleagues.

---

[2] We conjecture, but do not prove here, that in the case that we are given two global functions of the system, $g_1$ and $g_2$, if there does not exist a pathway in state space through the feasible region as defined by $g_1$ from the initial point $x_i$ to $x_f$, then this means that there is no pathway through the feasible region as defined by $g_2$. We call this the *Model Independence Conjecture*, which states that the specific function used to design the system is unimportant when determining whether a system can be built or not.

However, to date, no method has yet emerged that allows the global task to be accomplished without explicit global information.

This section begins with an investigation of the way in which microscopic hexagonal lattices may be built up from an originally randomly organized group of agents. These microscopic hexagonal lattices must then combine in such a way as to produce a macroscopic hexagonal lattice. We assume that the agents are capable of local sensing and carrying out moderate computation. The theory behind the system demonstrates that the final solution will be the desired macro-hexagonal state.

Application of this theory to real, physical systems should be direct. In order to make that happen, the agents should be restricted to actions and computations that could realistically occur on a mobile platform and with real local sensors. We restrict ourselves to sensors and actions that are realistic in terms of the actuators and sensors commonly available or likely to be available in the near future. Where reasonable, if precedents exist in the literature, we will omit the details of some group behaviors that we can reasonably assume can be done in tandem with the actions we are developing here.

In accordance with the *middle-meeting* methodology described in [112], we begin by first creating a global property using local agent-level properties. Once this has been built so that the desired global structure occurs when a specific unique value of the global property occurs, we continue by generating local behaviors that will yield the desired global property.

## 17.3.1 Microtheory

The first step is to list the various sensory capabilities of the agents. We assume that each agent is capable of measuring the distance and bearing from itself to any of its neighbors. That is, each agent $i$ has access to a measurement that we label as $r_{ij}$ which indicates the distance between agents $i$ and $j$. Note that each $r_{ij}$ can be calculated as

$$r_{ij} = |\boldsymbol{x}_i - \boldsymbol{x}_j| = \sqrt{(x_i - x_j)^2 + (y_i - y_j)^2}\,, \tag{17.4}$$

assuming that $\boldsymbol{x}_i$ and $\boldsymbol{x}_j$ represent the locations of agents $i$ and $j$, respectively, in some two-dimensional coordinate system.

It is interesting to note that in the following derivations, no other individual measurement is required to generate the micro- and macrobehaviors of the swarm. As a result, the agents do not need any other sensory capabilities. We also note that the distances are two-dimensional distances, indicating the tacit assumption that our agents are constrained to a two-dimensional surface. Such a limitation might be reasonable for a swarm constrained to act on the ground, or airborne swarms limited to a particular altitude by design or behavior.

In order to generate the hexagon, agents that are too close to one another must move away from each other, and those that are far away from one another must move closer to one another. As with the original artificial physics studies, we adopt the convention that the actions of the agents will be calculated and then implemented *as though external forces are at work on the agents*. In particular, we assume that each agent senses the distances to each of the other agents within sensor range $r_s$. Then, using this data, the agent calculates what its next move will be. The implementation is carried out by activating the agent's actuators, which simulate the effect one might expect if the motion came from a real physical system.

Using this micro-measurement, we now proceed to construct a global property. This global property has a well-defined and unique value when the desired global structure is achieved. Moreover, that value is not capable of being achieved in any other configuration. Generating that value, then, is a matter of undertaking behaviors that produce this value. The need for local properties derives from the requirement that the value be attainable using behaviors that are achievable by the individual agents.

We begin by defining a function known as the *force function*, $\mathbb{F}$. This function determines the virtual force caused by the proximity of two agents. Too close, and the force will be strongly repulsive; too far, and it will be attractive [240]. Assuming that $D$ is the desired separation distance between agents, $\mathbb{F}$ is defined as

$$\mathbb{F}(r_{ij}) = \begin{cases} -G_1 & : \ r_{ij} < D \ \text{ and } \ r_{ij} \leq r_s \, , \\ G_2 & : \ r_{ij} > D \ \text{ and } \ r_{ij} \leq r_s \ \text{ and} \\ 0 & \text{ otherwise} \, , \end{cases} \quad (17.5)$$

where $G_1$ and $G_2$ are two constants. (Be careful not to confuse these constants with the goal vector $\boldsymbol{G}$; they mean something quite different.) Recall that this is similar to the split Newtonian force law shown in Eq. 3.3, with $p = 0$. When many agents are involved, the effect of all the agents is cumulative. A single agent $j$ behaves as though it is reacting to the net vector force

$$\boldsymbol{F}_{ij} = \sum_{i \neq j}^{N} \left( \frac{\boldsymbol{x}_i - \boldsymbol{x}_j}{|\boldsymbol{x}_i - \boldsymbol{x}_j|} \right) \mathbb{F}(|\boldsymbol{x}_i - \boldsymbol{x}_j|) \, , \quad (17.6)$$

assuming a total of $N$ agents in the swarm.

The force equation defines the way in which the individual agent will behave at the microscopic level. However, we don't yet know whether or not this will lead to the desired global state—that of a single macroscopic hexagonal structure. We use the force to derive the global property *energy*:

$$E = \sum_{i \neq j}^{N} \int_{0}^{(D - r_{ij})} F_{ij}(x) \, dx$$

$$= \sum_{i \neq j}^{N} \mathbb{F}(|x_i - x_j|)(D - r_{ij}) \, . \tag{17.7}$$

Energy measures the total energy required to move all individuals from their current positions to their equilibrium positions.[3] It decreases as the system gets more perfect (stable) and increases as the system gets less perfect (unstable). Our goal is to achieve the smallest value of $E$ possible, because $E$ at its lowest value produces the most stable hexagonal lattice possible. We defer the proof of this fact to a later section of the chapter. Note that the abstract goal state for this problem, $G_f$, is the state of minimal energy of the lattice. Each state $G$ along the way to this objective measures one property, namely, the energy of the system. Equation 17.7 provides the problem-specific definition of the function $g$, which maps the swarm measurables, such as $x_i$, $x_j$, and $r_{ij}$, to the global property "energy."

In the following discussions, we approximate the force function for $r_{ij} \leq r_s$ as

$$\mathbb{F}(r_{ij}) = \lim_{k \to \infty} \left\{ G_1 + (G_2 - G_1) \left( \frac{\arctan[k(r_{ij} - D)]}{\pi} + \frac{1}{2} \right) \right\} . \tag{17.8}$$

Note that when $k \to \infty$, the function has the same behavior as that given above in Eq. 17.5.

In order for the system, which is likely to be in a random initial state, to continually evolve to a minimal energy, the time derivative of the energy must be negative. Therefore, we begin by computing the time derivative and work backward to the condition that each member of the swarm must obey in order to make the global task occur. Accordingly,

$$\frac{dE}{dt} = \sum_{i \neq j}^{N} \left( -\mathbb{F}(|x_i - x_j|) \left( \frac{\mathbb{D}}{r_{ij}} \right) \right)$$

$$+ \sum_{i \neq j}^{N} \left( \lim_{k \to \infty} \left( \frac{k(D - r_{ij})(G_2 - G_1)\mathbb{D}}{\pi r_{ij} \left( 1 + k^2 (|x_i - x_j| - D)^2 \right)} \right) \right) < 0 \, , \tag{17.9}$$

where $\mathbb{D} = \Delta x \frac{d\Delta x}{dt} + \Delta y \frac{d\Delta y}{dt}$. This is a problem-specific version of the generic Eq. 17.3. In Eq. 17.9, we can ignore the second term, since it is always zero except when $|x_i - x_j| = D$. This corresponds to a physical situation that will

---

[3] Hence, this is potential energy. See [240] for a similar derivation.

not generally occur and cannot be stable. As a result, we focus exclusively on the first term.

The energy is non-negative, and the minimal energy is the desired state. Thus, we are searching for all cases in which the first term of Eq. 17.9 is negative. This indicates that the system energy is decreasing, which is what we want it to do.

There are two cases leading to a negative value for the first term. First, if $r_{ij} < D$, $\mathbb{F}$ will be negative if $\mathbb{D} > 0$. This will happen if two agents are too close to one another and are therefore moving away from each other. On the other hand, if $r_{ij} > D$ then $\mathbb{D} < 0$. This means that agents that are too far apart must start coming closer to one another. Combining these two behaviors necessarily produces a behavior that makes the energy in the system non-increasing.

Despite the fact that the energy is non-increasing, it is still possible for the agents to be in a stable, low energy lattice that is not a perfect hexagon, e.g., by getting stuck in a "local minimum" of energy. In this situation, the energy of the hexagon is very low and the agents are very stable in their current positions. Although the global energy is smallest when a perfect hexagon is achieved, the stable lattices do not change because the energy required to move the misplaced agents out of their equilibrium positions is greater than the immediate difference in the global energy. In order to achieve the perfect hexagon, the lattice needs to be "shaken," or "pushed," to move the misplaced agents out of their equilibrium. Once those misplaced agents are out of their equilibrium positions, they will resume moving toward a stable position. They can either go back to the original equilibrium position or they can find a new equilibrium position that is more stable than the original equilibrium position.

This analysis has begun with the creation of a global property called "energy." This property has the requirement that the specific numerical value we are trying to achieve can only be attained with a unique system configuration. The fact that the system is nondegenerate in this sense means that once the specific value is achieved, the desired system configuration will be achieved.

The first swarm engineering step has just been shown in the context of a particular problem, namely, that of generating a hexagonal lattice. In particular, a global property called "energy" was defined and equated to measurables obtainable by agents in the swarm. We also prepared for the second step by defining the time derivative of the energy. However there is still one issue that remains. In order to generate a hexagonal lattice, we must first generate hexagonal sub-lattices. This can be accomplished by placing each individual agent at a specific distance from any of its neighbors. We've generated a method for achieving this, utilizing an analysis of our energy property. This has resulted in a very general method which may be applied in a variety of ways to individual agents.

The next two subsections will deal with the question of the minimal energy configuration of the larger lattice and with methods for moving the entire lattice into an energy configuration that is closer to this global energy configuration.

### 17.3.2 Lowest Energy Configuration

In previous sections, we have assumed that the lowest energy configuration of the lattice was a single perfect hexagon. We now prove this fact using a simple geometric argument.

In the lattice, the minimum distance between each pair of agents will always be greater than the equilibrium distance due to the high repulsion coefficient that we use for building hexagons:

$$\mathbb{F}(r_{ij}) = \begin{cases} -300 \ G & : \ r_{ij} < D \ \text{ and } \ r_{ij} \leq r_s \ , \\ G & : \ r_{ij} > D \ \text{ and } \ r_{ij} \leq r_s \ \text{ and} \\ 0 & \text{otherwise} \ . \end{cases} \quad (17.10)$$

Such a high repulsion coefficient of $300G$ makes it energetically impossible for any agent to get closer than the equilibrium distance other than transiently. The distance error, defined as $|(D - r_{ij})|$, is zero only when the agents are at their equilibrium distance. As a result, the only situation when the distance error of a pair of agents is nonzero is when the agents are farther away from one another than the equilibrium distance. The energy of a pair is given by

$$E = \mathbb{F}(|\boldsymbol{x}_i - \boldsymbol{x}_j|)(D - r_{ij}) \ . \quad (17.11)$$

This energy $E$ is always equal to the sum of all the energy that comes from the agents when they are more than the equilibrium distance apart:

$$E_{\text{excess}} = \sum_{i \neq j}^{N} (\mathbb{F}_{\text{Attraction}})(D - r_{ij}) \ . \quad (17.12)$$

The main factors in calculating the energy then, are the number of pairs that are more than the equilibrium distance apart and the magnitude of the distance. In hexagonal lattices, all six sides are completely symmetric. This symmetry minimizes the number of pairs of agents whose distances apart are greater than the equilibrium distance. Moreover, any agent moved out of this configuration will increase the sum of the distances between it and all other agents, thereby increasing the overall energy. Therefore, the lowest energy configuration will be the completely symmetric, perfect hexagonal lattice—because any other configuration will have a larger total distance error.

We have proved the following theorem:

**Theorem 17.1.** *The macroscopic state of the lattice of agents which minimizes the energy is the large (single) hexagonal state.*

### 17.3.3 Transitioning Between Minima

What we've demonstrated in the preceding subsections is that we can create a global property and use this property to motivate a class of behaviors, all of which are guaranteed to result in the desired global property being satisfied. Moreover, we've demonstrated that the system has a unique minimal value for this property, and that the minimal value occurs precisely at the desired system configuration. As a result, minimizing the value of the property is guaranteed to result in a system with the desired configuration.

Unfortunately, it is quite possible for the swarm, whether encountering an obstacle or not, to split. This happens regularly in the natural world with swarms, including birds, wildebeests, and schools of fish. We want to define a condition which would require that the swarm does not split, even under these circumstances.

Recall from Sec. 17.3.1 that our microtheory includes a mechanism that will move misplaced agents out of their temporary equilibrium positions and into their true equilibrium positions. The lattice as a whole moves in a hexagonal course, providing shocks at each turn. This gives the agents enough difference in energy to move out of a local minimum and possibly find an equilibrium position with even lower energy. Ultimately, the lattice will settle at the global energy minimum. This idea has its roots in *simulated annealing*, which finds the incorrect elements of a set and changes them, instead of replacing the whole set. With a decreasing energy function and implemented movements, the lattice will always achieve a perfect hexagon.

### 17.4 Simulation

In this section, we apply two swarm control methods derived from our understanding of the problem outlined in Sect. 17.3 to a virtual swarm, using a two-dimensional simulation. We begin by describing the simulation, focusing on the elements of the simulation which should make the control approaches easily portable to real robot platforms. Next we describe, in turn, two different solutions to the control problem given in Sect. 17.3.

## 17.4.1 Description of the Simulation

Our simulation is based on the artificial physics simulations of Spears et al. [235]. Briefly, those simulations focus on generating agent control using analogs of radial physical laws such as the law of gravity or electrostatics. Individual agents are represented as points and are capable of moving within the two-dimensional plane that makes up their environment. Each agent is assumed to have sensors capable of determining the distances and directions to its nearest neighbors. These distances and directions are then used to generate motion vectors. By utilizing a force function similar to that given in Sect. 17.3, Spears and his colleagues were able to have the individual agents arrange themselves in a locally hexagonal lattice (see also Chap. 3).

Our simulation is essentially identical to theirs. All agents are represented as points and are capable of movement within a two-dimensional plane, which is their environment. Agents are assumed to be equipped with directional sensors that provide them with the range and bearing to other agents. Each agent is assumed to have actuators which allow it to move, stop on a dime, and turn in place; inertia is not modeled.

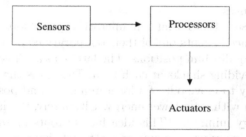

Fig. 17.1: This figure illustrates the data pathway of the agents in our system. All agents are reactive, and therefore data enters through the sensors, is processed directly in the processors, and is then translated to actuator commands. No memory is utilized in the control of these agents

Figure 17.1 illustrates the data pathway of the agents. In our system, all agents are reactive; memory is not required or utilized in the control algorithms. Thus, all sensory data is directly processed and translated to actuator commands in the same way that the original artificial physics simulations did. No special hardware is assumed here, and thus moderate timescales easily implemented on real agents may be used for real robotic implementations.

As indicated in Sect. 17.3, the basic behavior is nearly identical to the original behavior. This quickly changes a randomly organized group of agents to a relatively well-ordered group of agents whose energy value is significantly lower than the original energy level. However, it alone is incapable of generating a perfect hexagonal lattice, and so the perturbations described in

Sect. 17.3 must be applied. The basic behavior and energy minimization are illustrated in Figs. 17.2 and 17.3.

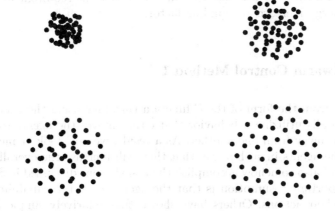

Fig. 17.2: Basic artificial physics behavior produces a semiordered lattice from a random lattice quickly

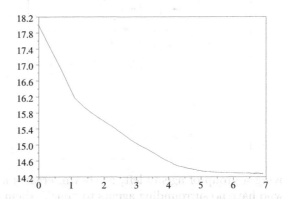

Fig. 17.3: As expected, with basic artificial physics, the energy decreases quickly. However, when the energy levels off, the system is caught in a higher energy state than the absolute minimum

We already noted that the basic behavior could be augmented by adding behaviors that encourage energy-requiring transitions between local minimum energy states. We shall see two different methods of achieving this in the next two subsections. Note that in both cases, the requirement of a temporary energy increase is a guiding factor.

## 17.4.2 Swarm Control Method 1

It is clear from the form of the $\mathbb{F}$ function that increasing the energy of the system entails utilizing a behavior that either moves the agents toward one another or away from one another. As a result, we now look at methods for moving the agents in such a way that they will increase the overall system's energy, and also somehow accomplish the translations discussed in Sect. 17.3.

An important assumption is that the agents have the capability to synchronize their actions. Others have shown that relatively simple behaviors may be utilized to obtain the decentralized synchronization of agents [95]. We assume that our agents have this added ability for the first solution.

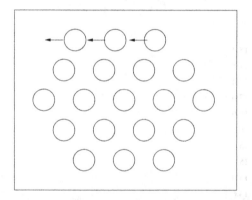

Fig. 17.4: Moving a group of agents left, as shown, creates a "drag" on the outer agents who have no surrounding agents to "push" them forward if they are left behind during an imperfectly synchronized turn. As a result, the outer agents' energy levels may be increased enough to shift

We also note, as depicted in Fig. 17.4, that when agents move in tandem and nearly perfect synchrony, the initial movement tends to create temporary movements of outer agents. This increases the outer agents' energies sufficiently to create a shift in agents as the entire group moves in a single direction.

As we've already noted, it is possible to generate a hexagon through energy-reducing shifts in the group of agents, thereby bringing the agents

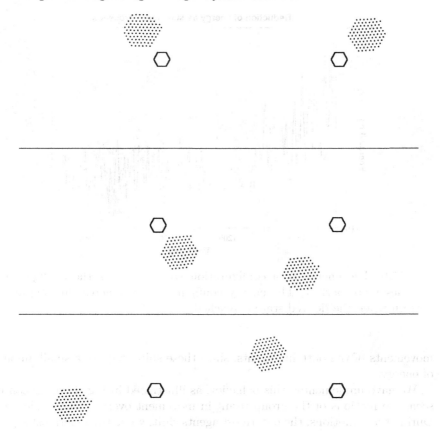

Fig. 17.5: This set of six images illustrates the "drifting" behavior that accomplishes the energy increase of the group of agents and the shifting of the outermost agents. The small hexagonal outline provides a frame of reference to aid in the visualization of the drifting, which is clockwise. As can be seen, this results, after several cycles if necessary, in the completion of the hexagonal structure

to more perfect positions and lowering the overall energy. However, in order to do this, one must move the group in such a way that the shifts happen along the outer edges of the large hexagon that is to be formed. Thus, the group movements must be aligned with the edges of the large hexagon. As a result, the group of agents must execute movements that are hexagonal, and this will both maintain the interior of the group's organization and affect the

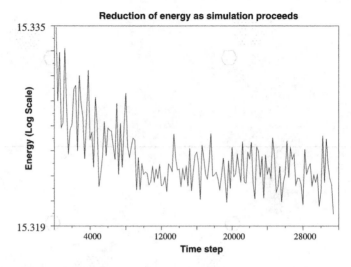

Fig. 17.6: The energy of a configuration steps down as the configuration becomes more perfect. The energy configuration completes this step-down pattern when the desired state is reached

movements of the outlying agents, since these shifts require a small amount of energy.

We have implemented this behavior, as illustrated in Fig. 17.5. As can be seen, the motions of the group result in movement over a rather wide area. During these motions, the outermost agents shift, while the innermost agents stay in formation. At the completion of sometimes several cycles, the agents gather into a very stable formation that does not support further shifts. This formation is the hexagonal formation, and it is the only one that does not allow shifts (see "moving.hexagon.avi").

It is interesting to examine the graph of energy versus time as the system settles into its lowest energy state. As the system shifts into a new state, although momentary variation continues, the base energy tends to step down until it reaches a minimal value. This is illustrated in Fig. 17.6.

The main drawback of this method is that the entire group must have an added layer of complexity as it executes the drifting behavior in tandem. Moreover, this behavior requires space, and the group may not be deployed until the behavior is complete. We have not explicitly built a behavior into the system that allows the agents to detect when they are in a large hexagon, since this added ability is beyond the scope of this study.

The next subsection examines a behavior that does not require the space of this method.

## 17.4.3 Swarm Control Method 2

As we saw in Sect. 17.4.2, it is possible to create group behaviors that accomplish the shifting of the agents by increasing the energy of the system. However, the first method had a few drawbacks that are difficult to envision for deployable swarms. In this section, we examine a more direct method of increasing the energy of a system in such a way that it goes into a shifting action that does not require synchrony among the agents or motion of the agents.

We now look at a second method, having the same objective as the first method, but that requires neither the coordination nor the space required by the first method. We start once again by considering what we can do to increase the energy of the system. As before, we realize that the energy can be increased by an individual agent by moving the agent out of equilibrium. A movement away from another agent will increase the energy much less than a movement towards the group. We therefore consider movements of agents towards one another, carefully attenuated so as to produce the small jumps required for the sliding action, but not for the rearrangement of the entire array.

The agents to which this behavior will be added must also be carefully controlled, since agents in the interior will generate larger energy changes with small movements than those on the exterior. Therefore the behavior requires the agents to consider their surroundings to enable the random triggering. Thus, the behavior's trigger is limited to those agents who have fewer than three neighbors. This will only occur with agents in external positions in the lattice, or to those at corners as a result of the agents' jostling around.

Figure 17.7 illustrates this behavior with an agent outside of a group of other agents. The exterior agent momentarily approaches the group. The approach increases the energy of the lattice, and the energy-reduction behavior moves the exterior agents in a translation motion. These actions can be initiated by single agents even though the entire group is in motion or static.

Figure 17.8 illustrates this behavior with a group of agents. The group begins in a state very far from the desired hexagonal state. However, as the translation movements continue, the system falls into a lower energy state, which eventually is that of the hexagonal organization. In all simulated runs, the system will settle into the hexagonal state eventually (see "stationary.hexagon.avi" for one example).[4]

---

[4] The simulation was run more than 10,000 times.

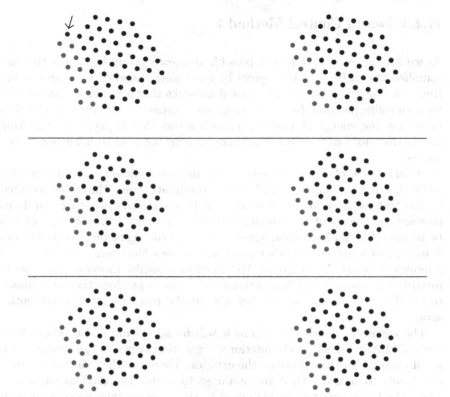

Fig. 17.7: The approach of a single agent can cause a set of other agents to slide into another position, with little change in the overall energy of the group

## 17.5 The Quark Model

Flocking has been very well studied in the swarm literature, beginning with the original Reynolds' boids [193] and continuing on through the years until today. In the natural world, animals flock together in ways that improve their survivability. The benefits include improved security, greater success in mating, and mutual predatory activity. The ways in which flocks stay together are varied, though in many natural flocks, they are mediated by flight. Despite the ability of the flocking group to hold their form, to turn and move together, and to make apparently seamless decisions, there are always cases in which the flock bifurcates and must somehow come back together.

For many tasks, bifurcation is an unacceptable property of man-made (synthetic) swarms. As a result, it's important to develop swarms of agents that, once connected, do not come apart. Moreover, it is also advantageous to

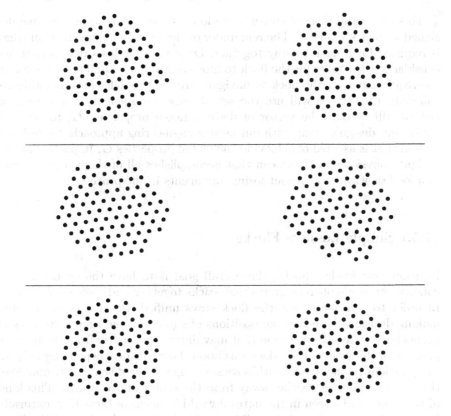

Fig. 17.8: This set of six images illustrates the "closing" behavior. The behavior initiates a number of different sliding behaviors that eventually cause the generation of a good hexagonal structure. As before, the energy inches down to a minimum under the action of the individual agent perturbations

investigate swarms that maintain a cohesive and regular global structure as a natural consequence of their cohesive behavior. The literature has several different approaches to this problem. These range from chorusing [95] to methods validated by temporal logic [272]. However, these methods have not yet been shown to be able to handle perturbations, such as global movement of the group or navigation through an obstacle course.

It is generally assumed that flocking is a function that is mediated entirely by the individuals using pairwise information that they can obtain by looking at their closest neighbors. Many authors have examined the problem and generated autonomous local-only information that can be obtained by using individual sensors. However, we claim that local-only information is insufficient to generate a control algorithm that prevents bifurcation of the swarm.

To address the issue of swarm cohesion without bifurcation, we have designed a "quark model."[5] The remainder of this section focuses first on what is required for a flock to stay together. Once this has been determined, we consider what it takes for the flock to move as a single unit. Then we explore the requirement for the flock to navigate around obstacles and maintain its cohesion. Finally, we will use the set of flock requirements just presented (which will become the vector of desired global properties, $G$, to be used for system design), along with our swarm engineering approach, to design a system that is assured of achieving the global properties $G$. In particular, we design a physicomimetic swarm that accomplishes all of these requirements in a lossless way, i.e., without losing any agents in the process.

## 17.5.1 Simple Lossless Flocks

In generating lossless flocks, the overall goal is to have the agents form a coherent or semi-coherent group that sticks together and loses no members. In order to guarantee that the flock stays unified, the agents must avoid making decisions based on the positions of subsets of the agents. Subsets of agents have an average position that may differ significantly from the average position of the swarm. If so, decisions about how to move, depending only on the positions of the agents within sensor range of the given agent, may lead the agent continually farther away from the center of the swarm. This kind of behavior can be seen in the natural world in flocks of birds that naturally break up into several parts, possibly coming back together later if the birds stay in the same general vicinity. Our strategy is to examine how to keep swarms of agents together once they have formed.

Put another way, each agent samples the flock using its sensors and makes decisions about what to do based on the sampling. In the case that the sampling is representative of the characteristics of the swarm and the behavior is intended to have a specific outcome relative to the characteristic, the agent can react appropriately. However, in the case that the sampling is not representative of the swarm, the behavior may or may not be appropriate.

We start by examining what each member of the flock can sense about the other agents in the flock. Each agent has a position; let the set of all positions of the agents be denoted by $\{x_i\}_{i=1}^{N}$ where $N$ represents the total number of agents. For simplicity, we will let $x_j$ represent the position of agent $j$. We assume that the agents have the ability to determine the relative positions of all other agents within a sensory range $r_s$.

Two groups of agents, $A$ and $B$, are detached if for all agents $a \in A$ and agents $b \in B$, $|x_a - x_b| > r_s$. Let's define an inter-group distance, $x_{AB} = x_m - x_n$, where $x_m$ and $x_n$ are chosen so that $|x_m - x_n|$ is minimal.

---

[5] The "quark model" is an analogy to the physical requirement that quarks only exist in pairs and triplets, but not alone.

Furthermore, let $\boldsymbol{x}_A = \sum_{i=1}^{N_A} \boldsymbol{x}_i / N_A$ and $\boldsymbol{x}_B = \sum_{i=1}^{N_B} \boldsymbol{x}_i\, N_B$ be the average positions of the sets of agents from group $A$ and group $B$, respectively. Note that group $A$ has $N_A$ agents while group $B$ has $N_B$. Then, in order for the flock to be truly lossless, it must be the case that any two groups of agents in the flock never become detached. That is, $|\boldsymbol{x}_{AB}| \le r_s$. It can be shown that in order for a group to be connected, any group of $N_s$ agents, where $N_s < N$, must be in sensory contact with at least one other agent outside of the group. If not, the group would be detached from the remainder of the agents. Therefore, the group of agents must be reactive to the agents not in the group in such a way that the group remains in contact with the agents external to the group.

Another condition for the flock to be truly lossless is that if an agent or a group of agents becomes detached, then the behavior of the detached agent(s) should be to reattach to the flock. That is, whenever $|\boldsymbol{x}_{AB}| > r_s$, it must be the case that $\frac{d|\boldsymbol{x}_{AB}|}{dt} < 0$.

Let us suppose that $A$ is a singleton group. Further suppose that $|\boldsymbol{x}_{AB}| > r_s$. Let

$$\widehat{\boldsymbol{x}_{A \to B}} = \frac{\boldsymbol{x}_B - \boldsymbol{x}_A}{|\boldsymbol{x}_B - \boldsymbol{x}_A|} \tag{17.13}$$

and

$$\widehat{\boldsymbol{x}_{A \to AB}} = \frac{\boldsymbol{x}_{AB} - \boldsymbol{x}_A}{|\boldsymbol{x}_{AB} - \boldsymbol{x}_A|} \,. \tag{17.14}$$

Further let

$$\phi = \cos^{-1}\left( \widehat{\boldsymbol{x}_{A \to B}} \cdot \widehat{\boldsymbol{x}_{A \to AB}} \right) \,.$$

Then it is straightforward to show that if

$$\cos^{-1}\left( \widehat{\boldsymbol{x}_{A \to B}} \cdot \frac{\boldsymbol{x}_A}{|\boldsymbol{x}_A|} \right) \le \frac{\pi}{2} - \phi \tag{17.15}$$

and

$$\cos^{-1}\left( \widehat{\boldsymbol{x}_{A \to AB}} \cdot \frac{\boldsymbol{x}_A}{|\boldsymbol{x}_A|} \right) \le \frac{\pi}{2} - \phi \tag{17.16}$$

then $\frac{d|\boldsymbol{x}_{AB}|}{dt} < 0$.

For larger groups of agents than singleton groups, the condition remains the same. The only caveat is that the entire group must move within these limits in order to recover the remainder of the swarm.

If we combine these requirements, letting the second requirement apply when the agent is "near" or beyond the edge of the swarm, this makes the swarm lossless and capable of reforming after perturbations of the spatial organization. Essentially, this is a simple geometric requirement for the agents to accomplish in order for the swarm to recover from any separation. However, the requirement clearly uses information that is not possible to obtain from simple pairwise sensing; the agent must be able to determine the global position, however noisily, of the swarm in order to reliably implement this behavior. As a result, global information must be available to the agents. Note that this result is independent of the model of the agents. Thus, the result holds for all swarms; lossless swarms can be developed if and only if individuals have access to global information about relative position.

## 17.5.2 Directional Coherent Flocking

In [115], Kazadi and Chang examined the general requirements for hijacking swarms. They determined that the general requirement for a swarm of agents to be "hijacked" was that the swarm property in question was not actively under the control of the swarm. That is, if a property is affected by the behaviors of the swarm, but the controller does not use this property's value in determining how the swarm will behave, then the property may be hijacked. In that case, the property can be affected by the smallest influence from any agent, internal or external.

In flocking systems, the swarm condition for lossless flocking centers around the use of the global information of the average *relative* position of the swarm and the agent(s) potentially leaving the swarm. The specific position of the swarm is not part of the control algorithm. This gives us the ability to control the exact position of the swarm *without interfering with the flocking of the swarm*. To see this, let

$$\bar{x} = \frac{1}{N_r} \sum_{i=1}^{N_r} x_i , \qquad (17.17)$$

where $N_r \leq N$. In this case,

$$\frac{\partial \cos^{-1}\left(\widehat{x_{A \to AB}} \cdot \frac{x_A}{|x_A|}\right)}{\partial \bar{x}} = \frac{\partial \cos^{-1}\left(\widehat{x_{A \to B}} \cdot \frac{x_A}{|x_A|}\right)}{\partial \bar{x}} = 0 . \qquad (17.18)$$

This verifies that the average position of the swarm has nothing to do with the control algorithm, and we are free to move it. Now, the question is, how can it be moved?

The researchers [28, 115, 229] all independently determined similar methods. In the first, the majority of the agents were designed to simply maintain the swarm in formation. A single "rogue" agent was added to the system whose controller included a directional "bias" (i.e., a "push" in a certain direction) to its direction of movement. This meant that while the main effect on the single agent was to keep it in the swarm, it tended to move in the bias direction, thereby dragging the entire swarm along with it. Celikkanat et al. designed a similar system in which a bias was added to a group of agents [28]. In this case, the idea was to have the agents not only move, but also to follow a path. Some agents were "informed" while others were not. The performance of the group improved with the number of informed agents.

In both systems, the additional behavior was to nudge movement in a particular direction, while maintaining the swarm cohesion. Another paper that examined this same problem in detail using different physicomimetics force laws and evolutionary computation-based optimization is [93]. In this paper, by Hettiarachchi and Spears, the swarm moved through an obstacle field while maintaining its form. The agents' controllers added a repulsive term which allowed the agents to avoid a collision when close to an obstacle, thus making a minimal perturbation of the system. This approach is also described in Chap. 14 of this book.

Let us suppose that we have a coherent swarm in the absence of obstacles. Suppose that a single robot $a_0$ is informed, which in this case means that a right-motion bias is added to its current direction of movement. In this case, the reaction to this bias will be for the swarm to follow the individual $a_0$, and move to the right. However, we have said that this is true as long as the angles are as given above in Eqs. 17.15 and 17.16. Let us understand when this is not so.

Suppose that the informed individual adds a directional bias to one robot's behavior. That is, the controller calculates a direction and speed to move, and the agent adds $\boldsymbol{v}_b$ to its current velocity vector, $\boldsymbol{v}_c$. The resultant magnitude and direction are given by

$$\boldsymbol{v}_T = \boldsymbol{v}_c + \boldsymbol{v}_b \, . \tag{17.19}$$

This means that

$$\cos^{-1}\left(\widehat{\boldsymbol{x}_{A \to AB}} \cdot \frac{\boldsymbol{v}_T}{|\boldsymbol{v}_T|}\right) \leq \frac{\pi}{2} - \phi \tag{17.20}$$

and

$$\cos^{-1}\left(\widehat{\boldsymbol{x}_{A \to B}} \cdot \frac{\boldsymbol{v}_T}{|\boldsymbol{v}_T|}\right) \leq \frac{\pi}{2} - \phi \, . \tag{17.21}$$

Inverting these equations obtains

$$\widehat{\boldsymbol{x}_{A \to B}} \cdot \boldsymbol{v}_c + \widehat{\boldsymbol{x}_{A \to B}} \cdot \boldsymbol{v}_b \leq |\boldsymbol{v}_c + \boldsymbol{v}_b| \sin(\phi) \leq (|\boldsymbol{v}_c| + |\boldsymbol{v}_b|) \sin(\phi) \tag{17.22}$$

and

$$\widehat{\boldsymbol{x}_{A \to AB}} \cdot \boldsymbol{v}_c + \widehat{\boldsymbol{x}_{A \to AB}} \cdot \boldsymbol{v}_b \le |\boldsymbol{v}_c + \boldsymbol{v}_b| \sin(\phi) \le (|\boldsymbol{v}_c| + |\boldsymbol{v}_b|) \sin(\phi) \ . \quad (17.23)$$

If we subtract Eq. 17.23 from Eq. 17.22 we obtain

$$\left( \widehat{\boldsymbol{x}_{A \to B}} - \widehat{\boldsymbol{x}_{A \to AB}} \right) \cdot \boldsymbol{v}_c \le \left( \widehat{\boldsymbol{x}_{A \to AB}} - \widehat{\boldsymbol{x}_{A \to B}} \right) \cdot \boldsymbol{v}_b \ . \quad\quad (17.24)$$

This requirement is also independent of the model of the agents. It gives limits on the magnitude and direction of $\boldsymbol{v}_b$ that must be obeyed in order to move the flock in a lossless way. Too great an added movement, and the swarm will lose the agent rather than being dragged along by it.[6]

## 17.5.3 Directional Coherent Flocking in the Presence of Obstacles

Flocks of individual agents often encounter obstacles, big and small. These obstacles can usually be overcome, with the entire flock swarming around a small obstacle. This can be achieved by small perturbations of the agent behaviors that allow agents to avoid obstacles in the way. This works fine when the obstacles are small, because the behaviors need only include an obstacle avoidance behavior (in behavior-based agents), a repulsion term in physicomimetic systems, or other biases that allow the agents to avoid the obstacles. The difficulty occurs when the obstacle is large in relation to the swarm and the agents in the swarm.

In the case that the obstacles are physically large with respect to the swarm, it is quite possible for the swarm to encounter the obstacle and be split. Upon encountering a large obstacle, the swarm splits into two separate groups which then move away from one another. Each agent is performing its behavior flawlessly, but the overall action still causes the swarm to pull apart into two groups. This occurs because the obstacle avoidance behavior and movement behavior eventually overwhelm the flocking behavior, essentially pulling the swarm apart a little at a time until the swarm has been split.

It can be shown, using a method similar to that in Sect. 17.5.2, that we require

$$\left( \widehat{\boldsymbol{x}_{A \to B}} - \widehat{\boldsymbol{x}_{A \to AB}} \right) \cdot \boldsymbol{v}_c \le \left( \widehat{\boldsymbol{x}_{A \to AB}} - \widehat{\boldsymbol{x}_{A \to B}} \right) \cdot (\boldsymbol{v}_b + \boldsymbol{v}_a) \ , \quad\quad (17.25)$$

where $\boldsymbol{v}_a$ represents the additional push due to obstacle avoidance. This is the additional requirement for the swarm of agents undergoing obstacle avoidance while maintaining a coherent swarm.

---

[6] Note that Chap. 6 also addresses a similar concern in the section about the theory for setting parameters. However those results are not model independent.

## 17.5.4 Quark Model

We have established the global-to-local engineering requirements (i.e., the desired global properties for the vector $G$) in the previous subsections. Now, we look at the local behaviors that accomplish these global objectives. In other words, the next swarm engineering task is to design the global function $g$. This entails engineering a swarm system that satisfies all of the requirements in Eqs. 17.15, 17.16, 17.24, and 17.25.

We can construct a physicomimetic system [76, 235] that satisfies the requirement of lossless flocks captured by Eqs. 17.15 and 17.16. This is done by creating a swarm in which each agent's behavior is most strongly affected by the agents that are within sensor range but are also furthest away. In this case, the direction of motion for agent $j$, obtained from the vector $f_j$, is determined as follows:

$$f_j = \sum_{i=1}^{N} [(1 - \delta_{ij})((a(r_{ij}) - b(r_{ij})) \hat{r}_{ij})] , \qquad (17.26)$$

where, as usual, $r_{ij}$ is the distance between agent $i$ and agent $j$, and $N$ is the total number of agents. Furthermore, $\delta_{ij}$ is 0 or 1, depending on whether agent $i$ is within sensing range $r_s$ of agent $j$ or not, respectively. Also, $\hat{r}_{ij}$ is a unit vector from agent $i$ to agent $j$, $a$ is a monotonically increasing real-valued function with a vertical asymptote at $x = r_s$, and $b$ is a monotonically decreasing real-valued function with a vertical asymptote at $x = 0$. In this case, the two limits behave like an infinite potential well with curved walls. In the physical analog, a particle with finite energy remains indefinitely trapped between 0 and $r_s$.

In our system, the vector $f_j$ gives the *direction* of motion rather than the magnitude of motion. We assume that the agent moves at its top speed in all cases, and the direction of motion for agent $j$ is given by the unit vector $u_j = f_j/|f_j|$. It is straightforward to see that this causes small groups of agents that have strayed from the main group to move directly toward the main group, and the main group will respond by moving toward the small group. This method clearly satisfies Eqs. 17.15 and 17.16 when approaching $r_s$. As a result, we expect it to generate coherent groups, despite the possibility of the group having a dynamically changing structure.

In order to validate this design, we simulate a circular (disk-shaped) swarm of agents in a two-dimensional plane. As before, the agents are capable of determining the positions of other agents within their sensor range $r_s$. Beyond this range, the agents cannot see the other agents. In our system,

$$f_j = \sum_{i=1}^{N} \left[ \cot\left(\frac{\pi r_{ij}}{r_s}\right) \hat{r}_{ij} \right] , \qquad (17.27)$$

which satisfies the definition given in Eq. 17.26 [113]. The sum of the different vectors produces a summed vector, which the agents then use to determine their direction of motion, and they head in this direction at their maximum speed. In our simulations, an individual agent's sensor range is 90 units, and agents are randomly dispersed around a 400 × 400 unbounded arena. We begin by examining the case where there are no obstacles or "informed" individuals.

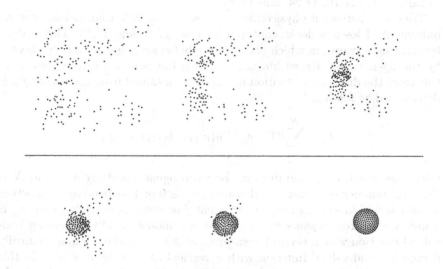

Fig. 17.9: These six images depict a typical swarm evolution of our unbounded, uninformed quark model. The swarm contracts efficiently and merges to form a single symmetric swarm

Figure 17.9 depicts an uninformed swarm contracting from random initial agent positions. The swarm quickly forms swarming centers, or areas of high density which might lead to the formation of a sub-swarm. The centers of high density merge together, eventually forming a single group 81% of the time. When the swarm does not form into a single group, this is the result of singleton agents failing to sense other agents when the simulation begins; no agents get lost along the way. See "quark.coalescing.avi" for an example simulation run.

We continue with an examination of the behavior of the swarm under the action of a single informed individual moving to the right, almost immediately when the simulation starts. Since this global movement does not affect the pairwise interactions, it is not expected to destabilize the swarm. Indeed, the swarm follows the informed agent losslessly, and still generates the same global behavior seen earlier. A typical run is depicted in Fig. 17.10 (see "quark.chasing.avi").

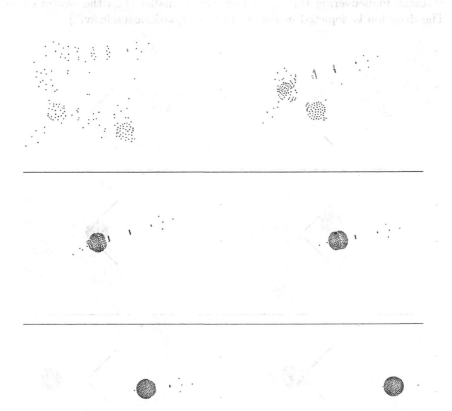

Fig. 17.10: These six images depict a typical swarm evolution of our unbounded, informed quark model with a single informed individual. The swarm begins moving as it contracts. The informed individual moves in its preferred direction, "pursued" by the bulk of the swarm, which eventually overtakes and absorbs the individual

In this model, the informed agent begins moving toward the right immediately. The other agents coalesce, but also begin following the moving agent, absorbing smaller groups into the larger aggregate. As the swarm congeals behind the moving agent, it remains connected by a few intermediate agents. The bulk of the swarm follows the moving agent and eventually absorbs it.

We also examine the behavior of the swarm in the presence of an obstacle. In this case, we have two "walls" generated by closely packed, deactivated agents. After initial random dispersion, one agent moves toward the right. The remainder of the swarm coalesces, and follows the agent through the obstacle, maneuvering through a hole much smaller than the swarm's size. The situation is depicted in Fig. 17.11 (see "quark.obstacle.avi").

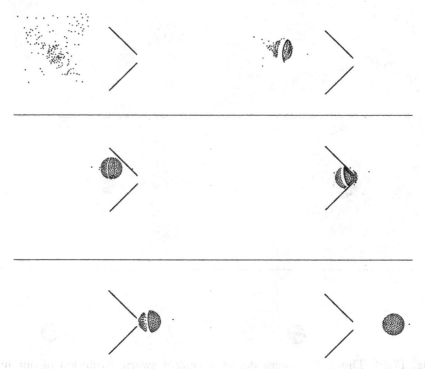

Fig. 17.11: These six images depict a typical swarm evolution of our unbounded, informed quark model with a single informed individual and an obstacle. The swarm begins moving as it contracts. The informed individual moves in its preferred direction, "pursued" by the bulk of the swarm, which eventually overtakes and absorbs the individual. When the swarm encounters the obstacle it temporarily loses the informed individual while it squeezes through, after which it recaptures the informed individual

The behavior of the quark model in the presence of obstacles very clearly depicts the goal of the lossless swarm. Not only does it not lose an individual leaving the swarm, but it is capable of getting through obstacles whose apertures are significantly smaller than the swarm size. The swarm, which is made up of autonomous purely reactive agents, manages to get through intact and resumes its normal behavior on the other side of the obstacle.

## 17.6 Discussion

The goal of the swarm engineer is the construction of swarms with predetermined global characteristics. These characteristics can be widely varied and include aspects of the swarm's spatial dispersion, information sharing, processing and group dynamics. What makes the design of swarms particularly difficult is the fact that an engineer's approach is typically very much tied to a particular model. This recasts the problem from "How can we build a swarm that accomplishes the task?" to the problem of "How do we build a swarm from agent type X that accomplishes the task?" Yet there is no a priori reason to believe that either X is the appropriate type to use, or that this particular agent type even has an appropriate solution associated with it.

Yet to be practical, the swarm engineer will become comfortable with a set of techniques that yield good results in terms of the design and simplicity of the swarm. Much of the criticism of swarm engineering in general comes from the inability of the community of swarm researchers to deliver predictable results that are simple, tunable, and yield robust swarms with control potential. Our combination of the swarm engineering methodology and the physicomimetic platforms does just this, since the global goal can be mathematically guaranteed as long as the swarm requirements are accomplished, and the physicomimetic system can accomplish the swarm requirements provably.

This combination of methods also has interesting applications beyond simply accomplishing a specific task. As a metaphor for living things, swarms can be remarkably like living creatures. Swarms of insects have processes that could be called *autonomic*, since they continue unabated at all times. They also have processes that are reactive and are initiated only when an event happens to trigger them. Swarms have economies of scale, and generally have an optimal size for tackling any particular task. Moreover, that size is regulated by some means either within or beyond the swarm's direct control.

One aspect of swarms that is typically stated, but not necessarily appreciated, is the fact that agents are really interchangeable—destroy or disable one of them, and another one can take over the task. The overall effectiveness of the swarm degrades gradually with the removal of individual agents. However, complex inter-agent interactions, such as those required to create, maintain and exploit computational centers specialized for handling of specific data, or those required to route data from one part of the swarm to another, may not be so indestructible. Yet it is precisely these kinds of complex swarms of differing and varied internal structures whose design we don't yet know how to generate, and that the current design work can help to bring forth.

Physicomimetics is a remarkably simple class of algorithms in terms of the requisite hardware and software, and also in terms of its understandability. It is also extremely applicable to a wide range of different problems. Because of their simplicity, physicomimetic algorithms are well-suited to autonomic swarm applications. These might be used to maintain cohesion, move the swarm, and adapt the swarm's global properties in response to external stimuli or control signals. On top of these algorithms, one might add sophisticated structures, such as the ability to process signals obtained by agents on the exterior of the swarm (as in data mining or small signal extraction). Internal special purpose networks, as in those one might find in the retinal ganglia or cortex, might be constructed so as to process these data. Moreover, removing a single agent due to failure could be easily adapted using algorithms that utilize proximity alone to generate these computational substructures. The result would be a swarm controlled and maintained by physicomimetics algorithms, and specialized for data manipulation using smaller sub-swarms within.

In order to achieve these kinds of tasks, however, the swarm engineer needs to have absolute predictability of the global structures upon which the local properties are built. As we've seen, this is achievable using our swarm engineering methods described above. As a result, we are now poised to begin the exciting task of extending the swarms into pseudo-organisms, including autonomic, control, and reaction algorithms, as well as structures analogous to a brain cortex, which is enabled with fault tolerance and network reconstruction. The result is a set of swarms that can be adaptive, controllable, reactive, and has significant and reconfigurable internal capabilities.

## 17.7 Conclusion

In this chapter, we've examined a combination of swarm engineering and physicomimetic models of swarm design. The two methods, when combined, represent a powerful aggregate methodology. What swarm engineering contributes is the global-to-local mapping of the global swarm goal(s). What physicomimetics achieves is provable fulfillment of the local requirements. Together, the methods generate a provable and mathematically rigorous approach to swarm design.

We examined the application of these methods to a hexagonal swarm problem in which a swarm achieves a hexagonal global structure and a hexagonal grid structure internally. We generated two different perturbations to the basic physicomimetics model, which yielded the global desired state. We also examined the creation of a completely lossless swarm. This swarm, when generated, exhibited the ability to coalesce into a single unit most of the time, and to successfully and losslessly move coherently in the presence or absence of obstacles.

Future work will continue to examine the theoretical aspects of generating swarms with potentially many global goals. Application of physicomimetics to autonomic swarm functions in the presence of non-global behaviors will also be examined. We will also examine mixed physicomimetic models in which one physicomimetic model is utilized in an interior space, and another is used externally.

# Part V
# Function Optimization

Part V

Function Optimization

# Chapter 18
# Artificial Physics Optimization Algorithm for Global Optimization

Liping Xie, Ying Tan, and Jianchao Zeng

## 18.1 Introduction

Global optimization problems, such as seeking the global minimum, are ubiquitous in many practical applications, e.g., in advanced engineering design, process control, biotechnology, data analysis, financial planning, risk management and so on. Some problems involve nonlinear functions of many variables which are irregular, high dimensional, non-differentiable, or complex multimodal. In general, the classical deterministic optimization techniques have difficulties in dealing with such kinds of global optimization problems. One of the main reasons for their failure is that they can easily get stuck in local minima and fail to obtain a reasonable solution. Moreover, these techniques cannot generate or even use the global information needed to find the global optimum for a function with multiple local optima.

Nature-inspired heuristics, such as evolutionary algorithms, swarm intelligence algorithms, bio-inspired methods, and physics-inspired algorithms have proven to be very effective in solving complex global optimization problems. Evolutionary algorithms heuristically "mimic" biological evolution, namely, using the process of natural selection and the "survival of the fittest" principle. The most popular type is the genetic algorithm (GA) [94], which uses techniques inspired by evolutionary biology such as inheritance, mutation,

Liping Xie
Taiyuan University of Science and Technology, Taiyuan, Shanxi, China, e-mail: xieliping1978@gmail.com

Ying Tan
Taiyuan University of Science and Technology, Taiyuan, Shanxi, China, e-mail: tanying1965@gmail.com

Jianchao Zeng
Taiyuan University of Science and Technology, Taiyuan, Shanxi, China, e-mail: zengjianchao@263.net

W.M. Spears, D.F. Spears (eds.), *Physicomimetics*,
DOI 10.1007/978-3-642-22804-9_18,

selection and crossover. Bio-inspired methodologies are even more elemental. The artificial immune system (AIS) heuristic, for example, is inspired by mammalian immunology at the cellular level. An important subset of AIS is the clonal selection algorithm (CSA) that mimics clonal selection in the natural immune response to the presence of "nonself" cells [46, 257].

Swarm intelligence algorithms are inspired by the collective behaviors of social insect colonies or animal societies; examples include ant and bee colonies, flocks of birds or schools of fish [18, 215]. In ecology systems, the swarm collective behaviors are complex, dynamic, and adaptive processes, which emerge from simple interactions and cooperation between individuals (e.g., self-organization). Several injured or dead individuals will hardly influence the swarm behavior—because they can be replaced with others and their task can be continued with swarm self-adaptation and self-repair. Similarly, self-organization, self-adaptation, and self-repair are precisely those principles exhibited by natural physical systems. Thus, many answers to the problems of swarm intelligence control can be found in the natural laws of physics.

As a well-known example of the swarm intelligence algorithm, particle swarm optimization (PSO) [119], is a population-based random search method motivated by social animal behaviors such as bird flocking, fish schooling, and insect herding. In the PSO algorithm, each individual, called a "particle," represents a potential solution, and it moves over the problem space (environment) to seek food (which is an optimum point) by adjusting its direction of motion according to its own experiences as well as those of the swarm. In an artificial physics sense, particles could be considered as searching the problem space, driven by two attractive forces based on the best history of their own position and that of the swarm. However, the standard PSO only provides a simple uniform control, omitting the differences and interactions among individuals, which are very important in a swarm foraging process.

Examples of physics-inspired algorithms include simulated annealing (SA) [126], the electromagnetism-like algorithm (EM) [15, 196], and central force optimization (CFO) [66]. SA performs optimization by analogizing the statistical mechanics of systems in thermal equilibrium. CFO analogizes gravitational kinematics using Newton's universal law of gravitation and his equations of motion. CFO searches for extrema by using a group of "probes" to "fly" through the decision space. The probes' trajectories are deterministically governed by the gravitational analogy. EM is inspired by Coulomb's force law for electrostatic charges.

All of these heuristics are subject to limitations that may result in their missing global minima. Getting trapped at local minima and failing to search the decision space completely are, perhaps, the main difficulties. Many algorithms are easily trapped in the local optima of highly multimodal or "deceptive" functions. Some algorithms initially converge very quickly, but are slow to fine-tune the solution for highly multimodal functions. Others converge slowly on very difficult problems (for example, the "needle-in-a-haystack" functions), or their computational efficiency is very low on problems of high dimensionality, i.e., they require too many calculations.

Artificial physics optimization (APO) [280, 285] was motivated by these observations and by the success of physicomimetics, or "artificial physics" (AP), as a metaphor for controlling multi-robot systems. The individual robots sense environmental parameters and respond to algorithmically created "virtual forces" that constitute the AP environment. The essential concept is that any system created according to the laws of physics can be controlled using the entire panoply of empirical, analytical, and theoretical tools available in modern physics and engineering.

Because the APO algorithm is a new swarm intelligence technique based on the AP framework for global optimization problems, there is considerable room for improvement. Since APO was first presented in 2009, many improvements to APO have been proposed, such as force law design [282, 283, 287], mass function selection [286], convergence analysis [279], an extended version of APO [281], its vector model [288], and local APO [161], which are briefly introduced in this chapter. In addition, APO has been straightforwardly applied to solve multi-objective optimization problems [258] and swarm robotic search [284], and it has been shown to be effective.

The rest of this chapter is organized as follows. Section 18.2 explains how to map the AP method to an optimization algorithm. Section 18.3 describes the general framework of the APO algorithm at length, designs three force laws and presents the selection principle of mass functions. Section 18.4 introduces the convergence conditions of APO and guidelines for selecting the gravitational constant $G$. Section 18.5 includes improvements on APO. Finally, the conclusions are presented, and further research directions are introduced.

## 18.2 Motivation

Swarm foraging and searching for optima in a global optimization problem are similar problems. Actually, the AP method for simulating swarm foraging can solve the global optimization problem. In the case of bird foraging, birds can communicate with each other, and each bird flies towards those that find more food, and away from those birds that find less food. In an artificial physics sense, if the communication between birds is regarded as a virtual force, that is, there exist virtual forces among birds, then each bird is attracted to others that find more food, and repelled by others that find less food. In the APO algorithm, each individual is called a "particle," and it represents a potential solution. Inspired by attractive–repulsive force laws between birds, the individual is attracted to others with a better fitness and repelled by others with a worse fitness. These virtual forces drive individuals to seek the optimum in a global optimization problem.

There are many similarities between the AP method and population-based optimization (PBO) algorithms, such as PSO, EM, CFO, etc. The mapping of the AP method to an optimization algorithm is shown in Fig. 18.1. In the PBO algorithm, each sample solution in the decision space is considered as a particle that has a fitness, position and/or velocity (for example, in PSO and EM). The decision space is a multidimensional search space, in which particles can be globally located. And particles have access to the information of the whole swarm. Various intelligent search strategies lead these particles to move towards the regions of the decision space with better fitness. Similarly, the AP method treats each entity as a physical individual that has a mass, position and velocity. The motion space is two- or three-dimensional space, in which individuals can only be locally located. Individuals only have access to information from their neighbors. An appropriate attraction–repulsion rule can lead APO's population to search regions of the decision space with better fitnesses.

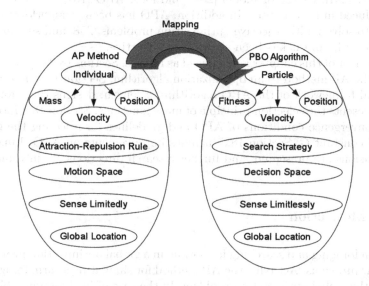

Fig. 18.1: The mapping of the AP method to the PBO algorithm

Obviously, the AP method can be easily mapped to the PBO algorithm. In the APO algorithm, we construct the relationship between an individual's mass and its fitness that we are trying to optimize. The better the objective (fitness) function value, the bigger the mass, and therefore the higher the magnitude of attraction. The individuals move towards the better fitness

regions, which can be mapped to individuals moving towards others with
bigger masses. The virtual forces drive each individual's motion. A bigger
mass implies a higher magnitude of attraction. In particular, an individual
is attracted to others with a better fitness and repelled by others with a
worse fitness than its own. Most importantly, the individual with the best
fitness attracts all the others, whereas it is never repelled or attracted by
others. The attractive–repulsive rule can be treated as a search strategy in
the optimization algorithm that will be used to lead the population to search
the better fitness regions of the problem.

To illustrate APO, the attractive–repulsive force law applied to three individuals (particles) in a two-dimensional decision space is shown in Fig. 18.2.
This system will be used to explain how APO works. The algorithm "moves"
a group of individuals through a decision space, driven by a virtual (user-
defined) force. Individuals $i$, $j$, and $l$, respectively, have objective function values $f(X_i)$, $f(X_j)$, and $f(X_l)$ and corresponding masses $m_i$, $m_j$, and $m_l$. When
performing minimization, if $f(X_l) < f(X_i) < f(X_j)$ then $m_l > m_i > m_j$. In
other words, if we are minimizing then "lower" is "better" when referring to
fitness. In Fig. 18.2, larger circles correspond to greater mass values (bigger
individuals with greater attraction). Under APO's novel attractive–repulsive
force law, individual $i$ is attracted to individual $l$ and repelled from individual
$j$, that is to say, an attractive force $F_{il}$ and a repulsive force $F_{ij}$ are exerted
on individual $i$. As in the physical universe, the total force $F_i$ exerted on individual $i$ is calculated by vectorially adding the forces $F_{il}$ and $F_{ij}$. Individual
$i$ moves through the decision space under the influence of this total force $F_i$
which then determines its "velocity." The velocity, in turn, is used to update
particle $i$'s position step-by-step throughout the optimization run.

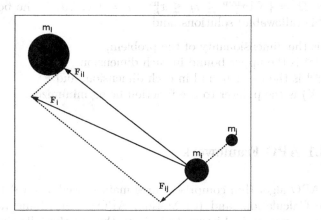

Fig. 18.2: A typical two-dimensional APO decision space

Listing 18.1: APO algorithmic framework

```
Begin
  Initialize population: both position x and velocity v;
  Set parameters, N, d, w and G, etc.;
  Iteration = 1;
While (termination criterion is not met) {
  Evaluate all individuals using corresponding fitness function;
  Update the global best position X_best;
  Calculate the mass using Eq. 18.2;
  Calculate the component force using Eq. 18.3;
  Calculate the total force using Eq. 18.4;
  Update the velocity v using Eq. 18.5;
  Update the position x using Eq. 18.6;
  Iteration = Iteration + 1;
}
End
```

## 18.3 General Framework of the APO Algorithm

The APO algorithm addresses the problem of locating the global minimum of a (nonlinear) objective function $f(x)$ defined on a bounded hyperspace,

$$\min\{f(x) : X \in \Omega \subset \mathbb{R}^d\}, \; f : \Omega \subset \mathbb{R}^d \to \mathbb{R}, \tag{18.1}$$

where $\Omega := \{X \mid x_k^{\min} \leq x_k \leq x_k^{\max}, \, k = 1, \ldots, d\}$ is the bounded region of feasible (allowable) solutions, and

- $d$ is the dimensionality of the problem,
- $x_k^{\max}$ is the upper bound in each dimension,
- $x_k^{\min}$ is the lower bound in each dimension, and
- $f(X)$ is the pointer to the function being minimized.

### 18.3.1 APO Framework

The APO algorithm comprises three main procedures: (a) Initialization, (b) Force Calculation, and (c) Motion. APO's algorithmic framework (pseudocode) appears in Listing 18.1. Note that in the following discussion the terms "time step," "step," "iteration," and "generation" are used interchangeably, as are the terms "individual" and "particle."

In the Initialization step, a swarm of individuals is randomly created in the $d$-dimensional decision space. $N$ denotes the number of individuals. The particles' velocities are initialized to random values in the region of decision space. The fitness (i.e., the value of the objective function) for each individual is calculated using the function pointer $f(x)$. The position vector of the particle with the best fitness at step $t$ is denoted $X_{\text{best}}$, which is the globally best position at that time step.

In the next step, Force Calculation, the total force exerted on each individual, is calculated based on the "masses" and distances between particles. The APO force law is an extension of the standard Newtonian physicomimetics force law shown earlier in this book, which first requires a definition of "mass" in APO space. Mass is a user-defined function of the value of the objective function to be minimized, that is,

$$m_i = g(f(X_i)) , \tag{18.2}$$

where $m_i \in (0, 1]$ and $g(\bullet) \geq 0$ is bounded and monotonically decreasing. Obviously, there are many functions meeting these requirements. The selection of the mass function is mentioned in Sect. 18.3.3.

Having developed an appropriate definition of mass, the next step is to compute the forces exerted on each individual by all other individuals, component by component, using the APO force law:

$$F_{ij,k} = \begin{cases} Gm_i m_j / |r_{ij,k}^p| & : f(X_j) < f(X_i) , \ \forall i \neq j, \ i \neq best , \\ - Gm_i m_j / |r_{ji,k}^p| & : f(X_j) \geq f(X_i) , \ \forall i \neq j, \ i \neq best , \end{cases} \tag{18.3}$$

where $F_{ij,k}$ is the $k$th component of force exerted on individual $i$ by individual $j$ , $r_{ij,k}$ is the distance from $x_{i,k}$ to $x_{j,k}$, which is computed by $r_{ij,k} = x_{j,k} - x_{i,k}$, and $x_{i,k}$ and $x_{j,k}$ represent the positions of individual $i$ and individual $j$ in the $k$th dimension, respectively. The exponent $p$ is user-defined, typically in the range $[-5,5]$. A properly chosen force law drives the AP particles to efficiently search the decision space by balancing global and local search. The result is an APO algorithm that performs very well. Different values of the exponent $p$ in Eq. 18.3 have been used to construct three distinct force laws, as described in Sect. 18.3.2.

The $k$th component of the total force $F_{i,k}$ exerted on individual $i$ by all other particles is obtained by summing over all other particles:

$$F_{i,k} = \sum_{j=1}^{N} F_{ij,k} , \ \forall i \neq best . \tag{18.4}$$

The total force on each particle is the vector sum of all the forces exerted by every other particle. Note that under the force law in Eq. 18.3, each individual neither attracts nor repels itself (i.e., the force is zero). Equation 18.4 therefore includes the force exerted on individual $i$ by itself by not excluding it from the summation—because the addend is zero. Note, too, that individual *best* cannot be attracted or repelled by other individuals—because it is excluded from the summation, which is equivalent to setting the total force exerted on the *best* particle to zero.

The third and final APO step is Motion, which refers to the movement of individuals through the decision space using the previously computed total force to calculate a "velocity," which is then used to update the particles' positions. The velocity and position of individual $i$ at time $t + 1$, the next iteration, are updated using Eqs. 18.5 and 18.6, respectively:

$$v_{i,k}(t+1) = w\, v_{i,k}(t) + \alpha \frac{F_{i,k}}{m_i} , \ \forall i \neq best , \qquad (18.5)$$

$$x_{i,k}(t+1) = x_{i,k}(t) + v_{i,k}(t+1) , \ \forall i \neq best , \qquad (18.6)$$

where $v_{i,k}(t)$ and $x_{i,k}(t)$, respectively, are the $k$th components of particle $i$'s velocity and coordinates at step $t$, and $\alpha$ is a random variable uniformly distributed on (0,1). The user-specified inertial weight $0 < w < 1$ determines how easily the previous velocity can be changed, with smaller values of $w$ corresponding to greater velocity changes. Each individual's movement is restricted to the domain of feasible solutions $x_{i,k} \in [x_k^{\min}, x_k^{\max}]$, and the velocities are similarly constrained, i.e., $v_{i,k} \in [v_k^{\min}, v_k^{\max}]$. It is important to note that under the scheme in Eqs. 18.5 and 18.6, the individual *best* does not move away from its current position, nor does its velocity change.

After updating the positions of all particles, the objective function fitnesses are updated at the particles' new positions. A new individual *best* is determined by the fitnesses, and its corresponding position vector replaces $X_{\text{best}}$. The two main processes Force Calculation and Motion are repeated until a termination criterion is satisfied. A variety of criteria may be used, and commonly employed ones include a predetermined maximum number of iterations, some successive number of iterations with no change in the *best* particle's position, and so on.

### 18.3.2 Force Law Design

The force law is key to understanding the performance of the APO algorithm. A well-designed force law can drive the individuals' search through the decision space intelligently and efficiently, which can make the APO algorithm perform very well. By choosing different values for the variable $p$ in the AP force law (Eq. 18.3), we construct three force laws: the *negative exponential force law*, the *unimodal force law* and the *linear force law* [283].

Initially, we set $p = 1$. The force is expressed as:

$$F_{ij,k} = \begin{cases} r_{ij,k}\, Gm_i m_j / r_{ij,k}^2 & : f(X_j) < f(X_i) , \ \forall i \neq j, \ i \neq best , \\ r_{ji,k}\, Gm_i m_j / r_{ji,k}^2 & : f(X_j) \geq f(X_i) , \ \forall i \neq j, \ i \neq best , \end{cases} \qquad (18.7)$$

where the gravitational constant $G = 1$. In this case, we use Eq. 18.2 to calculate the masses of all individuals, except for individual *best*. The mass of individual *best* is set to $K$, where $K$ is a positive constant and $K > 1$. We try to set the best individual's mass to a value that is sufficiently large that there will be enough attractive forces exerted on all other individuals.

According to Eq. 18.7, when the distance between individual $i$ and individual *best* is close to zero, the force exerted on individual $i$ via individual *best* is infinitely large. This implies that individual $i$ cannot carefully exploit the neighborhood of individual *best*. So, we restrict the magnitude of the total force exerted on each individual with $F_{i,k} \in [F_{\min}, F_{\max}]$ where $F_{\min} = -F_{\max}$. The magnitude of the negative exponential force is plotted in Fig. 18.3. The APO algorithm with the negative exponential force law (in Eq. 18.7) is called APO1.

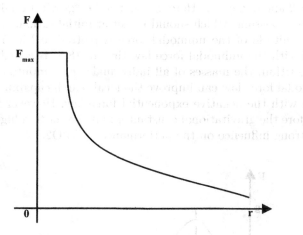

Fig. 18.3: The magnitude of the negative exponential force

APO1 with the negative exponential force law has two drawbacks. One is a poor local search capability, although $F_{i,k}$ is restricted in $[F_{\min}, F_{\max}]$ and $F_{\max}$ is set to a small value. The value $F_{\max}$ has a strong influence on the performance of APO1. To overcome this drawback, simulated annealing (SA) has been applied to local search [282]. The second drawback is a lack of globality, though the force law is global. According to the force law, each individual only attracts or repels its nearby neighbors, whereas it hardly attracts or repels others that are far away from it. Apparently, the force law is local.

If the magnitude of the force is decreased to zero when an individual gradually approaches another, this can make individual $i$ carefully exploit the neighborhood of individual $j$. This force law [287] can be set as follows:

$$F_{ij,k} = \begin{cases} \text{sgn}(r_{ij,k})G(r_{ij,k})m_im_j/r_{ij,k}^2 & : f(X_j) < f(X_i) \,,\ \forall i \neq j,\ i \neq best\,, \\ \text{sgn}(r_{ji,k})G(r_{ji,k})m_im_j/r_{ji,k}^2 & : f(X_j) \geq f(X_i) \,,\ \forall i \neq j,\ i \neq best\,, \end{cases}$$
$$(18.8)$$

where sgn() denotes the *signum function*, defined as:

$$\text{sgn}(r) = \begin{cases} 1 & : r \geq 0\,, \\ -1 & : r < 0\,. \end{cases} \tag{18.9}$$

Here, $G$ is not a constant and changes with the variable $r_{ij,k}$. It can be expressed as follows:

$$G(r) = \begin{cases} g|r|^h & : r \leq 1\,, \\ g|r|^q & : r > 1\,, \end{cases} \tag{18.10}$$

where $h > 2$ and $0 < q < 2$. Here, $g$ denotes the "gravitational constant" and is a positive constant, which should be set at initialization.

The magnitude of the unimodal force is plotted in Fig. 18.4. The APO algorithm with the unimodal force law (in Eq. 18.8) is called APO2. In the APO2 algorithm, the masses of all individuals are computed with Eq. 18.2. The unimodal force law can improve the local search capability of the APO1 algorithm with the negative exponential force law. However it is still local, and therefore the gravitational constant $g$ must be set to a high value so that it has a strong influence on the performance of APO2.

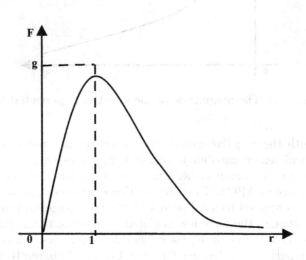

Fig. 18.4: The magnitude of the unimodal force

With the linear force law, the force increases as the distance increases. We set $p = -1$ in Eq. 18.3. The force can be expressed as:

$$F_{ij,k} = \begin{cases} Gm_im_jr_{ij,k} & : f(X_j) < f(X_i) \ , \ \forall i \neq j, \ i \neq best \ , \\ -Gm_im_jr_{ij,k} & : f(X_j) \geq f(X_i) \ , \ \forall i \neq j, \ i \neq best \ . \end{cases} \tag{18.11}$$

The magnitude of the linear force is plotted in Fig. 18.5. The APO algorithm with the linear force law (in Eq. 18.11) is called APO3. The linear force law is constructed to overcome the disadvantages of the former two force laws. APO3 with the linear force law has fewer parameters than APO1 and APO2, which can enhance the stability of the APO algorithm. Moreover, in the process of running experiments, we find that the gravitational constant $G$ has little effect on the performance of APO3.

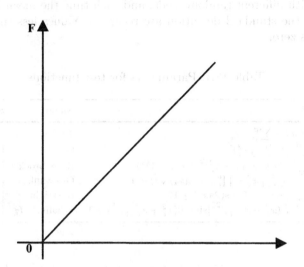

Fig. 18.5: The magnitude of the linear force

In order to evaluate the performance of the APO algorithms with the three force laws, six famous benchmark functions were chosen. These functions' names, formulation, and their parameters' ranges are listed in Table 18.1. Tablet and Quadric are unimodal functions. Rosenbrock is generally viewed as a unimodal function, although it can be treated as multimodal with two local optima. It has a narrow valley from the perceived local optima to the global optimum and provides little information to optimization algorithms by itself. Griewank, Rastrigin and Schaffer's $f_7$ functions are well-known multimodal nonlinear functions, which have wide search spaces and lots of local optima into which algorithms are easily trapped.

The details of the experimental environment and results are explained as follows. In each experiment with APO, the mass function is calculated with:

$$m_i = \exp\left(\frac{f(x_{\text{best}}) - f(x_i)}{f(x_{\text{worst}}) - f(x_{\text{best}})}\right), \tag{18.12}$$

where $best = \arg\{\min f(X_i), i \in S\}$ and $worst = \arg\{\max f(X_i), i \in S\}$. Equation 18.12 is one of the functions to compute the mass (see Eq. 18.2). The inertia weight $w$ in Eq. 18.5 is defined as $w = 0.9 - (0.5t/MAXITER)$ and is decreased linearly from 0.9 to 0.4, where $t$ denotes the current iteration. The velocity threshold $v_{\min}$ is set to the lower bound of the domain, and $v_{\max}$ is set to the upper bound of the domain. We terminate the APO procedure after a maximum number of iterations. The dimensionality $d$ and the population size $N$ are 50. The largest iteration is 2,500. Each of experiments was repeated 30 times with different random seeds, and each time the mean best function values and the standard deviation are recorded. Values less than $10^{-12}$ are regarded as zero.

Table 18.1: Parameters for test functions

| Function | Name | Search Range |
|---|---|---|
| $f_1(x) = 10^6 x_1^2 + \sum_{i=2}^{d} x_i^2$ | Tablet | $[-100, 100]^d$ |
| $f_2(x) = \sum_{i=1}^{d} (\sum_{j=2}^{i} x_j^2)^2$ | Quadric | $[-100, 100]^d$ |
| $f_3(x) = \sum_{i=1}^{d-1} (100(x_{i+1} - x_i^2)^2 + (x_i - 1)^2)$ | Rosenbrock | $[-50, 50]^d$ |
| $f_4(x) = \frac{1}{4000} \sum_{i=1}^{d} x_i^2 - \prod_{i=1}^{d} \cos(x_i/\sqrt{i}) + 1$ | Griewank | $[-600, 600]^d$ |
| $f_5(x) = \sum_{i=1}^{d} (x_i^2 - 10\cos(2\pi x_i) + 10)$ | Rastrigin | $[-5.12, 5.12]^d$ |
| $f_6(x) = \sum_{i=1}^{d-1} (x_i^2 + x_{i+1}^2)^{.25}[\sin(50(x_i^2 + x_{i+1}^2)^{.1}) + 1]$ | Schaffer's $f_7$ | $[-100, 100]^d$ |

The six functions in Table 18.1 are used to evaluate and compare the performances of the APO algorithms with the above three force laws. The functions all have a minimum value of 0.0 at $(0, \ldots, 0)$, except for Rosenbrock, where the minimum is at $(1, \ldots, 1)$. In the APO1 algorithm, $G = 1$, and $F_{\max}$ is an experiential value. In the APO2 algorithm, we set the parameters $h = 5$ and $q = 1$, the gravitational constant $g$ is an experimental value, $g = 2.0e + 4$ for Tablet, $g = 3.0e + 4$ for Quadric, $g = 5.0e + 3$ for Rosenbrock, $g = 5.0e + 5$ for Griewank, $g = 500$ for Rastrigin, and $g = 1.0e + 5$ for Shaffer's $f_7$, which makes APO2 perform better than it does with other values of $g$. In the APO3 algorithm, $G = 10$.

The comparison of the results of the six benchmarks in Table 18.1 is shown in Table 18.2, in which Min represents the best objective function value, Mean represents the mean best objective function value, and STD represents the standard deviation. To provide further analysis, the dynamic comparisons of these benchmarks with APOs in dimension 50 are plotted as Figs. 18.6–18.8. Figures 18.6–18.8 verify the dynamical behavior, and 20 sample points are selected within the same intervals. At these points, the average best fitness

Table 18.2: Performance comparison of APO1, APO2, and APO3 for benchmark problems with 50 dimensions

| Function | Algorithm | $G$ (or $g$) | Min | Mean | STD |
|---|---|---|---|---|---|
| | APO1 | 1 | 1.17 | 5.13 | 6.46e−01 |
| Tablet | APO2 | 2.0e+04 | 0 | 0 | 0 |
| | APO3 | 10 | 0 | 0 | 0 |
| | APO1 | 1 | 1.29e+01 | 1.98e+01 | 9.22e−01 |
| Quadric | APO2 | 3.0e+04 | 0 (53.3%) | 1.65e+04 | 6.84e+03 |
| | APO3 | 10 | 0 | 0 | 0 |
| | APO1 | 1 | 9.52e+01 | 1.63e+02 | 1.80e+01 |
| Rosenbrock | APO2 | 5.0e+03 | 4.90e+01 | 2.50e+07 | 9.16e+06 |
| | APO3 | 10 | 4.90e+01 | 4.90e+01 | 0 |
| | APO1 | 1 | 1.92e−01 | 2.20e−01 | 3.30e−03 |
| Griewank | APO2 | 5.0e+05 | 0 (40%) | 1.53e+01 | 3.70 |
| | APO3 | 10 | 0 | 0 | 0 |
| | APO1 | 1 | 3.98e+01 | 8.36e+01 | 3.47 |
| Rastrigin | APO2 | 5.0e+02 | 0 | 0 | 0 |
| | APO3 | 10 | 0 | 0 | 0 |
| | APO1 | 1 | 5.75 | 1.56e+01 | 1.00 |
| Schaffer's $f_7$ | APO2 | 1.0e+05 | 0 (40%) | 1.07 | 3.17e−01 |
| | APO3 | 10 | 0 | 0 | 0 |

of the historical best position of the population over all 30 runs is computed and plotted.

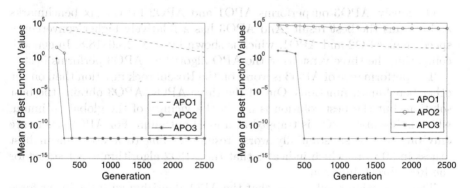

Fig. 18.6: Comparison of Tablet (left) and Quadric (right) functions with 50 dimensions

From the fourth column of Table 18.2, we can see that APO2 can find the global optimum of most functions with some probability, whereas APO1 only finds solutions in the vicinity of the global optimum. This indicates that APO1 has a poor local search capability, and this disadvantage can be overcome by APO2. From Table 18.2 and Figs. 18.6–18.8, we can see that the performances of APO2 are worse than those of APO1 for the Quadric, Rosenbrock, and Griewank functions in the early stage of the search. This is because APO2 has a slow convergence rate since the force of APO2 decreases as the distance increases when the distance is more than one. However, comparing the solutions for these functions (except Rosenbrock with APO2 and APO1), the simulation result shows that the global optimum can be found with APO2 rather than APO1, provided enough evolutionary generations have passed.

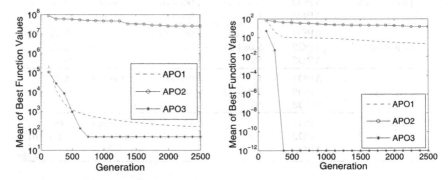

Fig. 18.7: Comparison of Rosenbrock (left) and Griewank (right) functions with 50 dimensions

Obviously, APO3 outperforms APO1 and APO2 for the six benchmarks and obtains the best result. And APO3 has a relatively higher convergence speed than APO2 and APO1, which is shown by Figs. 18.6–18.8. In general, comparing the three versions of the APO algorithm, APO3 performs best.

The performance of APO3 is worse on the Rosenbrock function than on the other benchmark functions. Out of the three APOs, APO3 obtains the best solution, but the best solution is not in the vicinity of the global optimum, which is because APO3 is trapped at a local optimum. For APO2, there are only one or two substantially worse results leading to a worse mean best fitness in 30 runs, which indicates that the APO2 algorithm is not stable for the Rosenbrock function.

The simulation results show that the APO algorithm with the linear force law (Eq. 18.11) is the most stable and effective among the three versions of the APO algorithm, and it is especially suitable for high-dimensional multimodal functions with many local optima.

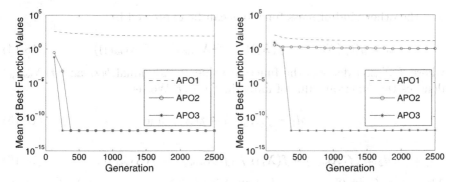

Fig. 18.8: Comparison of Rastrigin (left) and Schaffer's $f_7$ (right) functions with 50 dimensions

### 18.3.3 Mass Function Selection

Mass is an important parameter of the APO algorithm. In responding to virtual forces, APO's individuals move toward other particles possessing larger "masses" (with better fitnesses) and away from lower mass particles (with worse fitnesses). When APO is used to solve a maximization problem, its "mass" is a user-defined function that increases with increasing objective function value, whereas when performing minimization APO's mass increases with decreasing fitness. Moreover, in real-world physics, the value of an individual's mass is a positive constant. So the mass function can be expressed as $m = g(f(X))$ with the conditions that $m_i \in (0,1]$ and $g(\bullet)$ is a positive bounded monotonically increasing or decreasing function.

Obviously, there are many functions satisfying these requirements, and no doubt some will be better than others for specific optimization problems or certain classes of problems. The basic requirement is that the mass of the best individual has the largest (normalized) value, that is, $m_{\text{best}} = 1$, while all other individuals with worse fitnesses have smaller values. Typical mass functions include, as examples, $g_1(x) = e^x$, $g_2(x) = \arctan(x)$, $g_3(x) = \tanh(x) = (e^x - e^{-x})/(e^x + e^{-x})$, $x \in [\infty, -\infty]$ in which $g_k(x) \in [a,b]$ is mapped onto the interval $(0,1]$. The choice of mass function, however, does depend on whether maximization or minimization is the goal. Thus, an additional requirement is that the mass function be monotonically increasing or decreasing, respectively, for maximization or minimization, thereby guaranteeing greater mass, and hence greater attractive force, for the better fitness value.

For example, we construct two forms of mass functions. One kind of mass function is shown by

$$m_i = g(f(X_{\text{best}}) - f(X_i)), \tag{18.13}$$

and the other kind of mass functions can be expressed by

$$m_i = g((f(X_{\text{best}}) - f(X_i)) \, / \, (f(X_{\text{worst}}) - f(X_{\text{best}}))) \,, \qquad (18.14)$$

where $f(X_{\text{best}})$ denotes the function value of individual *best* and $f(X_{\text{worst}})$ denotes the function value of individual *worst*. We set

$$d_1 = f(X_{\text{best}}) - f(X_i) \quad \text{and} \qquad (18.15)$$

$$d_2 = (f(X_{\text{best}}) - f(X_i)) \, / \, (f(X_{\text{worst}}) - f(X_{\text{best}})) \,. \qquad (18.16)$$

Then $d_1 \in (-\infty, 0]$ and $d_2 \in [-1, 0]$. When $f(X_i) \in (-\infty, +\infty)$, $e^{d_1} \in (0, 1]$, $\arctan(d_1) \in (-\pi/2, 0]$, $\tanh(d_1) \in (-1, 0]$, $1 + d_2^{2n+1} \in [0, 1]$ for $n \geq 0$, $e^{d_2} \in (0, 1]$, $\arctan(d_2) \in (-\pi/2, 0]$, and $\tanh(d_2) \in (-1, 0]$. In the same way, $\arctan(d_1)$, $\tanh(d_1)$, $\arctan(d_2)$, and $\tanh(d_2)$ can also be mapped into $(0, 1)$ through elementary transformations. The above functions and their elementary transforms can be used to compute $m_i$, which should meet the condition of constructing the mass function.

According to their curvilinear styles, the above mass functions can be divided into *convex functions*, *linear functions*, and *concave functions*. In order to evaluate and compare the performances of these mass functions, we picked seven functions related to Eq. 18.16, which belong to three curve functions. The mass functions in Eqs. 18.17–18.19 are convex curves, those in Eq. 18.20 are straight lines, and those in Eqs. 18.21–18.23 are concave curves.

$$m_i = 1 + d_2^3 \,. \qquad (18.17)$$

$$m_i = 1 + d_2^5 \,. \qquad (18.18)$$

$$m_i = 1 + d_2^7 \,. \qquad (18.19)$$

$$m_i = 1 + d_2 \,. \qquad (18.20)$$

$$m_i = (\pi/2 + \arctan(d_2))/(\pi/2) \,. \qquad (18.21)$$

$$m_i = \tanh(d_2) + 1 \,. \qquad (18.22)$$

$$m_i = e^{d_2} \,. \qquad (18.23)$$

The three kinds of curves related to mass functions are shown as Fig. 18.9. From Fig. 18.9, we can see that mass functions with a concave curve can approach 1.0 more rapidly than those with a beeline or convex curve when the individuals have function values closer to $f(X_{\text{best}})$. However, mass functions

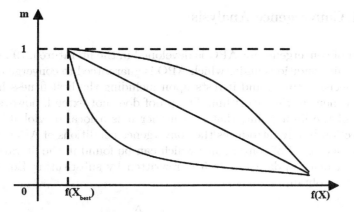

Fig. 18.9: Three kinds of curves related to mass functions

with a concave curve can approach 0.0 more slowly than those with a beeline or convex curve when the individuals have function values far from $f(X_{\text{best}})$.

In order to evaluate and compare the performance of the APO algorithm with different mass functions, the literature [286] suggests choosing seven mass functions belonging to three kinds of curves. The test results illustrate that a mass function with a convex curve has better global exploration capability than those with a concave curve or straight line, although the convergence rate is slow. Whereas the mass function with a concave curve maintains a powerful local exploitation capability and a relatively rapid convergence velocity, it is easily entrapped in local optima. In general, comparing the seven versions of the APO algorithm, mass functions with a concave curve are superior to those with a convex curve or straight line, and especially the mass function of Eq. 18.23 (namely, Eq. 18.12) yields better results.

## 18.4 Convergence Analysis of the APO Algorithm

As a new algorithm, APO lacks a deep theoretical foundation. Nonetheless, the convergence properties of APO and the selection strategies of the gravitational constant $G$ are briefly introduced in this section.

## 18.4.1 Convergence Analysis

A proof of convergence for APO is developed in the literature [279]. It reveals the specific conditions under which APO is guaranteed to converge to a point in the decision space, and it rests upon including the best fitness history as a key element in the algorithm. The proof does not extend, however, to the nature of the limit point, that is, whether it is a local or global optimum. This section briefly introduces the convergence conditions of APO and omits the detailed proof of convergence which can be found in the literature [279].

The velocity update Eq. 18.5 is rewritten by substituting Eq. 18.4 into Eq. 18.5 to obtain

$$v_{i,k}(t+1) = w\, v_{i,k}(t) + \alpha \sum_{j=1}^{N} F_{ij,k}/m_i \ , \ \forall i \ . \tag{18.24}$$

Without loss of generality, several assumptions are made to facilitate the convergence analysis by reducing the $d$-dimensional problem to a single dimension. $X_{\text{best}}$ is assumed to be constant over some number of iterations (of course, in operation $X_{\text{best}}$ is updated as APO progresses). If an individual $i$ is chosen arbitrarily, the results developed for that individual can be applied to all the others. It is evident from Eqs. 18.4–18.6, Eq. 18.11 and Eq. 18.24 that each vector component is updated independently of the others. As a result, only a single component need be analyzed without loss of generality, in effect reducing the convergence analysis to a one-dimensional problem. Thus, dropping the coordinate (dimension) index $k$ in Eqs. 18.6 and 18.24, the update equations become

$$V_i(t+1) = w\, V_i(t) + \sum_{j=1}^{N} \alpha_j F_{ij}/m_i \ \text{ and} \tag{18.25}$$

$$X_i(t+1) = X_i(t) + V_i(t+1) \ . \tag{18.26}$$

Define $N_i = \{j \,|\, f(X_j) < f(X_i), \forall j \in S\}$ and $M_i = \{j \,|\, f(X_j) \geq f(X_i), \forall j \in S\}$, where $S$ is the set of all individuals. Thus, for the single-particle one-dimensional APO algorithm, the velocity update Eq. 18.25 becomes

$$V_i(t+1) = w\, V_i(t) + \left( \sum_{j \in M_i} \alpha_j G m_j - \sum_{j \in N_i} \alpha_j G m_j \right) X_i(t) \tag{18.27}$$

$$+ \sum_{j \in N_i} \alpha_j G m_j X_j(t) - \sum_{j \in M_i} \alpha_j G m_j X_j(t) \ .$$

With the following definitions:

$$G_{N_i} = \sum_{j \in N_i} \alpha_j G m_j \, , \qquad (18.28)$$

$$G_{M_i} = \sum_{j \in M_i} \alpha_j G m_j \, , \qquad (18.29)$$

$$G_{NM_i} = G_{N_i} - G_{M_i} \quad \text{and} \qquad (18.30)$$

$$F_{G_i} = \sum_{j \in N_i} \alpha_j G m_j X_j(t) - \sum_{j \in M_i} \alpha_j G m_j X_j(t) \, , \qquad (18.31)$$

Equation 18.27 may be written as:

$$V_i(t+1) = w \, V_i(t) - G_{NM_i} X_i(t) + F_{G_i} \, . \qquad (18.32)$$

Substituting Eq. 18.32 into Eq. 18.26 yields the following non-homogeneous recurrence relation:

$$X_i(t+1) - (1 + w - G_{NM_i}) X_i(t) + w \, X_i(t-1) = F_{G_i} \, . \qquad (18.33)$$

Equation 18.33 is a second order difference equation with $F_{G_i}$ as the forcing function. By analyzing its properties, the convergence of the random variable sequence $\{E[X_i(t)]\}$ may be analyzed, where $E[X_i(t)]$ is the expectation of random variable $X_i(t)$. For the stochastic case, applying the expected value operator to Eq. 18.33 yields

$$E[X_i(t+1)] - (1 + w - \varphi_i) E[X_i(t)] + w \, E[X_i(t-1)] = \theta_i \, , \qquad (18.34)$$

where

$$\varphi_i = E[G_{NM_i}] = \frac{1}{2} G \left( \sum_{j \in N_i} m_j - \sum_{j \in M_i} m_j \right) \text{ and} \qquad (18.35)$$

$$\theta_i = E[F_{G_i}] = \frac{1}{2} G \left( \sum_{j \in N_i} m_j P_j - \sum_{j \in M_i} m_j P_j \right) .^{1} \qquad (18.36)$$

By inspection, the eigenvalue (characteristic) equation of the second order linear system described by Eq. 18.34 is

$$\lambda^2 - (1 + w + \varphi_i) \lambda + w = 0 \, . \qquad (18.37)$$

Stable solutions to Eq. 18.34 exist when its eigenvalues lie inside the unit circle in the complex $\lambda$-plane. These solutions, in turn, correspond to convergent sequences $\{E[X_i(t)]\}$, leading to the following result:

---

[1] The notation $P_j$ is defined in Sect. 18.5.1.

**Theorem 18.1.** *Given $w \geq 0$ and $G > 0$, if and only if $0 \leq w < 1$ and $0 < \varphi_i < 2(1+w)$, then $\{E[X_i(t)]\}$ converges to $X_{\text{best}}$.*

## 18.4.2 Guidelines for Selecting $G$

APO's convergence condition is $0 \leq w < 1$ and $0 < \varphi_i < 2(1+w)$, that is,

$$0 \leq w < 1 \quad \text{and} \quad 0 < G\left( \sum_{j \in N_i} m_j - \sum_{j \in M_i} m_j \right) < 4(1+w) . \qquad (18.38)$$

From this condition, it is evident that particle trajectories for which $\sum_{j \in N_i} m_j - \sum_{j \in M_i} m_j \leq 0$ or $\sum_{j \in N_i} m_j - \sum_{j \in M_i} m_j > 0$ and $G \geq 4(1+w)/(\sum_{j \in N_i} m_j - \sum_{j \in M_i} m_j)$ are *divergent*, while trajectories for which $\sum_{j \in N_i} m_j - \sum_{j \in M_i} m_j > 0$ and $0 < G < 4(1+w)/(\sum_{j \in N_i} m_j - \sum_{j \in M_i} m_j)$ are *convergent*.

Particles can be grouped according to their behavior by defining the following subsets:

**Definition 18.1.**

$$A \triangleq \left\{ i \mid \sum_{j \in N_i} m_j - \sum_{j \in M_i} m_j \leq 0 , \forall i \in S \right\} .$$

Particles in subset $A$ have the best fitnesses in the subset of particles whose trajectories are divergent.

**Definition 18.2.**

$$B \triangleq \left\{ i \mid \sum_{j \in N_i} m_j - \sum_{j \in M_i} m_j > 0 , \forall i \in S \right\} .$$

Particles in subset $B$ have worse fitness values than those in subset $A$.

**Definition 18.3.**

$$BD \triangleq \left\{ i \mid \sum_{j \in N_i} m_j - \sum_{j \in M_i} m_j > 0 \text{ and } G \geq \frac{4(1+w)}{\sum_{j \in N_i} m_j - \sum_{j \in M_i}} , \forall i \in S \right\} .$$

Trajectories of the individuals in subset $BD$ are divergent.

**Definition 18.4.**

$$BC \triangleq \left\{ i \mid \sum_{j \in N_i} m_j - \sum_{j \in M_i} m_j > 0 \text{ and } 0 < G < \frac{4(1+w)}{\sum_{j \in N_i} m_j - \sum_{j \in M_i}} , \forall i \in S \right\} .$$

Trajectories of the individuals in this subset are convergent. Note that $B = BD \cup BC$.

We can conclude Theorem 18.2 by reasoning:

**Theorem 18.2.** *A is a nonempty set.*

The gravitational constant $G$ is a particularly important parameter influencing the performance of the APO algorithm. Based on analyzing the convergence of the APO algorithm, all individuals of the swarm employed by APO can be divided into two subsets: a divergent subset $A \cup BD$ and a convergent subset $BC$. The divergent subset includes individuals which have divergent behaviors, and the convergent subset includes individuals which have convergent behaviors. There is a nonzero probability of APO's population having divergent behaviors in the whole search process. The value of $G$ determines the number of individuals in the two subsets. A constant $G$ and an adaptive $G$ are two selection strategies for $G$ in APO. A smaller constant value of $G$ makes almost all individuals satisfy APO's convergent conditions and limits the global search capability of individuals in the divergent subset. A bigger constant value of $G$ makes few individuals satisfy APO's convergent conditions and it impairs the local search capability of individuals in the convergent subset. An adaptive $G$ can not only tune the number of individuals in the two subsets but also improve the global search capability of individuals in the divergent subset and the local search capability of individuals in the convergent subset.

## 18.5 Improvements to APO

This section describes extensions to the basic APO algorithm that improve its performance.

## 18.5.1 An Extended APO

Swarm collective behaviors, such as bird flocking, insect foraging, and fish schooling, are complex, dynamic, and adaptive processes in which the experience of each individual is very helpful in advancing the search. That can be mapped onto each individual's best search history positions in the APO algorithm. The version of APO discussed so far in this chapter completely ignores this important information. APO is therefore modified to include the search history, in an extended version called EAPO [281].

$P_i = (p_{i,1}, p_{i,2}, \ldots, p_{i,d})$ is defined as the position vector of the best fitness found by individual $i$ over all iterations through the current iteration, $t$. At each successive iteration, the fitness history is updated as follows:

$$p_{i,k}(t+1) = \begin{cases} p_{i,k}(t) & : \ f(X_i(t+1)) > f(P_i(t)) \ , \ k = 1 \ldots n \ , \\ x_{i,k}(t+1) & : \ f(X_i(t+1)) \le f(P_i(t)) \ , \ k = 1 \ldots n \ . \end{cases} \quad (18.39)$$

The APO mass function in Eq. 18.12 is modified to include the fitness history as follows:

$$m_i = \exp\left(\frac{f(X_{\text{best}}) - f(P_i)}{f(X_{\text{worst}}) - f(X_{\text{best}})}\right) \ , \ \forall i \ . \quad (18.40)$$

The analogs of Eq. 18.11 and Eqs. 18.4–18.6 for computing the component-wise forces exerted on each individual by all others and the updated velocity and position components become

$$F_{ij,k} = \begin{cases} Gm_i m_j (p_{j,k} - x_{i,k}) & : \ f(P_j) \le f(X_i) \ , \ \forall i \ , \\ -Gm_i m_j (p_{j,k} - x_{i,k}) & : \ f(P_j) > f(X_i) \ , \ \forall i \ , \end{cases} \quad (18.41)$$

$$F_{i,k} = \sum_{j=1}^{N} F_{ij,k} \ , \ \forall i \ , \quad (18.42)$$

$$v_{i,k}(t+1) = w\, v_{i,k}(t) + \alpha F_{i,k}/m_i \ , \ \forall i \ , \ \text{and} \quad (18.43)$$

$$x_{i,k}(t+1) = x_{i,k}(t) + v_{i,k}(t+1) \ , \ \forall i \ . \quad (18.44)$$

These equations contain another important information departure from the original APO implementation. Whereas in APO the individual best was neither attracted nor repelled by the other particles, in EAPO the best is, in fact, attracted or repelled by other individuals. In EAPO the best individual's motion is determined by the total force exerted on it by all other particles. The pseudocode for the EAPO algorithm is described in Listing 18.2.

Obviously, EAPO is similar to particle swarm optimization (PSO) on each individual's velocity and position updated equations. The literature [281] suggests that PSO's velocity update equation can be reduced by transforming EAPO's. The PSO algorithm can be considered to be a special example of EAPO, while EAPO is a general form of PSO. The important difference between the two algorithms is that EAPO's force may be attractive or repulsive, whereas PSO's force is attractive. In addition, each particle's motion only uses the information of its personal best position and the global best position in PSO, whereas it utilizes the information of all other individuals' personal best positions in EAPO. The simulation results show that the EAPO algorithm has better diversity than the PSO algorithm. EAPO's performance has been compared to APO's and PSO's with numerical experiments which confirm that EAPO exhibits faster convergence and better diversity.

**Listing 18.2: The pseudocode of the EAPO algorithm**

```
Begin
   Initialize population: both position x and velocity v;
   Set parameters, N, d, w and G, etc.;
   Iteration = 1;
While (termination criterion is not met) {
   Evaluate all individuals using corresponding fitness function;
   Update the personal best position P_i using Eq. 18.39;
   Update the global best position X_best;
   Calculate the mass using Eq. 18.40;
   Calculate the component force using Eq. 18.41;
   Calculate the total force using Eq. 18.42;
   Update the velocity v using Eq. 18.43;
   Update the position x using Eq. 18.44;
   Iteration = Iteration + 1;
}
End
```

## 18.5.2 A Vector Model of APO

The vector model of APO (VM-APO) is now constructed to facilitate the analysis and application to solving constrained optimization problems. As mentioned in Sect. 18.4.1, two sets $N_i = \{j \mid f(X_j) < f(X_i), \forall j \in S\}$ and $M_i = \{j \mid f(X_j) \geq f(X_i), \forall j \in S\}$ are defined about individual $i$, where $X_i = (x_{i,1} \ldots, x_{i,k}, \ldots, x_{i,d})$ is the position vector of individual $i$.

The relative direction of individual $j$ towards individual $i$ is defined as:

$$\boldsymbol{r}_{ij} = \begin{pmatrix} r_{ij,1} \\ r_{ij,2} \\ \ldots \\ r_{ij,d} \end{pmatrix} . \tag{18.45}$$

$$r_{ij,k} = \begin{cases} 1 & : X_{j,k} > X_{i,k} , \\ 0 & : X_{j,k} = X_{i,k} , \\ -1 & : X_{j,k} < X_{i,k} . \end{cases} \tag{18.46}$$

$$\boldsymbol{F}_{ij} = \begin{pmatrix} F_{ij,1} \\ F_{ij,2} \\ \ldots \\ F_{ij,d} \end{pmatrix} = \begin{cases} G m_i m_j \|X_j - X_i\| \boldsymbol{r}_{ij} & : j \in N_i , \\ - G m_i m_j \|X_j - X_i\| \boldsymbol{r}_{ij} & : j \in M_i , \end{cases} \tag{18.47}$$

where $\|X_j - X_i\| = \sqrt{\sum_{k=1}^{d} (x_{j,k} - x_{i,k})^2}$.

The equation of motion that we consider for individual $i$ is given by Eqs. 18.48 and 18.49.

$$V_i(t+1) = w\,V_i(t) + \alpha \sum_{j=1, j \neq i}^{N} F_{ij}/m_i \,. \tag{18.48}$$

$$X_i(t+1) = X_i(t) + V_i(t+1) \,. \tag{18.49}$$

The procedure of VM-APO is summarized as follows:

**Step 1:** Initialize each coordinate $x_{i,k}$ and $v_{i,k}$ by random sampling within $[x_k^{\min}, x_k^{\max}]$ and $[v_k^{\min}, v_k^{\max}]$, respectively.
**Step 2:** Compute the force vectors exerted on each individual.
**Step 2.1:** Select the best and worst individuals at time $t$, then calculate the mass of each individual at time $t$ according to Eq. 18.12.
**Step 2.2:** Calculate the force vector exerted on each individual by all other individuals with Eq. 18.47.
**Step 3:** Update the velocity and position vectors of each individual with Eqs. 18.48 and 18.49, respectively.
**Step 4:** Compute the fitness value of each individual and update the global best position $X_{\text{best}}$.
**Step 5:** If the termination criteria are satisfied, output the best solution; otherwise, go to Step 2.

The above described VM-APO and APO have the same framework. The difference between VM-APO and APO is the equation for calculating the component force exerted on each individual by all other individuals.

In the literature [288], the updated velocity vector (Eq. 18.48) in VM-APO can be transformed into:

$$V_i(t+1) = \sum_{j=1}^{N} u_j r_{ij} \,, \tag{18.50}$$

and Eq. 18.49 can be shown as:

$$X_i(t+1) = X_i(t) + \sum_{j=1}^{N} u_j r_{ij} \,, \tag{18.51}$$

where $u_j$ is a vector of random numbers. If $r_{ij}$ is set to be linearly independent and $u_j$ is set to different values, both $V_i(t+1)$ and $X_i(t+1)$ can be denoted by any point in the problem space, which can indicate that the VM-APO algorithm performs well in terms of diversity.

### 18.5.3 Local APO

All individuals of a swarm can communicate with each other in the original APO algorithm, which can be also called "global APO" (GAPO). In real nature, animals have a limited ability to communicate. Inspired by this phenomenon, a "local APO" (LAPO) is presented with some simple topologies, in which individuals only interact with their neighbors [161]. These topologies, such as a ring and some transforming structures based on it, have a strong effect on the performances of the LAPO algorithm. The simulation results show that the LAPO algorithm is effective and that its parameter $G$ has a substantial effect on the performance of the LAPO algorithm with different topologies.

## 18.6 Conclusions and Future Work

The artificial physics optimization (APO) algorithm inspired by natural physical forces is a novel population-based stochastic algorithm based on the physicomimetics framework. In this chapter, the APO framework, with different force rules and mass functions, is introduced. Because APO is formulated in the physicomimetics framework, it is amenable to analysis using techniques from physics and engineering, as demonstrated. Discrete-time linear systems theory has therefore been applied to analyze APO's convergence properties. Explicit convergence conditions have been developed, demonstrating that APO can converge in an expected value sense, although the nature of the solution at the limit point has not been determined. The simulation results show that APO and its improved versions are effective and competitive.

The gravitational constant $G$ is a particularly important parameter. Future work will therefore focus on exploring an adaptive $G$'s influence on EAPO's performance, using the convergence criteria developed in Sect. 18.4. How to best set the velocity limits $v_k^{\min}$ and $v_k^{\max}$ is another open question that merits further study, as is the choice of the inertia weight $w$. The choice of APO mass function also remains an open question. The convergence proof presented here does not identify the nature of the limit point, that is, whether it corresponds to a local or global optimum. Future efforts will be directed at developing a proof of convergence that addresses this question. Finally, APO will be applied to solve multi-objective optimization problems, constrained optimization problems and swarm robotic search problems.

# Chapter 19
# Artificial Physics for Noisy Nonstationary Environments

William M. Spears

*"...science is in essence the pursuit of simplicity"* (1980) [248]

## 19.1 Artificial Physics Optimization and Tracking

Chapter 18 presented one way to apply artificial physics to function optimization, with good success. The authors note that their extended artificial physics optimizer (APO) is a general form of particle swarm optimization (PSO). Thus, this book has come full circle. In 1998 I wrote a paper with Jim Kennedy, one of the creators of PSO [120]. Soon thereafter I invented artificial physics [235]. Despite the fact that I was thinking of swarms of robots when I came up with artificial physics, I'm sure I must have been influenced by the particle nature of PSO. Scientific progress often occurs due to a confluence of two or more seemingly unrelated ideas.

This chapter presents an alternative artificial physics approach to function optimization. However, due to our work on chemical source localization in Chap. 8, our perspective is somewhat different from standard function optimization. The goal of chemical source localization is to surround the source of some chemical emission. This is not necessarily a static scenario. There are two mechanisms that can cause the environment to change (or appear to change). First, the amount of chemical is measured by hardware sensors. Sensors are not perfect and can produce erroneous results. This is referred to as "sensor noise." Second, the chemical emitter can move, producing an environment that is always changing.

We can think of the chemical emitter as an optimum in an environment. The environment is defined with a "fitness function" that produces a noisy value (the amount of chemical for example) at each location. The environment is "nonstationary" because it changes over time as the chemical emitter moves. Stated in terms of function optimization, we are examining a specific

William M. Spears
Swarmotics LLC, Laramie, Wyoming, USA, e-mail: wspears@swarmotics.com

W.M. Spears, D.F. Spears (eds.), *Physicomimetics*,
DOI 10.1007/978-3-642-22804-9_19,
© Springer-Verlag Berlin Heidelberg 2011

set of problems where the fitness function is noisy and nonstationary. This form of noisy function optimization is referred to as "tracking," because the goal is to continually track the position of the optimum as it moves.

Since artificial physics can create rigid formations it can perform tracking in noisy environments. Let's use an example to illustrate why this is the case. Suppose we want to find the lowest point of a crater. If the crater is smooth this would be a very simple task. You could take a ball bearing, place it at any point, and let it roll down to the lowest point. But what if the crater is not smooth? Perhaps smaller meteors and erosion created ruts and holes. Then the ball bearing could get stuck at a higher location. If you could view the whole crater you would understand immediately that although the large-scale structure of the crater has not changed, the roughness is a form of noise that changes the small-scale structure of the crater. These changes in the small-scale structure can cause the ball bearing to become stuck at a suboptimal location.

Now let's image a whole "population" of ball bearings. Both PSO and evolutionary algorithms (EAs) utilize the concept of a population to help overcome the problems caused by suboptima. The idea is that even if some ball bearings get stuck, the remainder of the population can find lower and lower positions in the crater. This idea works reasonably well, but unless great care is taken, such algorithms suffer from what is called "premature convergence." What this means is that all the ball bearings end up being nearby each other. They can no longer sense the large-scale structure. This becomes especially problematic if the large-scale structure is nonstationary and changes (e.g., a portion of the crater subsides further). It is impossible for the converged set of ball bearings to even notice the change, unless they are at precisely the correct location.

The lattice formations created by artificial physics, however, are a population of ball bearings that stay apart. They do not converge prematurely, because they continue to remain separated. Hence they can continue to sense the large-scale structure of an environment. As with chemical source localization, the goal is to *surround* an optimum. If the optimum moves, the surrounding ball bearings can move with the optimum.

The lattice of ball bearings also creates an elegant, yet implicit, mechanism for dealing with noise. Suppose that ball bearings are attracted to ball bearings that have lower fitness (since we are minimizing), while being repelled from ball bearings with higher fitness. Due to small-scale structural noise, individual pairs of attraction/repulsion interactions may be in the wrong direction (i.e., not towards the optimum). *But these small-scale structural interactions are largely canceled out by the rigid formation, allowing the large-scale structure to dominate. This is a form of implicit consensus.* To explain this more fully, let's use a different analogy. Suppose a set of people surround and lift a large table, in order to move it to another room. Each person pushes and pulls the table, according to where they think the table should go. As with any group of individuals that are not coordinating very well, they are not pushing and pulling in precisely the same direction. There is noise and disagreement in the decision making process of each person. Yet, the table will proceed in the *consensus* direction, despite the fact that an explicit consensus was never determined. The rigidity of the table implicitly canceled the disagreements without explicit agreement!

In summary, there are two inherent advantages to a formation of particles. First, since the formation does not collapse, it has a better chance of surrounding an area of interest *and* following that area if and when it moves. Note that no special trigger is required to detect the movement—the formation is always ready to respond. This is an important advantage because triggers (pieces of code that respond to special events) are extremely difficult to design. They either trigger too often, yielding false alarms, or too infrequently, missing the special events. It is especially difficult to create triggers that work well over a broad class of problems.

The second advantage of the formation is that it allows for implicit consensus in the face of noisy and/or contradictory data. No explicit agreement is required between the particles. This is important because explicit agreement requires yet another algorithm. As with triggers, it is difficult to design explicit consensus algorithms that perform properly over a broad class of problems.

Both triggers and explicit consensus algorithms require yet more parameters that must be adjusted by the user. It is generally best to avoid excess complexity in algorithms. APO is elegant because the use of the formation allows us to avoid this excess complexity.

## 19.2 NetLogo Implementation of APO

In order to test the ideas that we just presented, open the NetLogo simulation "apo.nlogo". The graphical user interface is shown in Fig. 19.1, after the Setup button has been clicked.

The graphical pane shows a depiction of the function being minimized, which is the Rastrigin function defined in Table 18.1. Brighter yellow represents higher fitness. This looks very much like the crater discussed above. The optimum (minimum) is shown with the blue circle at the lower left. The magenta APO particles start at the upper right, in a very small cluster. If you click Move Robots, the particles will self-organize into formation and move towards the optimum. An example is shown in Fig. 19.2.

The trail represents the path of the center of mass of the formation. The center of mass is defined as:

$$\frac{\sum_i m_i \boldsymbol{x}_i}{\sum_i m_i} , \tag{19.1}$$

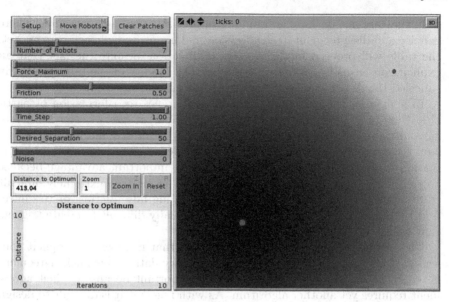

Fig. 19.1: A picture of the graphical user interface for artificial physics optimization

where $x_i$ represents the position of particle $i$. Since our goal is to surround the optimum, we want the center of mass to be as close to the optimum as possible. The Distance to Optimum monitor gives the distance of the center of mass from the optimum at each time step. The graph shows the distance as time increases. Both the trail and the graph indicate that the formation moves towards the goal very smoothly. This observation is not one that is typically made in function optimization, where all that matters is the best function value ever found. However, many real-world situations require not only a good solution, but a smooth trajectory towards that solution. An example would be a machine or process that needs to be optimized. It might not be possible to take large erratic steps towards the optimum. Gradual steps would be preferred.

The code to perform APO requires only a very simple modification to the code for triangular formations, as shown in Listing 19.1. Unlike as in Chap. 18 we assume that each particle has a mass of one. Each particle feels forces only from neighbors within sensor range $1.5D$, where $D$ is the desired separation between particles. The standard split Newtonian force law is used to create the formation, as described in Chap. 3. The new code provides a comparison of fitness values between pairs of particles within sensor range. Particles are attracted to neighbors that have lower fitness and are repelled from particles that have higher fitness.[1] The magnitude of this new force is set by the

---

[1] By changing the inequality to ">" the algorithm will maximize instead of minimize.

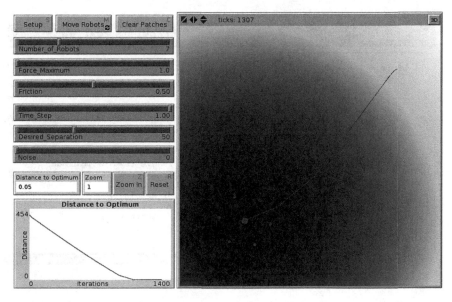

Fig. 19.2: A example of seven particles finding the optimum of the Rastrigin function

parameter apoF, which is determined by the gravitational constant $G$ used in the split Newtonian force law. We do not want particles to cluster, since that will reduce search efficacy. Equation 3.4 indicates that $G$ should be less than $F_{\max} D^p / 2\sqrt{3}$ to avoid clustering. Hence the NetLogo code automatically sets $G$ to $0.9 F_{\max} D^2 / 2\sqrt{3}$, since we assume $p = 2$ in this chapter. Under this assumption apoF should then be set to 0.10, as indicated by Eq. 6.2, and that is the value used in the NetLogo code.

Thus far we have dealt with the situation where there is no noise. The **Noise** slider allows you to introduce noise. This noise is generated by using a uniform distribution from $-\texttt{Noise}/2$ to $\texttt{Noise}/2$. For example, set the noise to 600,000. Then click Setup. Figure 19.3 shows what the fitness function looks like with the additional noise. Noise changes the small-scale structure of the environment. However, the location of the optimum is still visually discernible when the noise is at 600,000. This indicates that the large-scale structure is still intact. Now set the noise to 5,000,000 and click Setup. When the noise is 5,000,000 the graphics pane looks extremely random. Is the large-scale structure still there?

Try running APO with the increased levels of noise. As expected, the problem becomes more and more difficult, because the pairwise interactions that indicate the direction to move conflict more and more. Yet, even with a noise level of 5,000,000, seven particles are able to surround the optimum. It takes a long time, and the trajectory is not as straight as before, but the

Listing 19.1: The main piece of code for artificial physics optimization

```
if (r < 1.5 * D) [
    set F (G / (r ^ 2))
    if (F > FMAX) [set F FMAX]
    ifelse (r > D)
        [set Fx (Fx + F * (deltax / r))
         set Fy (Fy + F * (deltay / r))]
        [set Fx (Fx - F * (deltax / r))
         set Fy (Fy - F * (deltay / r))]

    ifelse ((rastrigin xcor ycor Noise) <  ; Move towards minimum
            (rastrigin ([xcor] of apo-robot ?)
                       ([ycor] of apo-robot ?) Noise))
        [set Fx (Fx - (apoF * (deltax / r))) ; Repulsive
         set Fy (Fy - (apoF * (deltay / r)))]
        [set Fx (Fx + (apoF * (deltax / r))) ; Attractive
         set Fy (Fy + (apoF * (deltay / r)))]
]
```

Fig. 19.3: The Rastrigin function when noise is 600,000 (left) and 5,000,000 (right)

task can still be performed. Although it is not visually apparent, enough of the large-scale structure remains for the formation to sense.

You can also change the number of robots and the desired separation. Since we want the formation to detect the large-scale structure, it seems reasonable to hypothesize that performance will be better when there are more particles and/or when the desired separation is larger. Larger formations should be able to follow large-scale structures more easily. To test this hypothesis we ran the NetLogo simulation for three, seven and 20 particles, where the desired separation $D$ varied from 20 to 50. Each experiment was performed 10 times (i.e., 10 runs), and the results were averaged. Noise was set to 5,000,000 and each run was terminated when the center of mass came within $D/4$ of the optimum. This ensured that the formation surrounded the optimum, as opposed to merely getting close to it.

Table 19.1: The average number of time steps required to surround the optimum, as the number of particles $N$ and desired separation $D$ change

|   | D | | | |
|---|---|---|---|---|
| N | 20 | 30 | 40 | 50 |
| 3 | 817015 | 433956 | 272401 | 198521 |
| 7 | 356632 | 293178 | 145085 | 129441 |
| 20 | 338822 | 148787 | 115346 | 97820 |

Table 19.2: The standard deviation in the number of time steps required to surround the optimum, as the number of particles $N$ and desired separation $D$ change

|   | D | | | |
|---|---|---|---|---|
| N | 20 | 30 | 40 | 50 |
| 3 | 454079 | 323352 | 107972 | 75647 |
| 7 | 168355 | 153214 | 53369 | 26617 |
| 20 | 105692 | 32279 | 36756 | 21945 |

The results are shown in Tables 19.1 and 19.2. The first table shows the average number of time steps to surround the optimum, while the second table gives the standard deviation of the results. A lower standard deviation means that the results are more consistent. One can see that the hypothesis is confirmed—larger formations find the optimum more quickly and more consistently.

The reason for the success of large formations on the Rastrigin function is that the large-scale structure guides the particles toward the optimum. In fact, there does not appear to be any small-scale structure when noise is zero. But is this really true? The Zoom In button in the simulation allows you to zoom into the fitness function by a factor of 10 each time you click. The Reset button resets the zoom back to one. Figure 19.4 shows what the Rastrigin function looks like when you zoom in by a factor of 10 and 100. If you look carefully at Fig. 19.4 (left) you will see a faint waffle board effect. The effect is pronounced in Fig. 19.4 (right). The Rastrigin function has an enormous number of local minima, just like the pitted crater we discussed earlier in this chapter.

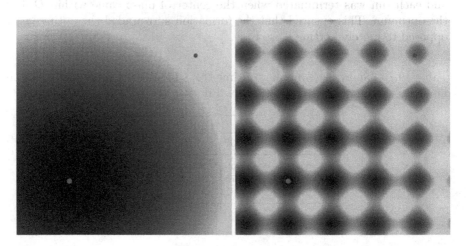

Fig. 19.4: The Rastrigin function when the zoom factor is 10 (left) and 100 (right)

If you try to run APO when the zoom factor is 10, it will usually succeed. However, it does not when the zoom factor is 100. The particles become stuck in the local optima. This is because the small-scale structure has now become large with respect to the desired separation $D$. This shows that if small-scale structures lead the search process to local optima, APO can compensate by using a desired separation $D$ that is large enough to make small-scale structures irrelevant.

As stated in this chapter, our emphasis is on tracking the optimum when it moves. The NetLogo simulation allows you to move the optimum by simply clicking anywhere in the graphics pane, while the robots are moving. You will notice a slight pause as the fitness of each patch in the graphics pane is recalculated. The blue circle moves to where you clicked the mouse, and the particles immediately respond to the changed environment by moving towards the optimum. This can be done in the zoomed environments also. Figure 19.5 shows an example where the optimum is moved several times, showing how the center of mass trail moves in response. The display of the fitness function was disabled for this illustration.

One interesting aspect of APO is that the minimum number of required particles depends on the dimensionality of the problem. For example, set the number of robots to two and run the NetLogo simulation. You will see that two particles will play "leader-follower" with each other (as we expect), but the trajectory is a straight line. Two particles can only create a one-dimensional object. But a one-dimensional object cannot search two dimen-

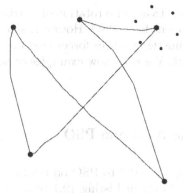

Fig. 19.5: An illustration of how a formation tracks a moving optimum

sions using APO. Hence, a lower bound on the number of particles needed to search a $d$-dimensional space is $d + 1$.

In the next section we will perform more rigorous experiments to see how well APO tracks a moving optimum. However, before this is done it is important to note that this version of APO differs in two important aspects from the APO algorithms presented in Chap. 18 as well as other algorithms, such as PSO and EAs. The first difference is the lack of randomness in the algorithm. Other than a random initial placement for the particles, the algorithm is purely deterministic. Randomness in search and function optimization algorithms is used to help the algorithms explore and avoid the premature convergence phenomenon described earlier. For evolutionary algorithms this randomness comes from selection, recombination, and mutation [233]. In PSO randomness is utilized to generate trajectories that provide more exploration. This is not to say that this version of APO would not benefit from the addition of randomness. However, the deterministic version presented here will serve as a benchmark for future research in artificial physics optimization.

The second difference is that this version of APO is rotationally invariant. To understand what this means, consider playing pool at your local pub. You are lined up for your shot. Now, assume that the whole building is magically rotated. You aren't even aware of the rotation. Will your shot be any different? No, of course not (assuming everything else remains constant, like the force of gravity, etc.). Physics is invariant to translation and rotation. This means that a sideways displacement of a system will make no difference to the system, and that a rotation of the system makes no difference.

Interestingly, both EAs and PSO are rotationally variant [34, 204, 270]. This means that if you rotate a fitness function, both algorithms perform differently. In fact, the rotational variance of PSO is tied to the form of randomness used by PSO to increase exploration [237]. Similarly, the versions of

APO presented in Chap. 18 are also rotationally variant, because they borrow their stochastic component from PSO. However, since the APO algorithm in this chapter is deterministic and uses forces that are circularly symmetric, it is rotationally invariant. We will show examples of how this affects behavior later in this chapter.

## 19.3 Comparison of APO and PSO

In this section we compare APO to PSO on tracking. First, we need to generalize the APO code shown in Listing 19.1 to $d$ dimensions. This is shown as pseudocode in Listing 19.2. One iteration of the algorithm means that all particles move once. The code in Listing 19.2 is executed by each particle to compute the movement. The code sums the component force interactions between two particles p and k, assuming that they are within sensor range $1.5D$. The split Newtonian force law is used to compute the magnitude of the "formation" force F. The components of the force are stored in the array Fx[], where index j refers to the dimension. Similarly the array delta[] stores the distance between the two particles, for each dimension.

After the formation force is computed the "fitness" force is calculated, where particle p compares its fitness with particle k. Note that the fitness of the particles is stored in the array fitness[]. This is done to minimize the number of calls to the fitness function. At every iteration the particles compute their fitness once, and store it in the array fitness[]. Hence, when there are $N$ particles, each iteration of APO consists of $N$ calls to the fitness function. This is also true for PSO.

PSO is summarized as follows [119]. The PSO population consists of $N$ particles. Each particle $i$ has a position at time $t$ denoted by $\boldsymbol{X}_i(t) = (X_{i,1}(t), \ldots, X_{i,d}(t))$, where $\boldsymbol{X}_i(t)$ is a $d$-dimensional vector. Each particle $i$ has a velocity $\boldsymbol{V}_i(t) = (V_{i,1}(t), \ldots, V_{i,d}(t))$. The equations of motion are given as [179]:

$$\boldsymbol{X}_i(t+1) = \boldsymbol{X}_i(t) + \boldsymbol{V}_i(t+1) \quad \text{and} \tag{19.2}$$

$$\boldsymbol{V}_i(t+1) = \omega \boldsymbol{V}_i(t) + c_1 \boldsymbol{r}_1 \odot (\boldsymbol{P}_i - \boldsymbol{X}_i(t)) + c_2 \boldsymbol{r}_2 \odot (\boldsymbol{G} - \boldsymbol{X}_i(t)), \tag{19.3}$$

where $\odot$ represents component-wise multiplication.

In this chapter we assume that all PSO particles can see each other (unlike in APO). Then $\boldsymbol{P}_i$ is the position of the best fitness ever encountered by particle $i$, while $\boldsymbol{G}$ is the best position ever found by all of the particles.[2] The best

---

[2] The vector $\boldsymbol{G}$ is not to be confused with the gravitational constant $G$ used elsewhere in this book.

> **Listing 19.2: The pseudocode for artificial physics optimization in $d$ dimensions**

```
if (r < 1.5*D) {                    // Note ``D'' is not the same as ``d''
    F = G / pow(r,POWER);
    if (F > FMAX) F = FMAX;
    if (r > D) {                    // Attractive
        for (j = 1; j <= d; j++) {
        Fx[j] = Fx[j] + (F * delta[j] / r);
    }
    }
    else {                          // Repulsive
        for (j = 1; j <= d; j++) {
        Fx[j] = Fx[j] - (F * delta[j] / r);
        }
    }

    // Compute the "fitness force", assume minimizing.
    if (fitness[p] < fitness[k]) {
    for (j = 1; j <= d; j++) {  // Repulsive
        Fx[j] = Fx[j] - (apoF * delta[j] / r);
    }
    }
    if (fitness[p] > fitness[k]) {
    for (j = 1; j <= d; j++) {  // Attractive
        Fx[j] = Fx[j] + (apoF * delta[j] / r);
    }
    }
}
```

positions are updated when particles find positions with better fitness. These updates are a form of memory for PSO, unlike APO, which maintains no memory. The $\omega$ term, an "inertial coefficient" from 0.0 to 1.0, was introduced in 1998 [219]. The "learning rates" $c_1$ and $c_2$ are non-negative constants. We use values of $\omega = 0.729$ and $c_1 = c_2 = 1.49$ as recommended [56]. Finally, $r_1$ and $r_2$ are vectors whose elements are random numbers generated in the range [0,1]. Listing 19.3 shows the implementation of Eq. 19.3, where U(0,1) is a uniform random generator in the range of [0,1].

## 19.3.1 Comparison of APO and PSO in Two Dimensions

We will first compare APO and PSO in two dimensions on the four fitness functions shown in Table 19.3. Each function is moved sideways so that the optimum is at (100,100) to avoid a PSO bias (preference) referred to as the "origin-seeking bias" [165]. Our experiments are performed in an environment

Listing 19.3: The velocity update code for particle swarm optimization

Listing 19.3: The velocity update code for particle swarm optimization

```
for (i = 1; i ≤ N; i++) {      // Loop through particles
    for (j = 1; j ≤ d; j++) { // Loop through dimensions
        r₁ = U(0,1);
        r₂ = U(0,1);
        V[i][j] = ω V[i][j] + c₁r₁ (P[i][j]-X[i][j])+c₂r₂ (G[j]-X[i][j]);
    }
}
```

that is $900 \times 900$, where the origin is at the upper left corner, using standard screen coordinates where positive $x$ is to the right, and positive $y$ is down. An example is shown in Fig. 19.6, which shows the optimum as the blue dot at (100,100).

Table 19.3: Four test functions and the amount of noise used

| Function | Name | Noise |
| --- | --- | --- |
| $20 + e - 20\exp(-.002\sqrt{\sum_{i=1}^{d} x_i^2/d}) - \exp(\sum_{i=1}^{d} \cos 2\pi x_i/d)$ | Ackley | 100 |
| $\frac{1}{4000}\sum_{i=1}^{d} x_i^2 - \prod_{i=1}^{d} \cos(x_i/\sqrt{i}) + 1$ | Griewank | 500 |
| $\sum_{i=1}^{d} (x_i^2 - 10\cos(2\pi x_i) + 10)$ | Rastrigin | 5000000 |
| $\sum_{i=1}^{d} (\lfloor x_i + 0.5 \rfloor)^2$ | Step | 5000000 |

The $900 \times 900$ size of the environment is important only with respect to the initialization of the particles. PSO particles are uniformly randomly placed within the $900 \times 900$ region. APO particles are initialized in a small cluster at some random location in the region. Figure 19.6 shows an example, with red PSO particles and green APO particles. Note that PSO has an immediate advantage simply because one or more of its particles are likely to be closer to the optimum than the APO cluster is.

PSO and APO particles are free to move beyond the $900 \times 900$ region if they wish. However, PSO particles are restricted in their velocity by the parameter $V_{max}$, which is often set according to the size of the initial region. As per the recommendation in [131] we set $V_{max} = 450$ because that is one-half the length of the search space. There is no corresponding $V_{max}$ parameter in APO.

As can be seen in Table 19.3, the amount of noise added to each fitness function is problem dependent. The objective was to make the problem hard without making it impossible. This required a different amount of noise for

Fig. 19.6: The initial particle locations for APO (green) and PSO (red) on the Rastrigin problem. The blue circle is the location of the optimum. The fitness function is not shown

each problem. The metric of performance is again the distance of the center of mass of the PSO and APO formations from the optimum. The APO desired separation $D$ was set to 300 for all of the two-dimensional experiments.

For the first experiment we ran APO and PSO on the four functions, for 100,000 iterations, with $N = 3$ and $N = 7$. The results were averaged over 50 independent runs. The optimum did not move, so although the environment was noisy, it was stationary. The results are shown in Fig. 19.7.

With only three particles APO does quite well, outperforming PSO. Performance improves for both algorithms when the number of particles is increased to seven, but the improvement is greater for PSO. This is because the additional particles aid in the initial exploration of the space by PSO. This is not true for APO since its particles are initialized in one small area. Hence PSO performs better than APO at the beginning of the runs. We will use this observation later in this chapter to improve APO.

For the next experiment the $x$-component and $y$-component of the optimum were each increased by 0.01 every iteration, after 10,000 iterations had elapsed. This allowed both APO and PSO a chance to get close to the optimum, before tracking commenced. Hence, after 10,000 iterations, the optimum moved a distance of $\sqrt{0.01^2 + 0.01^2} \approx 0.01414$ each iteration. The functions are now nonstationary. The results are shown in Fig. 19.8. It is clear that APO tracks the optimum quite easily, while PSO is unable to perform this task. Although it appears as if PSO has some initial success (the PSO curves drop for a while) visualization of the runs indicates that what is really happening is that because the original location of the optimum is in a

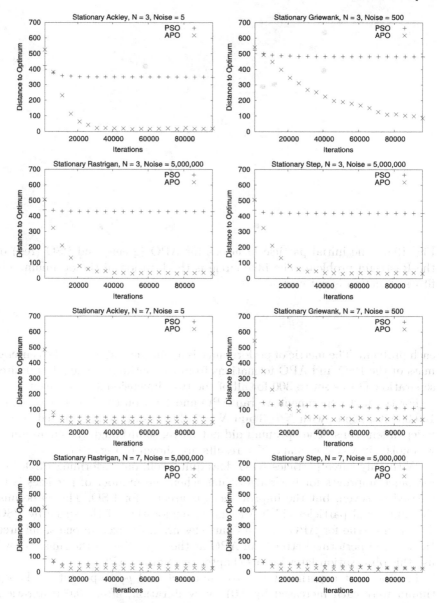

Fig. 19.7: Results of APO and PSO on four stationary fitness functions with three particles (top four graphs) and seven particles (bottom four graphs)

corner, the optimum first approaches the PSO particles and then leaves them behind. An additional experiment where the optimum moved (0.001,0.001) each iteration had similar results. In two dimensions PSO is unable to track even a very slowly changing environment.

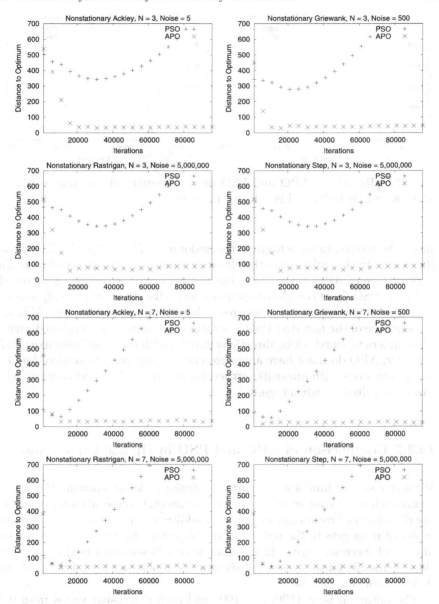

Fig. 19.8: Results of APO and PSO on four nonstationary fitness functions with three particles (top four graphs) and seven particles (bottom four graphs)

Before moving to a higher number of dimensions, it is instructive to perform one additional experiment on a stationary environment. In this case the fitness function is a highly eccentric ellipse as defined in [237]. Figure 19.9

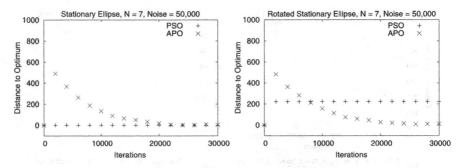

Fig. 19.9: Results of APO and PSO on a stationary ellipse that is aligned with the $x$-axis (left) and is rotated 45° (right)

shows the results. In the left graph the major axis of the ellipse is aligned with the $x$-axis. In the right graph the ellipse is rotated 45° counterclockwise. The performance of APO is the same, but the performance of PSO is extremely different. PSO finds the global optimum immediately when the ellipse is not rotated, but can't make any progress when the ellipse is rotated. This is a consequence of the fact that PSO is rotationally variant, as mentioned earlier. PSO prefers to search along directions parallel with the coordinate axes [237]. However, APO does not have any preference for any particular direction. For fitness functions with linear-like structures (narrow ridges and valleys), APO can have a distinct advantage.

## 19.3.2 Comparison of APO and PSO in Higher Dimensions

We now consider functions in thirty dimensions. Before showing the experiments with nonstationary environments, however, it is useful to continue with the discussion of rotational variance. In addition to measuring how close the center of mass gets to the optimum, the trajectory taken by a formation can also be of interest. Figure 19.10 shows a two-dimensional projection of the thirty-dimensional trajectories of the centers of mass for both PSO (red) and APO (green) formations.

The optimum is at (100, ..., 100) and each dimension varies from 0 to 900. The particles for both algorithms are initialized near the farthest corner of the hypercube from the optimum. The noisy stationary Rastrigin function is used for this example. What we see is that the preference of PSO for directions parallel to the coordinate axes has a very direct impact on the trajectory of the formation. Although both PSO and APO locate the optimum, PSO follows an unusual (and longer) trajectory. The APO trajectory is more natural. This is important if we are optimizing a real process or machine. Modifications can only be made gradually and incur an expense. In this situ-

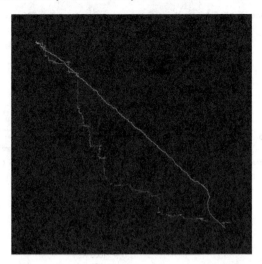

Fig. 19.10: The trajectory of APO (green) and PSO (red) on the Rastrigin function. The centers of mass of the thirty-dimensional APO and PSO formations are shown

ation the shortest trajectory is preferred. However, since the remainder of this chapter deals with standard test functions, rather than real-world processes, we will not focus on this issue any further.

Table 19.4: Eight test functions and the amount of noise used

| Function | Name | Noise |
|---|---|---|
| $20 + e - 20\exp(-.002\sqrt{\sum_{i=1}^{d} x_i^2/d}) - \exp(\sum_{i=1}^{d}\cos 2\pi x_i/d)$ | Ackley | 5 |
| $\sqrt{\sum_{i=1}^{d} x_i^2}$ | Distance | 500 |
| $\frac{1}{4000}\sum_{i=1}^{d} x_i^2 - \prod_{i=1}^{d}\cos(x_i/\sqrt{i}) + 1$ | Griewank | 500 |
| $\sum_{i=1}^{d-1}(100(x_{i+1} - x_i^2)^2 + (x_i - 1)^2)$ | Rosenbrock | 5 |
| $\sum_{i=1}^{d}(x_i^2 - 10\cos(2\pi x_i) + 10)$ | Rastrigin | 500000 |
| $\sum_{i=1}^{d-1}(x_i^2 + x_{i+1}^2)^{.25}[\sin(50(x_i^2 + x_{i+1}^2)^{.1}) + 1]$ | Schaffer's $f_7$ | 5 |
| $\sum_{i=1}^{d}\sum_{j=1}^{i} x_j^2$ | Schwefel | 500 |
| $\sum_{i=1}^{d}(\lfloor x_i + 0.5 \rfloor)^2$ | Step | 500000 |

For the experiment with the nonstationary optimum each component of the location of the optimum is again increased by 0.01 each iteration, after 10,000 iterations. Hence the optimum moved a distance of $\sqrt{30 \times 0.01^2} \approx 0.05477$ each iteration, which is considerably faster than the two-dimensional experiments. Since PSO generally performs better in the early stages of search we combined both PSO and APO to create a hybrid called "APOH." This

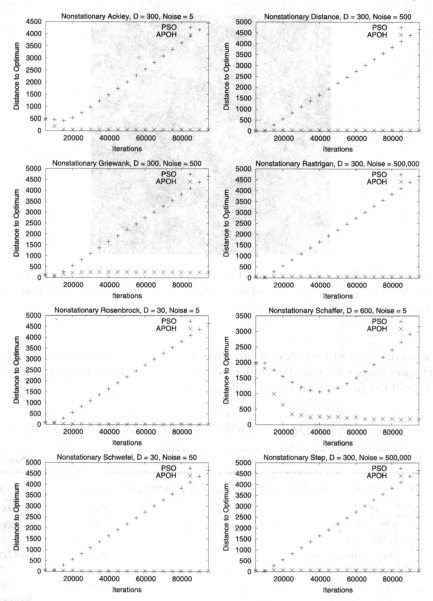

Fig. 19.11: Results of APOH and PSO on eight nonstationary fitness functions where the desired separation $D$ is set by the user

hybrid works in two stages. In stage one, APOH initializes particle locations in the same way that PSO does (uniformly randomly in the search space). Then for $I$ iterations standard PSO is run. The parameter $I$ should be high enough to allow PSO to converge. For these experiments we use 5,000 iterations.

This ends the first stage. In the second stage, the population is reinitialized as a small cluster around one randomly chosen particle from the converged population. Then the standard APO algorithm runs until termination. The motivation for this hybrid is to use the first stage (PSO) to get close to the optimum, while then using the second stage (APO) to track the moving optimum. The PSO parameters only affect the first stage of APOH. The APO parameters only affect the second stage of APOH. Hence the desired separation $D$ is used only by the second stage of APOH. We will see that both stages are complementary, and that the second stage can make further progress towards the optimum when the first stage does poorly.[3]

Four additional test functions were included, providing a suite of eight fitness functions, as shown in Table 19.4. Again, the amount of noise used is problem specific. Sometimes the noise is decreased from the amount used in the two-dimensional experiments, because the increased speed of the optimum, coupled with the higher dimensionality, makes the problems more difficult. The results are averaged over 10 independent runs. Although a population size of 31 is the minimum required for a thirty-dimensional problem, preliminary experiments indicate that a population of size $2d$ (where $d$ is the number of dimensions) works better. Our hypothesis is that this potentially allows there to be two particles per coordinate axis, allowing the formation to move along each axis without preference for any particular axis. Hence our experiments use a population size of $N = 60$ for both APOH and PSO.

The results are shown in Fig. 19.11. Once again it is clear that PSO cannot track the moving optimum, while APOH can.[4] However, one issue did arise for APOH. Although APOH worked well with $D = 300$ on a majority of problems, other problems required substantially different values. Rosenbrock and Schwefel required a small value of 30, while Schaffer required a higher value of 600.

The reason for this becomes clear when we examine pictures of these three functions. Figure 19.12 shows the Schaffer, Rosenbrock, and Schwefel functions, where the optimum has been moved to the center to aid visualization. Note that the Schaffer function is composed of circular ridges that are quite wide. A very large formation is needed to sense the global structure of the problem and move the population "over" the ridges. However, both the Rosenbrock and Schwefel functions have dark corridors that can be followed to the optimum. In this case small formations work better.

These results indicate that the desired separation $D$ is the crucial parameter in APOH, which is used in the second stage of the algorithm. The best $D$ depends on the structure of the problem. However, a nice feature of APOH is that the effect of $D$ is extremely simple to understand. This is in direct contrast to parameters in most other optimization algorithms, where

---

[3] It is very likely that an alternative hybrid that continually alternates between APO exploration and PSO exploitation might also work very well. We leave this alternative to future research.

[4] In fact, PSO performed no better when the experiments were rerun with zero noise!

Fig. 19.12: Pictures of the Schaffer (left), Rosenbrock (middle) and Schwefel (right) functions. The optimum location is shown with a center dot

the interplay between parameter settings and performance is complex and difficult to comprehend. Hence, although one can visualize fitness functions, this generally does not help the user to set parameter settings in most function optimization algorithms. However, visualization can be extremely useful in aiding the user to find the appropriate setting of $D$ for APOH.

As an alternative, the APOH algorithm can be modified to smoothly cycle the values of $D$. In our final experiment we modified APOH to cycle from 30 to 600 using the sin() function in the second stage of the algorithm. The results are shown in Fig. 19.13. The overall performance is extremely good. We calculated the mean distance of the center of mass from the optimum over all problems, runs, and iterations. When $D$ was set by the user, the mean distance was 118.73 with a standard deviation of 210.72. For the experiment with the cyclic $D$, the mean distance was 134.91 with a standard deviation of 223.42. In general, a fully automated system cannot perform as well as an algorithm with hand-optimized parameter settings. In this case, the mean from the automated system is only roughly 13% higher than the mean from the optimized system. These are very good results.

Let's see what the tracking behavior looks like. Figure 19.14 shows one run of APOH on four of the nonstationary problems. The optimum is moving down and to the right, while the path shows the movement of the center of mass of the APOH formation during the *second stage* of the algorithm (these are two-dimensional projections of the thirty-dimensional points). Note that although the environment is moving, the display of the environment is not updated, and what you are seeing is the environment at the beginning of each run. In these examples the first stage of APOH (PSO) has no difficulty finding the optimum on the Rosenbrock and Schwefel functions (right pictures), and the second stage easily tracks the optimum. On the Ackley and Schaffer functions (left pictures), the first stage was unable to reach the optimum. This is particularly problematic on the Schaffer function (lower left), where the first stage was unable to get any closer than the tip of the path at the lower left. In both cases the second stage had to continue to search for the

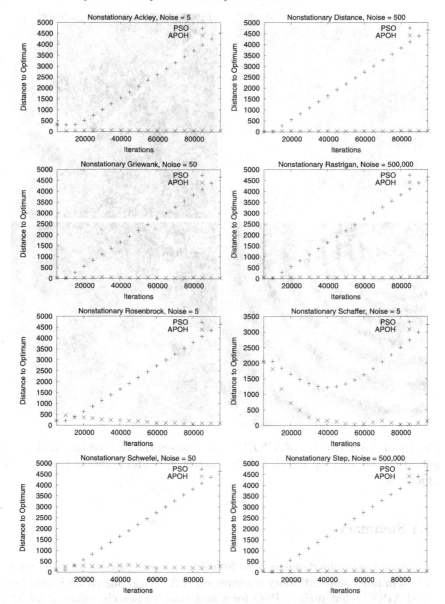

Fig. 19.13: Results of APOH and PSO on eight nonstationary fitness functions where the desired separation $D$ is cyclic

optimum. What is interesting is that the second stage succeeded when the first stage did not, indicating that both stages are complementary to each other. Finally, once APOH finds the optimum, tracking it is not difficult.

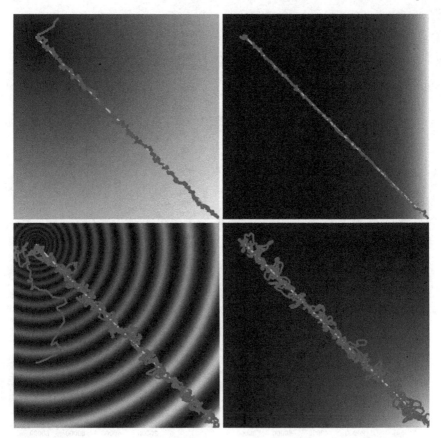

Fig. 19.14: Visualization of APOH (second stage) tracking on the Ackley (top left), Rosenbrock (top right), Schaffer (lower left) and Schwefel (lower right) functions

## 19.4 Summary

This chapter examined one method for using artificial physics to track moving optima in noisy nonstationary environments. A hybrid algorithm is proposed, called APOH, that utilizes PSO for a first stage of search, followed by APO for the second stage. In general, the two stages perform different tasks. PSO generally works well at finding a reasonable solution quickly, while APO excels at tracking, due to the APO formation. The key to the success is the fact that the APO formation serves two purposes. First, it can help to cancel noise, allowing the formation to move in a consensus direction. The consensus is achieved in an implicit manner, rather than explicitly. Second, the formation prevents premature convergence, allowing the algorithm to be continually ready for environment changes.

Not surprisingly, quite a bit of research has been performed in modifying PSO to perform tracking. As noted in [16], PSO has difficulty performing tracking due to outdated memory and premature convergence. In response, researchers have attempted to use multiple swarms simultaneously [141], triggers that rerandomize a portion of the population [100], and mechanisms that allow PSO to use "evaporation" to slowly forget [42]. None of these mechanisms are necessary for APO. Interestingly, more recent research has focused on physics-based additions to PSO to solve the tracking problem, including a "quantum atom" and "Coulomb repulsion" [16]. Gao et al. [74] have also proposed an extension to PSO that includes quantum behavior. This is not the first time that PSO has augmented its model with physics. As mentioned above, the "inertial coefficient" $\omega$ was added in 1998. This is identical to the friction used in artificial physics.

Other physics-based optimization techniques are currently being invented. Central force optimization (CFO) is a deterministic optimization algorithm, based on gravitational kinematics [66, 67]. The "Gravitational Search Algorithm" (GSA) is another approach, where the gravitational and inertial masses of particles are determined by the fitness function [186]. In contrast, Coulombs's law is utilized in the "Electromagnetism Algorithm" for optimization [15, 196]. Also, Coulomb's law is combined with the Newtonian laws of mechanics to create an optimization technique called "Charged System Search" [111]. Although quite different from APO, they are all physicomimetic algorithms.

When one examines the history of optimization, different metaphors appear at different times. First, there were the "mathematical" procedures, such as Fibonacci division, Newton–Raphson iteration, and Lagrangian interpolation. Such techniques often assume a problem is unimodal and attempt to "box in" the optimum. Alternatively, they required that the derivatives of the functions are known or they attempt to fit a polynomial to the fitness function. Schwefel [213] provides a summary of these earlier techniques, and compares them with evolution strategies. Newer techniques, such as EAs, neural networks, ant colony optimization, and PSO are based on biological metaphors. With the exception of simulated annealing [126], there was little in the way of physics-based algorithms.

The newest algorithms, however, are borrowing more and more from physics. The trend is clear. Physicomimetics is just beginning.

# Appendix A
# Anomalous Behavior of the `random()` Number Generator

William M. Spears and Derek T. Green

This appendix serves as a cautionary tale of trusting a random number generator too much, and illustrates the difficulty in conducting good science. As we were creating the artificial physics optimization (APO) algorithm in Chap. 19, we were simultaneously theoretically analyzing the particle swarm optimization (PSO) algorithm [237]. Unfortunately the theoretical results were not matching the experimental results. After checking it turned out that the theory was correct, but that the experimental results were wrong. Normally this would simply mean that an error was made in our experimental code. This turned out to not be the case. The cause was much more serious.

The experimental results were based on a Monte Carlo simulation that relied on the Linux/Unix "random()" random number generator used in the C and C++ languages.[1] Although it fails some tests [221] it is still commonly used by programmers. In this Appendix we expose anomalous behavior in random() that appears to be previously undocumented. Identical behavior has been shown under Ubuntu, Red Hat, Debian, and Mac OS X.

Rather than present you with the entirety of the experimental code, we first performed an ablation study. What is the smallest piece of code that exhibits the anomalous behavior? Surprisingly, it turned out to be an implementation of "rejection sampling" that is supposed to uniformly generate points in a unit circle [266]. The code is shown in Listing A.1. The portion of the code that generates numbers between zero and one is (double)random()/RAND_MAX. The distribution of numbers should be uniform (i.e., there should be no preference for any numbers or range of numbers). This is then converted to a number from negative one to one. Two

William M. Spears
Swarmotics LLC, Laramie, Wyoming, USA, e-mail: wspears@swarmotics.com

Derek T. Green
Department of Computer Science, University of Arizona, Tucson, Arizona, USA, e-mail: dtgreen@arizona.edu

---

[1] As of April 2006, rand() became the same as random() [6].

W.M. Spears, D.F. Spears (eds.), *Physicomimetics*,
DOI 10.1007/978-3-642-22804-9,
© Springer-Verlag Berlin Heidelberg 2011

Listing A.1: This code generates numbers in the unit circle and counts how many times each angle is generated

```
#include <stdio.h>
#include <stdlib.h>
#include <math.h>
int count[360];

int main(argc, argv)
int argc; char *argv[];
{
  int i, j, bin, S;
  double x, y;

  S = atoi(argv[1]);
  srandom(1);

  for (i = 0; i < 360; i++) count[i] = 0;

  for (i = 1; i <= 1000000000; i++) {
    for (j = 1; j <= S; j++) random();
    do {
      x = 2.0 * ((double)random()/RAND_MAX) - 1.0;
      y = 2.0 * ((double)random()/RAND_MAX) - 1.0;
    }
    while (sqrt(x * x + y * y) > 1);

    bin = (int)(180.5 + (atan2(y, x) * (180 / M_PI)));
    if (bin == 360) bin = 0;
    count[bin]++;
  }

  for (i = 0; i < 360; i++) printf("%d %d\n", i-180, count[i]);
  return(1);
}
```

such numbers are generated, creating a (x, y) point in the Cartesian coordinate system. If the distance of the point from the origin is greater than one, it is thrown away (i.e., rejected because it is not in the unit circle), and two more random numbers are generated. This loop continues until a point within the unit circle occurs. The code then calculates the angle of each point within the unit circle and counts how many times each angle occurs (using an array with 360 slots, one for each possible degree of angle). Since all points are supposed to be uniformly likely, all angles should also be uniformly likely. Note that the code takes one argument, which is assigned to the variable S. This variable controls the number of superfluous calls that are made to the random number generator (in our original experimental code the calls were not superfluous, but were used for other purposes). With a good random number generator it should not matter how many extraneous calls are made, because the assumption is that the calls are highly independent and without correlations or patterns.

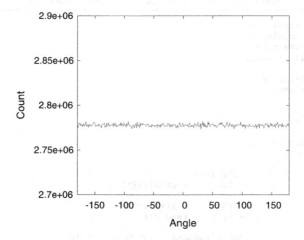

Fig. A.1: The distribution of angles is quite uniform when S is zero

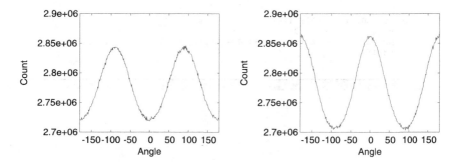

Fig. A.2: The distribution of angles is not uniform when S is 25 (left) and 27 (right)

If we compile and execute the code with S set to zero, the results are quite reasonable, as shown in Fig. A.1. The code generates 1,000,000,000 points in the unit circle. If every angle is uniformly likely, they should occur roughly 2,777,777 times. This is what we see. The amount of noise exhibited is normal for 1,000,000,000 sample points.

However, Fig. A.2 shows the results when S is 25 (left) and 27 (right). Clearly the results are no longer uniform! The left graph shows a preference for $-90°$ and $90°$. The right graph shows a preference for $0°$ and $180°$. The

```
#include <stdio.h>
#include <stdlib.h>
#include <math.h>
int grid[101][101];

int main(argc, argv)
int argc; char *argv[];
{
  int i, j, S;
  double x, y;

  S = atoi(argv[1]);
  srandom(1);

  for (i = 0; i <= 100; i++)
    for (j = 0; j <= 100; j++) grid[i][j] = 0;

  for (i = 1; i <= 1000000000; i++) {
    for (j = 1; j <= S; j++) random();
    do {
      x = 2.0 * ((double)random()/RAND_MAX) - 1.0;
      y = 2.0 * ((double)random()/RAND_MAX) - 1.0;
    }
    while (sqrt(x * x + y * y) > 1);

    grid[(int)(50*(x+1))][(int)(50*(y+1))]++;
  }

  for (i = 0; i <= 100; i++)
    for (j = 0; j <= 100; j++)
      printf("%f %f %d\n", (i/50.0 - 1), (j/50.0 - 1), grid[i][j]);
  return(1);
}
```

results are definitely problematic. To confirm these results we examined the
distribution of the points in the unit circle itself. A $100 \times 100$ grid of cells
was placed over the unit square that surrounds the unit circle. Each time a
point fell within a particular cell, the cell count was incremented. The code is
shown in Listing A.2. Figure A.3 shows the results when S is zero. The results
are quite reasonable, showing relative uniformity within the circle (there are
no obvious patterns). The cells that are totally outside the circle have zero
cell counts and are colored black.

Now let's redo the experiment with S set to 25 and 27. The results are
shown in Fig. A.4 and are quite striking. Lighter shades represent higher
counts. Hence the left graph shows a preference for $-90°$ and $90°$. The right

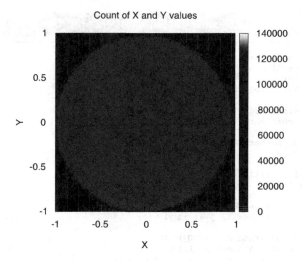

Fig. A.3: The distribution of points in the unit circle is quite uniform when
S is zero

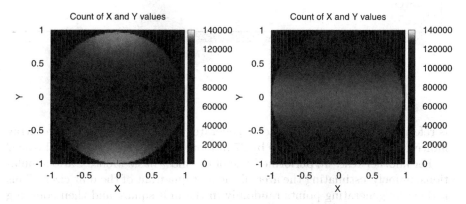

Fig. A.4: The distribution of angles is not uniform when S is 25 (left) and 27
(right)

graph shows a preference for 0° and 180°. This is perfectly consistent with
the results shown in Fig. A.2.

We reran our code using values of S from zero to 100. Although non-
uniform results appear for most values from one to 27, the results from 28 to
100 did not show any anomalies. The reason for these results is not at all clear.
We do note that it is claimed that `random()` is implemented via the C code
statement `r[i] = r[i-31] + r[i-3]`, which indicates that the current
random number is generated using the number generated three back in the
sequence, added to the number generated 31 back in the sequence [214]. This

Listing A.3: Code to estimate the area of the unit circle

```c
#include <stdio.h>
#include <stdlib.h>
#include <math.h>
#define TRIALS 1000000000

int main(argc, argv)
int argc; char *argv[];
{
  int i, j, success = 0, S;
  double x, y;

  S = atoi(argv[1]);
  srandom(1);

  for (i = 1; i <= TRIALS; i++) {
    for (j = 1; j <= S; j++) random();

    x = (double)random()/RAND_MAX;
    y = (double)random()/RAND_MAX;
    if (sqrt(x * x + y * y) <= 1) success++;
  }

  printf("%f\n", (double)(success)/TRIALS);
  return(1);
}
```

is referred as a "lagged-Fibonacci" generator [221]. Note that the two array indices (3 and 31) are separated by 27 random numbers. Is this a coincidence?

For another test, we performed one of the most basic Monte Carlo simulations, namely, estimating the area of the first quadrant of the unit circle. This is done by generating points randomly in the unit square and then counting the number of points that successfully fall within that quadrant of the circle. The ratio of the number of "successful" points to the total number of generated points yields an approximation of the area. Theoretically, the area of the quadrant should be $\pi r^2/4$, where $r = 1.0$. Hence $\pi/4 \approx 0.78539$. The code for this test is shown in Listing A.3. Note that once again we include the variable S to see if extraneous calls to random() affect the results. Our experiments showed that random() produces very accurate estimations of the area of the quadrant of the unit circle. Also, the value of the variable S had no noticeable effect.

Is it clear that the error lies with random() as opposed to our code? To double check, we substituted a different random number generator into the code in Listings A.1 and A.2. This generator is claimed to pass "all of the tests for random number generators" [21], and is discussed in more detail in [148]. This random generator performed much better, with almost no anomalies.

Also, we tested the rather old random number generators drand48() and erand48() [195]. These also performed very well.

In summary, we appear to have a puzzle. The random() random number generator works perfectly well at distributing points within the unit square, producing the proper estimation of the area of a quadrant of the unit circle. The number of superfluous calls to random() does not appear to have any effect. Yet, when generating points within the unit circle, the uniformity of the points within that circle depends on the number of superfluous calls to random(). The area test may not be as sensitive as our angle distribution test. Or there may be deeper issues. We leave the further exploration and explanation of this puzzle to the interested reader.

# Index

W.M. Spears, D.F. Spears (eds.), *Physicomimetics*,
DOI 10.1007/978-3-642-22804-9,
© Springer-Verlag Berlin Heidelberg 2011

# References

1. Abacom: AM-RTD-315 transceiver module. Tech. rep., ABACOM Technologies Inc (2008). URL http://www.abacom-tech.com/data_sheets/AM-RTD315.pdf. Accessed September 20, 2008
2. Agassounon, W., Spears, W.M., Welsh, R., Zarzhitsky, D., Spears, D.F.: Toxic plume source localization in urban environments using collaborating robots. In: IEEE Conference on Technologies for Homeland Security (2009)
3. Alur, R., Esposito, J., Kim, M., Kumar, J., Lee, I.: Formal modeling and analysis of hybrid systems: A case study in multi-robot coordination. Lecture Notes in Computer Science **1708**, 212–232 (1999)
4. AnalogDevices: AD7414/AD7415 ±0.5°C accurate, 10-bit digital temperature sensors in SOT-23. Tech. rep., Analog Devices (2005). URL http://www.analog.com/. Accessed October 13, 2008
5. Anderson, J.D.: Computational Fluid Dynamics. McGraw–Hill (1995)
6. Ashihara, H., Kuramoto, A., Wada, I., Matsumoto, M., Kikuchi, H., Kiyama, M.: Parallel computations reveal hidden errors of commonly used random number generators abstract. In: World Congress in Computer Science, Computer Engineering, and Applied Computing, International Conference on Scientific Computing (2006)
7. Balch, T., Arkin, R.: Behavior-based formation control for multi-robot teams. IEEE Transactions on Robotics and Automation **14**(6), 1–15 (1998)
8. Balch, T., Hybinette, M.: Social potentials for scalable multi-robot formations. IEEE Transactions on Robotics and Automation, **1**, 73–80 (2000)
9. Balkovsky, E., Shraiman, B.: Olfactory search at high Reynolds number. PNAS **99**(20), 12,589–12,593 (2002)
10. Batalin, M., Sukhatme, G.: Spreading out: A local approach to multi-robot coverage. In: Proceedings of the 6th International Symposium on Distributed Autonomous Robotics Systems, pp. 373–382 (2002)
11. Bayindir, L., Şahin, E.: A review of studies in swarm robotics. Turkish Journal of Electrical Engineering **15**(2), 115–147 (2007)
12. Beni, G., Hackwood, S.: Stationary waves in cyclic swarms. Intelligent Control, pp. 234–242 (1992)
13. Beni, G., Wang, J.: Swarm intelligence. In: Proceedings of the Seventh Annual Meeting of the Robotics Society of Japan, pp. 425–428. Tokyo, Japan (1989)
14. Bill, R.G., Herrnkind, W.F.: Drag reduction by formation movement in spiny lobsters. Science **193**, 1146–1148 (1976)
15. Birbil, S.I., Fang, S.C.: An electromagnetism-like mechanism for global optimization. Journal of Global Optimization **25**(3), 263–282 (2003)
16. Blackwell, T.: Particle swarm optimization in dynamic environments. In: S. Yang, Y. Ong, Y. Jin (eds.) Evolutionary Computation in Dynamic and Uncertain Environments, vol. 51, pp. 29–49. Springer (2007)

W.M. Spears, D.F. Spears (eds.), *Physicomimetics*, 629
DOI 10.1007/978-3-642-22804-9,
© Springer-Verlag Berlin Heidelberg 2011

17. Board on Atmospheric Sciences and Climate: Tracking and Predicting the Atmospheric Dispersion of Hazardous Material Releases: Implications for Homeland Security. National Academies Press (2003)

18. Bonabeau, E., Dorigo, M., Theraulaz, G.: Swarm Intelligence: From Natural to Artificial Intelligence. Oxford University Press, New York (1999)

19. Borenstein, J., Everett, H., Feng, L.: Where am I?: Sensors and methods for mobile robot positioning. In: Technical report, University of Michigan (1996)

20. Borenstein, J., Koren, Y.: Real-time obstacle avoidance for fast mobile robots. IEEE Transactions on Systems, Man, and Cybernetics **19**(5), 1179–1187 (1989)

21. Bourke, P.: Uniform random number generator (1998). URL http://paulbourke.net/miscellaneous/random/. Accessed March 4, 2011

22. Brogan, D., Hodgins, J.: Group behaviors for systems with significant dynamics. Autonomous Robots **4**, 137–153 (1997)

23. Bryant, B.D., Miikkulainen, R.: Neuroevolution for adaptive teams. In: Proceedings of the 2003 Congress on Evolutionary Computation, pp. 2194–2201 (2003)

24. Caldwell, S.L., D'Agostino, D.M., McGeary, R.A., Purdy, H.L., Schwartz, M.J., Weeter, G.K., Wyrsch, R.J.: Combating terrorism: Federal agencies' efforts to implement national policy and strategy. In: Congressional report GAO/NSIAD-97-254 (1997)

25. Camazine, S., Deneubourg, J.L., Franks, N.R., Sneyd, J., Theraulaz, G., Bonabeau, E.: Self-organization in biological systems. Princeton University Press (2003)

26. Campbell, A., Wu, A.S.: Multi-agent role allocation: Issues, approaches, and multiple perspectives. Journal of Autonomous Agents and Multi-Agent Systems **22**(2), 317–355 (2011)

27. Carlson, B., Gupta, V., Hogg, T.: Controlling agents in smart matter with global constraints. In: E.C. Freuder (ed.) AAAI-97 Workshop on Constraints and Agents—Technical Report WS-97-05 (1997)

28. Celikkanat, H., Turgut, A., Şahin, E.: Control of a mobile robot via informed robots. In: Proceedings of the 9th International Symposium on Distributed Autonomous Robotic Systems (2008)

29. Chalmers, A.: The Works of the English Poets. London (1810)

30. Chang, K., Hwang, J., Lee, E., Kazadi, S.: The application of swarm engineering technique to robust multi-chain robot system. In: IEEE Conference on Systems, Man, and Cybernetics, pp. 1429–1434. IEEE (2005)

31. Chen, Y.H.: Adaptive robust control of artificial swarm systems. Applied Mathematics and Computation **217**, 980–987 (2010)

32. Cheney, E.W., Kincaid, D.R.: Numerical Mathematics and Computing, fifth edn. Brooks Cole (2003)

33. Choset, H.: Coverage for robotics—A survey of recent results. Annals of Mathematics and Artificial Intelligence **31**(1–4), 113–126 (2001)

34. Clerc, M.: The old bias. Personal website p. 1 (2001). URL http://clerc.maurice.free.fr/pso/

35. Coe, G.: Ultrasonic rangefinder software. Tech. rep., Devantech Ltd (Robot Electronics) (2000). URL http://www.robot-electronics.co.uk/files/srf1.asm. Accessed October 12, 2008

36. Coirier, W.J., Fricker, D.M., Furmanczyk, M., Kim, S.: A computational fluid dynamics approach for urban area transport and dispersion modeling. Env Fluid Mech **5**(5), 443–479 (2005)

37. Cordesman, A.H.: Defending America: Asymmetric and terrorist attacks with chemical weapons (2001)

38. Cowen, E.: Special issue on chemical plume tracing. Environmental Fluid Mechanics **2** (2002)

39. Craig, J.J.: Introduction to Robotics: Mechanics and Control, second edn. Addison Wesley (1989)

40. Crimaldi, J., Koehl, M., Koseff, J.: Effects of the resolution and kinematics of olfactory appendages on the interception of chemical signals in a turbulent odor plume. Env Fluid Mech, Special Issue on Chemical Plume Tracing **2**, 35–63 (2002)

41. Cui, X., Hardin, C.T., Ragade, R.K., Elmaghraby, A.S.: A swarm approach for emission sources localization. In: Proceedings of the 16th IEEE International Conference on Tools with Artificial Intelligence, pp. 424–430 (2004)

42. Cui, X., Hardin, C.T., Ragade, R.K., Potok, T.E., Elmaghraby, A.S.: Tracking non-stationary optimal solution by particle swarm optimizer. In: Proceedings of the Sixth International Conference on Software Engineering, Artificial Intelligence, Networking and Parallel/Distributed Computing and First ACIS International Workshop on Self-Assembling Wireless Networks, pp. 133–138. IEEE Computer Society, Washington, DC, USA (2005). DOI 10.1109/SNPD-SAWN.2005.77. URL http://portal. acm.org/citation.cfm?id=1069813.1070325

43. Czirok, A., Vicsek, M., Vicsek, T.: Collective motion of organisms in three dimensions. Physica A **264**(299), 299–304 (1999)

44. D'Ambrosio, D.B.: Multiagent learning through indirect encoding. Ph.D. thesis, University of Central Florida (2011)

45. D'Ambrosio, D.B., Stanley, K.O.: Generative encoding for multiagent learning. In: Proceedings of the Genetic and Evolutionary Computation Conference (2008)

46. Dasgupta, D.: Advances in artificial immune systems. IEEE Computational Intelligence Magazine **1**(4), 40–49 (2006)

47. Decuyper, J., Keymeulen, D.: A reactive robot navigation system based on a fluid dynamics metaphor. Lecture Notes in Computer Science—Parallel Problem Solving from Nature (PPSN I) 496, 356–362 (1991)

48. Desai, J., Ostrowski, J., Kumar, V.: Controlling formations of multiple mobile robots. In: IEEE International Conference on Robotics and Automation, vol. 4, pp. 2864–2869. IEEE (1998)

49. Desai, J., Ostrowski, J., Kumar, V.: Modeling and control of formations of nonholonomic mobile robots. IEEE Transactions on Robotics and Automation **17**(6), 905–908 (2001)

50. Devantech: CMPS03—compass module. Tech. rep., Devantech Ltd (Robot Electronics) (2008). URL http://www.robot-electronics.co.uk/htm/ cmps3tech.htm. Accessed September 21, 2008

51. Devantech: LCD03—I2C/Serial LCD. Tech. rep., Devantech Ltd (Robot Electronics) (2008). URL http://www.robot-electronics.co.uk/htm/Lcd03tech. htm. Accessed September 21, 2008

52. Devantech: SRF04 ultrasonic ranger. Tech. rep., Devantech Ltd (Robot Electronics) (2008). URL http://www.robot-electronics.co.uk/htm/srf04tech. htm. Accessed September 20, 2008

53. Dewdney, A.K.: The Planiverse: Computer Contact with a Two-Dimensional World. Springer (2000)

54. DimensionEngineering: Sabertooth 2x10 user's guide. Tech. rep., Dimension Engineering (2007). URL http://www.dimensionengineering.com/. Accessed October 12, 2008

55. Ducatelle, F., Di Caro, G.A., Gambardella, L.M.: Cooperative self-organization in a heterogeneous swarm robotic system. In: Proceedings of the 12th Annual Conference on Genetic and Evolutionary Computation, pp. 87–94. ACM (2010)

56. Eberhart, R., Shi, Y.: Comparing inertia weights and constriction factors in particle swarm optimization. In: Proceedings of the Congress on Evolutionary Computation, pp. 84–88 (2000)

57. Elfes, A., Podnar, G., Dolan, J.M., Stancliff, S., Lin, E., Hosler, J.C., Ames, T.J., Moisan, J., Moisan, T.A., Higinbotham, J., Kulczycki, E.A.: The telesupervised adaptive ocean sensor fleet. Tech. rep., Carnegie–Mellon Robotics Institute, Paper 186 (2007)

58. Ellis, C., Wiegand, R.P.: Actuation constraints and artificial physics control. In: Proceedings of the Ninth International Conference on Parallel Problem Solving from Nature, pp. 389–398 (2008)
59. Farrell, J.A., Murlis, J., Li, W., Carde, R.T.: Filament-based atmospheric dispersion model to achieve short time-scale structure of odor plumes. In: E. Cowen (ed.) Environmental Fluid Mechanics, Special Issue on Chemical Plume Tracing, **2**, 143–169. Kluwer (2002)
60. Fax, J., Murray, R.: Information flow and cooperative control of vehicle formations. IEEE Transactions on Automatic Control **49**(9), 1465–1476 (2002)
61. Feddema, J.T., Robinett, R.D., Byrne, R.H.: An optimization approach to distributed controls of multiple robot vehicles. In: Proceedings of the IEEE/RSJ International Conference on Intelligent Robots and Systems (IROS'03) (2003)
62. Fernandes, J.C.A., Neves, J.A.B.C.: Using conical and spherical mirrors with conventional cameras for 360° panorama views in a single image. In: Mechatronics, 2006 IEEE International Conference on, pp. 157–160 (2006). DOI 10.1109/ICMECH.2006. 252515
63. Ferri, G., Jakuba, M.V., Yoerger, D.R.: A novel trigger-based method for hydrothermal vents prospecting using an autonomous underwater robot. Autonomous Robots **29**(1) (2010)
64. Fierro, R., Belta, C., Desai, J., Kumar, V.: On controlling aircraft formations. In: IEEE Conference on Decision and Control, vol. 2, pp. 1065–1070. Orlando, Florida (2001)
65. Fierro, R., Song, P., Das, A., Kumar, V.: Cooperative control of robot formations. In: R. Murphey, P. Pardalos (eds.) Cooperative Control and Optimization, vol. 66, pp. 73–93. Kluwer Academic Press, Hingham, MA (2002)
66. Formato, R.A.: Central force optimization: A new nature inspired computational framework for multidimensional search and optimization. In: Nature Inspired Cooperative Strategies for Optimization, vol. 129, pp. 221–238 (2008)
67. Formato, R.A.: Parameter-free deterministic global search with simplified central force optimization. In: Proceedings of the 6th International Conference on Advanced Intelligent Computing Theories and Applications: Intelligent Computing, ICIC'10, pp. 309–318. Springer–Verlag, Berlin, Heidelberg (2010). URL http://portal. acm.org/citation.cfm?id=1886896.1886946
68. Fox, D., Burgard, W., Thrun, S.: The dynamic window approach to collision avoidance (1995)
69. Fredslund, J., Matarić, M.: A general algorithm for robot formations using local sensing and minimal communication. IEEE Transactions on Robotics and Automation **18**(5) (2002)
70. Freescale: MC9S12DP256—advance information. Tech. rep., Freescale Semiconductor, Inc. (2000). Revision 1.1
71. Freund, J.E.: Modern Elementary Statistics, seventh edn. Prentice–Hall (1988)
72. Frey, C.L., Zarzhitsky, D., Spears, W.M., Spears, D.F., Karlsson, C., Ramos, B., Hamann, J., Widder, E.: A physicomimetics control framework for swarms of autonomous surface vehicles. Oceans'08 Conference (2008)
73. Gage, D.W.: Randomized search strategies with imperfect sensors. In: Proceedings of SPIE Mobile Robots VIII, pp. 270–279 (1993)
74. Gao, H., Xu, W., Sun, J., Tang, Y.: Multilevel thresholding for image segmentation through an improved quantum-behaved particle swarm algorithm. Transactions on Instrumentation and Measurement **59**(4), 934–946 (2010)
75. Garcia, A.L.: Numerical Methods for Physics, 2nd edn. Prentice Hall (2000)
76. Gordon, D., Spears, W., Sokolsky, O., Lee, I.: Distributed spatial control, global monitoring and steering of mobile physical agents. In: IEEE International Conference on Information, Intelligence, and Systems, pp. 681–688. Washington, DC (1999)
77. Grasso, F.: Invertebrate-inspired sensory-motor systems and autonomous, olfactory-guided exploration. Biology Bulletin **200**, 160–168 (2001)

78. Grasso, F., Atema, J.: Integration of flow and chemical sensing for guidance of autonomous marine robots in turbulent flows. Env Fluid Mech, Special Issue on Chemical Plume Tracing **2**(1–2), 95–114 (2002)

79. Greene, M., Gordon, D.F.: How patrollers set foraging direction in Harvester ants. American Naturalist **170**, 943–952 (2007)

80. Grefenstette, J.: Credit assignment in rule discovery systems based on genetic algorithms. Machine Learning **3**(2–3), 225–245 (1988)

81. Grinstead, C.M., Snell, J.L.: Introduction to Probability, 2nd edn., chap. 11, pp. 405–470. AMS (1997)

82. Guibas, L., Stolfi, J.: Primitives for the manipulation of general subdivisions and the computation of voronoi. ACM Trans. Graph. **4**(2), 74–123 (1985)

83. Guldner, J., Utkin, V.: Sliding mode control for gradient tracking and robot navigation using artificial potential fields. IEEE Transactions on Robotics and Automation **11**(2), 247–254 (1995)

84. Haeck, N.: Minimum distance between a point and a line (2002). URL http://www.simdesign.nl/tips/tip001.html. Accessed April 18, 2011

85. Halliday, D., Resnick, R.: Physics, third edn. John Wiley and Sons (1978)

86. Hayes, A., Martinoli, A., Goodman, R.: Swarm robotic odor localization. In: IEEE/RSJ International Conference on Intelligent Robots and Systems (2001)

87. Hazards, C., Poisons Division (London), H.P.A.: Chemical hazards and poisons report (2006)

88. Heil, R.: A trilaterative localization system for small mobile robots in swarms. Master's thesis, University of Wyoming, Laramie, WY (2004)

89. Helbing, D., Farkas, I., Vicsek, T.: Simulating dynamical features of escape panic. Nature **407**, 487–490 (2000)

90. Hettiarachchi, S.: Distributed evolution for swarm robotics. Ph.D. thesis, University of Wyoming, Laramie, WY (2007)

91. Hettiarachchi, S., Spears, W., Kerr, W., Zarzhitsky, D., Green, D.: Distributed agent evolution with dynamic adaptation to local unexpected scenarios. In: Second GSFC/IEEE Workshop on Radical Agent Concepts. Springer (2006)

92. Hettiarachchi, S., Spears, W.M.: Moving swarm formations through obstacle fields. In: International Conference on Artificial Intelligence, vol. 1, pp. 97–103. CSREA Press (2005)

93. Hettiarachchi, S., Spears, W.M.: DAEDALUS for agents with obstructed perception. In: IEEE Mountain Workshop on Adaptive and Learning Systems. IEEE Press (2006)

94. Holland, J.H.: Adaptation in Natural and Artificial Systems. University of Michigan Press, Ann Arbor, MI (1975)

95. Holland, O., Melhuish, C., Hoddell, S.: Chorusing and controlled clustering for minimal mobile agents. Robotics and Autonomous Systems **28**, 207–216 (1999)

96. Horling, B., Lesser, V.: A survey of multi-agent organizational paradigms. Knowledge Engineering Review **19**(4), 281–316 (2004)

97. How illuminating. The Economist (March 12–18, 2011)

98. Howard, A., Matarić, M., Sukhatme, G.: Mobile sensor network deployment using potential fields: A distributed, scalable solution to the area coverage problem. In: Sixth International Symposium on Distributed Autonomous Robotics Systems, pp. 299–308. ACM, Fukuoka, Japan (2002)

99. Hsu, S.S.: Sensors may track terror's fallout. In: Washington Post, p. A01 (2003)

100. Hu, X., Eberhart, R.C.: Adaptive particle swarm optimization: Detection and response to dynamic systems. In: Proceedings of the Congress on Evolutionary Computation, pp. 1666–1670. IEEE (2002)

101. Huang, W.H.: Optimal line-sweep-based decompositions for coverage algorithms. In: ICRA, pp. 27–32 (2001)

102. Hughes-Hallett, D., Gleason, A.M., McCallum, W.G.: Calculus. Single and Multivariable. John Wiley & Sons (1998)

103. Iacono, G.L., Reynolds, A.M.: Modelling of concentrations along a moving observer in an inhomogeneous plume. Biological application: Model of odour-mediated insect flights. Environmental Fluid Mechanics **8**(2), 147–168 (2008)
104. IBM: IEEE 802.3 local area network considerations. Tech. rep., IBM, Corp. (2008). IBM document GG22-9422-0
105. Ishida, H., Nakamoto, T., Moriizumi, T., Kikas, T., Janata, J.: Plume-tracking robots: A new application of chemical sensors. Biol Bull **200**, 222–226 (2001)
106. Ishida, H., Tanaka, H., Taniguchi, H., Moriizumi, T.: Mobile robot navigation using vision and olfaction to search for a gas/odor source. Autonomous Robots, Special Issue on Mobile Robot Olfaction **20**(3), 231–238 (2006)
107. Jantz, S., Doty, K.: Kinetics of robotics: The development of universal metrics in robotic swarms. Tech. rep., Dept of Electrical Engineering, University of Florida (1997)
108. Johnson, S.: Johnson's Dictionary. W. P. and L. Blake, Boston (1804)
109. Jones, C., Matarić, M.: Adaptive division of labor in large-scale multi-robot systems. In: Proceedings of the IEEE International Conference on Intelligent Robots and Systems, pp. 1969–1974. IEEE Press (2003)
110. Jones, C., Matarić, M.: From local to global behavior in intelligent self-assembly. In: Proceedings of the IEEE International Conference on Robotics and Automation, pp. 721–726. IEEE Press (2003)
111. Kaveh, A., Talatahari, S.: A novel heuristic optimization method: Charged system search. Acta Mechanica **213**, 267–289 (2010)
112. Kazadi, S.: On the development of a swarm engineering methodology. In: Proceedings of IEEE Conference on Systems, Man, and Cybernetics, pp. 1423–1428 (2005)
113. Kazadi, S.: Model independence in swarm robotics. International Journal of Intelligent Computing and Cybernetics **2**(4), 672–694 (2009)
114. Kazadi, S., Abdul-Khaliq, A., Goodman, R.: On the convergence of puck clustering systems. Robotics and Autonomous Systems **38**(2), 93–117 (2002)
115. Kazadi, S., Chang, J.: Hijacking swarms. In: Proceeding of the 11th IASTED International Conference on Intelligent Systems and Control, pp. 228–234 (2008)
116. Kazadi, S., Goodman, R., Tskikata, D., Green, D., Lin, H.: An autonomous water vapor plume tracking robot using passive resistance polymer sensors. Autonomous Robots **9**(2), 175–188 (2000)
117. Kazadi, S., Lee, J.R., Lee, J.: Artificial physics, swarm engineering, and the hamiltonian method. In: Proceedings from the 2007 World, Congress on Engineering and Computer Science, pp. 623–632 (2007)
118. Kelly, D., Keating, D.: Faster learning of control parameters through sharing experiences of autonomous mobile robots. International Journal of System Science **29**(7), 783–793 (1998)
119. Kennedy, J., Eberhart, R.: Particle swarm optimization. IEEE International Conference on Neural Networks **4**, 1942–1948 (1995)
120. Kennedy, J., Spears, W.M.: Matching algorithms to problems: An experimental test of the particle swarm and some genetic algorithms on the multimodal problem generator. In: IEEE International Conference on Evolutionary Computation, pp. 78–83. IEEE Press (1998)
121. Kerr, W., Spears, D.: Robotic simulation of gases for a surveillance task. In: IEEE/RSJ International Conference on Intelligent Robots and Systems (IROS'05) (2005)
122. Kerr, W., Spears, D., Spears, W., Thayer, D.: Two formal fluids models for multiagent sweeping and obstacle avoidance. Lecture Notes in Artificial Intelligence **3228** (2004)
123. Khalil, H.K.: Nonlinear Systems. Macmillan, New York (1992)
124. Khatib, O.: Real-time obstacle avoidance for manipulators and mobile robots. International Journal of Robotics Research **5**(1), 90–98 (1986)

125. Kim, J., Khosla, P.: Real-time obstacle avoidance using harmonic potential functions. In: Proc. IEEE International Conference on Robotics and Automation, pp. 790–796. IEEE Press (1991)

126. Kirkpatrick, S., Gelatt, C., Vecchi, M.: Optimization by simulated annealing. Science 220(4598), 671–680 (1983)

127. Koenig, S., Liu, Y.: Terrain coverage with ant robots: A simulation study. In: Agents, pp. 600–607 (2001)

128. Koenig, S., Szymanski, B., Liu, Y.: Efficient and inefficient ant coverage methods. Annals of Mathematics and Artificial Intelligence 31, 41–76 (2001)

129. Koren, Y., Borenstein, J.: Potential field methods and their inherent limitations for mobile robot navigation. In: Proc. IEEE International Conference on Robotics and Automation, pp. 1398–1404. IEEE Press (1991)

130. Kowadlo, G., Russell, R.A.: Robot odor localization: A taxonomy and survey. The International Journal of Robotics Research 27(8), 869–894 (2008)

131. Krink, T., Løvbjerg, M.: The lifecycle model: Combining particle swarm optimisation, genetic algorithms and hillclimbers. In: Proceedings of the 7th International Conference on Parallel Problem Solving from Nature, pp. 621–630 (2002)

132. Krishnanand, K.N., Ghose, D.: Glowworm swarm based optimization algorithm for multimodal functions with collective robotics applications. Multiagent and Grid Syst, Special Issue on Recent Progress in Distributed Intelligence 2(3), 209–222 (2006)

133. Kullback, S., Leibler, R.A.: On information and sufficiency. Annals of Mathematical Statistics, 22, 79–86 (1951)

134. Kunkel, T.M.: Hardware architecture of a swarm of robots. Master's thesis, University of Wyoming, Laramie, WY (2006)

135. Lemon, G.W.: English Etymology. London (1783)

136. Lennard-Jones, J.E.: On the determination of molecular fields. I. From the variation of the viscosity of a gas with temperature. Proceedings of the Royal Society of London. Series A, Containing Papers of a Mathematical and Physical Character 106(738), 441–462 (1924)

137. Leonard, N.E., Paley, D.A., Davis, R.E., Fratantoni, D.M., Lekien, F., Zhang, F.: Coordinated control of an underwater glider fleet in an adaptive ocean sampling field experiment in Monterey Bay. Journal of Field Robotics 27(6), 718–740 (2010)

138. Lerman, K., Galstyan, A.: A general methodology for mathematical analysis of multiagent systems. Tech. Rep. ISI-TR-529, USC Information Services (2001)

139. Levi, P., Kernbach, S.: Symbiotic Multi-Robot Organisms: Reliability, Adaptability, Evolution. Springer–Verlag (2010)

140. Lewis, M.A., Tan, K.H.: High precision formation control of mobile robots using virtual structures. Autonomous Robots 4(4), 387–403 (1997)

141. Li, C., Yang, S.: Fast multi-swarm optimization for dynamic optimization problems. In: Fourth International Conference on Natural Computation, pp. 624–628 (2008)

142. Li, W., Farrell, J., Pang, S., Arrieta, R.: Moth-inspired chemical plume tracing on an autonomous underwater vehicle. IEEE Transactions on Robotics 22(2) (2006)

143. Lilienthal, A., Duckett, T.: Approaches to gas source tracing and declaration by pure chemo-tropotaxis. In: Proceedings of AMS, pp. 161–171 (2003)

144. Lilienthal, A., Zell, A., Wandel, M., Weimar, U.: Experiences using gas sensors on an autonomous mobile robot. In: Proceedings of the 4th European Workshop on Advanced Mobile Robots (EUROBOT'01), pp. 1–8 (2001)

145. Lytridis, C., Kadar, E., Virk, G.: A systematic approach to the problem of odour source localization. Auton Robots, Special Issue on Mobile Robot Olfaction 20(3), 261–276 (2006)

146. Marques, L., de Almeida, A.: Special issue on mobile robot olfaction. Autonomous Robots 20(3) (2006)

147. Marques, L., Nunes, U., de Almeida, A.: Particle swarm-based olfactory guided search. Auton Robots, Special Issue on Mobile Robot Olfaction 20, 277–287 (2006)

148. Marsaglia, G., Zaman, A., Tsang, W.W.: Toward a universal random number generator. Statistics and Probability Letters **9**(1), 35–39 (1990). DOI 10.1016/ 0167-7152(90)90092-L. URL http://www.sciencedirect.com/science/ article/B6V1D-45DHJ42-S/2/65eff74c666aaddad67c802c9175ee62

149. Martinez, J.L., Mandow, A., Morales, J., Garcia-Cerezo, A., Pedraza, S.: Kinematic modelling of tracked vehicles by experimental identification. In: Proceedings of the 2004 IEEE/RSJ International Conference on Intelligent Robots and Systems (IROS 2004), vol. 2, pp. 1487–1492 (2004)

150. Martinoli, A.: Collective complexity out of individual simplicity. Artificial Life **7**(3), 315–319 (2001)

151. Martinson, E., Payton, D.: Lattice formation in mobile autonomous sensor arrays. Lecture Notes in Computer Science, vol. 3342, pp. 98–111. Springer–Verlag (2005)

152. Matarić, M.: Designing and understanding adaptive group behavior. Tech. rep., CS Dept, Brandeis Univ. (1995)

153. Matarić, M.J.: Designing emergent behaviors: From local interactions to collective intelligence. In: Proceedings of the Second International Conference on Simulation of Adaptive Behavior: From Animals to Animats, pp. 432–441. MIT Press, Cambridge, MA, USA (1993). URL http://portal.acm.org/citation.cfm?id= 171174.171225

154. Matarić, M.J.: Designing and understanding adaptive group behavior. Adaptive Behavior **4**(1), 51 (1995)

155. Maxim, P., Hettiarachchi, S., Spears, W.M., Spears, D.F., Hamann, J., Kunkel, T., Speiser, C.: Trilateration localization for multi-robot teams. In: Sixth International Conference on Informatics in Control, Automation and Robotics, Special Session on Multi-Agent Robotic Systems (2008)

156. Maxim, P., Spears, W.M.: Uniform coverage of arbitrary-shaped connected regions: Robot implementation. Carpathian Journal of Electronic and Computer Engineering **2**(1), 58–65 (2009)

157. McLurkin, J.: Hexagonal lattice formation in multi-robot systems. Tech. rep., Rice University (2011)

158. Megerian, S., Koushanfar, F., Potkonjak, M., Srivastava, M.: Worst and best-case coverage in sensor networks. IEEE Transactions on Mobile Computing **4**(1), 84–92 (2005)

159. Meng, Y., Gan, J.: A distributed swarm intelligence based algorithm for a cooperative multi-robot construction task. In: Swarm Intelligence Symposium, 2008. SIS 2008. IEEE, pp. 1–6. IEEE (2008)

160. Microchip: PIC16F767 data sheet. Tech. rep., Microchip Technology Inc. (2008)

161. Mo, S.M., Zeng, J.C.: Performance analysis of the artificial physics optimization algorithm with simple neighborhood topologies. In: International Conference on Computational Intelligence and Security, pp. 155–160 (2009)

162. Moline, M.: Bioluminescence potential in the transition zone to very shallow water (VSW). Tech. rep., California Polytechnic State University, San Luis Obispo Biological Sciences Dept. (2006)

163. Moline, M., Bissett, P., Blackwell, S., Mueller, J., Sevadjian, J., Trees, C., Zaneveld, R.: An autonomous vehicle approach for quantifying bioluminescence in ports and harbors. Photonics for Port and Harbor Security **5780**(13), 81–87 (2005)

164. Mondesire, S., Wiegand, R.P.: Evolving a non-playable character team with layered learning. In: Proceedings from the 2011 Conference on Multi-Criteria Decision Making (2011)

165. Monson, C., Seppi, K.: Exposing origin-seeking bias in PSO. Proceedings of the Conference on Genetic and Evolutionary Computation (2005). URL http://portal. acm.org/citation.cfm?id=1068009.1068045

166. More, G.M.: Using the I2C bus with HCS12 microcontrollers. Tech. rep., Freescale Semiconductor, Inc. (2002). AN2318/D Rev.0, Freescale Semiconductor, Inc.

167. Munson, B.R., Young, D.F., Okiishi, T.H.: Fundamentals of Fluid Mechanics. John Wiley & Sons Inc. (1990)

168. Navarro, I., Pugh, J., Martinoli, A., Matia, F.: A distributed scalable approach to formation control in multi-robot systems. In: Proceedings of the International Symposium on Distributed Autonomous Robotic Systems (2008). URL http://www.robot.t.u-tokyo.ac.jp/DARS2008/

169. Nokia: QT (2011). URL http://qt.nokia.com. Accessed March 18, 2011

170. NXP: UM10204 I2C-bus specification and user manual. Tech. Rep. Rev. 03, NXP B. V. (Philips Semiconductors) (2007). URL http://www.standardics.nxp.com/support/. Accessed October 12, 2008

171. O'Grady, R., Groß, R., Christensen, A.L., Mondada, F., Bonani, M., Dorigo, M.: Performance benefits of self-assembly in a swarm-bot. In: Intelligent Robots and Systems, 2007. IROS 2007. IEEE/RSJ International Conference on, pp. 2381–2387. IEEE (2007)

172. O'Hara, K.J., Bigio, V.L., Dodson, E.R., Irani, A., Walker, D.B., Balch, T.R.: Physical path planning using the gnats. In: IEEE International Conference on Robotics and Automation (2005)

173. Olszta, P.W.: The free OpenGL utility toolkit. Tech. rep., X.Org Foundation (2008). URL http://freeglut.sourceforge.net/. Accessed September 23, 2008

174. Panait, L., Luke, S.: Collaborative multi-agent learning: The state of the art. Autonomous Agents and Multi-agent Systems 11(3), 387–434 (2005)

175. Parker, L.: Toward the automated synthesis of cooperative mobile robot teams. In: SPIE Mobile Robots XIII, vol. 3525, pp. 82–93. Boston, MA (1998)

176. Passino, K.M.: Biomimicry for Optimization, Control, and Automation. Springer (2005)

177. Pearl, J.: Heuristics. Addison–Wesley, New York (1983)

178. Platts, J.: Platts' Synonyms. London (1845)

179. Poli, R.: Mean and variance of the sampling distribution of particle swarm optimizers during stagnation. IEEE Transactions on Evolutionary Computation 13(4), 712–721 (2009)

180. Potter, M.A., Meeden, L., Schultz, A.C.: Heterogeneity in the coevolved behaviors of mobile robots: The emergence of specialists. In: Proceedings from the International Joint Conference on Artificial Intelligence, pp. 1337–1343 (2001)

181. Potter, M.A., Wiegand, R.P., Blumenthal, H.J., Sofge, D.A.: Effects of experience bias when seeding with prior results. In: Congress on Evolutionary Computation, pp. 2730–2737. IEEE Press (2005)

182. Preiss, B., Aschuler, W.R.: The Microverse. Bantam Books (1989)

183. ProWave: Air ultrasonic ceramic transducers 400ST/R160. Tech. rep., ProWave Electronics Corporation (2008). URL http://www.prowave.com.tw/pdf/txall.pdf. Accessed September 20, 2008

184. Pugh, J., Martinoli, A.: Multi-robot learning with particle swarm optimization. In: International Conference on Autonomous Agents and Multiagent Systems, pp. 441–448 (2006)

185. Ramsey, A.: RF communications primer. Tech. rep., Ottawa Robotics Enthusiasts (2007). URL http://www.ottawarobotics.org/articles/rf/rf_article.pdf. Accessed September 16, 2008

186. Rashedi, E., Nezamabadi-pour, H., Saryazdi, S.: Gsa: A gravitational search algorithm. Information Science 179, 2232–2248 (2009)

187. Rebguns, A.: Using scouts to predict swarm success rate. Master's thesis, University of Wyoming (2008)

188. Rebguns, A., Anderson-Sprecher, R., Spears, D.F., Spears, W., Kletsov, A.: Using scouts to predict swarm success rate. Swarm Intelligence Symposium (SIS'08) (2008)

189. Rebguns, A., Spears, D.F., Anderson-Sprecher, R., Kletsov, A.: A theoretical framework for estimating swarm success probability using scouts. International Journal of Swarm Intelligence Research 1(4) (2010)

190. Reif, J.H., Wang, H.: Social potential fields: A distributed behavioral control for autonomous robots. Robotics and Autonomous Systems **27**(3), 171–194 (1999)

191. Rekleitis, I., New, A.P., Rankin, E.S., Choset, H.: Efficient boustrophedon multi-robot coverage: An algorithmic approach. Annals of Mathematics and Artificial Intelligence **52**(2–4), 109–142 (2008)

192. Reynolds, C.: Boids (1986). URL http://www.red3d.com/cwr/boids/

193. Reynolds, C.: Flocks, herds, and schools: A distributed behavioral model. In: Proceedings of SIGGRAPH'87, **21**(4), 25–34. ACM Computer Graphics, New York, NY (1987)

194. Rimon, E., Koditschek, D.E.: Exact robot navigation using artificial potential functions. Transactions on Robotics and Automation **8**(5), 501–518 (1992)

195. Roberts, C.S.: Implementing and testing new versions of a good, 48-bit, pseudo-random number generator. The Bell System Technical Journal **61**(8), 2053–2063 (1982). URL http://www.alcatel-lucent.com/bstj/vol61-1982/articles/bstj61-8-2053.pdf. Accessed March 5, 2011

196. Rocha, A.M.A.C., Fernandes, E.M.G.P.: On charge effects to the electromagnetism-like algorithm. In: The 20th International Conference, EURO Mini Conference: Continuous Optimization and Knowledge-Based Technologies, pp. 198–203. Vilnius Gediminas Technical University Publishing House: Technika (2008)

197. Rogge, J., Aeyels, D.: A novel strategy for exploration with multiple robots. In: Proceedings of 4th International Conference on Informatics in Control, Automation and Robotics, pp. 76–83 (2007)

198. Russell, R.A., Kleeman, L., Kennedy, S.: Using volatile chemicals to help locate targets in complex environments. In: Proceedings of the Australian Conference on Robotics and Automation (ACRA'00) (2000)

199. Russell, S., Norvig, P.: Artificial Intelligence: A Modern Approach. Prentice–Hall (2003)

200. Rybski, P.E., Burt, I., Drenner, A., Kratochvil, B., McMillen, C., Stoeter, S., Stubbs, K., Gini, M., Papanikolopoulos, N.: Evaluation of the scout robot for urban search and rescue. In: American Association for Artificial Intelligence (AAAI) Workshop (2001)

201. Şahin, E., Franks, N.R.: Simulation of nest assessment behavior by ant scouts. In: Proceedings of the Third International Workshop on Ant Algorithms (ANT'02), pp. 274–281 (2002)

202. Şahin, E., Spears, W.M. (ed.): Swarm Robotics. Lecture Notes in Computer Science State-of-the-Art Series, Springer–Verlag (2005)

203. Şahin, E., Spears, W.M., Winfield, A. (ed.): Swarm Robotics. Lecture Notes in Computer Science State-of-the-Art Series, Springer–Verlag (2007)

204. Salomon, R.: Reevaluating genetic algorithm performance under coordinate rotation of benchmark functions. BioSystems **39**(3), 263–278 (1995)

205. Sandini, G., Lucarini, G., Varoli, M.: Gradient driven self-organizing systems. In: Proceedings of the IEEE/RSJ International Conference on Intelligent Robots and Systems (IROS'93) (1993)

206. Sarma, J., De Jong, K.: Selection pressure and performance in spatially distributed evolutionary algorithms. In: IEEE World Congress on Computational Intelligence, pp. 553–557 (1998)

207. Schiller, C.: Motion Mountain, vol. 6. http://www.motionmountain.net (2009)

208. Schiller, C.: Motion Mountain, vol. 4. http://www.motionmountain.net (2009)

209. Schoenwald, D., Feddema, J., Oppel, F.: Decentralized control of a collective of autonomous robotic vehicles. In: American Control Conference, pp. 2087–2092. Arlington, VA (2001)

210. Schultz, A., Grefenstette, J., Adams, W.: Roboshepherd: Learning a complex behavior. In: Robotics and Manufacturing: Recent Trends in Research and Applications, vol. 6, pp. 763–768. ASME Press, New York (1996)

211. Schultz, A., Parker, L. (eds.): Multi-Robot Systems: From Swarms to Intelligent Automata, vol. 2. Kluwer (2002)

212. Schultz, A.C.: Using a genetic algorithm to learn strategies for collision avoidance and local navigation. In: International Symposium on Unmanned Untethered Submersible Technology, pp. 213–225 (1991)

213. Schwefel, H.P.: Numerical Optimization of Computer Models. John Wiley & Sons, Inc., New York, NY, USA (1981)

214. Selinger, P.: The glibc random number generator (2007). URL http://www.mscs.dal.ca/~selinger/random/. Accessed March 3, 2011

215. Shah-Hosseini, H.: The intelligent water drops algorithm: A nature-inspired swarm-based optimization algorithm. International Journal of Bio-Inspired Computation 1(1/2), 71–79 (2009)

216. SharpMicroElectronics: GP2D12 optoelectronic device. Tech. rep., Sharp Microelectronics of the Americas (2008). URL http://document.sharpsma.com/files/GP2D12-DATA-SHEET.PDF. Accessed September 20, 2008

217. Shehory, O., Kraus, S., Yadgar, O.: Emergent cooperative goal-satisfaction in large-scale automated-agent systems. Artificial Intelligence 110, 1–55 (1999)

218. Shen, W.M., Will, P., Galstyan, A., Chuong, C.M.: Hormone-inspired self-organization and distributed control of robotic swarms. Autonomous Robots 17(1), 93–105 (2004)

219. Shi, Y., Eberhart, R.: A modified particle swarm optimizer. IEEE World Congress on Computational Intelligence pp. 69–73 (1998)

220. Shinoda, T.: Koizumi's top-down leadership in the anti-terrorism legislation: The impact of political institutional changes. SAIS Review 23(1), 19–34 (2003)

221. Simard, R.: TestU01: A C library for empirical testing of random number generators. ACM Transactions on Mathematical Software p. 2007 (2007)

222. Simmons, R.: The curvature velocity method for local obstacle avoidance. In: Proceedings of the International Conference on Robotics and Automation, pp. 3375–3382 (1996)

223. Smith, B., Howard, A., McNew, J.M., Wang, J., Egerstedt, M.: Multi-robot deployment and coordination with Embedded Graph Grammars. Autonomous Robots 26(1), 79–98 (2009)

224. Smith, R.N., Chao, Y., Li, P.P., Caron, D.A., Jones, B.H., Sukhatme, G.S.: Planning and implementing trajectories for autonomous underwater vehicles to track evolving ocean processes based on predictions from a regional ocean model. The International Journal of Robotics Research 26(10) (2010)

225. Smith, R.N., Schwager, M., Smith, S.L., Rus, D., Sukhatme, G.S.: Persistent ocean monitoring with underwater gliders: Toward accurate reconstruction of dynamic ocean processes. In: IEEE International Conference on Robotics and Automation (ICRA'11) (2011)

226. Smullin, S.J., Geraci, A.A., Weld, D.M., Kaptitulnik, A.: Testing gravity at short distances. SLAC Summer Institute on Particle Physics (SSI04) (2004)

227. Sommerville, I.: Software engineering (5th ed.) Addison Wesley Longman Publishing Co., Inc., Redwood City, CA, USA (1995)

228. Song, Z., Chen, Y.: Challenges and some results for the MAS-net project. In: First Joint Emergency Preparation and Response Meeting, pp. 258–264 (2006)

229. Soysal, O., Şahin, E.: A macroscopic model for self-organized aggregation in swarm robotic systems. In: E. Şahin, W. Spears, A.F.T. Winfield (eds.) Swarm Robotics: Second SAB International Workshop, pp. 27–42. Springer (2007)

230. Spears, D.F.: Generating linear formations. Tech. rep., Naval Research Laboratory, Artificial Intelligence Center (2001)

231. Spears, D.F., Kerr, W., Spears, W.M.: Physics-based robot swarms for coverage problems. International Journal of Intelligent Control and Systems 11(3), 124–140 (2006)

232. Spears, D.F., Thayer, D., Zarzhitsky, D.: Foundations of swarm robotic chemical plume tracing from a fluid dynamics perspective. International Journal of Intelligent Computing and Cybernetics **2**(4), 745–785 (2009)

233. Spears, W.M.: Evolutionary Algorithms: The Role of Mutation and Recombination. Springer–Verlag (2000)

234. Spears, W.M., De Jong, K., Bäck, T., Fogel, D., de Garis, H.: An overview of evolutionary computation. In: European Conference on Machine Learning, vol. 667, pp. 442–459. Springer–Verlag (1993)

235. Spears, W.M., Gordon, D.F.: Using artificial physics to control agents. In: International Conference on Information, Intelligence, and Systems, pp. 281–288. IEEE (1999)

236. Spears, W.M., Gordon-Spears, D.F.: Analysis of a phase transition in a physics-based multiagent system. Lecture Notes in Artificial Intelligence 2699. Springer–Verlag (2003)

237. Spears, W.M., Green, D., Spears, D.F.: Biases in particle swarm optimization. International Journal of Swarm Intelligence Research **1**(2), 34–57 (2010)

238. Spears, W.M., Hamann, J., Maxim, P., Kunkel, T., Heil, R., Zarzhitsky, D., Spears, D.F., Karlsson, C.: Where are you? In: E. Şahin, W.M. Spears, A. Winfield (eds.) Second International Workshop on Swarm Robotics (2007)

239. Spears, W.M., Maxim, P.: Uniform coverage of arbitrary-shaped connected regions: Algorithm validation. Carpathian Journal of Electronic and Computer Engineering **2**(1), 50–57 (2009)

240. Spears, W.M., Spears, D.F., Hamann, J., Heil, R.: Distributed, physics-based control of swarms of vehicles. Autonomous Robots **17**(2–3) (2004)

241. Spears, W.M., Spears, D.F., Heil, R., Kerr, W.: An overview of physicomimetics. In: E. Şahin, W.M. Spears (eds.) Lecture Notes in Artificial Intelligence, State-of-the-Art Series (2005)

242. Spears, W.M., Spears, D.F., Zarzhitsky, D.: Physicomimetics positioning methodology for distributed, autonomous swarms. In: GOMACTech-05 Intelligent Technologies (2005)

243. Stark, H., Woods, J.W.: Probability, Random Processes, and Estimation Theory for Engineers. Prentice Hall (1986)

244. Stevens, B.L., Lewis, F.L.: Aircraft Control and Simulation. Whiley (2003)

245. Sugihara, K., Suzuki, I.: Distributed algorithms for formation of geometric patterns with many mobile robots. Journal of Robotic Systems **13**(3), 127–139 (1996)

246. Tan, M.: Multi-agent reinforcement learning: Independent vs. cooperative agents. In: Proceedings of the Tenth International Conference on Machine Learning, pp. 330–337 (1993)

247. Tannehill, J.C., Anderson, D.A., Pletcher, R.H.: Computational Fluid Mechanics and Heat Transfer. Taylor and Francis (1997)

248. Teller, E.: The Pursuit of Simplicity. Pepperdine University Press (1980)

249. TheMachineLab: MMP-5 mobile robot platform. Tech. rep., The Machine Lab, Inc. (2008). URL http://www.themachinelab.com/mmp5/MMP-5.pdf. Accessed September 20, 2008

250. Tikanmäki, A., Haverinen, J., Kemppainen, A., Röning, J.: Remote-operated robot swarm for measuring and environment. In: Proceedings of the International Conference on Machine Automation (ICMA'06) (2006)

251. Toner, J., Tu, Y.: Flocks, herds, and schools: A quantitative theory of flocking. Physical Review E **58**(4), 4828–4858 (1998)

252. Turgut, A.E., Çelikkanat, H., Gökçe, F., Şahin, E.: Self-organized flocking with a mobile robot swarm. In: Proceedings of the 7th international joint conference on Autonomous agents and multiagent systems—Volume 1, AAMAS '08, pp. 39–46. International Foundation for Autonomous Agents and Multiagent Systems, Richland, SC (2008). URL http://portal.acm.org/citation.cfm?id=1402383.1402394

253. Vail, D., Veloso, M.: Multi-robot dynamic role assignment and coordination through shared potential fields. In: A. Schultz, L. Parker, F. Schneider (eds.) Multi-Robot Systems, pp. 87–98. Kluwer, Hingham, MA (2003)

254. Varakantham, P., Maheswaran, R., Tambe, M.: Agent modelling in partially observable domains. In: Workshop on Modeling Other Agents from Observations, AAMAS04 (2004)

255. Veysal, G., Passino, K.M.: Swarm Stability and Optimization. Springer (2011)

256. Vicsek, T., Czirok, A., Ben-Jacob, E., Cohen, I., Shocher, O.: Novel type of phase transition in a system of self-driven particles. Physics Review Letters 75(6), 1226–1229 (1995)

257. Wang, X., Gao, X.Z., Ovaska, S.J.: A hybrid particle swarm optimization method. In: Proceeding of the IEEE International Conference on Systems, Man and Cybernetics, pp. 4151–4157 (2006)

258. Wang, Y., Zeng, J.C.: A multi-objective optimization algorithm based on artificial physics optimization. Control and Decision 25(7), 1040–1044 (2010)

259. Watson, R., Ficici, S., Pollack, J.: Embodied evolution: Distributing an evolutionary algorithm in a population of robots. Robotics and Autonomous Systems 39, 1–18 (2002)

260. Weissburg, M., Dusenbery, D., Ishida, H., Janata, J., Keller, T., Roberts, P., Webster, D.: A multidisciplinary study of spatial and temporal scales containing information in turbulent chemical plume tracking. Env Fluid Mech, Special Issue on Chemical Plume Tracing 2(1–2), 65–94 (2002)

261. Widder, E.A.: Bioluminescence and the plagic visual environment. Marine Fresh Behavior Physiology 35(1–2), 1–26 (2002)

262. Widder, E.A., Frey, C.L., Borne, L.J.: HIDEX Generation II: A new and improved instrument for measuring marine bioluminescence. Marine Technology Society of the IEEE (Oceans'03) 4 (2003)

263. Wiegand, R.P., Potter, M., Sofge, D., Spears, W.M.: A generalized graph-based method for engineering swarm solutions to multiagent problems. In: Parallel Problem Solving From Nature, pp. 741–750 (2006)

264. Wikipedia: Mean free path (2011). URL http://en.wikipedia.org/wiki/Mean_free_path. Accessed March 8, 2011

265. Wikipedia: Physics (2011). URL http://en.wikipedia.org/wiki/Physics. Accessed April 14, 2011

266. Wikipedia: Rejection sampling (2011). URL http://en.wikipedia.org/wiki/Rejection_sampling. Accessed March 3, 2011

267. Wilensky, U.: NetLogo termites model (1997). URL http://ccl.northwestern.edu/netlogo/models/Termites. Center for Connected Learning and Computer-Based Modeling, Northwestern University, Evanston, IL

268. Wilensky, U.: NetLogo flocking model (1998). URL http://ccl.northwestern.edu/netlogo/models/Flocking. Center for Connected Learning and Computer-Based Modeling, Northwestern University, Evanston, IL

269. Wilensky, U.: NetLogo (1999). URL http://ccl.northwestern.edu/netlogo/. Center for Connected Learning and Computer-Based Modeling, Northwestern University, Evanston, IL

270. Wilke, D.N.: Analysis of the particle swarm optimization algorithm. Master's thesis, University of Pretoria, Pretoria, South Africa (2005)

271. Winfield, A., Harper, C., Nembrini, J.: Towards dependable swarms and a new discipline of swarm engineering. In: 2004 SAB Swarm Robotics Workshop, pp. 133–148 (2004)

272. Winfield, A., Sa, J., Gago, M., Dixon, C., Fisher, M.: On formal specification of emergent behaviours in swarm robotics systems. International Journal of Advanced Robotic Systems 2(4), 363–370 (2005)

273. WirelessCables: AIRcable Serial3. Tech. rep., Wireless Cables Inc. (2008). URL
     http://www.aircable.net/support-serial3.html. Accessed September
     21, 2008
274. Witkin, A.: Physically based modeling: Principles and practice (particle system dy-
     namics). In: SIGGraph Course Notes C12. ACM Press (1997)
275. Wolfram, S.: A New Kind of Science. Wolfram Media (2002)
276. Wu, A., Schultz, A., Agah, A.: Evolving control for distributed micro air vehicles. In:
     IEEE Conference on Computational Intelligence in Robotics and Automation, pp.
     174–179. Belgium (1999)
277. Wytec: MiniDRAGON+ compact EVBplus MC9S12DP256 development board get-
     ting started manual ver. 1.22. Tech. rep., Wytec Company (2003)
278. Wytec: HCS12: MiniDRAGON+ development board. Tech. rep., Wytec Company
     (2008). URL http://www.evbplus.com/minidragonplus_hc12_68hc12_
     9s12_hcs12.html. Accessed September 20, 2008
279. Xie, L.P., Tan, Y., Zeng, J.C.: The convergence analysis and parameter selection of
     artificial physics optimization algorithm. In: International Conference on Modelling,
     Identification and Control, pp. 562–567 (2010)
280. Xie, L.P., Tan, Y., Zeng, J.C., Cui, Z.H.: Artificial physics optimization: A brief
     survey. International Journal of Bio-Inspired Computation 2(5), 291–302 (2010)
281. Xie, L.P., Zeng, J.C.: An extended artificial physics optimization algorithm for global
     optimization problem. In: Fourth International Conference on Innovative Computing
     (2009)
282. Xie, L.P., Zeng, J.C.: A global optimization based on physicomimetics framework.
     In: World Summit on Genetic and Evolutionary Computation, pp. 609–616 (2009)
283. Xie, L.P., Zeng, J.C.: The performance analysis of artificial physics optimization al-
     gorithm driven by different virtual forces. ICIC Express Letters 4(1), 239–244 (2009)
284. Xie, L.P., Zeng, J.C.: Physicomimetics for swarm robots search. Pattern Recognition
     and Artificial Intelligence 22(4), 647–652 (2009)
285. Xie, L.P., Zeng, J.C., Cui, Z.H.: General framework of artificial physics optimization
     algorithm. In: Proceedings of the World Congress on Nature and Biologically Inspired
     Computing, pp. 1321–1326 (2009)
286. Xie, L.P., Zeng, J.C., Cui, Z.H.: On mass effects to artificial physics optimization
     algorithm for global optimization problems. International Journal of Innovative Com-
     puting and Applications 2(2), 69–76 (2009)
287. Xie, L.P., Zeng, J.C., Cui, Z.H.: Using artificial physics to solve global optimization
     problems. In: The 8th IEEE International Conference on Cognitive Informatics, pp.
     502–508 (2009)
288. Xie, L.P., Zeng, J.C., Cui, Z.H.: The vector model of artificial physics optimization
     algorithm for global optimization problems. In: The 10th International Conference
     on Intelligent Data Engineering and Automated Learning, pp. 610–617 (2009)
289. Xiong, N., He, J., Park, J.H., Kim, T.H., He, Y.: Decentralized flocking algorithms
     for a swarm of mobile robots: Problem, current research and future directions. In:
     Consumer Communications and Networking Conference, pp. 1–6. IEEE (2009)
290. Yamaguchi, H., Burdick, J.W.: Asymptotic stabilization of multiple nonholonomic
     mobile robots forming group formations. In: Robotics and Automation, 1998. Pro-
     ceedings. 1998 IEEE International Conference on, vol. 4, pp. 3573–3580. IEEE (2002)
291. Zarzhitsky, D.: A physics-based approach to chemical source localization using mobile
     robotic swarms. Ph.D. thesis, University of Wyoming, Laramie, WY (2008)
292. Zarzhitsky, D., Spears, D.F.: Swarm approach to chemical source localization. In:
     Proceedings of the IEEE International Conference on Systems, Man and Cybernetics
     (SMC'05), pp. 1435–1440 (2005)
293. Zarzhitsky, D., Spears, D.F., Spears, W.M.: Distributed robotics approach to chemical
     plume tracing. In: IEEE/RSJ International Conference on Intelligent Robots and
     Systems (IROS'05) (2005)

294. Zarzhitsky, D., Spears, D.F., Thayer, D.: Experimental studies of swarm robotic chemical plume tracing using computational fluid dynamics simulations. International Journal of Intelligent Computing and Cybernetics $3(4)$, 631–671 (2010)

295. Zarzhitsky, D., Spears, D.F., Thayer, D., Spears, W.M.: Agent-based chemical plume tracing using fluid dynamics. Lecture Notes in Artificial Intelligence, vol. 3228. Springer–Verlag (2004)

296. Zarzhitsky, D., Spears, D.F., Thayer, D., Spears, W.M.: A fluid dynamics approach to multi-robot chemical plume tracing. In: Proceedings of the Third International Joint Conference on Autonomous Agents and Multi Agent Systems (2004)

297. Zwillinger, D.: CRC Standard Mathematical Tables and Formulae, thirty-first edn. Chapman and Hall/CRC (2002)